The Commonwealth Fund Book Program
gratefully acknowledges the assistance
of The Rockefeller University in the
administration of the program.

BLACK HOLES AND TIME WARPS

Einstein's Outrageous Legacy

BLACK HOLES

AND

TIME WARPS

———

Einstein's Outrageous Legacy

———

KIP S. THORNE

THE FEYNMAN PROFESSOR OF THEORETICAL PHYSICS
CALIFORNIA INSTITUTE OF TECHNOLOGY

A volume of
THE COMMONWEALTH FUND BOOK PROGRAM
under the editorship of Lewis Thomas, M.D.

W · W · NORTON & COMPANY

New York   London

The text of this book is composed in Walbaum,
with the display set in Walbaum display.
Composition and manufacturing by the Haddon Craftsmen, Inc.
Book design by Jacques Chazaud.
Illustrations by Matthew Zimet.

Library of Congress Cataloging-in-Publication Data

Thorne, Kip S.
From black holes to time warps : Einstein's outrageous legacy / Kip S. Thorne.
p.   cm.
Includes bibliographical references.
1. Physics—Philosophy. 2. Relativity (Physics) 3. Astrophysics. 4. Black holes (Astronomy)
I. Title.
QC6.T526 1993
530.1'1—dc20   93-2014

ISBN 0-393-03505-0

W. W. Norton & Company, Inc., 500 Fifth Avenue, New York, N.Y. 10110
W. W. Norton & Company Ltd., 10 Coptic Street, London WC1A 1PU

I dedicate this book to
JOHN ARCHIBALD WHEELER,
my mentor and friend.

Contents

11. What Is Reality?   *397*

*in which spacetime is viewed as curved on Sundays
and flat on Mondays, and horizons are made from vacuum
on Sundays and charge on Mondays, but Sunday's experiments
and Monday's experiments agree in all details*

12. Black Holes Evaporate   *412*

*in which a black-hole horizon is clothed in an atmosphere
of radiation and hot particles that slowly evaporate,
and the hole shrinks and then explodes*

13. Inside Black Holes   *449*

*in which physicists, wrestling with Einstein's equation,
seek the secret of what is inside a black hole: a route into
another universe? a singularity with infinite tidal gravity?
the end of space and time, and birth of quantum foam?*

14. Wormholes and Time Machines   *483*

*in which the author seeks insight into physical laws
by asking: can highly advanced civilizations build
wormholes through hyperspace for rapid interstellar
travel and machines for traveling backward in time?*

Epilogue   *523*

*an overview of Einstein's legacy, past and future,
and an update on several central characters*

Acknowledgments   *529*

*my debts of gratitude to friends and colleagues who
influenced this book*

Characters   *531*

*a list of characters who appear significantly
at several different places in the book*

Foreword

This book is about a revolution in our view of space and time, and its remarkable consequences, some of which are still being unraveled. It is also a fascinating account, written by someone closely involved, of the struggles and eventual success in a search for an understanding of what are possibly the most mysterious objects in the Universe—black holes.

It used to be thought obvious that the surface of the Earth was flat: It either went on forever or it had some rim that you might fall over if you were foolish enough to travel too far. The safe return of Magellan and other round-the-world travelers finally convinced people that the Earth's surface was curved back on itself into a sphere, but it was still thought self-evident this sphere existed in a space that was flat in the sense that the rules of Euclid's geometry were obeyed: Parallel lines never meet. However, in 1915 Einstein put forward a theory that combined space and time into something called spacetime. This was not flat but curved or warped by the matter and energy in it. Because spacetime is very nearly flat in our neighborhood, this curvature makes very little difference in normal situations. But the implications for the further reaches of the Universe were more surprising than even Einstein ever realized. One of these was the possibility that stars could collapse under their own gravity until the space around them became so curved that they cut themselves off from the rest of the Universe. Einstein himself didn't believe that such a collapse could ever occur, but a number of other people showed it was an inevitable consequence of his theory.

The story of how they did so, and how they found the peculiar properties of the black holes in space that were left behind, is the subject of this book. It is a history of scientific discovery in the making, written by one of the participants, rather like *The Double Helix* by James Watson about the discovery of the structure of DNA, which led to the understanding of the genetic code. But unlike the case of DNA, there were no experimental results to guide the investigators. Instead, the theory of black holes was developed before there was any indication from observations that they actually existed. I do not know any other example in science where such a great extrapolation was successfully made solely on the basis of thought. It shows the remarkable power and depth of Einstein's theory.

There is much we still don't know, such as what happens to objects and information that fall into a black hole. Do they reemerge elsewhere in the Universe, or in another universe? And can we warp space and time so much that one can travel back in time? These questions are part of our ongoing quest to understand the Universe. Maybe someone will come back from the future and tell us the answers.

STEPHEN HAWKING

Introduction

This book is based upon a combination of firmly established physical principles and highly imaginative speculation, in which the author attempts to reach beyond what is solidly known at present and project into a part of the physical world that has no known counterpart in our everyday life on Earth. His goal is, among other things, to examine both the exterior and interior of a black hole—a stellar body so massive and concentrated that its gravitational field prevents material particles and light from escaping in ways which are common to a star such as our own Sun. The descriptions given of events that would be experienced if an observer were to approach such a black hole from outside are based upon predictions of the general theory of relativity in a "strong-gravity" realm where it has never yet been tested. The speculations which go beyond that and deal with the region inside what is termed the black hole's "horizon" are based on a special form of courage, indeed of bravado, which Thorne and his international associates have in abundance and share with much pleasure. One is reminded of the quip made by a distinguished physicist, "Cosmologists are usually wrong but seldom in doubt." One should read the book with two goals: to learn some hard facts with regard to strange but real features of our physical Universe, and to enjoy informed speculation about what may lie beyond what we know with reasonable certainty.

As a preface to the work, it should be said that Einstein's general theory of relativity, one of the greatest creations of speculative science, was formulated just over three-quarters of a century ago. Its triumphs in the early 1920s in providing an explanation of the deviations of the motion of the planet Mercury from the predictions of the Newtonian theory of gravitation, and later an explanation of the redshift of distant nebulas discovered by Hubble and his colleagues at Mount Wilson Observatory, were followed by a period of relative quiet while the community of physical scientists turned much of its attention to the exploitation of quantum mechanics, as well as to nuclear physics, high-energy particle physics, and advances in observational cosmology.

The concept of black holes had been proposed in a speculative way soon after the discovery of Newton's theory of gravitation. With proper alterations, it was found to have a natural place in the theory of relativity if one was willing to extrapolate solutions of the basic equations to such strong gravitational fields—a procedure which Einstein regarded with skepticism at the time. Using the theory, however, Chandrasekhar pointed out in 1930 that, according to it, stars having a mass above a critical value, the so-called Chandrasekhar limit, should collapse to become what we now call black holes, when they have exhausted the nuclear sources of energy responsible for their high temperature. Somewhat later in the 1930s, this work was expanded by Zwicky and by Oppenheimer and his colleagues, who demonstrated that there is a range of stellar mass in which one would expect the star to collapse instead to a state in which it consists of densely packed neutrons, the so-called neutron star. In either case, the final implosion of the star when its nuclear energy is exhausted should be accompanied by an immense outpouring of energy in a relatively short time, an outpouring to be associated with the brilliance of the supernovae seen occasionally in our own galaxy as well as in more distant nebulas.

World War II interrupted such work. However, in the 1950s and 1960s the scientific community returned to it with renewed interest and vigor on both the experimental and theoretical frontiers. Three major advances were made. First, the knowledge gained from research in nuclear and high-energy physics found a natural place in cosmological theory, providing support for what is commonly termed the "big bang" theory of the formation of our Universe. Many lines of evidence now support the view that our Universe as we know it originated as the result of expansion from a small primordial soup of hot, densely packed particles, commonly called a fireball. The primary event occurred at some time between ten and twenty billion years ago. Perhaps the most dramatic support for the hypothesis was the discovery of the degraded remnants of the light waves that accompanied a late phase of the initial explosion.

Second, the neutron stars predicted by Zwicky and the Oppenheimer team were actually observed and behaved much as the theory predicted, giving full credence to the concept that the supernovae are associated with stars that have undergone what may be called a final gravitational collapse. If neutron stars can exist for a given range of stellar mass, it is not unreasonable to conclude that black holes will be produced by more massive stars, granting that much of the observational evidence will be indirect. Indeed, there is much such indirect evidence at present.

Finally, several lines of evidence have given additional support to the validity of the general theory of relativity. They include high-precision measurements of spacecraft and planetary orbits in our solar system, and observations of the "lensing" action of some galaxies upon light that reaches us from sources beyond those galaxies. Then, more recently, there is good evidence of the loss of energy of motion of mutually orbiting massive binary stars as a result of the generation of gravitational waves, a major prediction of the theory. Such observations give one courage to believe the untested predictions of the general theory of relativity in the proximity of a black hole and open the path to further imaginative speculation of the type featured here.

Several years ago the Commonwealth Fund decided at the suggestion of its president, Margaret E. Mahoney, to sponsor a Book Program in which working scientists of distinction were invited to write about their work for a literate lay audience. Professor Thorne is such a scientist, and the Book Program is pleased to offer his book as its ninth publication.

The advisory committee for the Commonwealth Fund Book Program, which recommended the sponsorship of this book, consists of the following members: Lewis Thomas, M.D., director; Alexander G. Bearn, M.D., deputy director; Lynn Margulis, Ph.D.; Maclyn McCarty,

M.D.; Lady Medawar; Berton Roueché; Frederick Seitz, Ph.D.; and
Otto Westphal, M.D. The publisher is represented by Edwin Barber,
vice-chairman and editor at W. W. Norton & Company, Inc.

FREDERICK SEITZ

Preface

*what this book is about,
and how to read it*

For thirty years I have been participating in a great quest: a quest to understand a legacy bequeathed by Albert Einstein to future generations—his relativity theory and its predictions about the Universe—and to discover where and how relativity fails and what replaces it.

This quest has led me through labyrinths of exotic objects: black holes, white dwarfs, neutron stars, singularities, gravitational waves, wormholes, time warps, and time machines. It has taught me epistemology: What makes a theory "good"? What transcending principles control the laws of nature? Why do we physicists think we know the things we think we know, even when technology is too weak to test our predictions? The quest has shown me how the minds of scientists work, and the enormous differences between one mind and another (say, Stephen Hawking's and mine) and why it takes many different types of scientists, each working in his or her own way, to flesh out our understanding of the Universe. Our quest, with its hundreds of participants scattered over the globe, has helped me appreciate the international character of science, the different ways the scientific enterprise is organized in different societies, and the intertwining of science with political currents, especially Soviet/American rivalry.

This book is my attempt to share these insights with nonscientists, and with scientists who work in fields other than my own. It is a book of interlocking themes held together by a thread of history: the history of our struggle to decipher Einstein's legacy, to discover its seemingly outrageous predictions about black holes, singularities, gravitational waves, wormholes, and time warps.

The book begins with a prologue: a science fiction tale that introduces the reader, quickly, to the book's physics and astrophysics concepts. Some readers may find this tale disheartening. The concepts (black holes and their horizons, wormholes, tidal forces, singularities, gravitational waves) fly by too fast, with too little explanation. My advice: Just let them fly by; enjoy the tale; get a rough impression. Each concept will be introduced again, in a more leisurely fashion, in the body of the book. After reading the body, return to the prologue and appreciate its technical nuances.

The body (Chapters 1 through 14) has a completely different flavor from the prologue. Its central thread is historical, and with this thread are interwoven the book's other themes. I pursue the historical thread for a few pages, then branch on to a tangential theme, and then another; then I return to the history for a while, and then launch on to another tangent. This branching, launching, and weaving expose the reader to an elegant tapestry of interrelated ideas from physics, astrophysics, philosophy of science, sociology of science, and science in the political arena.

Some of the physics may be tough going. As an aid, there is a glossary of physics concepts at the back of the book.

Science is a community enterprise. The insights that shape our view of the Universe come not from a single person or a small handful, but from the combined efforts of many. Therefore, this book has many characters. To help the reader remember those who appear several times, there is a list and a few words about each in the "Characters" section at the back of the book.

In scientific research, as in life, many themes are pursued simultaneously by many different people; and the insights of one decade may spring from ideas that are several decades old but were ignored in the intervening years. To make sense of it all, the book jumps backward and forward in time, dwelling on the 1960s for a while, then dipping back to the 1930s, and then returning to a main thread in the 1970s. Readers who get dizzy from all this time travel may find help in the chronology at the back of the book.

I do not aspire to a historian's standards of completeness, accuracy, or impartiality. Were I to seek completeness, most readers would drop by the wayside in exhaustion, as would I. Were I to seek much higher accuracy, the book would be filled with equations and would be unreadably technical. Although I have sought impartiality, I surely have failed; I am too close to my subject: I have been involved personally in its development from the early 1960s to the present, and several of my closest friends were personally involved from the 1930s onward. I have tried to balance my resulting bias by extensive taped interviews with other participants in the quest (see the bibliography) and by running chapters past some of them (see the acknowledgments). However, some bias surely remains.

As an aid to the reader who wants greater completeness, accuracy, and impartiality, I have listed in the notes at the back of the book the sources for many of the text's historical statements, and references to some of the original technical articles that the quest's participants have written to explain their discoveries to each other. The notes also contain more precise (and therefore more technical) discussions of some issues that are distorted in the text by my striving for simplicity.

Memories are fallible; different people, experiencing the same events, may interpret and remember them in very different ways. I have relegated such differences to the notes. In the text, I have stated my own final view of things as though it were gospel. May real historians forgive me, and may nonhistorians thank me.

John Wheeler, my principal mentor and teacher during my formative years as a physicist (and a central character in this book), delights in asking his friends, "What is the single most important thing you have learned about thus and so?" Few questions focus the mind so clearly. In the spirit of John's question, I ask myself, as I come to the end of fifteen years of on-and-off writing (mostly off), "What is the single most important thing that you want your readers to learn?"

My answer: the amazing power of the human mind—by fits and starts, blind alleys, and leaps of insight—to unravel the complexities of our Universe, and reveal the ultimate simplicity, the elegance, and the glorious beauty of the fundamental laws that govern it.

BLACK HOLES AND TIME WARPS

Einstein's Outrageous Legacy

Prologue:
A Voyage among the Holes

in which the reader,
in a science fiction tale,
encounters black holes
and all their strange properties
as best we understand them in the 1990s

Of all the conceptions of the human mind, from unicorns to gargoyles to the hydrogen bomb, the most fantastic, perhaps, is the black hole: a hole in space with a definite edge into which anything can fall and out of which nothing can escape; a hole with a gravitational force so strong that even light is caught and held in its grip; a hole that curves space and warps time.[1] Like unicorns and gargoyles, black holes seem more at home in the realms of science fiction and ancient myth than in the real Universe. Nonetheless, well-tested laws of physics predict firmly that black holes exist. In our galaxy alone there may be millions, but their darkness hides them from view. Astronomers have great difficulty finding them.[2]

Hades

Imagine yourself the owner and captain of a great spacecraft, with computers, robots, and a crew of hundreds to do your bidding. You

1. Chapters 3, 6, 7.
2. Chapter 8.

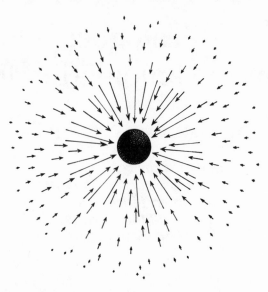

P.1 Atoms of gas, pulled by a black hole's gravity, stream toward the hole from all directions.

have been commissioned by the World Geographic Society to explore black holes in the distant reaches of interstellar space and radio back to Earth a description of your experiences. Six years into its voyage, your starship is now decelerating into the vicinity of the black hole closest to Earth, a hole called "Hades" near the star Vega.

On your ship's video screen you and your crew see evidence of the hole's presence: The atoms of gas that sparsely populate interstellar space, approximately one in each cubic centimeter, are being pulled by the hole's gravity (Figure P.1). They stream toward the hole from all directions, slowly at great distances where gravity pulls them weakly, faster nearer the hole where gravity is stronger, and extremely fast—almost as fast as light—close to the hole where gravity is strongest. If something isn't done, your starship too will be sucked in.

Quickly and carefully your first mate, Kares, maneuvers the ship out of its plunge and into a circular orbit, then shuts off the engines. As you coast around and around the hole, the centrifugal force of your circular motion holds your ship up against the hole's gravitational pull. Your ship is like a toy slingshot of your youth on the end of a whirling string, pushed out by its centrifugal force and held in by the string's tension, which is like the hole's gravity. As the starship

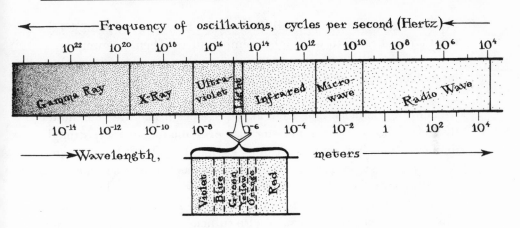

P.2   The spectrum of electromagnetic waves, running from radio waves at very long wavelengths (very low frequencies) to gamma rays at very short wavelengths (very high frequencies). For a discussion of the notation used here for numbers (10^{21}, 10^{-12}, etc.), see Box P.1 below.

coasts, you and your crew prepare to explore the hole.

At first you explore passively: You use instrumented telescopes to study the electromagnetic waves (the radiation) that the gas emits as it streams toward the hole. Far from the hole, the gas atoms are cool, just a few degrees above absolute zero. Being cool, they vibrate slowly; and their slow vibrations produce slowly oscillating electromagnetic waves, which means waves with long distances from one crest to the next— long wavelengths. These are radio waves; see Figure P.2. Nearer the hole, where gravity has pulled the atoms into a faster stream, they collide with each other and heat up to several thousand degrees. The heat makes them vibrate more rapidly and emit more rapidly oscillating, shorter wavelength waves, waves that you recognize as light of varied hues: red, orange, yellow, green, blue, violet (Figure P.2). Much closer to the hole, where gravity is much stronger and the stream much faster, collisions heat the atoms to several million degrees, and they vibrate very fast, producing electromagnetic waves of very short wavelength: X-rays. Seeing those X-rays pour out of the hole's vicinity, you are reminded that it was by discovering and studying just such X-rays that astrophysicists, in 1972, identified the first black hole in distant space: Cygnus X-1, 14,000 light-years from Earth.[3]

3. Chapter 8.

Turning your telescopes still closer to the hole, you see gamma rays from atoms heated to still higher temperatures. Then, looming up, at the center of this brilliant display, you see a large, round sphere, absolutely black; it is the black hole, blotting out all the light, X-rays, and gamma rays from the atoms behind it. You watch as superhot atoms stream into the black hole from all sides. Once inside the hole, hotter than ever, they must vibrate faster than ever and radiate more strongly than ever, but their radiation cannot escape the hole's intense gravity. Nothing can escape. That is why the hole looks black; pitch-black.[4]

With your telescope, you examine the black sphere closely. It has an absolutely sharp edge, the hole's surface, the location of "no escape." Anything just *above* this surface, with sufficient effort, can escape from gravity's grip: A rocket can blast its way free; particles, if fired upward fast enough, can escape; light can escape. But just *below* the surface, gravity's grip is inexorable; nothing can ever escape from there, regardless of how hard it tries: not rockets, not particles, not light, not radiation of any sort; they can never reach your orbiting starship. The hole's surface, therefore, is like the horizon on Earth, beyond which you cannot see. That is why it has been named *the horizon of the black hole.*[5]

Your first mate, Kares, measures carefully the circumference of your starship's orbit. It is 1 million kilometers, about half the circumference of the Moon's orbit around the Earth. She then looks out at the distant stars and watches them circle overhead as the ship moves. By timing their apparent motions, she infers that it takes 5 minutes and 46 seconds for the ship to encircle the hole once. This is the ship's *orbital period.*

From the orbital period and circumference you can now compute the mass of the hole. Your method of computation is the same as was used by Isaac Newton in 1685 to compute the mass of the Sun: The more massive the object (Sun or hole), the stronger its gravitational pull, and therefore the faster must an orbiting body (planet or starship) move to avoid being sucked in, and thus the shorter the body's orbital period must be. By applying Newton's mathematical version of this gravitational law[6] to your ship's orbit, you compute that the black hole Hades has a mass ten times larger than that of the sun ("10 solar masses").[7]

4. Chapters 3 and 6.
5. Chapter 6.
6. Chapter 2.
7. Readers who want to compute properties of black holes for themselves will find the relevant formulas in the notes at the end of the book.

You know that this hole was created long ago by the death of a star, a death in which the star, no longer able to resist the inward pull of its own gravity, imploded upon itself.[8] You also know that, when the star imploded, its mass did not change; the black hole Hades has the same mass today as its parent star had long ago—or almost the same. Hades' mass must actually be a little larger, augmented by the mass of everything that has fallen into the hole since it was born: interstellar gas, rocks, starships . . .

You know all this because, before embarking on your voyage, you studied the fundamental laws of gravity: laws that were discovered in an approximate form by Isaac Newton in 1687, and were radically revised into a more accurate form by Albert Einstein in 1915.[9] You learned that Einstein's gravitational laws, which are called *general relativity*, force black holes to behave in these ways as inexorably as they force a dropped stone to fall to earth. It is impossible for the stone to violate the laws of gravity and fall upward or hover in the air, and similarly it is impossible for a black hole to evade the gravitational laws: The hole must be born when a star implodes upon itself; the hole's mass, at birth, must be the same as the star's; and each time something falls into the hole, its mass must grow.[10] Similarly, if the star is spinning as it implodes, then the newborn hole must also spin; and the hole's *angular momentum* (a precise measure of how fast it spins) must be the same as the star's.

Before your voyage, you also studied the history of human understanding about black holes. Back in the 1970s Brandon Carter, Stephen Hawking, Werner Israel, and others, using Einstein's general relativistic description[11] of the laws of gravity, deduced that a black hole must be an exceedingly simple beast[12]: All of the hole's properties—the strength of its gravitational pull, the amount by which it deflects the trajectories of starlight, the shape and size of its surface—are determined by just three numbers: the hole's mass, which you now know; the angular momentum of its spin, which you don't yet know; and its electrical charge. You are aware, moreover, that no hole in interstellar space can contain much electrical charge; if it did, it quickly would pull

8. Chapters 3–5.
9. Chapter 2.
10. For further discussion of the concept that the laws of physics *force* black holes, and the solar system, and the Universe, to behave in certain ways, see the last few paragraphs of Chapter 1.
11. Chapter 2.
12. Chapter 7.

opposite charges from the interstellar gas into itself, thereby neutralizing its own charge.

As it spins, the hole should drag the space near itself into a swirling, tornado-like motion relative to space far away, much as a spinning airplane propeller drags air near itself into motion; and the swirl of space should cause a swirl in the motion of anything near the hole.[13]

To learn the angular momentum of Hades, you therefore look for a tornado-like swirl in the stream of interstellar gas atoms as they fall into the hole. To your surprise, as they fall closer and closer to the hole, moving faster and faster, there is no sign at all of any swirl. Some atoms circle the hole clockwise as they fall; others circle it counterclockwise and occasionally collide with clockwise-circling atoms; but on average the atoms' fall is directly inward (directly downward) with no swirl. Your conclusion: This 10-solar-mass black hole is hardly spinning at all; its angular momentum is close to zero.

Knowing the mass and angular momentum of the hole and knowing that its electrical charge must be negligibly small, you can now compute, using general relativistic formulas, all of the properties that the hole should have: the strength of its gravitational pull, its corresponding power to deflect starlight, and of greatest interest, the shape and size of its horizon.

If the hole were spinning, its horizon would have well-delineated north and south poles, the poles about which it spins and about which infalling atoms swirl. It would have a well-delineated equator halfway between the poles, and the centrifugal force of the horizon's spin would make its equator bulge out,[14] just as the equator of the spinning Earth bulges a bit. But Hades spins hardly at all, and thus must have hardly any equatorial bulge. Its horizon must be forced by the laws of gravity into an almost precisely spherical shape. That is just how it looks through your telescope.

As for size, the laws of physics, as described by general relativity, insist that the more massive the hole is, the larger must be its horizon. The horizon's circumference, in fact, must be 18.5 kilometers multiplied by the mass of the hole in units of the Sun's mass.[15] Since your

13. Chapter 7.
14. Ibid.
15. Chapter 3. The quantity 18.5 kilometers, which will appear many times in this book, is 4π (that is, 12.5663706 . . .) times Newton's gravitation constant times the mass of the Sun divided by the square of the speed of light. For this and other useful formulas describing black holes, see the notes to this chapter.

orbital measurements have told you that the hole weighs ten times as much as the Sun, its horizon circumference must be 185 kilometers—about the same as Los Angeles. With your telescopes you carefully measure the circumference: 185 kilometers; perfect agreement with the general relativistic formula.

This horizon circumference is minuscule compared to your starship's 1-million-kilometer orbit; and squeezed into that tiny circumference is a mass ten times larger than that of the Sun! If the hole were a solid body squeezed into such a small circumference, its average density would be 200 million (2×10^8) tons per cubic centimeter—2×10^{14} times more dense than water; see Box P.1. But the hole is not a solid body. General relativity insists that the 10 solar masses of stellar matter, which created the hole by imploding long ago, are now concentrated at the hole's very center—concentrated into a minuscule region of space called a *singularity*.[16] That singularity, roughly 10^{-33} centimeter in size (a hundred billion billion times smaller than an atomic nucleus), should be surrounded by pure emptiness, aside from the tenuous interstellar gas that is falling inward now and the radiation the gas emits. There should be near emptiness from the singularity out to the horizon, and near emptiness from the horizon out to your starship.

16. Chapter 13.

Box P.1

Power Notation for Large and Small Numbers

In this book I occasionally will use "power notation" to describe very large or very small numbers. Examples are 5×10^6, which means five million, or 5,000,000, and 5×10^{-6}, which means five millionths, or 0.000005.

In general, the power to which 10 is raised is the number of digits through which one must move the decimal point in order to put the number into standard decimal notation. Thus 5×10^6 means take 5 (5.00000000) and move its decimal point rightward through six digits. The result is 5000000.00. Similarly, 5×10^{-6} means take 5 and move its decimal point leftward through six digits. The result is 0.000005.

The singularity and the stellar matter locked up in it are hidden by the hole's horizon. However long you may wait, the locked-up matter can never reemerge. The hole's gravity prevents it. Nor can the locked-up matter ever send you information, not by radio waves, or light, or X-rays. For all practical purposes, it is completely gone from our Universe. The only thing left behind is its intense gravitational pull, a pull that is the same on your 1-million-kilometer orbit today as before the star imploded to form the hole, but a pull so strong at and inside the horizon that nothing there can resist it.

"What is the distance from the horizon to the singularity?" you ask yourself. (You choose not to measure it, of course. Such a measurement would be suicidal; you could never escape back out of the horizon to report your result to the World Geographic Society.) Since the singularity is so small, 10^{-33} centimeter, and is at the precise center of the hole, the distance from singularity to horizon should be equal to the horizon's radius. You are tempted to calculate this radius by the standard method of dividing the circumference by 2π (6.283185307 . . .). However, in your studies on Earth you were warned not to believe such a calculation. The hole's enormous gravitational pull completely distorts the geometry of space inside and near the hole,[17] in much the same manner as an extremely heavy rock, placed on a sheet of rubber, distorts the sheet's geometry (Figure P.3), and as a result the horizon's radius is not equal to its circumference divided by 2π.

"Never mind," you say to yourself. "Lobachevsky, Riemann, and other great mathematicians have taught us how to calculate the properties of circles when space is curved, and Einstein has incorporated those calculations into his general relativistic description of the laws of gravity. I can use these curved-space formulas to compute the horizon's radius."

But then you remember from your studies on Earth that, although a black hole's mass and angular momentum determine all the properties of the hole's horizon and exterior, they do not determine its interior. General relativity insists that the interior, near the singularity, should be chaotic and violently nonspherical,[18] much like the tip of the rubber sheet in Figure P.3 if the heavy rock in it is jagged and is bouncing up and down wildly. Moreover, the chaotic nature of the hole's core will depend not only on the hole's mass and angular momentum, but also on the details of the stellar implosion by which the hole was born, and

17. Chapters 3 and 13.
18. Chapter 13.

the details of the subsequent infall of interstellar gas—details that you do not know.

"So what," you say to yourself. "Whatever may be its structure, the chaotic core must have a circumference far smaller than a centimeter. Thus, I will make only a tiny error if I ignore it when computing the horizon's radius."

But then you remember that space can be so extremely warped near the singularity that the chaotic region might be millions of kilometers in radius though only a fraction of a centimeter in circumference, just as the rock in Figure P.3, if heavy enough, can drive the chaotic tip of the rubber sheet exceedingly far downward while leaving the circumference of the chaotic region extremely small. The errors in your cal-

P.3 A heavy rock placed on a rubber sheet (for example, a trampoline) distorts the sheet as shown. The sheet's distorted geometry is very similar to the distortions of the geometry of space around and inside a black hole. For example, the circumference of the thick black circle is far less than 2π times its radius, just as the circumference of the hole's horizon is far less than 2π times its radius. For further detail, see Chapters 3 and 13.

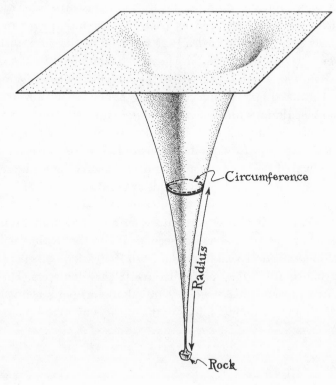

culated radius could thus be enormous. The horizon's radius is simply not computable from the meager information you possess: the hole's mass and its angular momentum.

Abandoning your musings about the hole's interior, you prepare to explore the vicinity of its horizon. Not wanting to risk human life, you ask a rocket-endowed, 10-centimeter-tall robot named Arnold to do the exploration for you and transmit the results back to your starship. Arnold has simple instructions: He must first blast his rocket engines just enough to halt the circular motion that he has shared with the starship, and then he must turn his engines off and let the hole's gravity pull him directly downward. As he falls, Arnold must transmit a brilliant green laser beam back to the starship, and on the beam's electromagnetic oscillations he must encode information about the distance he has fallen and the status of his electronic systems, much as a radio station encodes a newscast on the radio waves it transmits.

Back in the starship your crew will receive the laser beam, and Kares will decode it to get the distance and system information. She will also measure the beam's wavelength (or, equivalently, its color; see Figure P.2). The wavelength is important; it tells how fast Arnold is moving. As he moves faster and faster away from the starship, the green beam he transmits gets *Doppler-shifted,*[19] as received at the ship, to longer and longer wavelengths; that is, it gets more and more red. (There is an additional shift to the red caused by the beam's struggle against the hole's gravitational pull. When computing Arnold's speed, Kares must correct her calculations for this *gravitational redshift.*[20])

And so the experiment begins. Arnold blasts his way out of orbit and onto an infalling trajectory. As he begins to fall, Kares starts a clock to time the arrival of his laser signals. When 10 seconds have elapsed, the decoded laser signal reports that all his systems are functioning well, and that he has already fallen a distance of 2630 kilometers. From the color of the laser light, Kares computes that he is now moving inward with a speed of 530 kilometers per second. When the ticking clock has reached 20 seconds his speed has doubled to 1060 kilometers per second and his distance of fall has quadrupled to 10,500 kilometers. The clock ticks on. At 60 seconds his speed is 9700 kilometers per second, and he has fallen 135,000 kilometers, five-sixths of the way to the horizon.

19. See Box 2.3.
20. Chapters 2 and 3.

You now must pay very close attention. The next few seconds will be crucial, so Kares turns on a high-speed recording system to collect all details of the incoming data. At 61 seconds Arnold reports all systems still functioning normally; the horizon is 14,000 kilometers below him and he is falling toward it at 13,000 kilometers per second. At 61.7 seconds all is still well, 1700 kilometers more to go, speed 39,000 kilometers per second, or about one-tenth the speed of light, laser color beginning to change rapidly. In the next one-tenth of one second you watch in amazement as the laser color zooms through the electromagnetic spectrum, from green to red, to infrared, to microwave, to radiowave, to——. By 61.8 seconds it is all over. The laser beam is completely gone. Arnold has reached the speed of light and disappeared into the horizon. And in that last tenth of a second, just before the beam winked out, Arnold was happily reporting, "All systems go, all systems go, horizon approaching, all systems go, all systems go . . ."

As your excitement subsides, you examine the recorded data. There you find the full details of the shifting laser wavelength. You see that as Arnold fell, the wavelength of the laser signal increased very slowly at first, then faster and faster. But, surprisingly, after the wavelength had quadrupled, its rate of doubling became nearly constant; thereafter the wavelength doubled every 0.00014 second. After 33 doublings (0.0046 second) the wavelength reached 4 kilometers, the limit of your recording system's capabilities. Presumably the wavelength kept right on doubling thereafter. Since it takes an infinite number of doublings for the wavelength to become infinite, exceedingly faint, exceedingly long-wavelength signals might still be emerging from near the horizon!

Does this mean that Arnold has not yet crossed the horizon and never will? No, not at all. Those last, forever-doubling signals take forever long to climb out of the hole's gravitational grip. Arnold flew through the horizon, moving at the speed of light, many minutes ago. The weak remaining signals keep coming out only because their travel time is so long. They are relics of the past.[21]

After many hours of studying the data from Arnold's fall, and after a long sleep to reinvigorate yourself, you embark on the next stage of exploration. This time you, yourself, will probe the horizon's vicinity; but you will do it much more cautiously than did Arnold.

---

21. Chapter 6.

Bidding farewell to your crew, you climb into a space capsule and drop out of the belly of the starship and into a circular orbit alongside it. You then blast your rocket engines ever so gently to slow your orbital motion a bit. This reduces slightly the centrifugal force that holds your capsule up, and the hole's gravity then pulls you into a slightly smaller, coasting, circular orbit. As you again gently blast your engines, your circular orbit again gently shrinks. Your goal, by this gentle, safe, inward spiral, is to reach a circular orbit just above the horizon, an orbit with circumference just 1.0001 times larger than that of the horizon itself. There you can explore most of the horizon's properties, but still escape its fatal grip.

As your orbit slowly shrinks, however, something strange starts to happen. Already at a 100,000-kilometer circumference you feel it. Floating inside the capsule with your feet toward the hole and your head toward the stars, you feel a weak downward tug on your feet and upward tug on your head; you are being stretched like a piece of taffy candy, but gently. The cause, you realize, is the hole's gravity: Your feet are closer to the hole than your head, so the hole pulls on them harder than on your head. The same was true, of course, when you used to stand on the Earth; but the head-to-foot difference on Earth was so minuscule, less than one part in a million, that you never noticed it. By contrast, as you float in your capsule at a circumference of 100,000 kilometers, the head-to-foot difference is one-eighth of an Earth gravity ($\frac{1}{8}$ "g"). At the center of your body the centrifugal force of your orbital motion precisely counteracts the hole's pull. It is as though gravity did not exist; you float freely. But at your feet, the stronger gravity pulls down with an added $\frac{1}{16}$ g, and at your head the weaker gravity allows the centrifugal force to push up with an added $\frac{1}{16}$ g.

Bemused, you continue your inward spiral; but your bemusement quickly changes to worry. As your orbit grows smaller, the forces on your head and feet grow larger. At a circumference of 80,000 kilometers the difference is a $\frac{1}{4}$-g stretching force; at 50,000 kilometers it is a full Earth gravity stretch; at 30,000 kilometers it is 4 Earth gravities. Gritting your teeth in pain as your head and feet are pulled apart, you continue on in to 20,000 kilometers and a 15-g stretching force. More than this you cannot stand! You try to solve the problem by rolling up into a tight ball so your head and feet will be closer together and the difference in forces smaller, but the forces are so strong that they will not let you roll up; they snap you back out into a radial, head-to-foot stretch. If your capsule spirals in much farther, your body will give way; you will be torn apart! There is no hope of reaching the horizon's vicinity.

Frustrated and in enormous pain, you halt your capsule's descent, turn it around, and start carefully, gently, blasting your way back up through circular, coasting orbits of larger and larger circumference and then into the belly of the starship.

Entering the captain's chamber, you vent your frustrations on the ship's master computer, DAWN. "Tikhii, tikhii," she says soothingly (drawing words from the ancient Russian language). "I know you are upset, but it is really your own fault. You were told about those head-to-foot forces in your training. Remember? They are the same forces as produce the tides on the oceans of the Earth."[22]

Thinking back to your training, you recall that the oceans on the side of the Earth nearest the Moon are pulled most strongly by the Moon's gravity and thus bulge out toward the Moon. The oceans on the opposite side of the Earth are pulled most weakly and thus bulge out away from the Moon. The result is two oceanic bulges; and as the Earth turns, those bulges show up as two high tides every twenty-four hours. In honor of those tides, you recall, the head-to-foot gravitational force that you felt is called a *tidal force*. You also recall that Einstein's general relativity describes this tidal force as due to a curvature of space and warpage of time, or, in Einstein's language, a *curvature of spacetime*.[23] Tidal forces and spacetime distortions go hand in hand; one always accompanies the other, though in the case of ocean tides the distortion of spacetime is so tiny that it can be measured only with extremely precise instruments.

But what about Arnold? Why was he so blithely immune to the hole's tidal force? For two reasons, DAWN explains: first, because he was much smaller than you, only 10 centimeters high, and the tidal force, being the difference between the gravitational pulls at his head and his feet, was correspondingly smaller; and second, because he was made of a superstrong titanium alloy that could withstand the stretching force far better than your bones and flesh.

Then with horror you realize that, as he fell through the horizon and on in toward the singularity, Arnold must have felt the tidal force rise up in strength until even his superstrong titanium body could not resist it. Less than 0.0002 second after crossing the horizon, his disintegrat-

22. Chapter 2.
23. Ibid.

ing, stretching body must have neared the hole's central singularity. There, you recall from your study of general relativity back on Earth, the hole's tidal forces must come to life, dancing a chaotic dance, stretching Arnold's remains first in this direction, then in that, then in another, faster and faster, stronger and stronger, until even the individual atoms of which he was made are distorted beyond all recognition. That, in fact, is one essence of the singularity: It is a region where chaotically oscillating spacetime curvature creates enormous, chaotic tidal forces.[24]

Musing over the history of black-hole research, you recall that in 1965 the British physicist Roger Penrose used general relativity's description of the laws of physics to prove that a singularity must reside inside every black hole, and in 1969 the Russian troika of Lifshitz, Khalatnikov, and Belinsky used it to deduce that very near the singularity, tidal gravity must oscillate chaotically, like taffy being pulled first this way and then that by a mechanical taffy-pulling machine.[25] Those were the golden years of theoretical black-hole research, the 1960s and 1970s! But because the physicists of those golden years were not clever enough at solving Einstein's general relativity equations, one key feature of black-hole behavior eluded them. They could only conjecture that whenever an imploding star creates a singularity, it must also create a surrounding horizon that hides the singularity from view; a singularity can never be created "naked," for all the Universe to see. Penrose called this the "conjecture of cosmic censorship," since, if correct, it would censor all experimental information about singularities: One could never do experiments to test one's theoretical understanding of singularities, unless one were willing to pay the price of entering a black hole, dying while making the measurements, and not even being able to transmit the results back out of the hole as a memorial to one's efforts.

Although Dame Abygaile Lyman, in 2023, finally resolved the issue of whether cosmic censorship is true or not, the resolution is irrelevant to you now. The only singularities charted in your ship's atlases are those inside black holes, and you refuse to pay the price of death to explore them.

Fortunately, outside but near a black-hole horizon there are many phenomena to explore. You are determined to experience those phe-

24. Chapter 13.
25. Ibid.

nomena firsthand and report back to the World Geographic Society, but you cannot experience them near Hades' horizon. The tidal force there is too great. You must explore, instead, a black hole with weaker tidal forces.

General relativity predicts, DAWN reminds you, that as a hole grows more massive, the tidal forces at and above its horizon grow weaker. This seemingly paradoxical behavior has a simple origin: The tidal force is proportional to the hole's mass divided by the cube of its circumference; so as the mass grows, and the horizon circumference grows proportionally, the near-horizon tidal force actually decreases. For a hole weighing a million solar masses, that is, 100,000 times more massive than Hades, the horizon will be 100,000 times larger, and the tidal force there will be 10 billion (10^{10}) times weaker. That would be comfortable; no pain at all! So you begin making plans for the next leg of your voyage: a journey to the nearest million-solar-mass hole listed in Schechter's Black-Hole Atlas—a hole called Sagittario at the center of our Milky Way galaxy, 30,100 light-years away.

Several days later your crew transmit back to Earth a detailed description of your Hades explorations, including motion pictures of you being stretched by the tidal force and pictures of atoms falling into the hole. The description will require 26 years to cover the 26 light-year distance to Earth, and when it finally arrives it will be published with great fanfare by the World Geographic Society.

In their transmission the crew describe your plan for a voyage to the center of the Milky Way: Your starship's rocket engines will blast all the way with a 1-g acceleration, so that you and your crew can experience a comfortable 1-Earth-gravity force inside the starship. The ship will accelerate toward the galactic center for half the journey, then it will rotate 180 degrees and decelerate at 1 g for the second half. The entire trip of 30,100 light-years distance will require 30,102 years as measured on Earth; but as measured on the starship it will require only 20 years. In accordance with Einstein's laws of special relativity,[26] your ship's high speed will cause time, as measured on the ship, to "dilate"; and this *time dilation* (or *time warp*), in effect, will make the starship behave like a time machine, projecting you far into the Earth's future while you age only a modest amount.[27]

You explain to the World Geographic Society that your next trans-

26. Chapter 1.
27. Ibid.

mission will come from the vicinity of the galactic center, after you
have explored its million-solar-mass hole, Sagittario. Members of the
Society must go into deep-freeze hibernation for 60,186 years if they
wish to live to receive your transmission (30,102 − 26 = 30,076 years
from the time they receive your message until you reach the galactic
center, plus 30,110 years while your next transmission travels from the
galactic center to Earth).

Sagittario

After a 20-year voyage as measured in starship time, your ship decelerates into the Milky Way's center. There in the distance you see a
rich mixture of gas and dust flowing inward from all directions toward
an enormous black hole. Kares adjusts the rocket blast to bring the starship into a coasting, circular orbit well above the horizon. By measuring the circumference and period of your orbit and plugging the results
into Newton's formula, you determine the mass of the hole. It is 1
million times the mass of the Sun, just as claimed in Schechter's Black-
Hole Atlas. From the absence of any tornado-like swirl in the inflowing
gas and dust you infer that the hole is not spinning much; its horizon,
therefore, must be spherical and its circumference must be 18.5 million
kilometers, eight times larger than the Moon's orbit around the Earth.

After further scrutiny of the infalling gas, you prepare to descend
toward the horizon. For safety, Kares sets up a laser communication
link between your space capsule and your starship's master computer,
DAWN. You then drop out of the belly of the starship, turn your
capsule so its jets point in the direction of your circling orbital motion,
and start blasting gently to slow your orbital motion and drive yourself
into a gentle inward (downward) spiral from one coasting circular orbit
to another.

All goes as expected until you reach an orbit of circumference 55
million kilometers—just three times the circumference of the horizon.
There the gentle blast of your rocket engine, instead of driving you into
a slightly tighter circular orbit, sends you into a suicidal plunge toward
the horizon. In panic you rotate your capsule and blast with great force
to move back up into an orbit just outside 55 million kilometers.

"What the hell went wrong!?" you ask DAWN by laser link.

"Tikhii, tikhii," she replies soothingly. "You planned your orbit
using Newton's description of the laws of gravity. But Newton's de-

scription is only an approximation to the true gravitational laws that govern the Universe.[28] It is an excellent approximation far from the horizon, but bad near the horizon. Much more accurate is Einstein's general relativistic description; it agrees to enormous precision with the true laws of gravity near the horizon, and it predicts that, as you near the horizon, the pull of gravity becomes stronger than Newton ever expected. To remain in a circular orbit, with this strengthened gravity counterbalanced by the centrifugal force, you must strengthen your centrifugal force, which means you must increase your orbital speed around the black hole: As you descend through three horizon circumferences, you must rotate your capsule around and start blasting yourself forward. Because instead you kept blasting backward, slowing your motion, gravity overwhelmed your centrifugal force as you passed through three horizon circumferences, and hurled you inward."

"Damn that DAWN!" you think to yourself. "She always answers my questions, but she never volunteers crucial information. She never warns me when I'm going wrong!" You know the reason, of course. Human life would lose its zest and richness if computers were permitted to give warning whenever a mistake was being made. Back in 2032 the World Council passed a law that a Hobson block preventing such warnings must be embedded in all computers. As much as she might wish, DAWN cannot bypass her Hobson block.

Suppressing your exasperation, you rotate your capsule and begin a careful sequence of forward blast, inward spiral, coast, forward blast, inward spiral, coast, forward blast, inward spiral, coast, which takes you from 3 horizon circumferences to 2.5 to 2.0 to 1.6 to 1.55 to 1.51 to 1.505 to 1.501 to . . . What frustration! The more times you blast and the faster your resulting coasting, circular motion, the smaller becomes your orbit; but as your coasting speed approaches the speed of light, your orbit only approaches 1.5 horizon circumferences. Since you can't move faster than light, there is no hope of getting close to the horizon itself by this method.

Again you appeal to DAWN for help, and again she soothes you and explains: Inside 1.5 horizon circumferences there are no circular orbits at all. Gravity's pull there is so strong that it cannot be counteracted by any centrifugal forces, not even if one coasts around and around the hole at the speed of light. If you want to go closer, DAWN says, you must abandon your circular, coasting orbit and instead descend directly

28. Chapter 2.

toward the horizon, with your rockets blasting downward to keep you from falling catastrophically. The force of your rockets will support you against the hole's gravity as you slowly descend and then hover just above the horizon, like an astronaut hovering on blasting rockets just above the Moon's surface.

Having learned some caution by now, you ask DAWN for advice about the consequences of such a strong, steady rocket blast. You explain that you wish to hover at a location, 1.0001 horizon circumferences, where most of the effects of the horizon can be experienced, but from which you can escape. If you support your capsule there by a steady rocket blast, how much acceleration force will you feel? "One hundred and fifty million Earth gravities," DAWN replies gently.

Deeply discouraged, you blast and spiral your way back up into the belly of the starship.

After a long sleep, followed by five hours of calculations with general relativity's black-hole formulas, three hours of plowing through Schechter's Black-Hole Atlas, and an hour of consultation with your crew, you formulate the plan for the next leg of your voyage.

Your crew then transmit to the World Geographic Society, under the optimistic assumption that it still exists, an account of your experiences with Sagittario. At the end of their transmission your crew describe your plan:

Your calculations show that the larger the hole, the weaker the rocket blast you will need to support yourself, hovering, at 1.0001 horizon circumferences. For a painful but bearable 10-Earth-gravity blast, the hole must be 15 trillion (15×10^{12}) solar masses. The nearest such hole is the one called Gargantua, far outside the 100,000 (10^5) light-year confines of our own Milky Way galaxy, and far outside the 100 million (10^8) light-year Virgo cluster of galaxies, around which our Milky Way orbits. In fact, it is near the quasar 3C273, 2 billion (2×10^9) light-years from the Milky Way and 10 percent of the distance to the edge of the observable part of the Universe.

The plan, your crew explain in their transmission, is a voyage to Gargantua. Using the usual 1-g acceleration for the first half of the trip and 1-g deceleration for the second half, the voyage will require 2 billion years as measured on Earth, but, thanks to the speed-induced warpage of time, only 42 years as measured by you and your crew in the starship. If the members of the World Geographic Society are not willing to chance a 4-billion-year deep-freeze hibernation (2 billion years for the starship to reach Gargantua and 2 billion years for its transmission to return to Earth), then they will have to forgo receiving your next transmission.

Gargantua

Forty-two years of starship time later, your ship decelerates into the vicinity of Gargantua. Overhead you see the quasar 3C273, with two brilliant blue jets squirting out of its center[29]; below is the black abyss of Gargantua. Dropping into orbit around Gargantua and making your usual measurements, you confirm that its mass is, indeed, 15 trillion times that of the Sun, you see that it is spinning very slowly, and you compute from these data that the circumference of its horizon is 29 light-years. Here, at last, is a hole whose vicinity you can explore while experiencing bearably small tidal forces and rocket accelerations! The safety of the exploration is so assured that you decide to take the entire starship down instead of just a capsule.

Before beginning the descent, however, you order your crew to photograph the giant quasar overhead, the trillions of stars that orbit Gargantua, and the billions of galaxies sprinkled over the sky. They also photograph Gargantua's black disk below you; it is about the size of the sun as seen from Earth. At first sight it appears to blot out the light from all the stars and galaxies behind the hole. But looking more closely, your crew discover that the hole's gravitational field has acted like a lens[30] to deflect some of the starlight and galaxy light around the edge of the horizon and focus it into a thin, bright ring at the edge of the black disk. There, in that ring, you see several images of each obscured star: one image produced by light rays that were deflected around the left limb of the hole, another by rays deflected around the right limb, a third by rays that were pulled into one complete orbit around the hole and then released in your direction, a fourth by rays that orbited the hole twice, and so on. The result is a highly complex ring structure, which your crew photograph in great detail for future study.

The photographic session complete, you order Kares to initiate the starship's descent. But you must be patient. The hole is so huge that, accelerating and then decelerating at 1 g, it will require 13 years of starship time to reach your goal of 1.0001 horizon circumferences!

29. Chapter 9.
30. Chapter 8.

As the ship descends, your crew make a photographic record of the changes in the appearance of the sky around the starship. Most remarkable is the change in the hole's black disk below the ship: Gradually it grows larger. You expect it to stop growing when it has covered the entire sky below you like a giant black floor, leaving the sky overhead as clear as on Earth. But no; the black disk keeps right on growing, swinging up around the sides of your starship to cover everything except a bright, circular opening overhead, an opening through which you see the external Universe (Figure P.4). It is as though you had entered a cave and were plunging deeper and deeper, watching the cave's bright mouth grow smaller and smaller in the distance.

In growing panic, you appeal to DAWN for help: "Did Kares miscalculate our trajectory? Have we plunged through the horizon? Are we doomed?!"

P.4   The starship hovering above the black-hole horizon, and the trajectories along which light travels to it from distant galaxies (the light rays). The hole's gravity deflects the light rays downward ("gravitational lens effect"), causing humans on the starship to see all the light concentrated in a bright, circular spot overhead.

"Tikhii, tikhii," she replies soothingly. "We are safe; we are still outside the horizon. Darkness has covered most of the sky only because of the powerful lensing effect of the hole's gravity. Look there, where my pointer is, almost precisely overhead; that is the galaxy 3C295. Before you began your plunge it was in a horizontal position, 90 degrees from the zenith. But here near Gargantua's horizon the hole's gravity pulls so hard on the light rays from 3C295 that it bends them around from horizontal to nearly vertical. As a result 3C295 appears to be nearly overhead."

Reassured, you continue your descent. The console displays your ship's progress in terms of both the radial (downward) distance traveled and the circumference of a circle around the hole that passes through your location. In the early stages of your descent, for each kilometer of radial distance traveled, your circumference decreased by 6.283185307 . . . kilometers. The ratio of circumference decrease to radius decrease was 6.283185307 kilometers/1 kilometer, which is equal to 2π, just as Euclid's standard formula for circles predicts. But now, as your ship nears the horizon, the ratio of circumference decrease to radius decrease is becoming much smaller than 2π: It is 5.960752960 at 10 horizon circumferences; 4.442882938 at 2 horizon circumferences; 1.894451650 at 1.1 horizon circumferences; 0.625200306 at 1.01 horizon circumferences. Such deviations from the standard Euclidean geometry that teenagers learn in school are possible only in a curved space; you are seeing the curvature which Einstein's general relativity predicts must accompany the hole's tidal force.[31]

In the final stage of your ship's descent, Kares blasts the rockets harder and harder to slow its fall. At last the ship comes to a hovering rest at 1.0001 horizon circumferences, blasting with a 10-*g* acceleration to hold itself up against the hole's powerful gravitational pull. In its last 1 kilometer of radial travel the circumference decreases by only 0.062828712 kilometer.

Struggling to lift their hands against the painful 10-*g* force, your crew direct their telescopic cameras into a long and detailed photographic session. Except for wisps of weak radiation all around you from collisionally heated, infalling gas, the only electromagnetic waves to be photographed are those in the bright spot overhead. The spot is small, just 3 degrees of arc in diameter, six times the size of the Sun as seen from Earth. But squeezed into that spot are images of all the stars that

31. Chapters 2 and 3.

orbit Gargantua, and all the galaxies in the Universe. At the precise center are the galaxies that are truly overhead. Fifty-five percent of the way from the spot's center to its edge are images of galaxies like 3C295 which, if not for the hole's lens effect, would be in horizontal positions, 90 degrees from the zenith. Thirty-five percent of the way to the spot's edge are images of galaxies that you know are really on the opposite side of the hole from you, directly below you. In the outermost 30 percent of the spot is a second image of each galaxy; and in the outermost 2 percent, a third image!

Equally peculiar, the colors of all the stars and galaxies are wrong. A galaxy that you know is really green appears to be shining with soft X-rays: Gargantua's gravity, in pulling the galaxy's radiation downward to you, has made the radiation more energetic by decreasing its wavelength from 5×10^{-7} meter (green) to 5×10^{-9} meter (X-ray). And similarly, the outer disk of the quasar 3C273, which you know emits infrared radiation of wavelength 5×10^{-5} meter, appears to be shining with green 5×10^{-7} meter light.

After thoroughly recording the details of the overhead spot, you turn your attention to the interior of your starship. You half expect that here, so near the hole's horizon, the laws of physics will be changed in some way, and those changes will affect your own physiology. But no. You look at your first mate, Kares; she appears normal. You look at your second mate, Bret; he appears normal. You touch each other; you feel normal. You drink a glass of water; aside from the effects of the 10-*g* acceleration, the water goes down normally. Kares turns on an argon ion laser; it produces the same brilliant green light as ever. Bret pulses a ruby laser on and then off, and measures the time it takes for the pulse of light to travel from the laser to a mirror and back; from his measurement he computes the speed of the light's travel. The result is absolutely the same as in an Earth-based laboratory: 299,792 kilometers per second.

Everything in the starship is normal, absolutely the same as if the ship had been resting on the surface of a massive planet with 10-*g* gravity. If you did not look outside the starship and see the bizarre spot overhead and the engulfing blackness all around, you would not know that you were very near the horizon of a black hole rather than safely on the surface of a planet—or you almost wouldn't know. The hole curves spacetime inside your starship as well as outside, and with sufficiently accurate instruments, you can detect the curvature; for example, by its tidal stretch between your head and your feet. But whereas the curvature is enormously important on the scale of the horizon's 300-trillion-kilometer circumference, its effects are minuscule on the scale of your 1-kilometer starship; the curvature-produced tidal force between one end of the starship and the other is just one-hundredth of a trillionth of an Earth gravity (10^{-14} g), and between your own head and feet it is a thousand times smaller than that!

To pursue this remarkable normality further, Bret drops from the starship a capsule containing a pulsed-laser-and-mirror instrument for measuring the speed of light. As the capsule plunges toward the horizon, the instrument measures the speed with which light pulses travel from the laser in the capsule's nose to the mirror in its tail and back. A computer in the capsule transmits the result on a laser beam up to the ship: "299,792 kilometers per second; 299,792; 299,792; 299,792 . . ." The color of the incoming laser beam shifts from green to red to infrared to microwave to radio as the capsule nears the horizon, and still the message is the same: "299,792; 299,792; 299,792 . . ." And then the laser beam is gone. The capsule has pierced the horizon, and never once as it fell was there any change in the speed of light inside it, nor was there any change in the laws of physics that governed the workings of the capsule's electronic systems.

These experimental results please you greatly. In the early twentieth century Albert Einstein proclaimed, largely on philosophical grounds, that the local laws of physics (the laws in regions small enough that one can ignore the curvature of spacetime) should be the same everywhere in the Universe. This proclamation has been enshrined as a fundamental principle of physics, the *equivalence principle*.[32] Often in the ensuing centuries the equivalence principle was subjected to experimental test, but never was it tested so graphically and thoroughly as in your experiments here near Gargantua's horizon.

You and your crew are now tiring of the struggle with 10 Earth gravities, so you prepare for the next and final leg of your voyage, a return to our Milky Way galaxy. Your crew will transmit an account of your Gargantua explorations during the early stages of the voyage; and since your starship itself will soon be traveling at nearly the speed of light, the transmissions will reach the Milky Way less than a year before the ship, as measured from Earth.

As your starship pulls up and away from Gargantua, your crew make a careful, telescopic study of the quasar 3C273 overhead[33] (Figure P.5).

32. Chapter 2.
33. Chapter 9.

Its jets—thin spikes of hot gas shooting out of the quasar's core—are enormous: 3 million light-years in length. Training your telescopes on the core, your crew see the source of the jets' power: a thick, hot, doughnut of gas less than 1 light-year in size, with a black hole at its center. The doughnut, which astrophysicists have called an "accretion disk," orbits around and around the black hole. By measuring its rotation period and circumference, your crew infer the mass of the hole: 2 billion (2×10^9) solar masses, 7500 times smaller than Gargantua, but far larger than any hole in the Milky Way. A stream of gas flows from the doughnut to the horizon, pulled by the hole's gravity. As it nears the horizon the stream, unlike any you have seen before, swirls around and around the hole in a tornado-type motion. This hole must be spinning fast! The axis of spin is easy to identify; it is the axis about which the gas stream swirls. The two jets, you notice, shoot out along the spin axis. They are born just above the horizon's north and south poles, where they suck up energy from the hole's spin and from the doughnut,[34] much like a tornado sucks up dust from the earth.

The contrast between Gargantua and 3C273 is amazing: Why does Gargantua, with its 1000 times greater mass and size, not possess an encircling doughnut of gas and gigantic quasar jets? Bret, after a long telescopic study, tells you the answer: Once every few months a star in orbit around 3C273's smaller hole strays close to the horizon and gets ripped apart by the hole's tidal force. The star's guts, roughly 1 solar mass worth of gas, get spewed out and strewn around the hole. Gradually internal friction drives the strewn-out gas down into the doughnut. This fresh gas compensates for the gas that the doughnut is continually feeding into the hole and the jets. The doughnut and jets thereby are kept richly full of gas, and continue to shine brightly.

Stars also stray close to Gargantua, Bret explains. But because Gargantua is far larger than 3C273, the tidal force outside its horizon is too weak to tear any star apart. Gargantua swallows stars whole without spewing their guts into a surrounding doughnut. And with no doughnut, Gargantua has no way of producing jets or other quasar violence.

As your starship continues to rise out of Gargantua's gravitational grip, you make plans for the journey home. By the time your ship reaches the Milky Way, the Earth will be 4 billion years older than when you left. The changes in human society will be so enormous that you don't want to return there. Instead, you and your crew decide to

---

34. Chapters 9 and 11.

P.5   The quasar 3C273: a 2-billion-solar-mass black hole encircled by a dough-nut of gas ("accretion disk") and with two gigantic jets shooting out along the hole's spin axis.

colonize the space around a spinning black hole. You know that just as the spin energy of the hole in 3C273 helps power the quasar's jets, so the spin energy of a smaller hole can be used as a power source for human civilization.

You do not want to arrive at some chosen hole and discover that other beings have already built a civilization around it; so instead of aiming your starship at a rapidly spinning hole that already exists, you aim at a star system which will give birth to a rapidly spinning hole shortly after your ship arrives.

In the Milky Way's Orion nebula, at the time you left Earth, there was a *binary star system* composed of two 30-solar-mass stars orbiting each other. DAWN has calculated that each of those stars should have imploded, while you were outbound to Gargantua, to form a 24-solar-mass, nonspinning hole (with 6 solar masses of gas ejected during the implosion). Those two 24-solar-mass holes should now be circling around each other as a *black-hole binary,* and as they circle, they should emit ripples of tidal force (ripples of "spacetime curvature") called *gravitational waves.*[35] These gravitational waves should push back on the binary in much the same way as an outflying bullet pushes back on the gun that fires it, and this *gravitational-wave recoil* should drive the holes into a slow but inexorable inward spiral. With a slight adjustment of your starship's acceleration, you can time your arrival to coincide with the last stage of that inward spiral: Several days after you arrive, you will see the holes' nonspinning horizons whirl around and around each other, closer and closer, and faster and faster, until they coalesce to produce a single whirling, spinning, larger horizon.

Because the two parent holes do not spin, neither alone can serve as an efficient power source for your colony. However, the newborn, rapidly spinning hole will be ideal!

Home

After a 42-year voyage your starship finally decelerates into the Orion nebula, where DAWN predicted the two holes should be. There they are, right on the mark! By measuring the orbital motion of interstellar atoms as they fall into the holes, you verify that their horizons are not spinning and that each weighs 24 solar masses, just as DAWN

35. Chapter 10.

predicted. Each horizon has a circumference of 440 kilometers; they are 30,000 kilometers apart; and they are orbiting around each other once each 13 seconds. Inserting these numbers into the general relativity formulas for gravitational-wave recoil, you conclude that the two holes should coalesce seven days from now. There is just time enough for your crew to prepare their telescopic cameras and record the details. By photographing the bright ring of focused starlight that encircles each hole's black disk, they can easily monitor the holes' motions.

You want to be near enough to see clearly, but far enough away to be safe from the holes' tidal forces. A good location, you decide, is a starship orbit ten times larger than the orbit in which the holes circle each other—an orbital diameter of 300,000 kilometers and orbital circumference of 940,000 kilometers. Kares maneuvers the starship into that orbit, and your crew begin their telescopic, photographic observations.

Over the next six days the two holes gradually move closer to each other and speed up their orbital motion. One day before coalescence, the distance between them has shrunk from 30,000 to 18,000 kilometers and their orbital period has decreased from 13 to 6.3 seconds. One hour before coalescence they are 8400 kilometers apart and their orbital period is 1.9 seconds. One minute before coalescence: separation 3000 kilometers, period 0.41 second. Ten seconds before coalescence: separation 1900 kilometers, period 0.21 second.

Then, in the last ten seconds, you and your starship begin to shake, gently at first, then more and more violently. It is as though a gigantic pair of hands had grabbed your head and feet and were alternately compressing and stretching you harder and harder, faster and faster. And then, more suddenly than it started, the shaking stops. All is quiet.

"What was that?" you murmur to DAWN, your voice trembling.

"Tikhii, tikhii," she replies soothingly. "That was the undulating tidal force of gravitational waves from the holes' coalescence. You are accustomed to gravitational waves so weak that only very delicate instruments can detect their tidal force. However, here, close to the coalescing holes, they were enormously strong—strong enough that, had we parked our starship in an orbit 30 times smaller, it would have been torn apart by the waves. But now we are safe. The coalescence is complete and the waves are gone; they are on their way out into the Universe, carrying to distant astronomers a symphonic description of the coalescence."[36]

36. Chapter 10.

Training one of your crew's telescopes on the source of gravity below, you see that DAWN is right, the coalescence is complete. Where before there were two holes there now is just one, and it is spinning rapidly, as you see from the swirl of infalling atoms. This hole will make an ideal power generator for your crew and thousands of generations of their descendants.

By measuring the starship's orbit, Kares deduces that the hole weighs 45 solar masses. Since the parent holes totaled 48 solar masses, 3 solar masses must have been converted into pure energy and carried off by the gravitational waves. No wonder the waves shook you so hard!

As you turn your telescopes toward the hole, a small object unexpectedly hurtles past your starship, splaying brilliant sparks profusely in all directions, and then explodes, blasting a gaping hole in your ship's side. Your well-trained crew and robots rush to their battle stations, you search vainly for the attacking warship—and then, responding to an appeal for her help, DAWN announces soothingly over the ship's speaker system, "Tikhii, tikhii; we are not being attacked. That was just a freak primordial black hole, evaporating and then exploding."[37]

"A what?!" you cry out.

"A primordial black hole, evaporating and then destroying itself in an explosion," DAWN repeats.

"Explain!" you demand. "What do you mean by *primordial?* What do you mean by *evaporating and exploding?* You're not making sense. Things can fall into a black hole, but nothing can ever come out; nothing can 'evaporate.' And a black hole lives forever; it always grows, never shrinks. There is no way a black hole can 'explode' and destroy itself. That's absurd."

Patiently as always, DAWN educates you. "Large objects—such as humans, stars, and black holes formed by the implosion of a star—are governed by the *classical* laws of physics," she explains, "by Newton's laws of motion, Einstein's relativity laws, and so forth. By contrast, tiny objects—for example, atoms, molecules, and black holes smaller than an atom—are governed by a very different set of laws, the *quantum* laws of physics.[38] While the classical laws forbid a normal-sized black hole ever to evaporate, shrink, explode, or destroy itself, not so the quantum laws. They demand that any atom-sized black hole gradually evaporate and shrink until it reaches a critically small circumference,

37. Chapter 12.
38. Chapters 4–6, 10, 12–14.

about the same as an atomic nucleus. The hole, which despite its tiny size weighs about a billion tons, must then destroy itself in an enormous explosion. The explosion converts all of the hole's billion-ton mass into outpouring energy; it is a trillion times more energetic than the most powerful nuclear explosions that humans ever detonated on Earth in the twentieth century. Just such an explosion has now damaged our ship," DAWN explains.

"But you needn't worry that more explosions will follow," DAWN continues. "Such explosions are exceedingly rare because tiny black holes are exceedingly rare. The only place that tiny holes were ever created was in our Universe's big bang birth, twenty billion years ago; that is why they are called *primordial* holes. The big bang created only a few such primordial holes, and those few have been slowly evaporating and shrinking ever since their birth. Once in a great while one of them reaches its critical, smallest size and explodes.[39] It was only by chance—an extremely improbable occurrence—that one exploded while hurtling past our ship, and it is exceedingly unlikely that our starship will ever encounter another such hole."

Relieved, you order your crew to begin repairs on the ship while you and your mates embark on your telescopic study of the 45-solar-mass, rapidly spinning hole below you.

The hole's spin is obvious not only from the swirl of infalling atoms, but also from the shape of the bright-ringed black spot it makes on the sky below you: The black spot is squashed, like a pumpkin; it bulges at its equator and is flattened at its poles. The centrifugal force of the hole's spin, pushing outward, creates the bulge and flattening.[40] But the bulge is not symmetric; it looks larger on the right edge of the disk, which is moving away from you as the horizon spins, than on the left edge. DAWN explains that this is because the horizon can capture rays of starlight more easily if they move toward you along its right edge, against the direction of its spin, than along its left edge, with its spin.

By measuring the shape of the spot and comparing it with general relativity's black-hole formulas, Bret infers that the hole's spin angular momentum is 96 percent of the maximum allowed for a hole of its mass. And from this angular momentum and the hole's mass of 45 Suns you compute other properties of the hole, including the spin rate of its

39. Chapter 12.
40. Chapter 7.

horizon, 270 revolutions per second, and its equatorial circumference, 533 kilometers.

The spin of the hole intrigues you. Never before could you observe a spinning hole up close. So with pangs of conscience you ask for and get a volunteer robot, to explore the neighborhood of the horizon and transmit back his experiences. You give the robot, whose name is Kolob, careful instructions: "Descend to ten meters above the horizon and there blast your rockets to hold yourself at rest, hovering directly below the starship. Use your rockets to resist both the inward pull of gravity and the tornado-like swirl of space."

Eager for adventure, Kolob drops out of the starship's belly and plunges downward, blasting his rockets gently at first, then harder, to resist the swirl of space and remain directly below the ship. At first Kolob has no problems. But when he reaches a circumference of 833 kilometers, 56 percent larger than the horizon, his laser light brings the message, "I can't resist the swirl; I can't; I can't!" and like a rock caught up in a tornado he gets dragged into a circulating orbit around the hole.[41]

"Don't worry," you reply. "Just do your best to resist the swirl, and continue to descend until you are ten meters above the horizon."

Kolob complies. As he descends, he is dragged into more and more rapid circulating motion. Finally, when he stops his descent and hovers ten meters above the horizon, he is encircling the hole in near perfect lockstep with the horizon itself, 270 circuits per second. No matter how hard he blasts to oppose this motion, he cannot. The swirl of space won't let him stop.

"Blast in the other direction," you order. "If you can't circle more slowly than 270 circuits per second, try circling faster."

Kolob tries. He blasts, keeping himself always 10 meters above the horizon but trying to encircle it faster than before. Although he feels the usual acceleration from his blast, you see his motion change hardly at all. He still circles the hole 270 times per second. And then, before you can transmit further instructions, his fuel gives out; he begins to plummet downward; his laser light zooms through the electromagnetic spectrum from green to red to infrared to radio waves, and then turns black with no change in his circulating motion. He is gone, down the hole, plunging toward the violent singularity that you will never see.

41. Chapter 7.

After three weeks of mourning, experiments, and telescopic studies, your crew begin to build for the future. Bringing in materials from distant planets, they construct a girder-work ring around the hole. The ring has a circumference of 5 million kilometers, a thickness of 3.4 kilometers, and a width of 4000 kilometers. It rotates at just the right rate, two rotations per hour, for centrifugal forces to counterbalance the hole's gravitational pull at the ring's central layer, 1.7 kilometers from its inner and outer faces. Its dimensions are carefully chosen so that those people who prefer to live in 1 Earth gravity can set up their homes near the inner or outer face of the ring, while those who prefer weaker gravity can live nearer its center. These differences in gravity are due in part to the rotating ring's centrifugal force and in part to the hole's tidal force—or, in Einstein's language, to the curvature of spacetime.

The electric power that heats and lights this ring world is extracted from the black hole: Twenty percent of the hole's mass is in the form of energy that is stored in the tornado-like swirl of space outside but near the horizon.[42] This is 10,000 times more energy than the Sun will radiate as heat and light in its entire lifetime!—and being outside the horizon, it can be extracted. Never mind that the ring world's energy extractor is only 50 percent efficient; it still has a 5000 times greater energy supply than the Sun.

The energy extractor works on the same principle as do some quasars[43]: Your crew have threaded a magnetic field through the hole's horizon and they hold it on the hole, despite its tendency to pop off, by means of giant superconducting coils (Figure P.6). As the horizon spins, it drags the nearby space into a tornado-like swirl which in turn interacts with the threading magnetic field to form a gigantic electric power generator. The magnetic field lines act as transmission lines for the power. Electric current is driven out of the hole's equator (in the form of electrons flowing inward) and up the magnetic field lines to the ring world. There the current deposits its power. Then it flows out of the ring world on another set of magnetic field lines and down into the hole's north and south poles (in the form of positrons flowing inward). By adjusting the strength of the magnetic field, the world's inhabitants can adjust the power output: weak field and low power in the world's early years; strong field and high power in later years. Gradually as the power is extracted, the hole will slow its spin, but it

---

42. Chapters 7 and 11.
43. Chapters 9 and 11.

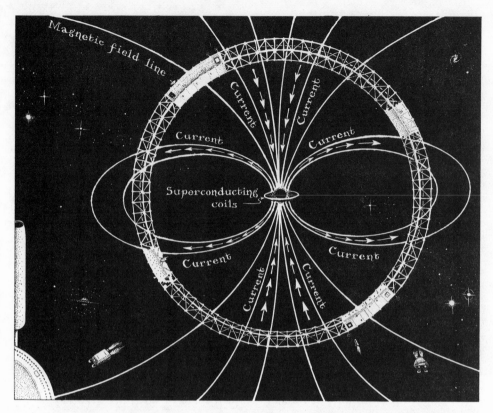

P.6 A city on a girder-work ring around a spinning black hole, and the electro-magnetic system by which the city extracts power from the hole's spin.

will take many eons to exhaust the hole's enormous store of spin energy.

Your crew and countless generations of their descendants can call this artificial world "home" and use it as a base for future explorations of the Universe. But not you. You long for the Earth and the friends whom you left behind, friends who must have been dead now for more than 4 billion years. Your longing is so great that you are willing to risk the last quarter of your normal, 200-year life span in a dangerous and perhaps foolhardy attempt to return to the idyllic era of your youth.

Time travel into the future is rather easy, as your voyage among the holes has shown. Not so travel into the past. In fact, such travel might be completely forbidden by the fundamental laws of physics. However, DAWN tells you of speculations, dating back to the twentieth century, that backward time travel might be achieved with the aid of a hypothetical space warp called a *wormhole*.[44] This space warp consists of two

44. Chapter 14.

P.7   The two mouths of a hypothetical wormhole. Enter either mouth, and you will emerge from the other, having traveled through a short tube (the wormhole's throat) that extends not through our Universe, but through hyperspace.

entrance holes (the wormhole's *mouths*), which look much like black holes but without horizons, and which can be far apart in the Universe (Figure P.7). Anything that enters one mouth finds itself in a very short tube (the wormhole's *throat*) that leads to and out of the other mouth. The tube cannot be seen from our Universe because it extends through *hyperspace* rather than through normal space. It might be possible for time to hook up through the wormhole in a different way than through our Universe, DAWN explains. By traversing the wormhole in one direction, say from the left mouth to the right, one might go backward in our Universe's time, while traversing in the opposite direction, from right to left, one would go forward. Such a wormhole would be a time warp, as well as a space warp.

The laws of quantum gravity demand that exceedingly tiny wormholes of this type exist,[45] DAWN tells you. These quantum wormholes must be so tiny, just 10^{-33} centimeter in size, that their existence is only fleeting—far too brief, 10^{-43} second, to be usable for time travel. They

45. Chapters 13 and 14.

must flash into existence and then flash out in a random, unpredictable manner—here, there, and everywhere. Very occasionally a flashing wormhole will have one mouth near the ring world today and the other near Earth in the era 4 billion years ago when you embarked on your voyage. DAWN proposes to try to catch such a wormhole as it flickers, enlarge it like a child blowing up a balloon, and keep it open long enough for you to travel through it to the home of your youth.

But DAWN warns you of great danger. Physicists have conjectured, though it has never been proved, that an instant before an enlarging wormhole becomes a time machine, the wormhole must self-destruct with a gigantic, explosive flash. In this way the Universe might protect itself from time-travel paradoxes, such as a man going back in time and killing his mother before he was conceived, thereby preventing himself from being born and killing his mother.[46]

If the physicists' conjecture is wrong, then DAWN might be able to hold the wormhole open for a few seconds, with a large enough throat for you to travel through. By waiting nearby as she enlarges the wormhole and then plunging through it, within a fraction of a second of your own time you will arrive home on Earth, in the era of your youth 4 billion years ago. But if the time machine self-destructs, you will be destroyed with it. You decide to take the chance . . .

The above tale sounds like science fiction. Indeed, part of it is: I cannot by any means guarantee that there exists a 10-solar-mass black hole near the star Vega, or a million-solar-mass hole at the center of the Milky Way, or a 15-trillion-solar-mass black hole anywhere at all in the Universe; they are all speculative but plausible fiction. Nor can I guarantee that humans will ever succeed in developing the technology for intergalactic travel, or even for interstellar travel, or for constructing ring worlds on girder-work structures around black holes. These are also speculative fiction.

On the other hand, I can guarantee with considerable but not complete confidence that black holes exist in our Universe and have the precise properties described in the above tale. If you hover in a blasting starship just above the horizon of a 15-trillion-solar-mass hole, I guarantee that the laws of physics will be the same inside your starship as on Earth, and that when you look out at the heavens around you, you will see the entire Universe shining down at you in a brilliant, small disk of light. I guarantee that, if you send a robot probe down near the horizon of a spinning hole, blast as it may it will never be able to move forward or backward at any speed other than the hole's own spin speed (270 circuits per second in my example). I guarantee that a rapidly spinning hole can store as much as 29 percent of its mass as spin energy, and that if one is clever enough, one can extract that energy and use it.

How can I guarantee all these things with considerable confidence? After all, I have never seen a black hole. Nobody has. Astronomers have found only indirect evidence for the existence of black holes[47] and no observational evidence whatsoever for their claimed detailed properties. How can I be so audacious as to guarantee so much about them? For one simple reason. Just as the laws of physics predict the pattern of ocean tides on Earth, the time and height of each high tide and each low tide, so also the laws of physics, if we understand them correctly, predict these black-hole properties, and predict them with no equivocation. From Newton's description of the laws of physics one can deduce, by mathematical calculations, the sequence of Earth tides for the year 1999 or the year 2010; similarly, from Einstein's general relativity description of the laws, one can deduce, by mathematical calculations, everything there is to know about the properties of black holes, from the horizon on outward.

And why do I believe that Einstein's general relativity description of the fundamental laws of physics is a highly accurate one? After all, we know that Newton's description ceases to be accurate near a black hole.

Successful descriptions of the fundamental laws contain within themselves a strong indication of where they will fail.[48] Newton's description tells us itself that it will probably fail near a black hole (though we only learned in the twentieth century how to read this out of Newton's description). Similarly, Einstein's general relativity description exudes confidence in itself outside a black hole, at the hole's horizon, and inside the hole all the way down almost (but not quite) to the singularity at its center. This is one thing that gives me confidence in general relativity's predictions. Another is the fact that, although general relativity's black-hole predictions have not yet been tested

47. Chapters 8 and 9.
48. Last section of Chapter 1.

directly, there have been high-precision tests of other features of general relativity on the Earth, in the solar system, and in binary systems that contain compact, exotic stars called pulsars. General relativity has come through each test with flying colors.

Over the past twenty years I have participated in the theoretical-physics quest which produced our present understanding of black holes and in the quest to test black-hole predictions by astronomical observation. My own contributions have been modest, but with my physicist and astronomer colleagues I have reveled in the excitement of the quest and have marveled at the insight it has produced. This book is my attempt to convey some sense of that excitement and marvel to people who are not experts in either astronomy or physics.

1

The Relativity
of Space and Time

*in which Einstein destroys
Newton's conceptions
of space and time as Absolute*

13 April 1901

Professor Wilhelm Ostwald
University of Leipzig
Leipzig, Germany

Esteemed Herr Professor!

Please forgive a father who is so bold as to turn to you, esteemed Herr
Professor, in the interest of his son.

I shall start by telling you that my son Albert is 22 years old, that
he studied at the Zurich Polytechnikum for 4 years, and that he
passed his diploma examinations in mathematics and physics with
flying colors last summer. Since then, he has been trying
unsuccessfully to obtain a position as Assistent, which would enable
him to continue his education in theoretical & experimental physics.
All those in position to give a judgment in the matter, praise his
talents; in any case, I can assure you that he is extraordinarily studious
and diligent and clings with great love to his science.

My son therefore feels profoundly unhappy with his present lack of

position, and his idea that he has gone off the tracks with his career &
is now out of touch gets more and more entrenched each day. In
addition, he is oppressed by the thought that he is a burden on us,
people of modest means.

Since it is you, highly honored Herr Professor, whom my son seems
to admire and esteem more than any other scholar currently active in
physics, it is you to whom I have taken the liberty of turning with the
humble request to read his paper published in the Annalen für
Physick and to write him, if possible, a few words of encouragement,
so that he might recover his joy in living and working.

If, in addition, you could secure him an Assistent's position for now
or the next autumn, my gratitude would know no bounds.

I beg you once again to forgive me for my impudence in writing to
you, and I am also taking the liberty of mentioning that my son does
not know anything about my unusual step.

I remain, highly esteemed Herr Professor, your devoted

Hermann Einstein

It was, indeed, a period of depression for Albert Einstein. He had been
jobless for eight months, since graduating from the Zurich Politechnikum at age twenty-one, and he felt himself a failure.

At the Politechnikum (usually called the "ETH" after its German-
language initials), Einstein had studied under several of the world's
most renowned physicists and mathematicians, but had not got on well
with them. In the turn-of-the-century academic world where most
Professors (with a capital P) demanded and expected respect, Einstein
gave little. Since childhood he had bristled against authority, always
questioning, never accepting anything without testing its truth himself. "Unthinking respect for authority is the greatest enemy of truth,"
he asserted. Heinrich Weber, the most famous of his two ETH physics
professors, complained in exasperation: "You are a smart boy, Einstein,
a very smart boy. But you have one great fault: you do not let yourself
be told anything." His other physics professor, Jean Pernet, asked him
why he didn't study medicine, law, or philology rather than physics.
"You can do what you like," Pernet said, "I only wish to warn you in
your own interest."

Einstein did not make matters better by his casual attitude toward
coursework. "One had to cram all this stuff into one's mind for the
examinations whether one liked it or not," he later said. His mathe-

matics professor, Hermann Minkowski, of whom we shall hear much in Chapter 2, was so put off by Einstein's attitude that he called him a "lazy dog."

But lazy Einstein was not. He was just selective. Some parts of the coursework he absorbed thoroughly; others he ignored, preferring to spend his time on self-directed study and thinking. Thinking was fun, joyful, and satisfying; on his own he could learn about the "new" physics, the physics that Heinrich Weber omitted from all his lectures.

Newton's Absolute Space and Time, and the Aether

The "old" physics, the physics that Einstein *could* learn from Weber, was a great body of knowledge that I shall call *Newtonian*, not because Isaac Newton was responsible for all of it (he wasn't), but because its foundations were laid by Newton in the seventeenth century.

By the late nineteenth century, all the disparate phenomena of the physical Universe could be explained beautifully by a handful of simple *Newtonian physical laws*. For example, all phenomena involving gravity could be explained by *Newton's laws of motion and gravity:*

- Every object moves uniformly in a straight line unless acted on by a force.
- When a force does act, the object's velocity changes at a rate proportional to the force and inversely proportional to its mass.
- Between any two objects in the Universe there acts a gravitational force that is proportional to the product of their masses and inversely proportional to the square of their separation.

By mathematically manipulating[1] these three laws, nineteenth-century physicists could explain the orbits of the planets around the Sun, the orbits of the moons around the planets, the ebb and flow of ocean tides, and the fall of rocks; and they could even learn how to weigh the Sun and the Earth. Similarly, by manipulating a simple set of electric and magnetic laws, the physicists could explain lightning, magnets, radio waves, and the propagation, diffraction, and reflection of light.

---

1. Readers who wish to understand what is meant by *"mathematically manipulating"* the laws of physics will find a discussion in the notes section at the end of the book.

Fame and fortune awaited those who could harness the Newtonian laws for technology. By mathematically manipulating the Newtonian laws of heat, James Watt figured out how to convert a primitive steam engine devised by others into the practical device that came to bear his name. By leaning heavily on Joseph Henry's understanding of the laws of electricity and magnetism, Samuel Morse devised his profitable version of the telegraph.

Inventors and physicists alike took pride in the perfection of their understanding. Everything in the heavens and on Earth seemed to obey the Newtonian laws of physics, and mastery of the laws was bringing humans a mastery of their environment—and perhaps one day would bring mastery of the entire Universe.

All the old, well-established Newtonian laws and their technological applications Einstein could learn in Heinrich Weber's lectures, and learn well. Indeed, in his first several years at the ETH, Einstein was enthusiastic about Weber. To the sole woman in his ETH class, Mileva Marić (of whom he was enamored), he wrote in February 1898, "Weber lectured masterfully. I eagerly anticipate his every class."

But in his fourth year at the ETH Einstein became highly dissatisfied. Weber lectured only on the *old* physics. He completely ignored some of the most important developments of recent decades, including James Clerk Maxwell's discovery of a new set of elegant electromagnetic laws from which one could deduce *all* electromagnetic phenomena: the behaviors of magnets, electric sparks, electric circuits, radio waves, light. Einstein had to teach himself Maxwell's unifying laws of electromagnetism by reading up-to-date books written by physicists at other universities, and he presumably did not hesitate to inform Weber of his dissatisfaction. His relations with Weber deteriorated.

In retrospect it is clear that of all things Weber ignored in his lectures, the most important was the mounting evidence of cracks in the foundation of Newtonian physics, a foundation whose bricks and mortar were Newton's concepts of space and time as absolute.

Newton's *absolute space* was the space of everyday experience, with its three dimensions: east–west, north–south, up–down. It was obvious from everyday experience that there is one and only one such space. It is a space shared by all humanity, by the Sun, by all the planets and the stars. We all move through this space in our own ways and at our own speeds, and regardless of our motion, we experience the space in the same way. This space gives us our sense of length and breadth and height; and according to Newton, we all, regardless of our motion, will agree on the length, breadth, and height of an object, so long as we make sufficiently accurate measurements.

Newton's *absolute time* was the time of everyday experience, the time that flows inexorably forward as we age, the time measured by high-quality clocks and by the rotation of the Earth and motion of the planets. It is a time whose flow is experienced in common by all humanity, by the Sun, by all the planets and the stars. According to Newton we all, regardless of our motion, will agree on the period of some planetary orbit or the duration of some politician's speech, so long as we all use sufficiently accurate clocks to time the orbit or speech.

If Newton's concepts of space and time as absolute were to crumble, the whole edifice of Newtonian physical laws would come tumbling down. Fortunately, year after year, decade after decade, century after century, Newton's foundational concepts had stood firm, producing one scientific triumph after another, from the domain of the planets to the domain of electricity to the domain of heat. There was no sign of any crack in the foundation—until 1881, when Albert Michelson started timing the propagation of light.

It seemed obvious, and the Newtonian laws so demanded, that if one measures the speed of light (or of anything else), the result must depend on how one is moving. If one is at rest in absolute space, then one should see the same light speed in all directions. By contrast, if one is moving through absolute space, say eastward, then one should see eastward-propagating light slowed and westward-propagating light speeded up, just as a person on an eastbound train sees eastward-flying birds slowed and westward-flying birds speeded up.

For the birds, it is the air that regulates their flight speed. Beating their wings against the air, the birds of each species move at the same maximum speed through the air regardless of their flight direction. Similarly, for light it was a substance called the *aether* that regulated the propagation speed, according to Newtonian physical laws. Beating its electric and magnetic fields against the aether, light propagates always at the same universal speed through the aether, regardless of its propagation direction. And since the aether (according to Newtonian concepts) is at rest in absolute space, anyone at rest will measure the same light speed in all directions, while anyone in motion will measure different light speeds.

Now, the Earth moves through absolute space, if for no other reason than its motion around the Sun; it moves in one direction in January,

then in the opposite direction six months later, in June. Correspondingly, we on Earth should measure the speed of light to be different in
different directions, and the differences should change with the seasons—though only very slightly (about 1 part in 10,000), because the
Earth moves so slowly compared to light.

To verify this prediction was a fascinating challenge for experimental physicists. Albert Michelson, a twenty-eight-year-old American,
took up the challenge in 1881, using an exquisitely accurate experimental technique (now called "Michelson interferometry"[2]) that he
had invented. But try as he might, Michelson could find no evidence
whatsoever for any variation of light speed with direction. The speed
turned out to be the same in *all* directions and at *all* seasons in his
initial 1881 experiments, and the same to much higher precision in
later 1887 experiments that Michelson performed in Cleveland, Ohio,
jointly with a chemist, Edward Morley. Michelson reacted with a mixture of elation at his discovery and dismay at its consequences. Heinrich Weber and most other physicists of the 1890s reacted with skepticism.

It was easy to be skeptical. Interesting experiments are often terribly
difficult—so difficult, in fact, that regardless of how carefully they are
carried out, they can give wrong results. Just one little abnormality in
the apparatus, or one tiny uncontrolled fluctuation in its temperature,
or one unexpected vibration of the floor beneath it, might alter the
experiment's final result. Thus, it is not surprising that physicists of
today, like physicists of the 1890s, are occasionally confronted by terribly difficult experiments which conflict with each other or conflict
with our deeply cherished beliefs about the nature of the Universe and
its physical laws. Recent examples are experiments that purported to
discover a "fifth force" (one not present in the standard, highly successful physical laws) and other experiments denying that such a force
exists; also experiments claiming to discover "cold fusion" (a phenomenon forbidden by the standard laws, if physicists understand those laws
correctly) and other experiments denying that cold fusion occurs. Almost always the experiments that threaten our deeply cherished beliefs
are wrong; their radical results are artifacts of experimental error.
However, occasionally they are right and point the way toward a revolution in our understanding of nature.

One mark of an outstanding physicist is an ability to "smell" which

2. Chapter 10.

experiments are to be trusted, and which not; which are to be worried about, and which ignored. As technology improves and the experiments are repeated over and over again, the truth ultimately becomes clear; but if one is trying to contribute to the progress of science, and if one wants to place one's own imprimatur on major discoveries, then one needs to divine early, not later, which experiments to trust.

Several outstanding physicists of the 1890s examined the Michelson–Morley experiment and concluded that the intimate details of the apparatus and the exquisite care with which it was executed made a strongly convincing case. This experiment "smells good," they decided; something might well be wrong with the foundations of Newtonian physics. By contrast, Heinrich Weber and most others were confident that, given time and further experimental effort, all would come out fine; Newtonian physics would triumph in the end, as it had so many times before. It would be inappropriate to even mention this experiment in one's university lectures; one should not mislead young minds.

The Irish physicist George F. Fitzgerald was the first to accept the Michelson–Morley experiment at face value and speculate about its implications. By comparing it with other experiments, he came to the radical conclusion that the fault lies in physicists' understanding of the concept of "length," and correspondingly there might be something wrong with Newton's concept of absolute space. In a short 1889 article in the American journal *Science*, he wrote in part:

> I have read with much interest Messrs. Michelson and Morley's wonderfully delicate experiment. . . . Their result seems opposed to other experiments. . . . I would suggest that almost the only hypothesis that can reconcile this opposition is that the length of material bodies changes, according as they are moving through the aether [through absolute space] or across it, by an amount depending on the square of the ratio of their velocities to that of light.

A tiny (five parts in a billion) contraction of length along the direction of the Earth's motion could, indeed, account for the null result of the Michelson–Morley experiment. But this required a repudiation of physicists' understanding of the behavior of matter: No known force could make moving objects contract along their direction of motion, not even by so minute an amount. If physicists understood correctly the nature of space and the nature of the molecular forces inside solid bodies, then uniformly moving solid bodies would always have to retain their same shape and size relative to absolute space, regardless of how fast they moved.

Hendrik Lorentz in Amsterdam also believed the Michelson–Morley experiment, and he took seriously Fitzgerald's suggestion that moving objects contract. Fitzgerald, upon learning of this, wrote to Lorentz expressing delight, since "I have been rather laughed at for my view over here." In a search for deeper understanding, Lorentz—and independently Henri Poincaré in Paris, France, and Joseph Larmor in Cambridge, England—reexamined the laws of electromagnetism, and noticed a peculiarity that dovetailed with Fitzgerald's length-contraction idea:

If one expressed Maxwell's electromagnetic laws in terms of the electric and magnetic fields measured at rest in absolute space, the laws took on an especially simple and beautiful mathematical form. For example, one of the laws said, simply, "As seen by anyone at rest in absolute space, magnetic field lines have no ends" (see Figure 1.1a,b). However, if one expressed Maxwell's laws in terms of the slightly different fields measured by a moving person, then the laws looked far more complicated and ugly. In particular, the "no ends" law became, "As seen by someone in motion, most magnetic field lines are endless, but a few get cut by the motion, thereby acquiring ends. Moreover, when the moving person shakes the magnet, new field lines get cut, then heal, then get cut again, then reheal" (see Figure 1.1c).

The new mathematical discovery by Lorentz, Poincaré, and Larmor was a way to make the moving person's electromagnetic laws look beautiful, and in fact look identical to the laws used by a person at rest in absolute space: "Magnetic field lines never end, under any circumstances whatsoever." One could make the laws take on this beautiful form by pretending, contrary to Newtonian precepts, that all moving objects get contracted along their direction of motion by precisely the amount that Fitzgerald needed to explain the Michelson–Morley experiment!

If the Fitzgerald contraction had been the only "new physics" that one needed to make the electromagnetic laws universally simple and beautiful, Lorentz, Poincaré, and Larmor, with their intuitive faith that the laws of physics *ought to be* beautiful, might have cast aside Newtonian precepts and believed firmly in the contraction. However, the contraction by itself was not enough. To make the laws beautiful, one also had to pretend that time flows more slowly as measured by someone moving through the Universe than by someone at rest; motion "dilates" time.

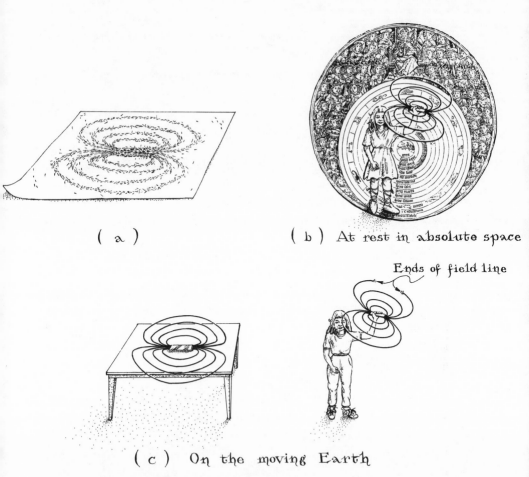

(a)

(b) At rest in absolute space

Ends of field line

(c) On the moving Earth

1.1 One of Maxwell's electromagnetic laws, as understood within the framework of nineteenth-century, Newtonian physics: (a) The concept of a magnetic field line: When one places a bar magnet under a sheet of paper and scatters iron filings on top of the sheet, the filings mark out the magnet's field lines. Each field line leaves the magnet's north pole, swings around the magnet and reenters it at the south pole, and then travels through the magnet to the north pole, where it attaches onto itself. The field line is therefore a closed curve, somewhat like a rubber band, without any ends. The statement that "magnetic field lines never have ends" is Maxwell's law in its simplest, most beautiful form. (b) According to Newtonian physics, this version of Maxwell's law is correct no matter what one does with the magnet (for example, even if one shakes it wildly) *so long as one is at rest in absolute space.* No magnetic field line *ever* has any ends, from the viewpoint of someone at rest. (c) When studied by someone riding on the surface of the Earth as it moves through absolute space, Maxwell's law is much more complicated, according to Newtonian physics. If the moving person's magnet sits quietly on a table, then a few of its field lines (about one in a hundred million) will have ends. If the person shakes the magnet wildly, additional field lines (one in a trillion) will get cut temporarily by the shaking, and then will heal, then get cut, then reheal. Although one field line in a hundred million or a trillion with ends was far too few to be discerned in any nineteenth-century physics experiment, the fact that Maxwell's laws predicted such a thing seemed rather complicated and ugly to Lorentz, Poincaré, and Larmor.

Now, the Newtonian laws of physics were unequivocal: Time is *absolute*. It flows uniformly and inexorably at the same universal rate, independently of how one moves. If the Newtonian laws were correct, then motion cannot cause time to dilate any more than it can cause lengths to contract. Unfortunately, the clocks of the 1890s were far too inaccurate to reveal the truth; and, faced with the scientific and techno-logical triumphs of Newtonian physics, triumphs grounded firmly on the foundation of absolute time, nobody was willing to assert with conviction that time really does dilate. Lorentz, Poincaré, and Larmor waffled.

Einstein, as a student in Zurich, was not yet ready to tackle such heady issues as these, but already he was beginning to think about them. To his friend Mileva Marić (with whom romance was now budding) he wrote in August 1899, "I am more and more convinced that the electrodynamics of moving bodies, as presented today, is not correct." Over the next six years, as his powers as a physicist matured, he would ponder this issue and the reality of the contradiction of lengths and dilation of time.

Weber, by contrast, showed no interest in such speculative issues. He kept right on lecturing about Newtonian physics as though all were in perfect order, as though there were no hints of cracks in the foundation of physics.

As he neared the end of his studies at the ETH, Einstein naively assumed that, because he was intelligent and had not really done all that badly in his courses (overall mark of 4.91 out of 6.00), he would be offered the position of "Assistent" in physics at the ETH under Weber, and could use it in the usual manner as a springboard into the academic world. As an Assistent he could start doing research of his own, leading in a few years to a Ph.D. degree.

But such was not to be. Of the four students who passed their final exams in the combined physics–mathematics program in August 1900, three got assistantships at the ETH working under mathematicians; the fourth, Einstein, got nothing. Weber hired as Assistents two engi-neering students rather than Einstein.

Einstein kept trying. In September, one month after graduation, he applied for a vacant Assistent position in mathematics at the ETH. He was rejected. In winter and spring he applied to Wilhelm Ostwald in Leipzig, Germany, and Heike Kamerlingh Onnes in Leiden, the Neth-erlands. From them he seems never to have received even the courtesy of a reply—though his note to Onnes is now proudly displayed in a museum in Leiden, and though Ostwald ten years later would be the first to nominate Einstein for a Nobel Prize. Even the letter to Ostwald from Einstein's father seems to have elicited no response.

To the saucy and strong-willed Mileva Marić, with whom his romance had turned intense, Einstein wrote on 27 March 1901, "I'm absolutely convinced that Weber is to blame. . . . it doesn't make any sense to write to any more professors, because they'll surely turn to Weber for information about me at a certain point, and he'll just give me another bad recommendation." To a close friend, Marcel Grossmann, he wrote on 14 April 1901, "I could have found [an Assistent position] long ago had it not been for Weber's underhandedness. All the same, I leave no stone unturned and do not give up my sense of humor . . . God created the donkey and gave him a thick hide."

A thick hide he needed; not only was he searching fruitlessly for a job, but his parents were vehemently opposing his plans to marry Mileva, and his relationship to Mileva was growing turbulent. Of Mileva his mother wrote, "This Miss Marić is causing me the bitterest hours of my life, if it were in my power, I would make every effort to banish her from our horizon, I really dislike her." And of Einstein's mother, Mileva wrote, "That lady seems to have made it her life's goal to embitter as much as possible not only my life but also that of her son. . . . I wouldn't have thought it possible that there could exist such heartless and outright wicked people!"

Einstein wanted desperately to escape his financial dependence on his parents, and to have the peace of mind and freedom to devote most of his energy to physics. Perhaps this could be achieved by some means other than an Assistent position in a university. His degree from the ETH qualified him to teach in a *gymnasium* (high school), so to this he turned: He managed in mid-May 1901 to get a temporary job at a technical high school in Winterthur, Switzerland, substituting for a mathematics teacher who had to serve a term in the army.

To his former history professor at the ETH, Alfred Stern, he wrote, "I am beside myself with joy about [this teaching job], because today I received the news that everything has been definitely arranged. I have not the slightest idea as to who might be the humanitarian who recommended me there, because from what I have been told, I am not in the good books of any of my former teachers." The job in Winterthur, followed in autumn 1901 by another temporary high school teaching job in Schaffhausen, Switzerland, and then in June 1902 by a job as "technical expert third class" in the Swiss Patent Office in Bern, gave him independence and stability.

Despite continued turbulence in his personal life (long separations from Mileva; an illegitimate child with Mileva in 1902, whom they seem to have put up for adoption, perhaps to protect Einstein's career possibilities in staid Switzerland; his marriage to Mileva a year later in spite of his parents' violent opposition), Einstein maintained an optimistic spirit and remained clear-headed enough to think, and think deeply about physics: From 1901 through 1904 he seasoned his powers as a physicist by theoretical research on the nature of the forces between molecules in liquids, such as water, and in metals, and research on the nature of heat. His new insights, which were substantial, were published in a sequence of five articles in the most prestigious physics journal of the early 1900s: the *Annalen der Physik.*

The patent office job in Bern was well suited to seasoning Einstein's powers. On the job he was challenged to figure out whether the inventions submitted would work—often a delightful task, and one that sharpened his mind. And the job left free half his waking hours and all weekend. Most of these he spent studying and thinking about physics, often in the midst of family chaos.

His ability to concentrate despite distractions was described by a student, who visited him at home several years after his marriage to Mileva: "He was sitting in his study in front of a heap of papers covered with mathematical formulas. Writing with his right hand and holding his younger son in his left, he kept replying to questions from his elder son Albert who was playing with his bricks. With the words, 'Wait a minute, I've nearly finished,' he gave me the children to look after for a few moments and went on working."

In Bern, Einstein was isolated from other physicists (though he did have a few close non-physicist friends with whom he could discuss science and philosophy). For most physicists, such isolation would be disastrous. Most require continual contact with colleagues working on similar problems to keep their research from straying off in unproductive directions. But Einstein's intellect was different; he worked more fruitfully in isolation than in a stimulating milieu of other physicists.

Sometimes it helped him to talk with others—not because they offered him deep new insights or information, but rather because by explaining paradoxes and problems to others, he could clarify them in his own mind. Particularly helpful was Michele Angelo Besso, an Italian engineer who had been a classmate of Einstein's at ETH and now

Left: Einstein seated at his desk in the patent office in Bern, Switzerland, ca. 1905. *Right:* Einstein with his wife, Mileva, and their son Hans Albert, ca. 1904. [Left: courtesy the Albert Einstein Archives of the Hebrew University of Jerusalem; right: courtesy Schweizerisches Literaturachiv/Archiv der Einstein-Gesellschaft, Bern.]

was working beside Einstein in the patent office. Of Besso, Einstein said, "I could not have found a better sounding board in the whole of Europe."

Einstein's Relative Space and Time, and Absolute Speed of Light

Michele Angelo Besso was especially helpful in May 1905, when Einstein, after focusing for several years on other physics issues, returned to Maxwell's electrodynamic laws and their tantalizing hints of length contraction and time dilation. Einstein's search for some way to make sense of these hints was impeded by a mental block. To clear the block, he sought help from Besso. As he recalled later, "That was a very beautiful day when I visited [Besso] and began to talk with him as follows: 'I have recently had a question which was difficult for me to understand. So I came here today to bring with me a battle on the question.' Trying a lot of discussions with him, I could suddenly comprehend the matter. The next day I visited him again and said to him without greeting: 'Thank you. I've completely solved the problem.' "

Einstein's solution: *There is no such thing as absolute space. There is no such thing as absolute time. Newton's foundation for all of physics was flawed. And as for the aether: It does not exist.*

By rejecting absolute space, Einstein made absolutely meaningless the notion of "being at rest in absolute space." There is no way, he asserted, to ever measure the Earth's motion through absolute space, and that is why the Michelson–Morley experiment turned out the way it did. One can measure the Earth's velocity only *relative to other physical objects* such as the Sun or the Moon, just as one can measure a train's velocity only relative to physical objects such as the ground and the air. For neither Earth nor train nor anything else is there any standard of absolute motion; motion is purely "relative."

By rejecting absolute space, Einstein also rejected the notion that everyone, regardless of his or her motion, must agree on the length, height, and width of some table or train or any other object. On the contrary, Einstein insisted, *length, height, and width are "relative" concepts.* They depend on the relative motion of the object being measured and the person doing the measuring.

By rejecting absolute time, Einstein rejected the notion that everyone, regardless of his or her motion, must experience the flow of time in the same manner. *Time is relative,* Einstein asserted. Each person traveling in his or her own way must experience a different time flow than others, traveling differently.

It is hard not to feel queasy when presented with these assertions. If correct, not only do they cut the foundations out from under the entire edifice of Newtonian physical law, they also deprive us of our common-sense, everyday notions of space and time.

But Einstein was not just a destroyer. He was also a creator. He offered us a new foundation to replace the old, a foundation just as firm and, it has turned out, in far more perfect accord with the Universe.

Einstein's new foundation consisted of two new fundamental principles:

- *The principle of the absoluteness of the speed of light:* Whatever might be their nature, space and time must be so constituted as to make the speed of light absolutely the same in all directions, and absolutely independent of the motion of the person who measures it.

This principle is a resounding affirmation that the Michelson–Morley experiment was correct, and that regardless of how accurate light-measuring devices may become in the future, they must always continue to give the same result: a universal speed of light.

· *The principle of relativity:* Whatever might be their nature, the laws of physics must treat all states of motion on an equal footing.

This principle is a resounding rejection of absolute space: If the laws of physics did not treat all states of motion (for example, that of the Sun and that of the Earth) on an equal footing, then using the laws of physics, physicists would be able to pick out some "preferred" state of motion (for example, the Sun's) and define it as the state of "absolute rest." Absolute space would then have crept back into physics. We shall return to this later in the chapter.

From the absoluteness of the speed of light, Einstein deduced, by an elegant logical argument described in Box 1.1 below, that if you and I move relative to each other, *what I call space must be a mixture of your space and your time, and what you call space must be a mixture of my space and my time.*

This "mixing of space and time" is analogous to the mixing of directions on Earth. Nature offers us two ways to reckon directions, one tied to the Earth's spin, the other tied to its magnetic field. In Pasadena, California, magnetic north (the direction a compass needle points) is offset eastward from true north (the direction toward the Earth's spin axis, that is, toward the "North Pole") by about 20 degrees; see Figure 1.2. This means that in order to travel in the magnetic north direction, one must travel partly (about 80 percent) in the true north direction and partly (about 20 percent) toward true east. In this sense, *magnetic north is a mixture of true north and true east;* similarly, true north is a mixture of magnetic north and magnetic west.

To understand the analogous mixing of space and time *(your space is a mixture of my space and my time, and my space is a mixture of your space and your time),* imagine yourself the owner of a powerful sports car. You like to drive your car down Colorado Boulevard in Pasadena, California, at extremely high speed in the depths of the night, when I, a policeman, am napping. To the top of your car you attach a series of firecrackers, one over the front of the hood, one over the rear of the trunk, and many in between; see Figure 1.3a. You set the firecrackers to detonate simultaneously as seen by you, just as you are passing my police station.

Figure 1.3b depicts this from your own viewpoint. Drawn vertically is the flow of time, as measured by you ("your time"). Drawn horizontally is distance along your car, from back to front, as measured by you ("your space"). Since the firecrackers are all at rest in your space (that is, as seen by you), with the passage of your time they all remain at the same horizontal locations in the diagram. The dashed lines, one for each firecracker, depict this. They extend vertically upward in the diagram, indicating no rightward or leftward motion in space whatsoever as time passes—and they then terminate abruptly at the moment the firecrackers detonate. The detonation events are depicted by asterisks.

This figure is called a *spacetime diagram* because it plots space horizontally and time vertically; the dashed lines are called *world lines* because they show where in the world the firecrackers travel as time passes. We shall make extensive use of spacetime diagrams and world lines later in this book.

If one moves horizontally in the diagram (Figure 1.3b), one is moving through space at a fixed moment of your time. Correspondingly, it is convenient to think of each horizontal line in the diagram as depicting space, as seen by you ("your space"), at a specific moment of your time. For example, the dotted horizontal line is your space at the moment of firecracker detonation. As one moves vertically upward in the diagram, one is moving through time at a fixed location in your

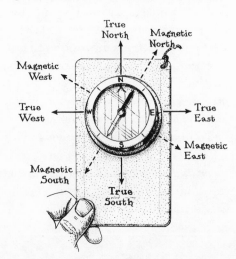

1.2   Magnetic north is a mixture of
true north and true east, and
true north is a mixture of magnetic
north and magnetic west.

space. Correspondingly, it is convenient to think of each vertical line in the spacetime diagram (for example, each firecracker world line) as depicting the flow of your time at a specific location in your space.

I, in the police station, were I not napping, would draw a rather different spacetime diagram to depict your car, your firecrackers, and the detonation (Figure 1.3c). I would plot the flow of time, as measured by me, vertically, and distance along Colorado Boulevard horizontally. As time passes, each firecracker moves down Colorado Boulevard with your car at high speed, and correspondingly, the firecracker's world line tilts rightward in the diagram: At the time of its detonation, the firecracker is farther to the right down Colorado Boulevard than at earlier times.

Now, the surprising conclusion of Einstein's logical argument (Box 1.1) is that the absoluteness of the speed of light requires the firecrackers *not* to detonate simultaneously as seen by me, even though

1.3 (a) Your sports car speeding down Colorado Boulevard with firecrackers attached to its roof. (b) Spacetime diagram depicting the firecrackers' motion and detonation from your viewpoint (riding in the car). (c) Spacetime diagram depicting the same firecracker motion and detonation from my viewpoint (at rest in the police station).

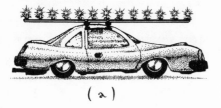

(a)

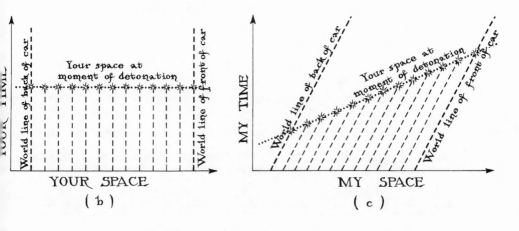

they detonate simultaneously as seen by you. From my viewpoint the rearmost firecracker on your car detonates first, and the frontmost one detonates last. Correspondingly, the dotted line that we called "your space at moment of detonation" (Figure 1.3b) is tilted in my spacetime diagram (Figure 1.3c).

From Figure 1.3c it is clear that, to move through your space at your moment of detonation (along the dotted detonation line), I must move through both my space and my time. In this sense, your space is a mixture of my space and my time. This is just the same sense as the statement that magnetic north is a mixture of true north and true east (compare Figure 1.3c with Figure 1.2).

You might be tempted to assert that this "mixing of space and time" is nothing but a complicated, jingoistic way of saying that "simultaneity depends on one's state of motion." True. However, physicists, building on Einstein's foundations, have found this way of thinking to be powerful. It has helped them to decipher Einstein's legacy (his new laws of physics), and to discover in that legacy a set of seemingly outrageous phenomena: black holes, wormholes, singularities, time warps, and time machines.

From the absoluteness of the speed of light and the principle of relativity, Einstein deduced other remarkable features of space and time. In the language of the above story:

- Einstein deduced that, as you speed eastward down Colorado Boulevard, I must see your space and everything at rest in it (your car, your firecrackers, and you) contracted along the east–west direction, but not north–south or up–down. This was the contraction inferred by Fitzgerald, but now put on a firm foundation: The contraction is caused by the peculiar nature of space and time, and not by any physical forces that act on moving matter.
- Similarly, Einstein deduced that, as you speed eastward, you must see my space and everything at rest in it (my police station, my desk, and me) contracted along the east–west direction, but not north–south or up–down. That you see me contracted and I see you contracted may seem puzzling, but in fact it could not be otherwise: It leaves your state of motion and mine on an equal footing, in accord with the principle of relativity.
- Einstein also deduced that, as you speed past, I see your flow of time slowed, that is, dilated. The clock on your car's dashboard appears to tick more slowly than my clock on the police station

Box 1.1
Einstein's Proof of the Mixing of Space and Time

Einstein's principle of the absoluteness of the speed of light enforces the mixing of space and time; in other words, it enforces the relativity of simultaneity: Events that are simultaneous as seen by you (that lie in your space at a specific moment of your time), as your sports car speeds down Colorado Boulevard, are not simultaneous as seen by me, at rest in the police station. I shall prove this using descriptive words that go along with the spacetime diagrams shown below. This proof is essentially the same as the one devised by Einstein in 1905.

Place a flash bulb at the middle of your car. Trigger the bulb. It sends a burst of light forward toward the front of your car, and a burst backward toward the back of your car. Since the two bursts are emitted simultaneously, and since they travel the same distance as measured by you in your car, and since they travel at the same speed (the speed of light is absolute), they must arrive at the front and back of your car simultaneously from your viewpoint; see the left diagram, below. The two events of burst arrival (call them *A* at your car's front and *B* at its back) are thus simultaneous from your viewpoint, and they happen to coincide with the firecracker detonations of Figure 1.4, as seen by you.

Next, examine the light bursts and their arrival events *A* and *B* from my viewpoint as your car speeds past me; see the right diagram, below. From my viewpoint, the back of your car is moving forward, toward the backward-directed burst of light, and they thus meet each other (event *B*) sooner as seen by me than as seen by you. Similarly, the front of your car is moving forward, away from the frontward-directed burst, and they thus meet each other (event *A*) later as seen by me than as seen by you. (These conclusions rely crucially on the fact that the speeds of the two light bursts are the same as seen by me; that is, they rely on the absoluteness of the speed of light.) Therefore, I regard event *B* as occurring before event *A*; and similarly, I see the firecrackers near the back of your car detonate before those near the front.

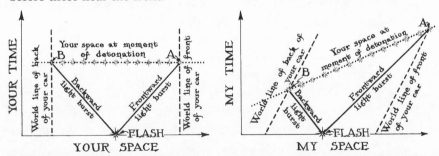

Note that the locations of the detonations (your space at a specific moment of your time) are the same in the above spacetime diagrams as in Figure 1.4. This justifies the asserted mixing of space and time discussed in the text.

wall. You speak more slowly, your hair grows more slowly, you age more slowly than I.

· Similarly, in accord with the principle of relativity, as you speed past me, you see my flow of time slowed. You see the clock on my station wall tick more slowly than the one on your dashboard. To you I seem to speak more slowly, my hair grows more slowly, and I age more slowly than you.

How can it possibly be that I see your time flow slowed, while you see mine slowed? How is that logically possible? And how can I see your space contracted, while you see my space contracted? The answer lies in the relativity of simultaneity. You and I disagree about whether events at different locations in our respective spaces are simultaneous, and this disagreement turns out to mesh with our disagreements over the flow of time and the contraction of space in just such a way as to keep everything logically consistent. To demonstrate this consistency, however, would take more pages than I wish to spend, so I refer you, for a proof, to Chapter 3 of Taylor and Wheeler (1992).

How is it that we as humans have never noticed this weird behavior of space and time in our everyday lives? The answer lies in our slowness. We always move relative to each other with speeds far smaller than that of light (299,792 kilometers per second). If your car zooms down Colorado Boulevard at 150 kilometers per hour, I should see your time flow dilated and your space contracted by roughly one part in a hundred trillion (1×10^{-14})—far too little for us to notice. By contrast, if your car were to move past me at 87 percent the speed of light, then (using instruments that respond very quickly) I should see your time flow twice as slowly as mine, while you see my time flow twice as slowly as yours; similarly, I should see everything in your car half as long, east–west, as normal, and you should see everything in my police station half as long, east–west, as normal. Indeed, a wide variety of experiments in the late twentieth century have verified that space and time do behave in just this way.

How did Einstein arrive at such a radical description of space and time?

Not by examining the results of experiments. Clocks of his era were too inaccurate to exhibit, at the low speeds available, any time dilation or disagreements about simultaneity, and measuring rods were too inaccurate to exhibit length contraction. The only relevant experiments were those few, such as Michelson and Morley's, which suggested that the speed of light on the Earth's surface might be the same in all directions. These were very skimpy data indeed on which to base such a radical revision of one's notions of space and time! Moreover, Einstein paid little attention to these experiments.

Instead, Einstein relied on his own innate intuition as to how things *ought to* behave. After much reflection, it became *intuitively obvious* to him that the speed of light must be a universal constant, independent of direction and independent of one's motion. Only then, he reasoned, could Maxwell's electromagnetic laws be made uniformly simple and beautiful (for example, "magnetic field lines never ever have any ends"), and he was firmly convinced that the Universe in some deep sense insists on having simple and beautiful laws. He therefore introduced, as a new principle on which to base all of physics, his principle of the absoluteness of the speed of light.

This principle by itself, without anything else, already guaranteed that the edifice of physical laws built on Einstein's foundation would differ profoundly from that of Newton. *A Newtonian physicist, by presuming space and time to be absolute, is forced to conclude that the speed of light is relative—it depends on one's state of motion (as the bird and train analogy earlier in this chapter shows). Einstein, by presuming the speed of light to be absolute, was forced to conclude that space and time are relative—they depend on one's state of motion. Having deduced that space and time are relative, Einstein was then led onward by his quest for simplicity and beauty to his principle of relativity: No one state of motion is to be preferred over any other; all states of motion must be equal, in the eyes of physical law.*

Not only was experiment unimportant in Einstein's construction of a new foundation for physics, the ideas of other physicists were also unimportant. He paid little attention to others' work. He seems not even to have read any of the important technical articles on space, time, and the aether that Hendrik Lorentz, Henri Poincaré, Joseph Larmor, and others wrote between 1896 and 1905.

In their articles, Lorentz, Poincaré, and Larmor were groping toward the same revision of our notions of space and time as Einstein, but they were groping through a fog of misconceptions foisted on them by Newtonian physics. Einstein, by contrast, was able to cast off the Newtonian misconceptions. His conviction that the Universe loves simplicity and beauty, and his willingness to be guided by this conviction, even if it meant destroying the foundations of Newtonian physics, led him, with a clarity of thought that others could not match, to his new description of space and time.

The principle of relativity will play an important role later in this book. For this reason I shall devote a few pages to a deeper explanation of it.

A deeper explanation requires the concept of a *reference frame*. A reference frame is a laboratory that contains all the measuring apparatus one might need for whatever measurements one wishes to make. The laboratory and all its apparatus must move through the Universe together; they must all undergo the same motion. In fact, the motion of the reference frame is really the central issue. When a physicist speaks of "different reference frames," the emphasis is on different states of motion and not on different measuring apparatuses in the two laboratories.

A reference frame's laboratory and its apparatus need not be real. They perfectly well can be imaginary constructs, existing only in the mind of the physicist who wants to ask some question such as, "If I were in a spacecraft floating through the asteroid belt, and I were to measure the size of some specific asteroid, what would the answer be?" Such physicists imagine themselves as having a reference frame (laboratory) attached to their spacecraft and as using that frame's apparatus to make the measurement.

Einstein expressed his principle of relativity not in terms of arbitrary reference frames, but in terms of rather special ones: frames (laboratories) that move freely under their own inertia, neither pushed nor pulled by any forces, and that therefore continue always onward in the same state of uniform motion as they began. Such frames Einstein called *inertial* because their motion is governed solely by their own inertia.

A reference frame attached to a firing rocket (a laboratory inside the rocket) is *not* inertial, because its motion is affected by the rocket's thrust as well as by its inertia. The thrust prevents the frame's motion from being uniform. A reference frame attached to the space shuttle as it reenters the Earth's atmosphere also is not inertial, because friction between the shuttle's skin and the Earth's air molecules slows the shuttle, making its motion nonuniform.

Most important, near any massive body such as the Earth, *all* reference frames are pulled by gravity. There is no way whatsoever to shield a reference frame (or any other object) from gravity's pull. Therefore, by restricting himself to inertial frames, Einstein prevented himself from considering, in 1905, physical situations in which gravity is important[3]; in effect, he idealized our Universe as one in which there is no gravity at all. Extreme idealizations like this are central to progress in physics; one throws away, conceptually, aspects of the Universe that are difficult to deal with, and only after gaining intellectual control over the remaining, easier aspects does one return to the harder ones. Einstein gained intellectual control over an idealized universe without gravity in 1905. He then turned to the harder task of understanding the nature of space and time in our real, gravity-endowed Universe, a task that eventually would force him to conclude that gravity warps space and time (Chapter 2).

With the concept of an inertial reference frame understood, we are now ready for a deeper, more precise formulation of Einstein's principle of relativity: *Formulate any law of physics in terms of measurements made in one inertial reference frame. Then, when restated in terms of measurements in any other inertial frame, that law of physics must take on precisely the same mathematical and logical form as in the original frame.* In other words, the laws of physics must not provide any means to distinguish one inertial reference frame (one state of uniform motion) from any other.

Two examples of physical laws will make this more clear:

- "Any free object (one on which no forces act) that initially is at rest in an inertial reference frame will always remain at rest; and any free object that initially is moving through an inertial reference frame will continue forever forward, along a straight line with constant speed." If (as is the case) we have strong reason to believe that this relativistic version of Newton's first law of motion is true in at least one inertial reference frame, then the principle of relativity insists that it must be true in all inertial reference frames regardless of where they are in the Universe and regardless of how fast they are moving.
- Maxwell's laws of electromagnetism must take on the same mathematical form in all reference frames. They failed to do so, when physics was built on Newtonian foundations (magnetic field lines could have ends in some frames but not in others), and this failure was deeply disturbing to Lorentz, Poincaré, Larmor, and Einstein.

---

3. This means that it was a bit unfair of me to use a high-speed sports car, which feels the Earth's gravity, in my example above. However, it turns out that because the Earth's gravitational pull is perpendicular to the direction of the car's motion (downward versus horizontal), it has no effect on any of the issues discussed in the sports-car story.

In Einstein's view it was utterly unacceptable that the laws were simple and beautiful in one frame, that of the aether, but complex and ugly in all frames that moved relative to the aether. By reconstructing the foundations of physics, Einstein enabled Maxwell's laws to take on one and the same simple, beautiful form (for example, "magnetic field lines never ever have any ends") in each and every inertial reference frame—in accord with his principle of relativity.

The principle of relativity is actually a *metaprinciple* in the sense that it is not itself a law of physics, but instead is a pattern or rule which (Einstein asserted) must be obeyed by *all* laws of physics, no matter what those laws might be, no matter whether they are laws governing electricity and magnetism, or atoms and molecules, or steam engines and sports cars. The power of this metaprinciple is breathtaking. Every new law that is proposed must be tested against it. If the new law passes the test (if the law is the same in every inertial reference frame), then the law has some hope of describing the behavior of our Universe. If it fails the test, then it has no hope, Einstein asserted; it must be rejected.

All of our experience in the nearly 100 years since 1905 suggests that Einstein was right. All new laws that have been successful in describing the real Universe have turned out to obey Einstein's principle of relativity. This metaprinciple has become enshrined as a governor of physical law.

In May 1905, once his discussion with Michele Angelo Besso had broken his mental block and enabled him to abandon absolute time and space, Einstein needed only a few weeks of thinking and calculating to formulate his new foundation for physics, and to deduce its consequences for the nature of space, time, electromagnetism, and the behaviors of high-speed objects. Two of the consequences were spectacular: mass can be converted into energy (which would become the foundation for the atomic bomb; see Chapter 6), and the inertia of every object must increase so rapidly, as its speed approaches the speed of light, that no matter how hard one pushes on the object, one can never make it reach or surpass the speed of light ("nothing can go faster than light").[4]

4. But see Chapter 14 for a caveat.

In late June, Einstein wrote a technical article describing his ideas and their consequences, and mailed it off to the *Annalen der Physik*. His article carried the somewhat mundane title "On the Electrodynamics of Moving Bodies." But it was far from mundane. A quick perusal showed Einstein, the Swiss Patent Office's "technical expert third class," proposing a whole new foundation for physics, proposing a metaprinciple that all future physical laws must obey, radically revising our concepts of space and time, and deriving spectacular consequences. Einstein's new foundation and its consequences would soon come to be known as *special relativity* ("special" because it correctly describes the Universe only in those special situations where gravity is unimportant).

Einstein's article was received at the offices of the *Annalen der Physik* in Leipzig on 30 June 1905. It was perused for accuracy and importance by a referee, was passed as acceptable, and was published.

In the weeks after publication, Einstein waited expectantly for a response from the great physicists of the day. His viewpoint and conclusions were so radical and had so little experimental basis that he expected sharp criticism and controversy. Instead, he was met with stony silence. Finally, many weeks later, there arrived a letter from Berlin: Max Planck wanted clarification of some technical issues in the paper. Einstein was overjoyed! To have the attention of Planck, one of the most renowned of all living physicists, was deeply satisfying. And when Planck went on, the following year, to use Einstein's principle of relativity as a central tool in his own research, Einstein was further heartened. Planck's approval, the gradual approval of other leading physicists, and most important his own supreme self-confidence held Einstein firm throughout the following twenty years as the controversy he had expected did, indeed, swirl around his relativity theory. The controversy was still so strong in 1922 that, when the secretary of the Swedish Academy of Sciences informed Einstein by telegram that he had won the Nobel Prize, the telegram stated explicitly that relativity was *not* among the works on which the award was based.

The controversy finally died in the 1930s, as technology became sufficiently advanced to produce accurate experimental verifications of special relativity's predictions. By now, in the 1990s, there is absolutely no room for doubt: Every day more than 10^{17} electrons in particle accelerators at Stanford University, Cornell University, and elsewhere are driven up to speeds as great as 0.9999999995 of the speed of light—and their behaviors at these ultra-high speeds are in complete accord with Einstein's special relativistic laws of physics. For example, the electrons' inertia increases as they near the speed of light, preventing them from ever reaching it; and when the electrons collide with targets, they produce high-speed particles called mu mesons that live for only 2.22 microseconds as measured by their own time, but because of time dilation live for 100 microseconds or more as measured by the physicists' time, at rest in the laboratory.

The Nature of Physical Law

Does the success of Einstein's special relativity mean that we must totally abandon the Newtonian laws of physics? Obviously not. The Newtonian laws are still used widely in everyday life, in most fields of science, and in most technology. We don't pay attention to time dilation when planning an airplane trip, and engineers don't worry about length contraction when designing an airplane. The dilation and contraction are far too small to be of concern.

Of course, if we wished to, we *could* use Einstein's laws rather than Newton's in everyday life. The two give almost precisely the same predictions for all physical effects, since everyday life entails relative speeds that are very small compared to the speed of light.

Einstein's and Newton's predictions begin to diverge strongly only at relative speeds approaching the speed of light. Then and only then must one abandon Newton's predictions and adhere strictly to Einstein's.

This is an example of a very general pattern, one that we shall meet again in future chapters. It is a pattern that has been repeated over and over in the history of twentieth-century physics: One set of laws (in our case the *Newtonian laws*) is widely accepted at first, because it accords beautifully with experiment. But then experiments become more accurate and this first set of laws turns out to work well only in a limited domain, its *domain of validity* (for Newton's laws, the domain of speeds small compared to the speed of light). Physicists then struggle, experimentally and theoretically, to understand what is going on at the boundary of that domain of validity, and they finally formulate a new set of laws which is highly successful inside, near, and beyond the boundary (in Newton's case, *Einstein's special relativity*, valid at speeds approaching light as well as at low speeds). Then the process repeats. We shall meet the repetition in coming chapters: The failure of special relativity when gravity becomes important, and its replacement by a new set of laws called *general relativity* (Chapter 2); the failure of general relativity near the singularity inside a black hole, and its replacement by a new set of laws called *quantum gravity* (Chapter 13).

There has been an amazing feature of each transition from an old set of laws to a new one: In each case, physicists (if they were sufficiently clever) did not need any experimental guidance to tell them where the old set would begin to break down, that is, to tell them the boundary of its domain of validity. We have seen this already for Newtonian physics: Maxwell's laws of electrodynamics did not mesh nicely with the absolute space of Newtonian physics. At rest in absolute space (in the frame of the aether), Maxwell's laws were simple and beautiful—for example, magnetic field lines have no ends. In moving frames, they became complicated and ugly—magnetic field lines sometimes have ends. However, the complications had negligible influence on the outcome of experiments when the frames moved, relative to absolute space, at speeds small compared to light; then almost all field lines are endless. Only at speeds approaching light were the ugly complications predicted to have a big enough influence to be measured easily: lots of ends. Thus, it was reasonable to suspect, even without the Michelson–Morley experiment, that the domain of validity of Newtonian physics might be speeds small compared to light, and that the Newtonian laws might break down at speeds approaching light.

In Chapter 2 we shall see, similarly, how special relativity predicts its own failure in the presence of gravity; and in Chapter 13, how general relativity predicts its own failure near a singularity.

When contemplating the above sequence of sets of laws (Newtonian physics, special relativity, general relativity, quantum gravity)—and a similar sequence of laws governing the structure of matter and elementary particles—most physicists are driven to believe that these sequences are converging toward a set of ultimate laws that truly governs the Universe, laws that *force* the Universe to behave the way it does, that *force* rain to condense on windows, *force* the Sun to burn nuclear fuel, *force* black holes to produce gravitational waves when they collide, and so on.

One might object that each set of laws in the sequence "looks" very different from the preceding set. (For example, the absolute time of Newtonian physics looks very different from the many different time flows of special relativity.) In the "looks" of the laws, there is no sign whatsoever of convergence. Why, then, should we expect convergence? The answer is that one must distinguish sharply between the predictions made by a set of laws and the mental images that the laws convey (what the laws "look like"). I expect convergence only in terms of predictions, but that is all that ultimately counts. The mental images (one absolute time in Newtonian physics versus many time flows in relativistic physics) are not important to the ultimate nature of *reality*. In fact, it is possible to change completely what a set of laws "looks like" without changing its predictions. In Chapter 11, I shall discuss this remarkable fact and give examples, and shall explain its implications for the nature of reality.

Why do I expect convergence in terms of predictions? Because all the evidence we have points to it. Each set of laws has a larger domain of validity than the sets that preceded it: Newton's laws work throughout the domain of everyday life, but not in physicists' particle accelerators and not in exotic parts of the distant Universe, such as pulsars, quasars, and black holes; Einstein's general relativity laws work everywhere in our laboratories, and everywhere in the distant Universe, except deep inside black holes and in the big bang where the Universe was born; the laws of quantum gravity (which we do not yet understand at all well) may turn out to work absolutely everywhere.

Throughout this book, I shall adopt, without apology, the view that there *does* exist an ultimate set of physical laws (which we do not as yet know but which might be quantum gravity), and that those laws truly *do* govern the Universe around us, everywhere. They *force* the Universe to behave the way it does. When I am being extremely accurate, I shall say that the laws we now work with (for example, general relativity) are "an approximation to" or "an approximate description of" the true laws. However, I shall usually drop the qualifiers and not distinguish between the true laws and our approximations to them. At these times I shall assert, for example, that "the general relativistic laws [rather than the true laws] *force* a black hole to hold light so tightly in its grip that the light cannot escape from the hole's horizon." This is how my colleagues and I as physicists think, when struggling to understand the Universe. It is a fruitful way to think; it has helped produce deep new insights into imploding stars, black holes, gravitational waves, and other phenomena.

This viewpoint is incompatible with the common view that physicists work with *theories* which try to describe the Universe, but which are only human inventions and have no real power over the Universe. The word *theory*, in fact, is so ladened with connotations of tentativeness and human quirkiness that I shall avoid using it wherever possible. In its place I shall use the phrase *physical law* with its firm connotation of truly ruling the Universe, that is, truly forcing the Universe to behave as it does.

2

The Warping
of Space and Time

*in which Hermann Minkowski
unifies space and time,
and Einstein warps them*

Minkowski's Absolute Spacetime

The views of space and time which I wish to lay before you have
sprung from the soil of experimental physics, and therein lies their
strength. They are radical. Henceforth, space by itself, and time by
itself, are doomed to fade away into mere shadows, and only a kind
of union of the two will preserve an independent reality.

With these words Hermann Minkowski revealed to the world, in
September 1908, a new discovery about the nature of space and time.

Einstein had shown that space and time are "relative." The length of
an object and the flow of time are different when viewed from different
reference frames. My time differs from yours if I move relative to you,
and my space differs from yours. My time is a mixture of your time and
your space; my space is a mixture of your space and your time.

Minkowski, building on Einstein's work, had now discovered that
the Universe is made of a four-dimensional "spacetime" fabric that is
absolute, not relative. This four-dimensional fabric is the same as seen

from all reference frames (if only one can learn how to "see" it); it exists independently of reference frames.

The following tale (adapted from Taylor and Wheeler, 1992) illustrates the idea underlying Minkowski's discovery.

Once upon a time, on an island called Mledina in a far-off Eastern sea, there lived a people with strange customs and taboos. Each June, on the longest day of the year, all the Mledina men journeyed in a huge sailing vessel to a distant, sacred island called Serona, there to commune with an enormous toad. All night long the toad would enchant them with marvelous tales of stars and galaxies, pulsars and quasars. The next day the men would sail back to Mledina, filled with inspiration that sustained them for the whole of the following year.

Each December, on the longest night of the year, the Mledina women sailed to Serona, communed with the same toad all the next day, and returned the next night, inspired with the toad's visions of stars and galaxies, quasars and pulsars.

Now, it was absolutely taboo for any Mledina woman to describe to any Mledina man her journey to the sacred island of Serona, or any details of the toad's tales. The Mledina men were ruled by the same taboo. Never must they expose to a woman anything about their annual voyage.

In the summer of 1905 a radical Mledina youth named Albert, who cared little for the taboos of his culture, discovered and exposed to all the Mledinans, female and male, two sacred maps. One was the map by which the Mledina priestess guided the sailing vessel on the women's midwinter journey. The other was the map used by the Mledina priest on the men's midsummer voyage. What shame the men felt, having their sacred map exposed. The women's shame was no less. But there the maps were, for everyone to see—and they contained a great shock: They disagreed about the location of Serona. The women were sailing eastward 210 furlongs, then northward 100 furlongs, while the men were sailing eastward 164.5 furlongs, then northward 164.5 furlongs. How could this be? Religious tradition was firm; the women and the men were to seek their annual inspiration from the same sacred toad on the same sacred island of Serona.

Most of the Mledinans dealt with their shame by pretending the exposed maps were fakes. But a wise old Mledina man named Hermann believed. For three years he struggled to understand the mystery of the maps' discrepancy. Finally, one autumn day in 1908, the truth came to him: The Mledina men must be navigating by magnetic compass, and the Mledina women by the stars (Figure 2.1). The Mledina men reckoned north and east magnetically, the Mledina women reckoned them by the rotation of the Earth which makes the stars turn overhead, and the two methods of reckoning differed by 20 degrees. When the men sailed northward, as reckoned by them, they were actually sailing "north 20 degrees east," or about 80 percent north and 20 percent east, as reckoned by the women. In this sense, the men's north was a mixture of the women's north and east, and similarly the women's north was a mixture of the men's north and west.

The key that led Hermann to this discovery was the formula of Pythagoras: Take two legs of a right triangle; square the length of one leg, square the length of the other, add them, and take the square root. The result should be the length of the triangle's hypotenuse.

The hypotenuse was the straight-line path from Mledina to Serona. The absolute distance along that straight-line path was $\sqrt{210^2 + 100^2}$ = 232.6 furlongs as reckoned using the women's map with its legs

2.1   The two maps of the route from Mledina to Serona superimposed on each other, together with Hermann's notations of magnetic north, true north, and the absolute distance.

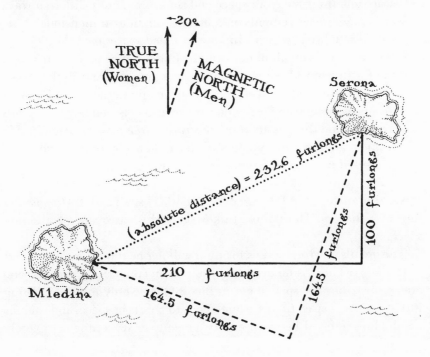

along true east and true north. As reckoned using the men's map with
its legs along magnetic east and magnetic north, the absolute distance
was $\sqrt{164.5^2 + 164.5^2} = 232.6$ furlongs. The eastward distance and
the northward distance were "relative"; they depended on whether the
map's reference frame was magnetic or true. But from either pair of
relative distances one could compute the same, absolute, straight-line
distance.

History does not record how the people of Mledina, with their culture of taboos, responded to this marvelous discovery.

Hermann Minkowski's discovery was analogous to the discovery by
Hermann the Mledinan: Suppose that you move relative to me (for
example, in your ultra-high-speed sports car). Then:

- Just as magnetic north is a mixture of true north and true east, so
  also my time is a mixture of your time and your space.
- Just as magnetic east is a mixture of true east and true south, so
  also my space is a mixture of your space and your time.
- Just as magnetic north and east, and true north and east, are
  merely different ways of making measurements on a preexisting,
  two-dimensional surface—the surface of the Earth—so also my
  space and time, and your space and time, are merely different ways
  of making measurements on a preexisting, four-dimensional "surface" or "fabric," which Minkowski called *spacetime.*
- Just as there is an absolute, straight-line distance on the surface of
  the Earth from Mledina to Serona, computable from Pythagoras's
  formula using either distances along magnetic north and east or
  distances along true north and east, so also between any two *events*
  in spacetime there is an *absolute straight-line interval,* computable
  from an analogue of Pythagoras's formula using lengths and times
  measured in either reference frame, mine or yours.

It was this analogue of Pythagoras's formula (I shall call it *Minkowski's
formula*) that led Hermann Minkowski to his discovery of absolute
spacetime.

The details of Minkowski's formula will *not* be important in the rest
of this book. There is no need to master them (though for readers who
are curious, they are spelled out in Box 2.1). The only important thing
is that events in spacetime are analogous to points in space, and there is
an absolute interval between any two events in spacetime completely

Box 2.1
Minkowski's Formula

You zoom past me in a powerful, 1-kilometer-long sports car, at a speed of 162,000 kilometers per second (54 percent of the speed of light); recall Figure 1.3. Your car's motion is shown in the following spacetime diagrams. Diagram (a) is drawn from your viewpoint; (b) from mine. As you pass me, your car backfires, ejecting a puff of smoke from its tailpipe; this backfire event is labeled *B* in the diagrams. Two microseconds (two-millionths of a second) later, as seen by you, a firecracker on your front bumper detonates; this detonation event is labeled *D*.

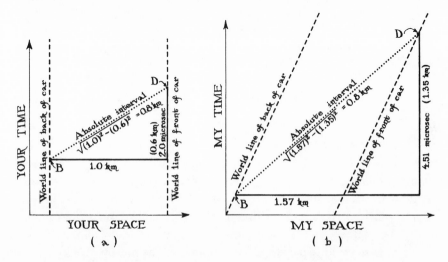

Because space and time are relative (your space is a mixture of my space and my time), you and I disagree about the time separation between the backfire event *B* and the detonation event *D*. They are separated by 2.0 microseconds of your time, and by 4.51 microseconds of mine. Similarly, we disagree about the events' spatial separation; it is 1.0 kilometer in your space and 1.57 kilometers in mine. Despite these temporal and spatial disagreements, we *agree* that the two events are separated by a straight line in four-dimensional spacetime, and we *agree* that the "absolute interval" along that line (the spacetime length of the line) is 0.8 kilometer. (This is analogous to the Mledinan men and women agreeing on the straight-line distance between Mledina and Serona.)

We can use Minkowski's formula to compute the absolute interval: We each multiply the events' time separation by the speed of light (299,792 kilometers per second), getting the rounded-off numbers shown in the diagrams (0.600 kilometer for you, 1.35 kilometers for me). We then square the events' time and space separations, we *subtract* the squared

(continued next page)

(Box 2.1 continued)

time separation from the squared space separation, and we take the square root. (This is analogous to the Mledinans squaring the eastward and northward separations, *adding* them, and taking the square root.) As is shown in the diagrams, although your time and space separations differ from mine, we get the same final answer for the absolute interval: 0.8 kilometer.

There is only one important difference between Minkowski's formula, which you and I follow, and Pythagoras's formula, which the Mledinans follow: Our squared separations are to be subtracted rather than added. This subtraction is intimately connected to the physical difference between spacetime, which you and I are exploring, and the Earth's surface, which the Mledinans explore—but at the risk of infuriating you, I shall forgo explaining the connection, and simply refer you to the discussions in Taylor and Wheeler (1992).

analogous to the straight-line distance between any two points on a flat sheet of paper. The absoluteness of this interval (the fact that its value is the same, regardless of whose reference frame is used to compute it) demonstrates that spacetime has an absolute reality; it is a four-dimensional fabric with properties that are independent of one's motion.

As we shall see in the coming pages, gravity is produced by a curvature (a warpage) of spacetime's absolute, four-dimensional fabric, and black holes, wormholes, gravitational waves, and singularities are all constructed wholly and solely from that fabric; that is, each of them is a specific type of spacetime warpage.

Because the absolute fabric of spacetime is responsible for such fascinating phenomena, it is frustrating that you and I do not experience it in our everyday lives. The fault lies in our low-velocity technology (for example, sports cars that travel far more slowly than light). Because of our low velocities relative to each other, we experience space and time solely as separate entities, we never notice the discrepancies between the lengths and times that you and I measure (we never notice that space and time are relative), and we never notice that our relative spaces and times are unified to form spacetime's absolute, four-dimensional fabric.

Minkowski, you may recall, was the mathematics professor who had labeled Einstein a lazy dog in his student days. In 1902 Minkowski, a

Russian by birth, had left the ETH in Zurich to take up a more attractive professorship in Göttingen, Germany. (Science was as international then as it is now.) In Göttingen, Minkowski studied Einstein's article on special relativity, and was impressed. That study led him to his 1908 discovery of the absolute nature of four-dimensional spacetime.

When Einstein learned of Minkowski's discovery, he was *not* impressed. Minkowski was merely rewriting the laws of special relativity in a new, more mathematical language; and, to Einstein, the mathematics obscured the physical ideas that underlie the laws. As Minkowski continued to extol the beauties of his spacetime viewpoint, Einstein began to make jokes about Göttingen mathematicians describing relativity in such complicated language that physicists wouldn't be able to understand it.

The joke, in fact, was on Einstein. Four years later, in 1912, he would realize that Minkowski's absolute spacetime is an essential foundation for incorporating gravity into special relativity. Sadly, Minkowski did not live to see this; he died of appendicitis in 1909, at age forty-five.

I shall return to Minkowski's absolute spacetime later in this chapter. First, however, I must develop another thread of my story: Newton's law of gravity and Einstein's first steps toward reconciling it with special relativity, steps he took before he began to appreciate Minkowski's breakthrough.

Newton's Gravitational Law, and Einstein's First Steps to Marry It to Relativity

Newton conceived of gravity as a force that acts between every pair of objects in the Universe, a force that pulls the objects toward each other. The larger the objects' masses and the closer they are together, the stronger the force. Stated more precisely, the force is proportional to the product of the objects' masses and inversely proportional to the square of the distance between them.

This gravitational law was an enormous intellectual triumph. When combined with Newton's laws of motion, it explained the orbits of the planets around the Sun, and the moons around the planets, the ebb and flow of ocean tides, and the fall of rocks; and it taught Newton and his seventeenth-century compatriots how to weigh the Sun and the Earth.[1]

During the two centuries that separated Newton and Einstein, astronomers' measurements of celestial orbits improved manyfold, putting Newton's gravitational law to ever more stringent tests. Occasionally new astronomical measurements disagreed with Newton's law, but in due course the observations or their interpretation turned out to be wrong. Time after time Newton's law triumphed over experimental or intellectual error. For example, when the motion of the planet Uranus (which had been discovered in 1781) appeared to violate the predictions of Newton's gravitational law, it seemed likely that the gravity of some other, undiscovered planet must be pulling on Uranus, perturbing its orbit. Calculations, based solely on Newton's laws of gravity and motion and on the observations of Uranus, predicted where in the sky that new planet should be. In 1846, when U. J. J. Leverrier trained his telescope on the spot, there the predicted planet was, too dim to be seen by the naked eye but bright enough for his telescope. This new planet, which vindicated Newton's gravitational law, was given the name "Neptune."

In the early 1900s, there remained two other exquisitely small, but puzzling discrepancies with Newton's gravitational law. One, a peculiarity in the orbit of the planet Mercury, would ultimately turn out to herald a failure of Newton's law. The other, a peculiarity in the Moon's orbit, would ultimately go away; it would turn out to be a misinterpretation of the astronomers' measurements. As is so often the case with exquisitely precise measurements, it was difficult to discern which of the two discrepancies, if either, should be worried about.

Einstein correctly suspected that Mercury's peculiarity (an anomalous shift of its *perihelion;* Box 2.2) was real and the Moon's peculiarity was not. Mercury's peculiarity "smelled" real; the Moon's did not. However, this suspected disagreement of experiment with Newton's gravitational law was far less interesting and important to Einstein than his conviction that Newton's law would turn out to violate his newly formulated principle of relativity (the "metaprinciple" that all the laws of physics must be the same in every inertial reference frame). Since Einstein believed firmly in his principle of relativity, such a violation would mean that Newton's gravitational law must be flawed.[2]

---

1. See the note to page 61 for details.

2. It was not completely obvious that Newton's gravitational law violated Einstein's principle of relativity, because Einstein, in formulating his principle, had relied on the concept of an

Box 2.2

The Perihelion Shift of Mercury

Kepler described the orbit of Mercury as an ellipse with the Sun at one focus (left diagram, in which the elliptical elongation of the orbit is exaggerated). However, by the late 1800s astronomers had deduced from their observations that Mercury's orbit is not quite elliptical. After each trip around its orbit, Mercury fails by a tiny amount to return to the same point as it started. This failure can be described as a shift, with each orbit, in the location of Mercury's closest point to the Sun (a shift of its *perihelion*). Astronomers measured a perihelion shift of 1.38 seconds of arc during each orbit (right diagram, in which the shift is exaggerated).

Newton's law of gravity could account for 1.28 arc seconds of this 1.38-arc-second shift: It was produced by the gravitational pull of Jupiter and the other planets on Mercury. However, there remained a 0.10-arc-second discrepancy: an *anomalous 0.10-arc-second shift of Mercury's perihelion during each orbit.* The astronomers claimed that the errors and uncertainties in their measurement were only 0.01 arc second in size, but considering the tiny angles being measured (0.01 arc second is equivalent to the angle subtended by a human hair at a distance of 10 kilometers), it is not surprising that many physicists of the late nineteenth and early twentieth centuries were skeptical, and expected Newton's laws to triumph in the end.

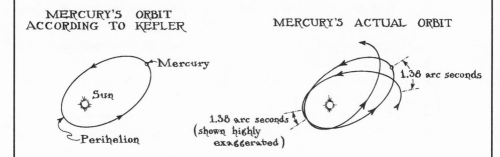

MERCURY'S ORBIT ACCORDING TO KEPLER

MERCURY'S ACTUAL ORBIT

Mercury

Sun

Perihelion

1.38 arc seconds
(shown highly exaggerated)

1.38 arc seconds

---

inertial reference frame, and this concept could not be used in the presence of gravity. (There is no way to shield a reference frame from gravity and thereby permit it to move solely under the influence of its own inertia.) However, Einstein was convinced that there must be some way to extend the sway of his relativity principle into the realm of gravity (some way to "generalize" it to include gravitational effects), and he was convinced that Newton's gravitational law would violate that yet-to-be-formulated "generalized principle of relativity."

Einstein's reasoning was simple: According to Newton, the gravitational force depends on the *distance* between the two gravitating objects (for example, the Sun and Mercury), but according to relativity, that distance is different in different reference frames. For example, Einstein's relativity laws predict that the distance between the Sun and Mercury will differ by about a part in a billion, depending on whether one is riding on Mercury's surface when measuring it or riding on the surface of the Sun. If both reference frames, Mercury's and the Sun's, are equally good in the eyes of the laws of physics, then which frame should be used to measure the distance that appears in Newton's gravitational law? Either choice, Mercury's frame or the Sun's, would violate the principle of relativity. This quandary convinced Einstein that Newton's gravitational law must be flawed.

Einstein's audacity is breathtaking. Having discarded Newton's absolute space and absolute time with almost no experimental justification, he was now inclined to discard Newton's enormously successful law of gravity, and with even less experimental justification. However, he was motivated not by experiment, but by his deep, intuitive insight into how the laws of physics *ought* to behave.

Einstein began his search for a new law of gravity in 1907. His initial steps were triggered and guided by a writing project: Although the patent office now classified him as only a "technical expert second class" (recently promoted from third class), he was sufficiently respected by the world's great physicists to be invited to write a review article for the annual publication *Jahrbuch der Radioaktivität und Elektronik* about his special relativistic laws of physics and their consequences. As he worked on his review, Einstein discovered a valuable strategy for scientific research: The necessity to lay out a subject in a self-contained, coherent, pedagogical manner forces one to think about it in new ways. One is driven to examine all the subject's gaps and flaws, and seek cures for them.

Gravity was his subject's biggest gap; special relativity, with its inertial frames on which no gravitational force can act, was totally ignorant of gravity. So while Einstein wrote, he kept looking for ways to incorporate gravity into his relativistic laws. As happens to most people immersed in a puzzle, even when Einstein wasn't thinking directly about this problem, the back of his mind mulled it over. Thus it was that one day in November 1907, in Einstein's own words, "I was sitting in a chair in the patent office at Bern, when all of a sudden a thought occurred to me: 'If a person falls freely, he will not feel his own weight.' "

Now you or I could have had that thought, and it would not have led anywhere. But Einstein was different. He pursued ideas to their ultimate ends; he wrung from them every morsel of insight that he could. And this idea was key; it pointed toward a revolutionary new view of gravity. He later called it "the happiest thought of my life."

The consequences of this thought tumbled forth quickly, and were immortalized in Einstein's review article. If you fall freely (for example, by jumping off a cliff), not only will you not feel your own weight, it will seem to you, in all respects, as though gravity had completely disappeared from your vicinity. For example, if you drop some rocks from your hand as you fall, you and the rocks will then fall together, side by side. If you look at the rocks and ignore your other surroundings, you cannot discern whether you and the rocks are falling together toward the ground below or are floating freely in space, far from all gravitating bodies. In fact, Einstein realized, in your immediate vicinity, gravity is so irrelevant, so impossible to detect, that *all* the laws of physics, in a small reference frame (laboratory) that you carry with you as you fall, must be the same as if you were moving freely through a universe without gravity. In other words, your small, freely falling reference frame is "equivalent to" an inertial reference frame in a gravity-free universe, and the laws of physics that you experience are the same as those in a gravity-free inertial frame; they are the laws of special relativity. (We shall learn later why the reference frame must be kept small, and that "small" means very small compared to the size of the Earth—or, more generally, very small compared to the distance over which the strength and direction of gravity change.)

As an example of the equivalence between a gravity-free inertial frame and your small, freely falling frame, consider the special relativistic law that describes the motion of a freely moving object (let it be a cannonball) in a universe without gravity. As measured in any inertial frame in that idealized universe, the ball must move along a straight line and with uniform velocity. Compare this with the ball's motion in our real, gravity-endowed Universe: If the ball is fired from a cannon on a grassy meadow on Earth and is watched by a dog who sits on the grass, the ball arcs up and over and falls back to Earth (Figure 2.2). It moves along a parabola (solid black curve) as measured in the dog's reference frame. Einstein asks that you view this same cannonball from a small, freely falling reference frame. This is easiest if the meadow is at the edge of a cliff. Then you can jump off the cliff just as the cannon is fired, and watch the ball as you fall.

As an aid in depicting what you see as you fall, imagine that you hold in front of yourself a window with twelve panes of glass, and you watch the ball through your window (middle segment of Figure 2.2). As you fall, you see the clockwise sequence of scenes shown in Figure 2.2. In looking at this sequence, ignore the dog, cannon, tree, and cliff; focus solely on your windowpanes and the ball. As seen by you, relative to your windowpanes, the ball moves along the straight dashed line with constant velocity.

Thus, in the dog's reference frame the ball obeys Newton's laws; it moves along a parabola. In your small, freely falling reference frame it obeys the laws of gravity-free special relativity; it moves along a straight line with constant velocity. And what is true in this example must be true in general, Einstein realized in a great leap of insight:

In any small, freely falling reference frame anywhere in our real, gravity-endowed Universe, the laws of physics must be the same as they are in an inertial reference frame in an idealized, gravity-free universe. Einstein called this the *principle of equivalence*, because it asserts that small, freely falling frames in the presence of gravity are equivalent to inertial frames in the absence of gravity.

This assertion, Einstein realized, had an enormously important consequence: It implied that, if we merely give the name "inertial reference frame" to every small, freely falling reference frame in our real, gravity-endowed Universe (for example, to a little laboratory that you carry as you fall over the cliff), then everything that special relativity says about inertial frames in an idealized universe without gravity will automatically also be true in our real Universe. Most important, the *principle of relativity* must be true: All small, inertial (freely falling) reference frames in our real, gravity-endowed Universe must be "created equal"; none can be preferred over any other in the eyes of the laws of physics. Or, stated more precisely (see Chapter 1):

Formulate any law of physics in terms of measurements made in one small, inertial (freely falling) reference frame. Then, when restated in terms of measurements in any other small inertial (freely falling) frame, that law of physics must take on precisely the same mathematical and logical form as in the original frame. And this must be true whether the (freely falling) inertial frame is in gravity-free intergalactic space, or is falling off a cliff on Earth, or is at the center of our galaxy, or is falling through the horizon of a black hole.

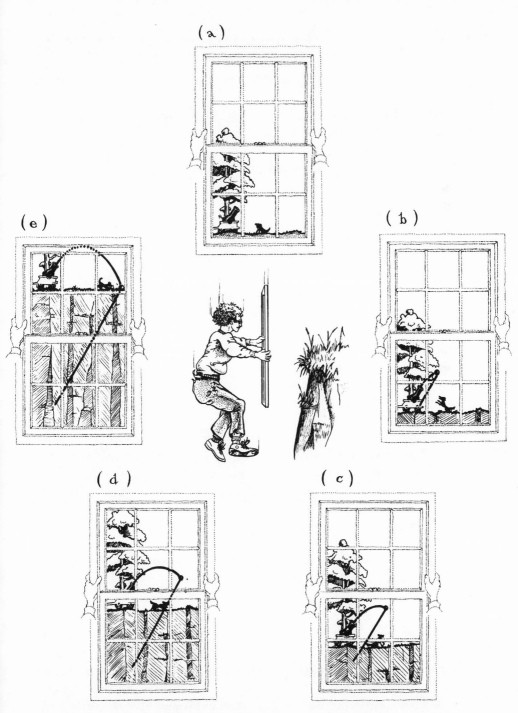

2.2 Center: You jump off a cliff holding a twelve-paned window in front of yourself. Remainder of figure, clockwise from the top: What you see through the window when a cannon is fired. Relative to the falling window frame, the ball's trajectory is the straight, dashed line; relative to the dog and the Earth's surface, it is the solid parabola.

With this extension of his principle of relativity to include gravity, Einstein took his first step toward a new set of gravitational laws—his first step from *special* relativity to *general* relativity.

Be patient, dear reader. This chapter is probably the most difficult one in the book. My story will get less technical in the next chapter, when we start exploring black holes.

Within days after formulating his equivalence principle, Einstein used it to make an amazing prediction, called *gravitational time dilation: If one is at rest relative to a gravitating body, then the nearer one is to the body, the more slowly one's time must flow.* For example, in a room on Earth, time must flow more slowly near the floor than near the ceiling. This Earthly difference turns out to be so minuscule, however (only 3 parts in 10^{16}; that is, 300 parts in a billion billion), that it is exceedingly difficult to detect. By contrast (as we shall see in the next chapter), near a black hole gravitational time dilation is enormous: If the hole weighs 10 times as much as the Sun, then time will flow 6 million times more slowly at 1 centimeter height above the hole's horizon than far from its horizon; and right at the horizon, the flow of time will be completely stopped. (Imagine the possibilities for time travel: If you descend to just above a black hole's horizon, hover there for one year of near-horizon time flow, and then return to Earth, you will find that during that one year of your time, millions of years have flown past on Earth!)

Einstein discovered gravitational time dilation by a somewhat complicated argument, but later he produced a simple and elegant demonstration of it, one that illustrates beautifully his methods of physical reasoning. That demonstration is presented in Box 2.4, and the *Doppler shift* of light, on which it relies, is explained in Box 2.3.

When starting to write his 1907 review article, Einstein expected it to describe relativity in a universe without gravity. However, while writing, he had discovered three clues to the mystery of how gravity might mesh with his relativity laws—the equivalence principle, gravitational time dilation, and the extension of his principle of relativity to include gravity—so he incorporated those clues into his article. Then, around the beginning of December, he mailed the article off to the editor of the *Jahrbuch der Radioaktivität und Elektronik* and turned his attention full force to the challenge of devising a complete, relativistic description of gravity.

Box 2.3
Doppler Shift

Whenever an emitter and a receiver of waves are moving toward each other, the receiver sees the waves shifted to higher frequency—that is, shorter period and shorter wavelength. If the emitter and receiver are moving apart, then the receiver sees the waves shifted to lower frequency—that is, longer period and longer wavelength. This is called the *Doppler shift*, and it is a property of all types of waves: sound waves, waves on water, electromagnetic waves, and so forth.

For sound waves, the Doppler shift is a familiar everyday phenomenon. One hears it in the sudden lowering of the sound's pitch when a speeding ambulance passes with siren screeching (drawing b), or when a landing airplane passes overhead. One can understand the Doppler shift by thinking about the diagrams below.

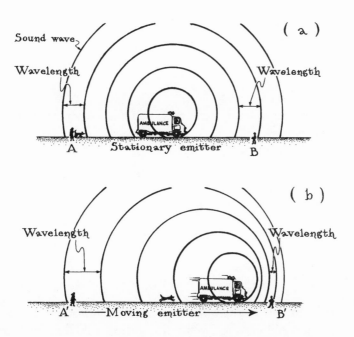

What is true of waves is also true of pulses. If the emitter transmits regularly spaced pulses of light (or of anything else), then the receiver, as the emitter moves toward it, will encounter the pulses at a higher frequency (a shorter time between pulses) than the frequency with which they were emitted.

BOX 2.4
Gravitational Time Dilation

Take two identical clocks. Place one on the floor of a room beside a hole into which it later will fall, and attach the other to the room's ceiling by a string. The ticking of the floor clock is regulated by the flow of time near the floor, and the ticking of the ceiling clock is regulated by the flow of time near the ceiling.

Let the ceiling clock emit a very short pulse of light whenever it ticks, and direct the pulses downward, toward the floor clock. Immediately before the ceiling clock emits its first pulse, cut the string that holds it, so it is falling freely. If the time between ticks is very short, then at the moment it next ticks and emits its second pulse, the clock will have fallen only imperceptibly and will still be very nearly at rest with respect to the ceiling (diagram a). This in turn means that the clock is still feeling the same flow of time as does the ceiling itself; that is, the interval between its pulse emissions is governed by the ceiling's time flow.

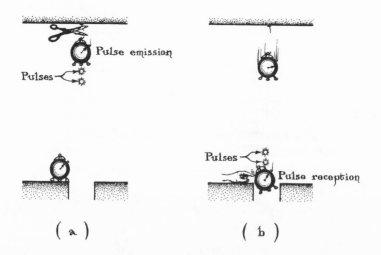

(a) (b)

Immediately before the first pulse of light reaches the floor, drop the floor clock into the hole (diagram b). The second pulse arrives so soon afterward that the freely falling floor clock has moved imperceptibly between pulses, and is still very nearly at rest with respect to the floor, and therefore is still feeling the same flow of time as does the floor itself.

In this way, Einstein converted the problem of comparing the flow of time as felt by the ceiling and the floor into the problem of comparing the ticking rates of two freely falling clocks: the falling ceiling clock which

(continued next page)

(Box 2.4 continued)

feels ceiling time, and the falling floor clock which feels floor time. Einstein's equivalence principle then permitted him to compare the ticks of the freely falling clocks with the aid of his special relativistic laws.

Because the ceiling clock was dropped before the floor clock, its downward speed is always greater than that of the floor clock (diagram b); that is, it moves toward the floor clock. This implies that the floor clock will see the ceiling clock's light pulses *Doppler-shifted* (Box 2.3); that is, it will see them arrive more closely spaced in time than the time between its own ticks. Since the time between pulses was regulated by the ceiling's time flow, and the time between floor-clock ticks is regulated by the floor's time flow, this means that time must flow more slowly near the floor than near the ceiling; in other words, *gravity must dilate the flow of time.*

On December 24, he wrote to a friend saying, "At this time I am busy with considerations on relativity theory in connection with the law of gravitation . . . I hope to clear up the so-far unexplained secular changes of the perihelion shift of Mercury . . . but thus far it does not seem to work." By early 1908, frustrated by no real progress, Einstein gave up, and turned his attention to the realm of atoms, molecules, and radiation (the "realm of the small"), where the unsolved problems for the moment seemed more tractable and interesting.[3]

Through 1908 (while Minkowski unified space and time, and Einstein pooh-poohed the unification), and through 1909, 1910, and 1911, Einstein stayed with the realm of the small. These years also saw him move from the patent office in Bern to an associate professorship at the University of Zurich, and a full professorship in Prague—a center of the Austro-Hungarian empire's cultural life.

Einstein's life as a professor was not easy. He found it irritating to have to give regular lectures on topics not close to his research. He could summon neither the energy to prepare such lectures well nor the enthusiasm to make them scintillate, even though when lecturing on topics dear to his heart, he was brilliant. Einstein was now a full-fledged member of Europe's academic circle, but he was paying a price. Despite this price, his research in the realm of the small moved forward impressively, producing insights that later would win him the Nobel Prize (see Box 4.1).

Then, in mid-1911, Einstein's fascination with the small waned and his attention returned to gravity, with which he would struggle almost

3. Chapter 4 and especially Box 4.1.

full time until his triumphant formulation of general relativity in November 1915.

The initial focus of Einstein's gravitational struggle was *tidal gravitational forces.*

Tidal Gravity and Spacetime Curvature

Imagine yourself an astronaut out in space, far above the Earth's equator, and falling freely toward it. Although, as you fall, you will not feel your own weight, you will, in fact, feel some tiny, residual effects of gravity. Those residuals are called "tidal gravity," and they can be understood by thinking about the gravitational forces you feel, first from the viewpoint of someone watching you from the Earth below, and then from your own viewpoint.

2.3   As you fall toward Earth, tidal gravitational forces stretch you from head to foot and squeeze you from the sides.

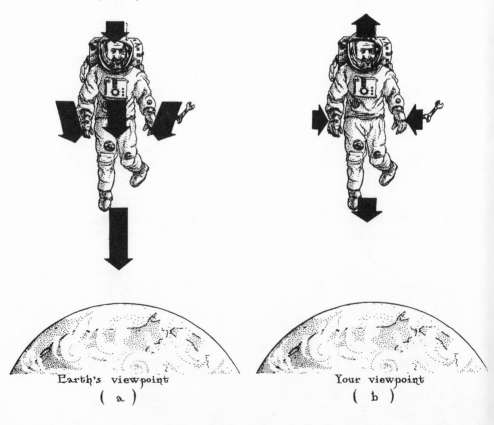

Earth's viewpoint
(a)

Your viewpoint
(b)

As seen from Earth (Figure 2.3a), the gravitational pull is slightly different on various parts of your body. Because your feet are closer to the Earth than your head, gravity pulls more strongly on them than on your head, so it stretches you from foot to head. And because gravity pulls always toward the Earth's center, a direction that is slightly left-ward on your right side and slightly rightward on your left side, the pull is slightly leftward on your right and slightly rightward on your left; that is, it squeezes your sides inward.

From your viewpoint (Figure 2.3b), the large, downward force of gravity is gone, vanished. You feel weightless. However, the vanished piece of gravity is only the piece that pulled you downward. The head-to-foot stretch and side-to-side squeeze remain. They are caused by the *differences* between gravity on the outer parts of your body and gravity at your body's center, differences that you cannot get rid of by falling freely.

The vertical stretch and lateral squeeze that you feel, as you fall, are called tidal gravity or tidal gravitational forces, because, when the Moon is their source rather than the Earth and when the Earth is feeling them rather than you, they produce the ocean tides. See Box 2.5.

In deducing his principle of equivalence, Einstein ignored tidal gravitational forces; he pretended they do not exist. (Recall the essence of his argument: As you fall freely, you "will not feel your own weight" and "it will seem to you, in all respects, as though gravity has disap-peared from your vicinity.") Einstein justified ignoring tidal forces by imagining that you (and your reference frame) are very small. For example, if you are the size of an ant or smaller, then your body parts will all be very close to each other, the direction and strength of grav-ity's pull will therefore be very nearly the same on the outer parts of your body as at its center, and the *difference* in gravity between your outer parts and your center, which causes the tidal stretch and squeeze, will be extremely small. On the other hand, if you are a 5000-kilome-ter-tall giant, then the direction and strength of the Earth's gravita-tional pull will differ greatly between the outer parts of your body and its center; and correspondingly, as you fall, you will experience a huge tidal stretch and squeeze.

This reasoning convinced Einstein that, in a sufficiently small, freely falling reference frame (a frame very small compared to the distance over which gravity's pull changes), one should not be able to detect any influences of tidal gravity whatsoever; that is, small, freely

Ocean Tides Produced by Tidal Forces

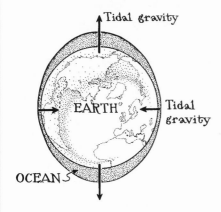

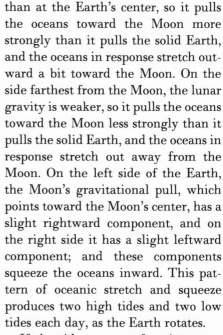

On the side of the Earth nearest the Moon, the lunar gravity is stronger than at the Earth's center, so it pulls the oceans toward the Moon more strongly than it pulls the solid Earth, and the oceans in response stretch outward a bit toward the Moon. On the side farthest from the Moon, the lunar gravity is weaker, so it pulls the oceans toward the Moon less strongly than it pulls the solid Earth, and the oceans in response stretch out away from the Moon. On the left side of the Earth, the Moon's gravitational pull, which points toward the Moon's center, has a slight rightward component, and on the right side it has a slight leftward component; and these components squeeze the oceans inward. This pattern of oceanic stretch and squeeze produces two high tides and two low tides each day, as the Earth rotates.

If the tides at your favorite ocean beach do not behave in precisely this way, it is not the fault of the Moon's gravity; rather, it is because of two effects: (1) There is a lag in the water's response to the tidal gravity. It takes time for the water to move in and out of bays, harbors, river channels, fjords, and other indentations in the coastline. (2) The Sun's gravitational stretch and squeeze are almost as strong on the Earth as the Moon's, but are oriented differently because the Sun's position in the sky is (usually) different from the Moon's. The Earth's tides are a result of the combined tidal gravity of the Sun and the Moon.

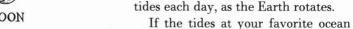

falling reference frames in our gravity-endowed Universe are equivalent to inertial frames in a universe without gravity. But not so for large frames. And the tidal forces felt in large frames seemed to Einstein, in 1911, to be a key to the ultimate nature of gravity.

It was clear how Newton's gravitational law explains tidal forces: They are produced by a difference in the strength and direction of gravity's pull, from one place to another. But Newton's law, with its gravitational force that depends on distance, had to be wrong; it violated the principle of relativity ("in whose frame was the distance to be measured?"). Einstein's challenge was to formulate a completely new gravitational law that is simultaneously compatible with the principle of relativity and explains tidal gravity in some new, simple, compelling way.

From mid-1911 to mid-1912, Einstein tried to explain tidal gravity by assuming that time is warped, but space is flat. This radical-sounding idea was a natural outgrowth of gravitational time dilation: The different rates of flow of time near the ceiling and the floor of a room on Earth could be thought of as a warpage of time. Perhaps, Einstein speculated, a more complicated pattern of time warpage might produce all known gravitational effects, from tidal gravity to the elliptical orbits of the planets to even the anomalous perihelion shift of Mercury.

After a twelve-month pursuit of this intriguing idea, Einstein abandoned it, and for a good reason. Time is relative. Your time is a mixture of my time and my space (if we move with respect to each other), and therefore, if your time is warped but your space is flat, then my time and my space must both be warped, as must be everybody else's. You and only you will have a flat space, so the laws of physics must be picking out your reference frame as fundamentally different from all others—in violation of the principle of relativity.

Nevertheless, time warpage "smelled right" to Einstein, so perhaps—he reasoned—everybody's time is warped and, inevitably alongside that, everybody's space is warped. Perhaps these combined warpages could explain tidal gravity.

The idea of a warpage of *both* time and space was rather daunting. Since the Universe admits an infinite number of different reference frames, each moving with a different velocity, there would have to be an infinity of warped times and an infinity of warped spaces! Fortunately, Einstein realized, Hermann Minkowski had provided a powerful tool for simplifying such complexity: "Henceforth, space by itself, and time by itself, are doomed to fade away into mere shadows, and only a kind of union of the two will preserve an independent reality." There is just one, unique, absolute, four-dimensional spacetime in our Universe; and a warpage of everyone's time and everyone's space must show up as *a warpage of Minkowski's single, unique, absolute spacetime*.

This was the conclusion to which Einstein was driven in the summer of 1912 (though he preferred to use the word "curvature" rather than "warpage"). After four years of ridiculing Minkowski's idea of absolute spacetime, Einstein had finally been driven to embrace it, and warp it.

What does it mean for spacetime to be curved (or warped)? For clarity, ask first what it means for a two-dimensional surface to be curved (or warped). Figure 2.4 shows a flat surface and a curved surface. On the flat surface (an ordinary sheet of paper) are drawn two absolutely straight lines. The lines start out side by side and parallel. The ancient Greek mathematician Euclid, who created the subject now called "Euclidean geometry," used as one of his geometric postulates the demand that two such initially parallel lines never cross. This non-crossing is an unequivocal test for the flatness of the surface on which the lines are drawn. If space is flat, then initially parallel straight lines can never cross. If we ever find a pair of initially parallel straight lines that do cross, then we will know that space is not flat.

The curved surface in Figure 2.4 is a globe of the Earth. Locate on that globe the city of Quito, Equador; it sits on the equator. Send out a precisely straight line from Quito, directed northward. The line will travel northward, at constant longitude, through the North Pole.

In what sense is this a straight line? In two senses. One is the sense so crucial to airlines: It is a great circle, and the great circles on the Earth's

2.4   Two straight lines, initially parallel, never cross on a flat surface such as the sheet of paper shown on the left. Two straight lines, initially parallel, will typically cross on a curved surface such as the globe of the world shown on the right.

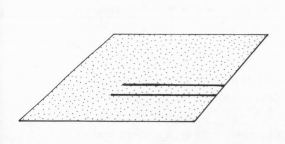
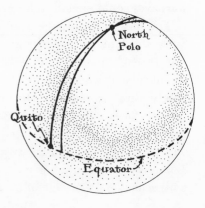

globe are the shortest routes between two points and thus are the kinds of routes along which airlines like to fly. Construct any other line connecting Quito to the North Pole; it will necessarily be longer than the great circle.

The second sense of straightness is the one that we shall use below, when discussing spacetime: In sufficiently small regions on the globe along the great circle's route, the globe's curvature can hardly be noticed. In such a region, the great circle looks straight in the usual flat-sheet-of-paper sense of straightness—the sense of straightness used by professional surveyors, who lay out boundaries of property using transits or laser beams. The great circle is straight, in this surveyors' sense, in each and every small region along its route.

Mathematicians use the name *geodesic* for any line, on a curved or warped surface, that is straight in these two senses: the airlines' "shortest route" sense, and the surveyors' sense.

Now move eastward on the globe from Quito by a few centimeters, and construct a new straight line (great circle; geodesic) that is precisely parallel, at the equator, to the one through Quito. This straight line, like the first one, will pass through the globe's North Pole. *It is the curvature of the globe's surface that forces the two straight lines, initially parallel, to cross at the North Pole.*

With this understanding of the effects of curvature in two-dimensional surfaces, we can return to four-dimensional spacetime and ask about curvature there.

In an idealized universe without gravity, there is no warpage of space, no warpage of time; spacetime has no curvature. In such a universe, according to Einstein's special relativity laws, freely moving particles must travel along absolutely straight lines. They must maintain constant direction and constant velocity, as measured in any and every inertial reference frame. This is a fundamental tenet of special relativity.

Now, Einstein's equivalence principle guarantees that gravity cannot change this fundamental tenet of free motion: Whenever a freely moving particle, in our real, gravity-endowed Universe, enters and passes through a small, inertial (freely falling) reference frame, the particle must move along a straight line through that frame. Straight-line motion through a small inertial frame, however, is the obvious analogue of straight-line behavior as measured by surveyors in a small region of the Earth's surface; and just as such straight-line behavior in small regions on Earth implies that a line is actually a geodesic of the Earth's surface, so also the particle's straight-line motion in small regions of spacetime implies that the particle moves along a geodesic of spacetime. And what is true of this particle must be true of all particles: *Every freely moving particle (every particle on which no forces, except gravity, act) travels along a geodesic of spacetime.*

As soon as Einstein realized this, it became obvious to him that *tidal gravity is a manifestation of spacetime curvature.*

To understand why, imagine the following thought experiment (mine, not Einstein's). Stand on the ice sheet at the North Pole, holding two small balls, one in each hand (Figure 2.5). Throw the balls into the air side by side, so they rise upward along precisely parallel trajectories,

2.5 Two balls thrown into the air on precisely parallel trajectories, if able to pass unimpeded through the Earth, will collide near the Earth's center.

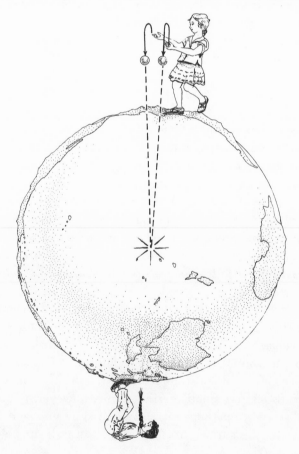

and then watch them fall back to Earth. Now, in a thought experiment such as this, you can do anything you wish so long as it does not violate the laws of physics. You wish to watch the trajectories of the balls as they fall under the action of gravity, not only above the Earth's surface, but also below. For this purpose, you can pretend that the balls are made of a material that falls through the Earth's soil and rock without being slowed at all (tiny black holes would have this property), and you can pretend that you and a friend on the opposite side of the Earth, who also watches, can follow the balls' motion inside the Earth via "X-ray vision."

As the balls fall into the Earth, the Earth's tidal gravity squeezes them together in the same way as it squeezes your sides if you are a falling astronaut (Figure 2.3). The strength of the tidal gravity is just right to make both balls fall almost precisely toward the Earth's center, and hit each other there.

Now comes the payoff of this thought experiment: Each ball moved along a precisely straight line (a geodesic) through spacetime. Initially the two straight lines were parallel. Later they crossed (the balls collided). This crossing of initially parallel, straight lines signals a curvature of spacetime. From Einstein's viewpoint, spacetime curvature *causes* the crossing, that is, causes the balls' collision, just as the curvature of the globe caused straight lines to cross in Figure 2.4. From Newton's viewpoint, tidal gravity causes the crossing.

Thus, Einstein and Newton, with their very different viewpoints on the nature of space and time, give very different names to the agent that causes the crossing. Einstein calls it spacetime curvature; Newton calls it tidal gravity. But there is just one agent acting. Therefore, *spacetime curvature and tidal gravity must be precisely the same thing, expressed in different languages.*

Our human minds have great difficulty visualizing curved surfaces with more than two dimensions; therefore, it is nearly impossible to visualize the curvature of four-dimensional spacetime. Some insight can be gained, however, by looking at various two-dimensional pieces of spacetime. Figure 2.6 uses two such pieces to explain how spacetime curvature creates the tidal stretch and squeeze that produce the ocean tides.

Figure 2.6a depicts one piece of spacetime in the vicinity of Earth, a piece that includes time, plus space along the direction toward the Moon. The Moon curves this piece of spacetime, and the curvature stretches apart two geodesics, in the manner shown. Correspondingly, we humans see two freely moving particles, which travel along the geodesics, get stretched apart as they travel, and we interpret that stretching as a tidal gravitational force. This stretching tidal force (spacetime curvature) affects not only freely moving particles, but also the Earth's oceans; it stretches the oceans in the manner shown in Box 2.5, producing oceanic bulges on the sides of the Earth nearest and farthest from the Moon. The two bulges are trying to travel along geodesics of the curved spacetime (Figure 2.6a), and therefore are trying to fly apart; but the Earth's gravity (the spacetime curvature produced by the Earth; not shown in the diagram) is counteracting that flight, so the ocean merely bulges.

Figure 2.6b is a different piece of spacetime near Earth, a piece that includes time, plus space along a direction transverse to the Moon's direction. The Moon curves this piece of spacetime, and the curvature squeezes geodesics together in the manner shown. Correspondingly, we humans see freely moving particles that travel along geodesics transverse to the Moon's direction get squeezed together by the curvature (by the Moon's tidal gravity), and similarly we see the Earth's oceans get squeezed along directions transverse to the direction of the Moon. This tidal squeeze produces the transverse oceanic compressions shown in Box 2.5.

2.6 Two two-dimensional pieces of curved spacetime, in the vicinity of the Earth. The curvature is produced by the Moon. The curvature creates a tidal stretch along the direction toward the Moon (a), and a tidal squeeze along the direction transverse to the Moon (b), and this stretch and squeeze produce the ocean's tides in the manner discussed in Box 2.5, above.

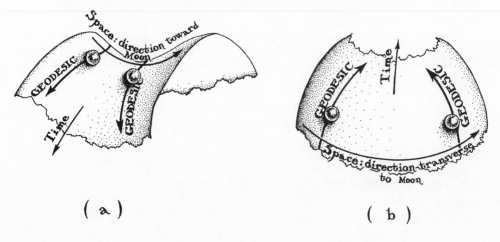

(a)   (b)

Einstein was a professor in Prague in the summer of 1912, when he realized that tidal gravity and spacetime curvature are one and the same thing. It was a wonderful revelation—though he was not yet certain of it and did not yet understand it as fully as I have described it, and it did not provide a complete explanation of gravity. It told Einstein that spacetime curvature dictates the motion of free particles and raises the tides on the ocean, but it did not tell him how the curvature is produced. Einstein believed that the matter inside the Sun and Earth and other planets is somehow responsible for the curvature. But how? *How does matter warp spacetime, and what are the details of the warpage?* A quest for the *law of warpage* became Einstein's central concern.

A few weeks after "discovering" spacetime curvature, Einstein moved from Prague back to Zurich, to take up a professorship at his alma mater, the ETH. Upon arriving in Zurich in August 1912, Einstein sought advice from an old classmate, Marcel Grossmann, who was now a professor of mathematics there. Einstein explained his idea that tidal gravity is spacetime curvature, and then asked whether any mathematician had ever developed a set of mathematical equations that could help him figure out the law of warpage, that is, the law that describes how matter forces spacetime to curve. Grossmann, whose specialty was other aspects of geometry, wasn't sure, but after browsing in the library he came back with an answer: Yes, the necessary equations did exist. They had been invented largely by the German mathematician Bernhard Riemann in the 1860s, the Italian Gregorio Ricci in the 1880s, and Ricci's student Tullio Levi-Civita in the 1890s and 1900s; they were called the "absolute differential calculus" (or, in physicists' language of 1915–1960, "tensor analysis," or in the language of 1960 to the present, "differential geometry"). But, Grossmann told Einstein, this differential geometry is a terrible mess which physicists should not be involved with. Were there any other geometries that could be used to figure out the law of warpage? No.

And so, with much help from Grossmann, Einstein set out to master the intricacies of differential geometry. As Grossmann taught mathematics to Einstein, Einstein taught something of physics to Grossmann. Einstein later quoted Grossmann as saying, "I concede that I did after all gain something rather important from the study of physics. Before, when I sat on a chair and felt a trace of heat left by my 'pre-sitter,' I used to shudder a little. That is completely gone, for on this point physics has taught me that heat is something completely impersonal."

Learning differential geometry was not an easy task for Einstein. The spirit of the subject was alien to the intuitive physical arguments that he found so natural. In late October 1912 he wrote to Arnold Sommerfeld, a leading German physicist: "I am now occupying myself exclusively with the problem of gravitation and believe that, with the aid of a local mathematician [Grossmann] who is a friend of mine I'll now be able to master all the difficulties. But one thing is certain, that in all my life I have never struggled so hard, and that I have been infused with great respect for mathematics the subtler parts of which, in my simple-mindedness, I had considered pure luxury up to now! Compared to this problem the original relativity theory [special relativity] is child's play."

Together Einstein and Grossmann struggled through the autumn and into the winter with the puzzle of how matter forces spacetime to curve. But despite their all-out effort, the mathematics could not be brought into accord with Einstein's vision. The law of warpage eluded them.

Einstein was convinced that the law of warpage should obey a *generalized (enlarged) version of his principle of relativity:* It should look the same in every reference frame—not just inertial (freely falling) frames, but non-inertial frames as well. The law of warpage should not rely for its formulation on any special reference frame or any special class of reference frames whatsoever.[4] Sadly, the equations of differential geometry did not seem to admit such a law. Finally, in late winter, Einstein and Grossmann gave up the search and published the best law of warpage they could find—a law that relied for its definition on a special class of reference frames.

Einstein, eternally the optimist, managed to convince himself, briefly, that this was no catastrophe. To his physicist friend Paul Ehrenfest he wrote in early 1913, "What can be more beautiful than that this necessary specialization follows from [the mathematical equations for the conservation of energy and momentum]?" But after further thought he regarded it a disaster. He wrote to Lorentz in August 1913: "My faith in the reliability of the theory [the "law of warpage"] still fluctuates. . . . [Because of the failure to obey the generalized principle of relativity,] the theory contradicts its own starting point and all is up in the air."

---

4. Einstein used the new phrase "general covariance" for this property, although it was just a natural extension of his principle of relativity.

As Einstein and Grossmann struggled with spacetime curvature, other physicists scattered over the European continent took up the challenge of uniting the laws of gravity with special relativity. But none of them—Gunnar Nordström in Helsinki, Finland; Gustav Mie in Greifswald, Germany; Max Abraham in Milano, Italy—adopted Einstein's spacetime curvature viewpoint. Instead they treated gravity, like electromagnetism, as due to a force field which lives in Minkowski's flat, special relativistic spacetime. And no wonder they took this approach: The mathematics used by Einstein and Grossmann was horrendously complex, and it had produced a law of warpage that violated its authors' own precepts.

Controversy swirled among the proponents of the various viewpoints. Wrote Abraham, "Someone who, like this author, has had to warn repeatedly against the siren song of [the principle of relativity] will greet with satisfaction the fact that its originator has now convinced himself of its untenability." Wrote Einstein in reply, "In my opinion the situation does not indicate the failure of the relativity principle. . . . There is not the slightest ground to doubt its validity." And privately he described Abraham's theory of gravity as "a stately horse which lacks three legs." Writing to friends in 1913 and 1914 Einstein said of the controversy, "I enjoy it that this affair is at least taken up with the requisite animation. I enjoy the controversies. Figaro mood: I'll play him a tune." "I enjoy it that colleagues occupy themselves at all with the theory [developed by Grossmann and me], although for the time being with the purpose of killing it. . . . On the face of it, Nordstrom's theory . . . is much more plausible. But it, too, is built on [flat, Minkowskian spacetime], the belief in which amounts, I feel, to something like a superstition."

In April 1914 Einstein left the ETH for a professorship in Berlin which carried no teaching duties. At last he could work on research as much as he wished, and even do so in the stimulating vicinity of Berlin's great physicists, Max Planck and Walther Nernst. In Berlin, despite the June 1914 outbreak of the First World War, Einstein continued his quest for an acceptable description of how matter curves spacetime, a description that did not rely on any special class of reference frames—an improved law of warpage.

A three-hour train ride from Berlin, in the university village of Göttingen where Minkowski had worked, there lived one of the greatest mathematicians of all time: David Hilbert. During 1914 and 1915

Hilbert pursued a passionate interest in physics. Einstein's published ideas fascinated him, so in late June of 1915 he invited Einstein down for a visit. Einstein stayed for about a week and gave six two-hour lectures to Hilbert and his colleagues. Several days after the visit Einstein wrote to a friend, "I had the great joy of seeing in Göttingen that everything [about my work] is understood to the last detail. With Hilbert I am just enraptured."

Several months after returning to Berlin, Einstein became more deeply distressed than ever with the Einstein–Grossmann law of warpage. Not only did it violate his vision that the laws of gravity should be the same in all reference frames, but also, he discovered after arduous calculation, it gave a wrong value for the anomalous perihelion shift of Mercury's orbit. He had hoped his theory would explain the perihelion shift, thereby triumphantly resolving the shift's discrepancy with Newton's laws. Such an achievement would give at least some experimental confirmation that his laws of gravity were right and Newton's wrong. However, his calculation, based on the Einstein–Grossmann law of warpage, gave a perihelion shift half as large as was observed.

Pouring over his old calculations with Grossmann, Einstein discovered a few crucial mistakes. Feverishly he worked through the month of October, and on 4 November he presented, at the weekly plenary session of the Prussian Academy of Sciences in Berlin, an account of his mistakes and a revised law of warpage—still slightly dependent on a special class of reference frames, but less so than before.

Remaining dissatisfied, Einstein struggled all the next week with his 4 November law, found mistakes, and presented yet another proposal for the law of warpage at the Academy meeting of 11 November. But still the law relied on special frames; still it violated his principle of relativity.

Resigning himself to this violation, Einstein struggled during the next week to compute consequences of his new law that could be observed with telescopes. It predicted, he found, that starlight passing the limb of the Sun should be deflected gravitationally by an angle of 1.7 seconds of arc (a prediction that would be verified four years later by careful measurements during a solar eclipse). More important to Einstein, the new law yielded the correct perihelion shift for Mercury! He was beside himself with joy; for three days he was so excited that he couldn't work. This triumph he presented at the next meeting of the Academy on 18 November.

But his law's violation of the relativity principle still troubled him.

So during the next week Einstein poured back over his calculations and found another mistake—the crucial one. At last everything fell into place. The entire mathematical formalism was now free of any dependence on special reference frames: It had the same form when expressed in each and every reference frame (see Box 2.6 below) and thus obeyed the principle of relativity. Einstein's vision of 1914 was fully vindicated! And the new formalism still gave the same predictions for the shift of Mercury's perihelion and for the gravitational deflection of light, and it incorporated his 1907 prediction of gravitational time dilation. These conclusions, and the final definitive form of his *general relativity* law of warpage, Einstein presented to the Prussian Academy on 25 November.

Three days later Einstein wrote to his friend Arnold Sommerfeld: "During the past month I had one of the most exciting and strenuous times of my life, but also one of the most successful." Then, in a January letter to Paul Ehrenfest: "Imagine my joy [that my new law of warpage obeys the principle of relativity] and at the result that the [law predicts] the correct perihelion motion of Mercury. I was beside myself with ecstasy for days." And, later, speaking of the same period: "The years of searching in the dark for a truth that one feels but cannot express, the intense desire and the alternations of confidence and misgiving until one breaks through to clarity and understanding, are known only to him who has himself experienced them."

Remarkably, Einstein was not the first to discover the correct form of the law of warpage, the form that obeys his relativity principle. Recognition for the first discovery must go to Hilbert. In autumn 1915, even as Einstein was struggling toward the right law, making mathematical mistake after mistake, Hilbert was mulling over the things he had learned from Einstein's summer visit to Göttingen. While he was on an autumn vacation on the island of Rugen in the Baltic the key idea came to him, and within a few weeks he had the right law—derived not by the arduous trial-and-error path of Einstein, but by an elegant, succinct mathematical route. Hilbert presented his derivation and the resulting law at a meeting of the Royal Academy of Sciences in Göttingen on 20 November 1915, just five days before Einstein's presentation of the same law at the Prussian Academy meeting in Berlin.

Quite naturally, and in accord with Hilbert's own view of things, the resulting law of warpage was quickly given the name the *Einstein field equation* (Box 2.6) rather than being named after Hilbert. Hilbert had

Box 2.6

The Einstein Field Equation:
Einstein's Law of Spacetime Warpage

Einstein's law of spacetime warpage, the *Einstein field equation*, states that "mass and pressure warp spacetime." More specifically:

At any location in spacetime, choose an arbitrary reference frame. In that reference frame, explore the curvature of spacetime by studying how the curvature (that is, tidal gravity) pushes freely moving particles together or pulls them apart along each of the three directions of the chosen frame's space: the east–west direction, the north–south direction, and the up–down direction. The particles move along geodesics of spacetime (Figure 2.6), and the rate at which they are pushed together or pulled apart is proportional to the strength of the curvature along the direction between them. If they are pushed together as in diagrams (a) and (b), the curvature is said to be positive; if they are pulled apart as in (c), the curvature is negative.

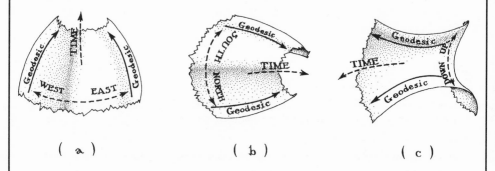

(a)   (b)   (c)

Add together the strengths of the curvatures along all three directions, east–west [diagram (a)], north–south [diagram (b)], and up–down [diagram (c)]. Einstein's field equation states that *the sum of the strengths of these three curvatures is proportional to the density of mass in the particle's vicinity (multiplied by the speed of light squared to convert it into a density of energy; see Box 5.2), plus 3 times the pressure of matter in the particles' vicinity.*

Even though you and I may be at the same location in spacetime (say, flying over Paris, France, at noon on 14 July 1996), if we move relative to each other, your space will be different from mine and similarly the density of mass (for example, the mass of the air around us) that you measure will be different from the density that I measure, and the pressure of matter (for example, the air pressure) that we measure will differ. Similarly, it turns out, the sum of the three curvatures of spacetime that

(continued next page)

(Box 2.6 continued)

you measure will be different from the sum that I measure. However, you and I must each find that the sum of the curvatures we measure is proportional to the density of mass we measure plus 3 times the pressure we measure. In this sense, the Einstein field equation is the same in every reference frame; it obeys Einstein's principle of relativity.

Under most circumstances (for example, throughout the solar system), the pressure of matter is tiny compared to its mass density times the speed of light squared, and therefore the pressure is an unimportant contributor to spacetime curvature; *the spacetime warpage is due almost solely to mass.* Only deep inside neutron stars (Chapter 5), and in a few other exotic places, is pressure a significant contributor to the warpage.

By mathematically manipulating the Einstein field equation, Einstein and other physicists have not only explained the deflection of starlight by the Sun and the motions of the planets in their orbits, including the mysterious perihelion shift of Mercury, they have also predicted the existence of black holes (Chapter 3), gravitational waves (Chapter 10), singularities of spacetime (Chapter 13), and perhaps the existence of wormholes and time machines (Chapter 14). The remainder of this book is devoted to this legacy of Einstein's genius.

carried out the last few mathematical steps to its discovery independently and almost simultaneously with Einstein, but Einstein was responsible for essentially everything that preceded those steps: the recognition that tidal gravity must be the same thing as a warpage of spacetime, the vision that the law of warpage must obey the relativity principle, and the first 90 percent of that law, the Einstein field equation. In fact, without Einstein the general relativistic laws of gravity might not have been discovered until several decades later.

As I browse through Einstein's published scientific papers (a browsing which, unfortunately, I must do in the 1965 Russian edition of his collected works because I read no German and most of his papers have not as of 1993 been translated into English!), I am struck by the profound change of character of Einstein's work in 1912. Before 1912 his papers are fantastic for their elegance, their deep intuition, and their modest use of mathematics. Many of the arguments are the same as those which I and my friends use in the 1990s when we teach courses on relativity. Nobody has learned to improve on those arguments. By contrast, after 1912, complex mathematics abounds in Einstein's papers—though usually in combination with insights about physical laws. This combination of mathematics and physical insight, which only Einstein among all physicists working on gravity had in the period 1912–1915, ultimately led Einstein to the full form of his gravitational laws.

But Einstein wielded his mathematical tools with some clumsiness. As Hilbert was later to say, "Every boy in the streets of Göttingen understands more about four-dimensional geometry than Einstein. Yet, in spite of that, Einstein did the work [formulated the general relativistic laws of gravity] and not the mathematicians." He did the work because mathematics was not enough; Einstein's unique physical insight was also needed.

Actually, Hilbert exaggerated. Einstein was a rather good mathematician, though in mathematical technique he was not the towering figure that he was in physical insight. As a result, few of Einstein's post-1912 arguments are presented today in the way Einstein presented them. People have learned improvements. And, with the quest to understand the laws of physics becoming more and more mathematical as the years after 1915 passed, Einstein became less and less the dominant figure he had been. The torch was passed to others.

3

Black Holes
Discovered and
Rejected

*in which Einstein's laws
of warped spacetime
predict black holes,
and Einstein rejects the prediction*

"The essential result of this investigation," Albert Einstein wrote in a technical paper in 1939, "is a clear understanding as to why the 'Schwarzschild singularities' do not exist in physical reality." With these words, Einstein made clear and unequivocal his rejection of his own intellectual legacy: the black holes that his general relativistic laws of gravity seemed to be predicting.

Only a few features of black holes had as yet been deduced from Einstein's laws, and the name "black holes" had not yet been coined; they were being called "Schwarzschild singularities." However, it was clear that anything that falls into a black hole can never get back out and cannot send light or anything else out, and this was enough to convince Einstein and most other physicists of his day that black holes are outrageously bizarre objects which surely should not exist in the real Universe. Somehow, the laws of physics must protect the Universe from such beasts.

What was known about black holes, when Einstein so strongly rejected them? How firm was general relativity's prediction that they do exist? How could Einstein reject that prediction and still maintain confidence in his general relativistic laws? The answers to these questions have their roots in the eighteenth century.

Throughout the 1700s, scientists (then called natural philosophers) believed that gravity was governed by Newton's laws, and that light was made of corpuscles (particles) that are emitted by their sources at a very high, universal speed. That speed was known to be about 300,000 kilometers per second, thanks to telescopic measurements of light emitted by Jupiter's moons as they orbit around their parent planet.

In 1783 John Michell, a British natural philosopher, dared to combine the corpuscular description of light with Newton's gravitation laws and thereby predict what very compact stars should look like. He did this by a thought experiment which I repeat here in modified form:

Launch a particle from the surface of a star with some initial speed, and let it move freely upward. If the initial speed is too low, the star's gravity will slow the particle to a halt and then pull it back to the star's surface. If the initial speed is high enough, gravity will slow the particle but not stop it; the particle will manage to escape. The dividing line, the minimum initial speed for escape, is called the "escape velocity." For a particle ejected from the Earth's surface, the escape velocity is 11 kilometers per second; for a particle ejected from the Sun's surface, it is 617 kilometers per second, or 0.2 percent of the speed of light.

Michell could compute the escape velocity using Newton's laws of gravity, and could show that it is proportional to the square root of the star's mass divided by its circumference. Thus, for a star of fixed mass, the smaller the circumference, the larger the escape velocity. The reason is simple: The smaller the circumference, the closer the star's surface is to its center, and thus the stronger is gravity at its surface, and the harder the particle has to work to escape the star's gravitational pull.

There is a *critical circumference*, Michell reasoned, for which the escape velocity is the speed of light. If corpuscles of light are affected by gravity in the same manner as other kinds of particles, then light can barely escape from a star that has this critical circumference. For a star a bit smaller, light cannot escape at all. When a corpuscle of light is launched from such a star with the standard light velocity of 299,792 kilometers per second, it will fly upward at first, then slow to a halt and fall back to the star's surface; see Figure 3.1.

Michell could easily compute the critical circumference; it was 18.5 kilometers, if the star had the same mass as the Sun, and proportionately larger if the mass were larger.

Nothing in the eighteenth-century laws of physics prevented so

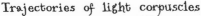

Trajectories of light corpuscles

STAR

3.1 The behavior of light emitted from a star that is smaller than the critical circumference, as computed in 1783 by John Michell using Newton's laws of gravity and corpuscular description of light.

compact a star from existing. Thus, Michell was led to speculate that the Universe might contain a huge number of such dark stars, each living happily inside its own critical circumference, and each invisible from Earth because the corpuscles of light emitted from its surface are inexorably pulled back down. Such *dark stars* were the eighteenth-century versions of black holes.

Michell, who was Rector of Thornhill in Yorkshire, England, reported his prediction that dark stars might exist to the Royal Society of London on 27 November 1783. His report made a bit of a splash among British natural philosophers. Thirteen years later, the French natural philosopher Pierre Simon Laplace popularized the same prediction in the first edition of his famous work *Le Systeme du Monde*, without reference to Michell's earlier work. Laplace kept his dark-star prediction in the second (1799) edition, but by the time of the third (1808) edition, Thomas Young's discovery of the interference of light with itself[1] was forcing natural philosophers to abandon the corpuscular description of light in favor of a wave description devised by Christiaan Huygens—and it was not at all clear how this wave description should be meshed with Newton's laws of gravity so as to compute the effect of a star's gravity on the light it emits. For this reason, presumably, Laplace deleted the concept of a dark star from the third and subsequent editions of his book.

1. Chapter 10.

Only in November 1915, after Einstein had formulated his general relativistic laws of gravity, did physicists once again believe they understood gravitation and light well enough to compute the effect of a star's gravity on the light it emits. Only then could they return with confidence to the dark stars (black holes) of Michell and Laplace.

The first step was made by Karl Schwarzschild, one of the most distinguished astrophysicists of the early twentieth century. Schwarzschild, then serving in the German army on the Russian front of World War I, read Einstein's formulation of general relativity in the 25 November 1915 issue of the *Proceedings of the Prussian Academy of Sciences.* Almost immediately he set out to discover what predictions Einstein's new gravitation laws might make about stars.

Since it would be very complicated, mathematically, to analyze a star that spins or is nonspherical, Schwarzschild confined himself to stars that do not spin at all and that are precisely spherical, and to ease his calculations, he sought first a mathematical description of the star's exterior and delayed its interior until later. Within a few days he had the answer. He had calculated, in exact detail, from Einstein's new field equation, the curvature of spacetime outside *any* spherical, non-spinning star. His calculation was elegant and beautiful, and the curved spacetime geometry that it predicted, the *Schwarzschild geometry* as it soon came to be known, was destined to have enormous impact on our understanding of gravity and the Universe.

Schwarzschild mailed to Einstein a paper describing his calculations, and Einstein presented it in his behalf at a meeting of the Prussian Academy of Sciences in Berlin on 13 January 1916. Several weeks later, Einstein presented the Academy a second paper by Schwarzschild: an exact computation of the spacetime curvature *inside* the star. Only four months later, Schwarzschild's remarkable productivity was halted: On 19 June, Einstein had the sad task of reporting to the Academy that Karl Schwarzschild had died of an illness contracted on the Russian front.

The Schwarzschild geometry is the first concrete example of spacetime curvature that we have met in this book. For this reason, and because it is so central to the properties of black holes, we shall examine it in detail.

If we had been thinking all our lives about space and time as an absolute, unified, four-dimensional spacetime "fabric," then it would

Karl Schwarzschild in his academic robe in Göttingen, Germany. [Courtesy AIP Emilio Segrè Visual Archives.]

be appropriate to describe the Schwarzschild geometry immediately in the language of curved (warped), four-dimensional spacetime. However, our everyday experience is with three-dimensional space and one-dimensional time, un-unified; therefore, I shall give a description in which warped spacetime is split up into warped space plus warped time.

Since space and time are "relative" (my space differs from your space and my time from yours, if we are moving relative to each other[2]), such a split requires first choosing a reference frame—that is, choosing a state of motion. For a star, there is a natural choice, one in which the star is at rest; that is, the star's own reference frame. In other words, it is natural to examine the star's own space and the star's own time rather than the space and time of someone moving at high speed through the star.

As an aid in visualizing the curvature (warpage) of the star's space, I shall use a drawing called an *embedding diagram*. Because embedding diagrams will play a major role in future chapters, I shall introduce the concept carefully, with the help of an analogy.

Imagine a family of human-like creatures who live in a universe with only two spatial dimensions. Their universe is the curved, bowl-like surface depicted in Figure 3.2. They, like their universe, are two-dimensional; they are infinitesimally thin perpendicular to the surface. Moreover, they cannot see out of the surface; they see by means of light rays that move in the surface and never leave it. Thus, these "2D beings," as I shall call them, have no method whatsoever to get any information about anything outside their two-dimensional universe.

These 2D beings can explore the geometry of their two-dimensional universe by making measurements on straight lines, triangles, and circles. Their straight lines are the "geodesics" discussed in Chapter 2 (Figure 2.4 and associated text): the straightest lines that exist in their two-dimensional universe. In the bottom of their universe's "bowl," which we see in Figure 3.2 as a segment of a sphere, their straight lines are segments of great circles like the equator of the Earth or its lines of constant longitude. Outside the lip of the bowl their universe is flat, so their straight lines are what we would recognize as ordinary straight lines.

If the 2D beings examine any pair of parallel straight lines in the outer, flat part of their universe (for example, L1 and L2 of Figure 3.2),

---

2. Figure 1.3, and the lessons of the tale of Mledina and Serona in Chapter 2.

then no matter how far the beings follow those lines, they will never see them cross. In this way, the beings discover the flatness of the outer region. On the other hand, if they construct the parallel straight lines L3 and L4 outside the bowl's lip, and then follow those lines into the bowl region, keeping them always as straight as possible (keeping them geodesics), they will see the lines cross at the bottom of the bowl. In this way, they discover that the inner, bowl region of their universe is curved.

The 2D beings can also discover the flatness of the outer region and the curvature of the inner region by measuring circles and triangles (Figure 3.2). In the outer region, the circumferences of all circles are equal to π (3.14159265 . . .) times their diameters. In the inner region, circumferences of circles are less than π times their diameters; for example, the large circle drawn near the bowl's bottom in Figure 3.2 has a circumference equal to 2.5 times its diameter. When the 2D beings construct a triangle whose sides are straight lines (geodesics) and then add up the triangle's interior angles, they obtain 180 degrees in the outer, flat region, and more than 180 degrees in the inner, curved region.

Having discovered, by such measurements, that their universe is curved, the 2D beings might begin to speculate about the existence of a three-dimensional space in which their universe resides—in which it is *embedded*. They might give that three-dimensional space the name

3.2 A two-dimensional universe peopled by 2D beings.

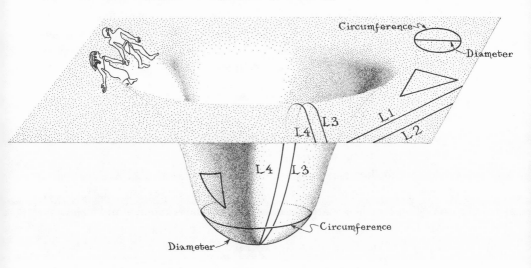

hyperspace, and speculate about its properties; for example, they might presume it to be "flat" in the Euclidean sense that straight, parallel lines in it never cross. You and I have no difficulty visualizing such a hyperspace; it is the three-dimensional space of Figure 3.2, the space of our everyday experience. However, the 2D beings, with their limited two-dimensional experience, would have great difficulty visualizing it. Moreover, there is no way that they could ever learn whether such a hyperspace really exists. They can never get out of their two-dimensional universe and into hyperspace's third dimension, and because they see only by means of light rays that stay always in their universe, they can never see into hyperspace. For them, hyperspace would be entirely hypothetical.

The third dimension of hyperspace has nothing to do with the 2D beings' "time" dimension, which they might also think of as a third dimension. When thinking about hyperspace, the beings would actually have to think in terms of four dimensions: two for the space of their universe, one for its time, and one for the third dimension of hyperspace.

W e are three-dimensional beings, and we live in a curved three-dimensional space. If we were to make measurements of the geometry of our space inside and near a star—the *Schwarzschild geometry*—we would discover it to be curved in a manner closely analogous to that of the 2D beings' universe.

We can speculate about a higher-dimensional, flat hyperspace in which our curved, three-dimensional space is embedded. It turns out that such a hyperspace must have six dimensions in order to accommodate curved three-dimensional spaces like ours inside itself. (And when we remember that our Universe also has a time dimension, we must think in terms of seven dimensions in all.)

Now, it is even harder for me to visualize our three-dimensional space embedded in a six-dimensional hyperspace than it would be for 2D beings to visualize their two-dimensional space embedded in a three-dimensional hyperspace. However, there is a trick that helps enormously, a trick depicted in Figure 3.3.

Figure 3.3 shows a thought experiment: A thin sheet of material is inserted through a star in its equatorial plane (upper left), so the sheet bisects the star leaving precisely identical halves above and below it. Even though this equatorial sheet looks flat in the picture, it is not really flat. The star's mass warps three-dimensional space inside and around the star in a manner that the upper left picture cannot convey, and that warpage curves the equatorial sheet in a manner the picture does not show. We can discover the sheet's curvature by making geometric measurements on it in our real, physical space, in precisely the same way as the 2D beings make measurements in the two-dimensional space of their universe. Such measurements will reveal that straight lines which are initially parallel cross near the star's center, the circumference of any circle inside or near the star is less than π times its diameter, and the sums of the internal angles of triangles are greater than 180 degrees. The details of these curved-space distortions are predicted by Schwarzschild's solution of Einstein's equation.

To aid in visualizing this Schwarzschild curvature, we, like the 2D beings, can imagine extracting the equatorial sheet from the curved, three-dimensional space of our real Universe, and embedding it in a fictitious, flat, three-dimensional hyperspace (lower right in Figure 3.3). In the uncurved hyperspace, the sheet can maintain its curved geometry only by bending downward like a bowl. Such diagrams of two-dimensional sheets from our curved Universe, embedded in a hypothetical, flat, three-dimensional hyperspace, are called *embedding diagrams.*

3.3 The curvature of the three-dimensional space inside and around a star (upper left), as depicted by means of an *embedding diagram* (lower right). This is the curvature predicted by Schwarzschild's solution to Einstein's field equation.

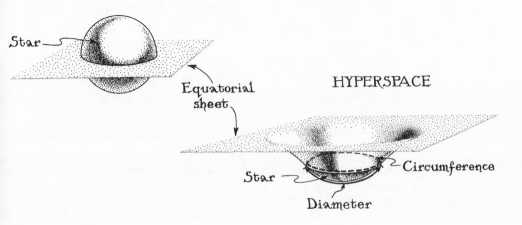

PHYSICAL SPACE

Star

Equatorial sheet

HYPERSPACE

Star

Circumference

Diameter

It is tempting to think of hyperspace's third dimension as being the same as the third spatial dimension of our own Universe. We must avoid this temptation. Hyperspace's third dimension has nothing whatsoever to do with any of the dimensions of our own Universe. It is a dimension into which we can never go and never see, and from which we can never get any information; it is purely hypothetical. Nonetheless, it is useful. It helps us visualize the Schwarzschild geometry, and later in this book it will help us visualize other curved-space geometries: those of black holes, gravitational waves, singularities, and wormholes (Chapters 6, 7, 10, 13, and 14).

As the embedding diagram in Figure 3.3 shows, the Schwarzschild geometry of the star's equatorial sheet is qualitatively the same as the geometry of the 2D beings' universe: Inside the star, the geometry is bowl-like and curved; far from the star it becomes flat. As with the large circle in the 2D beings' bowl (Figure 3.2), so also here (Figure 3.3), the star's circumference divided by its diameter is less than π. For our Sun, the ratio of circumference to diameter is predicted to be less than π by several parts in a million; in other words, inside the Sun, space is flat to within several parts in a million. However, if the Sun kept its same mass and were made smaller and smaller in circumference, then the curvature inside it would become stronger and stronger, the downward dip of the bowl in the embedding diagram of Figure 3.3 would become more and more pronounced, and the ratio of circumference to diameter would become substantially less than π.

Because space is different in different reference frames ("your space is a mixture of my space and my time, if we move relative to each other"), the details of the star's spatial curvature will be different as measured in a reference frame that moves at high speed relative to the star than as measured in a frame where the star is at rest. In the space of the high-speed reference frame, the star is somewhat squashed perpendicular to its direction of motion, so the embedding diagram looks much like that of Figure 3.3, but with the bowl compressed transversely into an oblong shape. This squashing is the curved-space variant of the contraction of space that Fitzgerald discovered in a universe without gravity (Chapter 1).

Schwarzschild's solution to the Einstein field equation describes not only this curvature (or warpage) of space, but also a warpage of time near the star—a warpage produced by the star's strong gravity. In a reference frame that is at rest with respect to the star, and not flying past it at high speed, this time warpage is precisely the *gravitational time dilation* discussed in Chapter 2 (Box 2.4 and associated discussion): Near the star's surface, time flows more slowly than far away, and at the star's center, it flows slower still.

In the case of the Sun, the time warpage is small: At the Sun's surface, time should flow more slowly by just 2 parts in a million (64 seconds in one year) than far from the Sun, and at the Sun's center it should flow more slowly than far away by about 1 part in 100,000 (5 minutes in one year). However, if the Sun kept its same mass and were made smaller in circumference so its surface was closer to its center, then its gravity would be stronger, and correspondingly its gravitational time dilation—its warpage of time—would become larger.

One consequence of this time warpage is the *gravitational redshift* of light emitted from a star's surface. Since the light's frequency of oscillation is governed by the flow of time at the place where the light is emitted, light emerging from atoms on the star's surface will have a lower frequency when it reaches Earth than light emitted by the same kinds of atoms in interstellar space. The frequency will be lowered by precisely the same amount as the flow of time is slowed. A lower frequency means a longer wavelength, so light from the star must be shifted toward the red end of the spectrum by the same amount as time is dilated on the star's surface.

At the Sun's surface the time dilation is 2 parts in a million, so the gravitational redshift of light arriving at the Earth from the Sun should also be 2 parts in a million. This was too small a redshift to be measured definitively in Einstein's day, but in the early 1960s, technology began to catch up with Einstein's laws of gravity: Jim Brault of Princeton University, in a very delicate experiment, measured the redshift of the Sun's light, and obtained a result in nice agreement with Einstein's prediction.

Within a few years after Schwarzschild's untimely death, his spacetime geometry became a standard working tool for physicists and astrophysicists. Many people, including Einstein, studied it and computed its implications. All agreed and took seriously the conclusion that, if the star were rather large in circumference, like the Sun, then spacetime inside and around it should be very slightly curved, and light emitted from its surface and received at Earth should be shifted in color, ever so slightly, toward the red. All also agreed that the more compact the star, the greater must be the warpage of its spacetime and the larger the gravitational redshift of light from its surface. However, few were

willing to take seriously the extreme predictions that the Schwarzschild geometry gave for highly compact stars (Figure 3.4):

The Schwarzschild geometry predicted that for each star there is a
critical circumference, which depends on the star's mass—the same
critical circumference as had been discovered by John Michell and
Pierre Simon Laplace more than a century earlier: 18.5 kilometers
times the mass of the star in units of the mass of the Sun. If the star's
actual circumference is larger than this critical one by a factor of 4
(upper part of Figure 3.4), then the star's space will be moderately
curved as shown, time at its surface will flow 15 percent more slowly
than far away, and light emitted from its surface will be shifted toward
the red end of the spectrum by 15 percent. If the star's circumference is
smaller, just twice the critical one (middle part of Figure 3.4), its space
will be more strongly curved, time at its surface will flow 41 percent

3.4   General relativity's predictions for the curvature of space and the redshift
of light from three highly compact stars with the same mass but different circumferences. The first is four times larger than the critical circumference, the
second is twice as large as critical, and the third has its circumference precisely
critical. In modern language, the surface of the third star is a black-hole horizon.

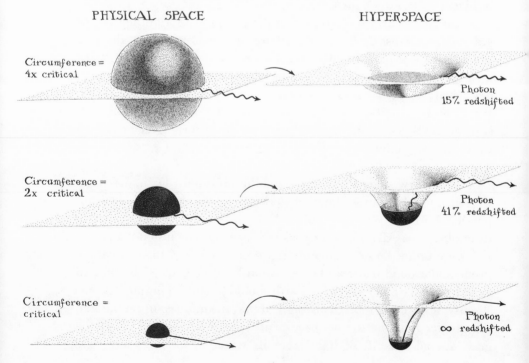

more slowly than far away, and light from its surface will be redshifted by 41 percent. These predictions seemed acceptable and reasonable. What did not seem at all reasonable to physicists and astrophysicists of the 1920s, or even as late as the 1960s, was the prediction for a star whose actual circumference was the same as its critical one (bottom part of Figure 3.4). For such a star, with its more strongly curved space, the flow of time at the star's surface is infinitely dilated; time does not flow at all—it is frozen. And correspondingly, no matter what may be the color of light when it begins its journey upward from the star's surface, it must get shifted beyond the red, beyond the infrared, beyond radio wavelengths, all the way to infinite wavelengths; that is, all the way out of existence. In modern language, the star's surface, with its critical circumference, is precisely at the horizon of a black hole; the star, by its strong gravity, is creating a black-hole horizon around itself.

The bottom line of this Schwarzschild-geometry discussion is the same as that found by Michell and Laplace: A star as small as the critical circumference must appear completely dark, when viewed from far away; it must be what we now call a black hole. The bottom line is the same, but the mechanism is completely different:

Michell and Laplace, with their Newtonian view of space and time as absolute and the speed of light as relative, believed that for a star just a bit smaller than the critical circumference, corpuscles of light would very nearly escape. They would fly up to great heights above the star, higher than any orbiting planet; but as they climbed, they would be slowed by the star's gravity, then halted somewhere short of interstellar space, then turned around and pulled back down to the star. Though creatures on an orbiting planet could see the star by its slow-moving light (to them it would not be dark), we, living far away on Earth, could not see it at all. The star's light could not reach us. For us the star would be totally black.

By contrast, Schwarzschild's spacetime curvature required that light always propagate with the same universal speed; it can never be slowed. (The speed of light is absolute, but space and time are relative.) However, if emitted from the critical circumference, the light must get shifted in wavelength an infinite amount, while traveling upward an infinitesimal distance. (The wavelength shift must be infinite because the flow of time is infinitely dilated at the horizon, and the wavelength always shifts by the same amount as time is dilated.) This infinite shift of wavelength, in effect, removes all the light's energy; and the light, thereupon, ceases to exist! Thus, no matter how close a planet might be to the critical circumference, creatures on it cannot see any light at all emerging from the star.

In Chapter 7, we shall study how the light behaves as seen from inside a black hole's critical circumference, and shall discover that it does not cease to exist after all. Rather, it simply is unable to escape the critical circumference (the hole's horizon) even though it is moving outward at the standard, universal speed of 299,792 kilometers per second. But this early in the book, we are not yet ready to comprehend such seemingly contradictory behavior. We must first build up our understanding of other things, as did physicists during the decades between 1916 and 1960.

During the 1920s and into the 1930s, the world's most renowned experts on general relativity were Albert Einstein and the British astrophysicist Arthur Eddington. Others understood relativity, but Einstein and Eddington set the intellectual tone of the subject. And, while a few others were willing to take black holes seriously, Einstein and Eddington were not. Black holes just didn't "smell right"; they were outrageously bizarre; they violated Einstein's and Eddington's intuitions about how our Universe ought to behave.

In the 1920s Einstein seems to have dealt with the issue by ignoring it. Nobody was pushing black holes as a serious prediction, so there was not much need on that score to straighten things out. And since other mysteries of nature were more interesting and puzzling to Einstein, he put his energies elsewhere.

Eddington in the 1920s took a more whimsical approach. He was a bit of a ham, he enjoyed popularizing science, and so long as nobody was taking black holes too seriously, they were a playful thing to dangle in front of others. Thus, we find him writing in 1926 in his book *The Internal Constitution of the Stars* that no observable star can possibly be more compact than the critical circumference: "Firstly," he wrote, "the force of gravitation would be so great that light would be unable to escape from it, the rays falling back to the star like a stone to the Earth. Secondly, the redshift of the spectral lines would be so great that the spectrum would be shifted out of existence. Thirdly, the mass would produce so much curvature of the space-time metric that space would close up round the star, leaving us outside (i.e. nowhere)." The first conclusion was the Newtonian version of light not escaping; the second was a semi-accurate, relativistic description; and the third was typical Eddingtonian hyperbole. As one sees clearly from the embedding diagrams of Figure 3.4, when a star is as small as the critical circumference, the curvature of space is strong but not infinite, and space is definitely not wrapped up around the star. Eddington may have known this, but his description made a good story, and it captured in a whimsical way the spirit of Schwarzschild's spacetime curvature.

In the 1930s, as we shall see in Chapter 4, the pressure to take black holes seriously began to mount. As the pressure mounted, Eddington, Einstein, and others among the "opinion setters" began to express unequivocal opposition to these outrageous objects.

In 1939, Einstein published a general relativistic calculation that he interpreted as an example of why black holes cannot exist. His calculation analyzed the behavior of an idealized kind of object which one might have thought could be used to make a black hole. The object was a cluster of particles that pull on each other gravitationally and thereby

3.5 Einstein's evidence that no object can ever be as small as its critical circumference. *Left:* If Einstein's spherical cluster of particles is smaller than 1.5 critical circumferences, then the particles' speeds must exceed the speed of light, which is impossible. *Right:* If a star with constant density is smaller than 9/8 = 1.125 critical circumferences, then the pressure at the star's center must be infinite, which is impossible.

EINSTEIN'S CLUSTER of PARTICLES   SCHWARZSCHILD'S STAR with
                                              CONSTANT DENSITY

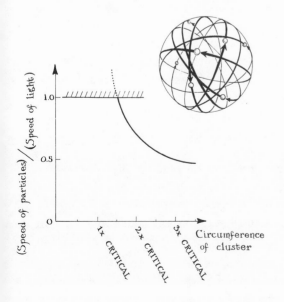

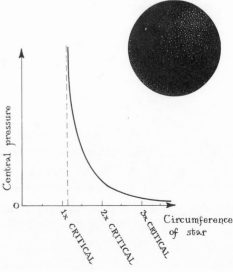

hold the cluster together, in much the same way as the Sun holds the solar system together by pulling gravitationally on its planets. The particles in Einstein's cluster all moved in circular orbits around a common center; their orbits formed a sphere, with particles on one side of the sphere pulling gravitationally on those on the other side (left half of Figure 3.5).

Einstein imagined making this cluster smaller and smaller, trying to drive its actual circumference down toward the critical circumference. As one might expect, his calculation showed that the more compact the cluster, the stronger the gravity at its spherical surface and the faster the particles must move on its surface to prevent themselves from being pulled in. If the cluster were smaller than 1.5 times the critical circumference, Einstein's calculations showed, then its gravity would be so strong that the particles would have to move faster than the speed of light to avoid being pulled in. Since nothing can move faster than light, there was no way the cluster could ever be smaller than 1.5 times critical. "The essential result of this investigation," Einstein wrote, "is a clear understanding as to why the 'Schwarzschild singularities' do not exist in physical reality."

As backing for his view, Einstein could also appeal to the internal structure of an idealized star made of matter whose density is constant throughout the stellar interior (right half of Figure 3.5). Such a star was prevented from imploding by the pressure of the gas inside it. Karl Schwarzschild had used general relativity to derive a complete mathematical description of such a star, and his formulas showed that, if one makes the star more and more compact, then in order to counteract the increased strength of its internal gravity, the star's internal pressure must rise higher and higher. As the star's shrinking circumference nears $9/8 = 1.125$ times its critical circumference, Schwarzschild's formulas show the central pressure becoming infinitely large. Since no real gas can ever produce a truly infinite pressure (nor can any other kind of matter), such a star could never get as small as 1.125 times critical, Einstein believed.

Einstein's calculations were correct, but his reading of their message was not. The message he extracted, that no object can ever become as small as the critical circumference, was determined more by Einstein's intuitive opposition to Schwarzschild singularities (black holes) than by the calculations themselves. The correct message, we now know in retrospect, was this:

Einstein's cluster of particles and the constant-density star could never be so compact as to form a black hole because Einstein demanded that some kind of force inside them counterbalance the squeeze of gravity: the force of gas pressure in the case of the star; the centrifugal force due to the particles' motions in the case of the cluster. In fact, it is true that no force whatsoever can resist the squeeze of gravity when an object is very near the critical circumference. But this does not mean the object can never get so small. Rather, it means that, if the object does get that small, then *gravity necessarily overwhelms all other forces inside the object, and squeezes the object into a catastrophic implosion, which forms a black hole.* Since Einstein's calculations did not include the possibility of implosion (he left it out of all his equations), he missed this message.

We are so accustomed to the idea of black holes today that it is hard not to ask, "How could Einstein have been so dumb? How could he leave out the very thing, implosion, that makes black holes?" Such a reaction displays our ignorance of the mindset of nearly *everybody* in the 1920s and 1930s.

General relativity's predictions were poorly understood. Nobody realized that a sufficiently compact object *must* implode, and that the implosion will produce a black hole. Rather, Schwarzschild singularities (black holes) were imagined, incorrectly, to be objects that are hovering at or just inside their critical circumference, supported against gravity by some sort of internal force; Einstein therefore thought he could debunk black holes by showing that nothing supported by internal forces can be as small as the critical circumference.

If Einstein had suspected that "Schwarzschild singularities" can really exist, he might well have realized that implosion is the key to forming them and internal forces are irrelevant. But he was so firmly convinced they cannot exist (they "smelled wrong"; terribly wrong) that he had an impenetrable mental block against the truth—as did nearly all his colleagues.

In T. H. White's epic novel *The Once and Future King* there is a society of ants which has the motto, "Everything not forbidden is compulsory." That is *not* how the laws of physics and the real Universe work. Many of the things permitted by the laws of physics are so highly improbable that in practice they never happen. A simple and time-worn example is the spontaneous reassembly of a whole egg from fragments splattered on the floor: Take a motion picture of an egg as it falls to the floor and splatters into fragments and goo. Then run the motion picture backward, and watch the egg spontaneously regenerate itself and fly up into the air. The laws of physics permit just such a regeneration with time going forward, but it never happens in practice because it is highly improbable.

Physicists' studies of black holes during the 1920s and 1930s, and even on into the 1940s and 1950s, dealt only with the issue of whether the laws of physics *permit* such objects to exist—and the answer was equivocal: At first sight, black holes seemed to be permitted; then Einstein, Eddington, and others gave (incorrect) arguments that they are forbidden. In the 1950s, when those arguments were ultimately disproved, many physicists turned to arguing that black holes might be permitted by the laws of physics, but are so highly improbable that (like the reassembling egg) they never occur in practice.

In reality, black holes, unlike the reassembling egg, are compulsory in certain common situations; but only in the late 1960s, when the evidence that they *are* compulsory became overwhelming, did most physicists begin to take black holes seriously. In the next three chapters I shall describe how that evidence mounted from the 1930s through the 1960s, and the widespread resistance it met.

This widespread and almost universal twentieth-century resistance to black holes is in marked contrast to the enthusiasm with which black holes were met in the eighteenth-century era of John Michell and Pierre Simon Laplace. Werner Israel, a modern-day physicist at the University of Alberta who has studied the history in depth, has speculated on the reasons for this difference.

"I am sure [that the eighteenth-century acceptance of black holes] was not just a symptom of the revolutionary fervour of the 1790s," Israel writes. "The explanation must be that Laplacian dark stars [black holes] posed no threat to our cherished faith in the permanence and stability of matter. By contrast, twentieth-century black holes are a great threat to that faith."

Michell and Laplace both imagined their dark stars as made from matter with about the same density as water or earth or rock or the Sun, about 1 gram per cubic inch. With this density, a star, to be dark (to be contained within its critical circumference), must have a mass about 400 million times greater than the Sun's and a circumference about 3 times larger than the Earth's orbit. Such stars, governed by Newton's laws of physics, might be exotic, but they surely were no threat to any cherished beliefs about nature. If one wanted to see the star, one need only land on a planet near it and look at its light corpuscles as they rose in their orbits, before plummeting back to the star's surface. If one wanted a sample of the material from which the star was made, one need only fly down to the star's surface, scoop some up, and bring it back to Earth for laboratory study. I do not know whether Michell, Laplace, or others of their day speculated about such things, but it is clear that if they did, there was no reason for concern about the laws of nature, about the permanence and stability of matter.

The critical circumference (horizon) of a twentieth-century black hole presents quite a different challenge. At no height above the horizon can one see any emerging light. Anything that falls through the horizon can never thereafter escape; it is lost from our Universe, a loss that poses a severe challenge to physicists' notions about the conservation of mass and energy.

"There is a curious parallel between the histories of black holes and continental drift [the relative drifting motion of the Earth's continents]," Israel writes. "Evidence for both was already non-ignorable by 1916, but both ideas were stopped in their tracks for half a century by a resistance bordering on the irrational. I believe the underlying psychological reason was the same in both cases. Another coincidence: resistance to both began to crumble around 1960. Of course, both fields [astrophysics and geophysics] benefitted from postwar technological developments. But it is nonetheless interesting that this was the moment when the Soviet H-bomb and Sputnik swept away the notion of Western science as engraved in stone and beyond challenge, and, perhaps, instilled the suspicion that there might be more in heaven and earth than Western science was prepared to dream of."

4

The Mystery of the White Dwarfs

*in which Eddington and Chandrasekhar do battle
over the deaths of massive stars;
must they shrink when they die,
creating black holes?
or will quantum mechanics save them?*

The year was 1928; the place, southeast India, the city of Madras on the Bay of Bengal. There, at the University of Madras, a seventeen-year-old Indian boy named Subrahmanyan Chandrasekhar was immersed in the study of physics, chemistry, and mathematics. Chandrasekhar was tall and handsome, with regal bearing and pride in his academic achievements. He had recently read Arnold Sommerfeld's classic textbook *Atomic Structure and Spectral Lines* and was now overjoyed that Sommerfeld, one of the world's great theoretical physicists, had come from his home in Munich to visit Madras.

Eager for personal contact, Chandrasekhar went to Sommerfeld's hotel room and asked for an interview. Sommerfeld granted an appointment for several days hence.

On the day of his appointment Chandrasekhar, filled with pride and confidence in his mastery of modern physics, walked up to Sommerfeld's hotel room and knocked on the door. Sommerfeld greeted him politely, inquired about his studies, then deflated him. "The physics you have been studying is a thing of the past. Physics has all changed in the five years since my textbook was written," he explained. He went on to describe a revolution in physicists' understanding of the laws that govern the realm of the small: the realm of atoms, molecules, electrons, and protons. In this realm, the Newtonian laws had been found to fail in ways that relativity had not anticipated. Their replacement was a radically new set of physical laws—laws that were called *quantum mechanics*[1] because they deal with the behavior (the "mechanics") of particles of matter ("quanta"). Though only two years old, the new quantum mechanical laws were already having great success in explaining how atoms and molecules behave.

Chandrasekhar had read in Sommerfeld's book about the first, tentative version of the new laws. But the tentative quantum laws had been unsatisfactory, Sommerfeld explained to him. Although they agreed well with experiments on simple atoms and molecules such as hydrogen, the tentative laws could not account for the behaviors of more complicated atoms and molecules, and they did not mesh in a logically consistent way with each other or with the other laws of physics. They were little more than a mishmash of unaesthetic, ad-hoc rules of computation.

The new version of the laws, though radical in form, looked far more promising. It explained complicated atoms and complicated molecules, and it seemed to be meshing quite nicely with the rest of physics.

Chandrasekhar listened to the details, entranced.

Quantum Mechanics and the Guts of White Dwarfs

When they parted, Sommerfeld gave Chandrasekhar the galley proofs of a technical article that he had just written. It contained a derivation of the quantum mechanical laws that govern large collections of electrons squeezed together into small volumes, in a metal for instance.

Chandrasekhar read Sommerfeld's galley proofs with fascination, understood them, and then spent many days in the University library studying all the research articles he could find relating to them. Especially interesting was an article entitled "On Dense Matter" by the English physicist R. H. Fowler, published in the 10 December 1926

1. For a clear discussion of the laws of quantum mechanics, see *The Cosmic Code* by Heinz Pagels (Simon and Schuster, 1982).

issue of *Monthly Notices of the Royal Astronomical Society*. Fowler's article pointed Chandrasekhar to a most fascinating book, *The Internal Constitution of the Stars*, by the eminent British astrophysicist Arthur S. Eddington, in which Chandrasekhar found a description of the *mystery of white-dwarf stars*.

White dwarfs were a type of star that astronomers had discovered through their telescopes. The mysterious thing about white dwarfs was the extremely high density of the matter inside them, a density far greater than humans had ever before encountered. Chandrasekhar had no way of knowing it when he opened Eddington's book, but the struggle to unravel the mystery of this high density would ultimately force him and Eddington to confront the possibility that massive stars, when they die, might shrink to form black holes.

"White dwarfs are probably very abundant," Chandrasekhar read in Eddington's book. "Only three are definitely known, but they are all within a small distance of the Sun. . . . The most famous of these stars is the Companion of [the ordinary star] Sirius," which has the name *Sirius B*. Sirius and Sirius B are the sixth and seventh nearest stars to the Earth, 8.6 light-years away, and Sirius is the brightest star in our sky. Sirius B orbits Sirius just as the Earth orbits the Sun, but Sirius B requires 50 years to complete an orbit, the Earth only one.

Eddington described how astronomers had estimated, from telescopic observations, the mass and circumference of Sirius B. The mass was that of 0.85 Sun; the circumference, 118,000 kilometers. This meant that the mean density of Sirius B was 61,000 grams per cubic centimeter—61,000 times greater density than water and just about a ton to the cubic inch. "This argument has been known for some years," Eddington wrote. "I think it has generally been considered proper to add the conclusion 'which is absurd.' " Most astronomers could not take seriously a density so much greater than ever encountered on Earth— and had they known the real truth, as revealed by more modern astronomical observations (a mass of 1.05 Suns, a circumference of 31,000 kilometers, and a density of 4 million grams per cubic centimeter or 60 tons per cubic inch), they would have considered it even more absurd; see Figure 4.1.

Eddington went on to describe a key new observation that reinforced the "absurd" conclusion. If Sirius B were, indeed, 61,000 times denser than water, then according to Einstein's laws of gravity, light climbing out of its intense gravitational field would be shifted to the red by 6 parts in 100,000—a shift 30 times greater than for light emerging from

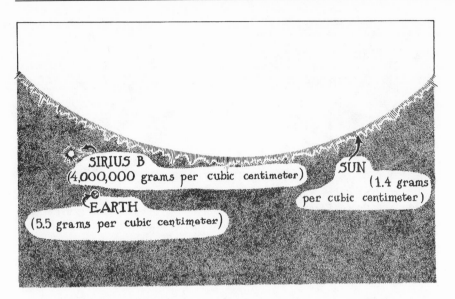

4.1 Comparison of the sizes and mean densities of the Sun, the Earth, and the white-dwarf star Sirius B, using modern values.

the Sun, and therefore easier to measure. This redshift prediction, it seemed, had been tested and verified just before Eddington's book went to press in 1925, by the astronomer W. S. Adams at Mount Wilson Observatory on a mountaintop above Pasadena, California.[2] "Professor Adams has killed two birds with one stone," Eddington wrote; "he has carried out a new test of Einstein's general theory of relativity and he has confirmed our suspicion that matter 2000 times denser than platinum is not only possible, but is actually present in the Universe."

Further on in Eddington's book, Chandrasekhar found a description of how the internal structure of a star, such as the Sun or Sirius B, is governed by the balance of internal pressure against gravitational squeeze. This squeeze/pressure balance can be understood (though this was not Eddington's way) by analogy with squeezing a balloon in your hands (left half of Figure 4.2): The inward force of your squeezing hands is precisely counterbalanced by the outward force of the balloon's air pressure—air pressure that is created by air molecules inside the balloon bombarding the balloon's rubber wall.

2. It is dangerously easy, in a delicate measurement, to get the result that one thinks one is supposed to get. Adams's gravitational redshift measurement is an example. His result agreed with the predictions, but the predictions were severely wrong (five times too small) due to errors in astronomers' estimates of the mass and circumference of Sirius B.

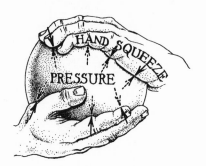 

4.2 *Left:* The balance between the squeeze of your hands and the pressure inside a balloon. *Right:* The analogous balance between the gravitational squeeze (weight) of an outer shell of stellar matter and the pressure of an inner ball of stellar matter.

For a star (right half of Figure 4.2) the analogue of your squeezing hands is the weight of an outer shell of stellar matter, and the analogue of the air in the balloon is the spherical ball of matter inside that shell. The boundary between the outer shell and inner ball can be chosen anywhere one wishes—a meter deep into the star, a kilometer deep, a thousand kilometers deep. Wherever one chooses the boundary, it must fulfill the requirement that the weight of the outer shell squeezing on the inner ball (the outer shell's "gravitational squeeze") is precisely counterbalanced by the pressure of the inner ball's molecules bombarding the outer shell. This balance, enforced at each and every location inside the star, determines the star's *structure;* that is, it determines the details of how the star's pressure, gravity, and density vary from the star's surface down to its center.

Eddington's book also described a troubling paradox in what was then known about the structures of white-dwarf stars. Eddington believed—indeed all astronomers believed in 1925—that the pressure of white-dwarf matter, like that in your balloon, must be caused by its heat. Heat makes the matter's atoms fly about inside the star at high speed, bombarding each other and bombarding the interface between the star's outer shell and its inner ball. If we take a "macroscopic" viewpoint, too coarse to detect the individual atoms, then all we can measure is the total bombardment force of all the atoms that hit, say, one square centimeter of the interface. That total force is the star's pressure.

As the star cools by emitting radiation into space, its atoms will fly about more slowly, their pressure will go down, and the weight of the star's outer shell will then squeeze its inner ball into a smaller volume. This compression of the ball, however, heats it up again, raising its pressure so a new squeeze/pressure balance can be achieved—one with the star slightly smaller than before. Thus, as Sirius B continues gradually to cool by radiating heat into interstellar space, it must gradually shrink in size.

How does this gradual shrinkage end? What will be the ultimate fate of Sirius B? The most obvious (but wrong) answer, that the star will shrink until it is so small that it becomes a black hole, was anathema to Eddington; he refused even to consider it. The only reasonable answer, he asserted, was that the star must ultimately turn cold and then support itself not by thermal pressure (that is, heat-induced pressure), but rather by the only other type of pressure known in 1925: the pressure that one finds in solid objects like rocks, a pressure due to repulsion between adjacent atoms. But such "rock pressure" was only possible, Eddington believed (incorrectly), if the star's matter has a density something like that of a rock, a few grams per cubic centimeter—10,000 times less than the present density of Sirius B.

This line of argument led to Eddington's paradox. In order to reexpand to the density of rock and thereby be able to support itself when it turns cold, Sirius B would have to do enormous work against its own gravity, and physicists did not know of any energy supply inside the star adequate for such work. "Imagine a body continually losing heat but with insufficient energy to grow cold!" Eddington wrote. "It is a curious problem and one may make many fanciful suggestions as to what actually will happen. We here leave aside the difficulty as not necessarily fatal."

Chandrasekhar had found the resolution of this 1925 paradox in R. H. Fowler's 1926 article "On Dense Matter." The resolution lay in the failure of the laws of physics that Eddington used. Those laws had to be replaced by the new quantum mechanics, which described the pressure inside Sirius B and other white dwarfs as due not to heat, but instead to a new, quantum mechanical phenomenon: the *degenerate motions of electrons,* also called *electron degeneracy.*[3]

---

3. This usage of the word "degenerate" does not have its origins in the concept of "moral degeneracy" (the *lowest possible level of morality*), but rather in the concept of the electrons having reached their *lowest possible levels of energy.*

Electron degeneracy is somewhat like human claustrophobia. When matter is squeezed to a density 10,000 times higher than that of rock, the cloud of electrons around each of its atomic nuclei gets squashed 10,000-fold. Each electron thereby gets confined to a "cell" with 10,000 times smaller volume than the one it previously was allowed to move around in. With so little space available to it, the electron, like a claustrophobic human, starts to shake uncontrollably. It flies about its tiny cell at high speed, kicking with great force against adjacent electrons in their cells. This *degenerate motion*, as physicists call it, cannot be stopped by cooling the matter. Nothing can stop it; it is forced on the electron by the laws of quantum mechanics, even when the matter is at absolute zero temperature.

This degenerate motion is a consequence of a feature of matter that Newtonian physicists never dreamed of, a feature called *wave/particle duality*: Every kind of particle, according to quantum mechanics, sometimes behaves like a wave, and every kind of wave sometimes behaves like a particle. Thus, waves and particles are really the same thing, a "thing" that sometimes behaves like a wave and sometimes like a particle; see Box 4.1.

Electron degeneracy is easily understood in terms of wave/particle duality. When matter is compressed to high densities, and each electron inside the matter gets confined to an extremely small cell squeezed up against neighboring electrons' cells, the electron begins to behave in part like a wave. The wavelength of the electron wave (the distance between its crests) cannot be larger than the electron's cell; if it were, the wave would extend beyond the cell. Now, particles with very short wavelengths are necessarily highly energetic. (A common example is the particle associated with an electromagnetic wave, the photon. An X-ray photon has a wavelength far shorter than that of a photon of light, and as a result X-ray photons are far more energetic than photons of light. Their higher energies enable the X-ray photons to penetrate human flesh and bone.)

In the case of an electron inside very dense matter, the electron's short wavelength and accompanying high energy imply rapid motion, and this means that the electron must fly around inside its cell, behaving like an erratic, high-speed mutant: half particle, half wave. Physicists say that the electron is "degenerate," and they call the pressure that its erratic high speed motion produces "electron degeneracy pressure." There is no way to get rid of this degeneracy pressure; it is an

Box 4.1

A Brief History of Wave/Particle Duality

Already in Isaac Newton's time (the late 1600s), physicists were struggling over the issue of whether *light* is made of particles or waves. Newton, though equivocal about the issue, leaned toward particles and called them *corpuscles*, while Christiaan Huygens argued for waves. Newton's particle view prevailed until the early 1800s, when the discovery that light can interfere with itself (Chapter 10) converted physicists to Huygens' wave viewpoint. In the mid-1800s, James Clerk Maxwell put the wave description on a firm footing with his unified laws of electricity and magnetism, and physicists then thought the issue had finally been settled. However, that was before quantum mechanics.

In the 1890s Max Planck noticed hints, in the shape of the spectrum of the light emitted by very hot objects, that something might be missing in physicists' understanding of light. Einstein, in 1905, showed what was missing: Light sometimes behaves like a wave and sometimes like a particle (now called a *photon*). It behaves like a wave, Einstein explained, when it interferes with itself; but it behaves like a particle in the *photoelectric effect:* When a dim beam of light shines on a piece of metal, the beam ejects electrons from the metal one at a time, precisely as though individual particles of light (individual photons) were hitting the electrons and knocking them out of the metal's surface one by one. From the electrons' energies, Einstein inferred that the photon energy is always inversely proportional to the light's wavelength. Thus, the photon and wave properties of light are intertwined; the wavelength is inexorably tied to the photon energy. Einstein's discovery of the wave/particle duality of light, and the tentative quantum mechanical laws of physics that he began to build around this discovery, won him the 1921 Nobel Prize in 1922.

Although Einstein almost single-handedly formulated general relativity, he was only one among many who contributed to the laws of quantum mechanics—the laws of the "realm of the small."

When Einstein discovered the wave/particle duality of light, he did not realize that an electron or proton might also behave sometimes like a particle and sometimes like a wave. Nobody recognized it until the mid-1920s when Louis de Broglie raised it as a conjecture and then Erwin Schrödinger used it as a foundation for a full set of quantum mechanical laws, laws in which an electron is a wave of probability. Probability for what? For the location of a particle. These "new" quantum mechanical laws (which have been enormously successful in explaining how electrons, protons, atoms, and molecules behave) will not concern us much in this book. However, from time to time a few of their features will be important. In this chapter, the important feature is electron degeneracy.

inevitable consequence of confining the electron to such a small cell. Moreover, the higher the matter's density, the smaller the cell, the shorter the electron wavelength, the higher the electron energy, the faster the electron's motion, and thus the larger its degeneracy pressure. In ordinary matter with ordinary densities, the degeneracy pressure is so tiny that one never notices it, but at the huge densities of white dwarfs it is enormous.

When Eddington wrote his book, electron degeneracy had not yet been predicted, and it was not possible to compute correctly how rock or other materials will respond if compressed to the ultra-high densities of Sirius B. With the laws of electron degeneracy in hand such computations were now possible, and they had been conceived and carried out by R. H. Fowler in his 1926 article.

According to Fowler's computations, because the electrons in Sirius B and other white-dwarf stars have been compressed into such tiny cells, their degeneracy pressure is far larger than their thermal (heat-induced) pressure. Accordingly, when Sirius B cools off, its minuscule thermal pressure will disappear, but its enormous degeneracy pressure will remain and will continue to support it against gravity.

Thus, the resolution of Eddington's white-dwarf paradox was two-fold: (1) Sirius B is not supported against its own gravity primarily by thermal pressure as everyone had thought before the advent of the new quantum mechanics; rather, it is supported primarily by degeneracy pressure. (2) When Sirius B cools off, it need not reexpand to the density of rock in order to support itself; rather, it will continue to be supported quite satisfactorily by degeneracy pressure at its present density of 4 million grams per cubic centimeter.

Chandrasekhar, reading these things and studying their mathematical formulations in the library in Madras, was enchanted. This was his first contact with modern astronomy, and he was finding here, side by side, deep consequences of the two twentieth-century revolutions in physics: Einstein's general relativity, with its new viewpoints on space and time, was showing up in the gravitational redshift of light from Sirius B; and the new quantum mechanics, with its wave/particle duality, was responsible for Sirius B's internal pressure. This astronomy was a fertile field in which a young man could make his mark.

As he continued his university studies in Madras, Chandrasekhar explored further the consequences of quantum mechanics for the astronomical Universe. He even wrote a small article on his ideas, mailed it to England to R. H. Fowler, whom he had never met, and Fowler arranged for it to be published.

Finally, in 1930 at age nineteen, Chandrasekhar completed the Indian equivalent of an American bachelor's degree, and in the last week of July he boarded a steamer bound for far-off England. He had been accepted for graduate study at Cambridge University, the home of his heroes, R. H. Fowler and Arthur Eddington.

The Maximum Mass

Those eighteen days at sea, steaming from Madras to Southampton, were Chandrasekhar's first opportunity in many months to think quietly about physics without the distraction of formal studies and examinations. The solitude of the sea was conducive to thought, and Chandrasekhar's thoughts were fertile. So fertile, in fact, that they would help to win him the Nobel Prize, but only fifty-four years later, and only after a great struggle to get them accepted by the world's astronomical community.

Aboard the steamer, Chandrasekhar let his mind reminisce over white dwarfs, Eddington's paradox, and Fowler's resolution. Fowler's resolution almost certainly had to be correct; there was none other in sight. However, Fowler had not worked out the full details of the balance between degeneracy pressure and gravity in a white-dwarf star, nor had he computed the star's resulting internal structure—the manner in which its density, pressure, and gravity change as one goes from its surface down to its center. Here was an interesting challenge to help ward off boredom during the long voyage.

As a tool in working out the star's structure, Chandrasekhar needed to know the answer to the following question: Suppose that white-dwarf matter has already been compressed to some density (for example, a density of a million grams per cubic centimeter). Compress the matter (that is, reduce its volume and increase its density) by an additional 1 percent. The matter will protest against this additional compression by raising its pressure. By what percentage will its pressure go up? Physicists use the name *adiabatic index* for the percentage increase in pressure that results from a 1 percent additional compression. In this book I shall use the more graphic name *resistance to compression* or simply *resistance*. (This "resistance to compression" should not be confused with "electrical resistance"; they are completely different concepts.)

Chandrasekhar worked out the resistance to compression by examining step by step the consequences of a 1 percent increase in the density of white-dwarf matter: the resulting decrease in electron cell size, the decrease in electron wavelength, the increase in electron energy and speed, and finally the increase in pressure. The result was clear: A 1 percent increase in density produced ⅗ of a percent (1.667 percent) increase in pressure. The resistance of white-dwarf matter, therefore, was ⅗.

Many decades before Chandrasekhar's voyage, astrophysicists had computed the details of the balance of gravity and pressure inside any star whose matter has a resistance to compression that is independent of depth in the star—that is, a star whose pressure and density increase in step with each other, as one moves deeper and deeper into the star, with a 1 percent increase in density always accompanied by the same fixed percentage increase in pressure. The details of the resulting stellar structures were contained in Eddington's book *The Internal Constitution of the Stars,* which Chandrasekhar had brought on board the ship because he treasured it so much. Thus, when Chandrasekhar discovered that white-dwarf matter has a resistance to compression of ⅗, independent of its density, he was pleased. He could now go directly to Eddington's book to discover the star's internal structure: the manner in which its density and pressure changed from surface to center.

Among the things that Chandrasekhar discovered, by combining the formulas in Eddington's book with his own formulas, were the density at the center of Sirius B, 360,000 grams per cubic centimeter (6 tons per cubic inch), and the speed of the electrons' degeneracy motion there, 57 percent of the speed of light.

This electron speed was disturbingly large. Chandrasekhar, like R. H. Fowler before him, had computed the resistance of white-dwarf matter using the laws of quantum mechanics, but ignoring the effects of relativity. However, when any object moves at almost the speed of light, even a particle obeying the laws of quantum mechanics, the effects of special relativity must become important. At 57 percent of the speed of light, relativity's effects might not be too terribly big, but a more massive white dwarf with its stronger gravity would require a larger central pressure to support itself, and the random speeds of its electrons would be correspondingly higher. In such a white dwarf the effects of relativity surely could not be ignored. So Chandrasekhar returned to the starting point of his analysis, the calculation of the resistance to compression for white-dwarf matter, vowing to include the effects of relativity this time around.

To include relativity in the computation would require meshing the laws of special relativity with the laws of quantum mechanics—a mesh that the great minds of theoretical physics were only then working out. Alone on the steamer and barely graduated from university, Chandrasekhar could not produce that full mesh. However, he was able to produce enough to indicate the principal effects of high electron speeds.

Quantum mechanics insists that when already dense matter is compressed a bit, making each electron's cell smaller than it was, the electron's wavelength must decrease and correspondingly the energy of its degeneracy motion must increase. However, Chandrasekhar realized, the nature of the additional electron energy is different, depending on whether the electron is moving slowly compared to light or at close to light speed. If the electron's motion is slow, then, as in everyday life, an increase of energy means more rapid motion, that is, higher speed. If the electron is already moving at close to light speed, however, there is no way its speed can go up much (if it did, it would exceed the speed limit!), so the energy increase takes a different form, one unfamiliar in everyday life: The additional energy goes into inertia; that is, it increases the electron's resistance to being speeded up—it makes the electron behave as though it had become a bit heavier. These two different fates of the added energy (added speed versus added inertia) produce different increases in the electron's pressure, and thus different resistances to compression, Chandrasekhar deduced: at low electron speeds, a resistance of ⁵⁄₃, the same as he had computed before; at high speeds, a resistance of ⁴⁄₃.

By combining his ⁴⁄₃ resistance for *relativistically degenerate matter* (that is, matter so dense that the degenerate electrons move at nearly the speed of light) with the formulas given in Eddington's book, Chandrasekhar then deduced the properties of high-density, high-mass white dwarfs. The answer was astonishing: The high-density matter would have difficulty supporting itself against gravity—sufficient difficulty that *only if the star's mass were less than that of 1.4 Suns could the squeeze of gravity be counterbalanced.* This meant that no white dwarf could ever have a mass exceeding 1.4 solar masses!

With his limited knowledge of astrophysics, Chandrasekhar was deeply puzzled about the meaning of this strange result. Time and again Chandrasekhar checked his calculations, but he could find no error. So, in the last few days of his voyage, he wrote two technical manuscripts for publication. In one he described his conclusions about the structure of low-mass, low-density white dwarfs such as Sirius B. In the other he explained, very briefly, his conclusion that no white dwarf can ever be heavier than 1.4 Suns.

When Chandrasekhar arrived in Cambridge, Fowler was out of the country. In September, when Fowler returned, Chandrasekhar eagerly went to his office and gave him the two manuscripts. Fowler approved the first one and sent it to *Philosophical Magazine* for publication, but the second one, the white-dwarf maximum mass, puzzled him. He could not understand Chandrasekhar's proof that no white dwarf can be heavier than 1.4 Suns; but then, he was a physicist rather than an astronomer, so he asked his colleague, the famous astronomer E. A. Milne, to look at it. When Milne couldn't understand the proof either, Fowler declined to send it for publication.

Chandrasekhar was annoyed. Three months had passed since his arrival in England, and Fowler had been sitting on his paper for two months. This was too long to wait for approval to publish. So, piqued, Chandrasekhar abandoned his attempts to publish in Britain and mailed the manuscript to the *Astrophysical Journal* in America.

After some weeks there came a response from the editor at the University of Chicago: The manuscript had been sent to the American physicist Carl Eckart for refereeing. In the manuscript Chandrasekhar stated, without explanation, the result of his relativistic and quantum mechanical calculation, that the resistance to compression is $4/3$ at ultra-high densities. This $4/3$ resistance was essential to the limit on how heavy a white dwarf can be. If the resistance were larger than $4/3$, then white dwarfs could be as heavy as they wished—and Eckart thought it should be larger. Chandrasekhar fired off a reply containing a mathematical derivation of the $4/3$ resistance; Eckart, reading the details, conceded that Chandrasekhar was right and approved his paper for publication. Finally, a full year after Chandrasekhar had written it, his paper got published.[4]

The response of the astronomical community was deafening silence. Nobody seemed interested. So Chandrasekhar, wanting to complete his Ph.D. degree, turned to other, more acceptable research.

4. In the meantime, Edmund C. Stoner had independently derived and published the existence of the white-dwarf maximum mass, but his derivation was rather less convincing than Chandrasekhar's because it pretended the star had a constant density throughout its interior.

Three years later, with his Ph.D. finished, Chandrasekhar visited Russia to exchange research ideas with Soviet scientists. In Leningrad a young Armenian astrophysicist, Viktor Amazapovich Ambartsumian, told Chandrasekhar that the world's astronomers would not believe his strange limit on the masses of white dwarfs unless he computed, from the laws of physics, the masses of a representative sample of white dwarfs and demonstrated explicitly that they were all below his claimed limit. It was not enough, Ambartsumian asserted, that Chandrasekhar had analyzed white dwarfs with rather low densities and resistances of $5/3$, and white dwarfs with extremely high densities and resistances of $4/3$; he needed also to analyze a goodly sample of white dwarfs with densities in between and show that they, too, always have masses below 1.4 Suns. Upon returning to Cambridge, Chandrasekhar took up Ambartsumian's challenge.

One foundation that Chandrasekhar would need was the *equation of state* of white-dwarf matter over the entire range of densities, running from low to extremely high. (By the "state" of the matter, physicists mean the matter's density and pressure—or equivalently its density and its resistance to compression, since from the resistance and the density one can compute the pressure. By "equation of state" is meant the relationship between the resistance and the density, that is, the resistance *as a function of* density.)

In late 1934, when Chandrasekhar took up Ambartsumian's challenge, the equation of state for white-dwarf matter was known, thanks to calculations by Edmund Stoner of Leeds University in England and Wilhelm Anderson of Tartu University in Estonia. The Stoner–Anderson equation of state showed that, as the density of the white-dwarf matter is squeezed higher and higher, moving from the nonrelativistic regime of low densities and low electron speeds into the relativistic domain of extremely high densities and electron speeds near the speed of light, the matter's resistance to compression decreases smoothly from $5/3$ to $4/3$ (left half of Figure 4.3). The resistance could not have behaved more simply.

To meet Ambartsumian's challenge, Chandrasekhar had to combine this equation of state (this dependence of resistance on density) with the star's law of balance between gravity and pressure, and thereby obtain a *differential equation*[5] describing the star's internal structure—

5. A differential equation is one that combines in a single formula various functions and their rates of change, that is, the functions and their "derivatives." In Chandrasekhar's differential equation, the functions were the star's density and pressure and the strength of its

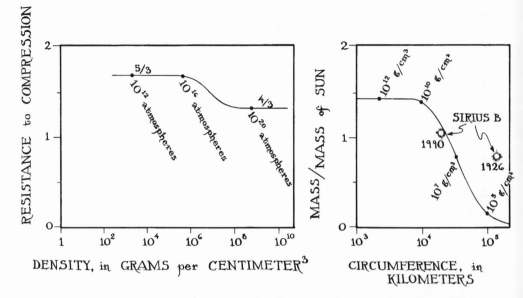

4.3 *Left:* The Stoner–Anderson *equation of state* for white-dwarf matter, that is, the relationship between the matter's density and its resistance to compression. Plotted horizontally is the density to which the matter has been squeezed. Plotted vertically is its resistance (the percentage increase of pressure that accompanies a 1 percent increase of density). Along the curve is marked the squeeze pressure (equal to internal pressure), in multiples of the pressure of the Earth's atmosphere. *Right:* The circumferences (plotted horizontally) and masses (plotted vertically) of white-dwarf stars as computed by Chandrasekhar using Eddington's Braunschweiger mechanical calculator. Along the curve is marked the density of the matter at the center of the star, in grams per cubic centimeter.

that is, describing the variation of its density with distance from the star's center. He then had to solve that differential equation for a dozen or so stars that have central densities spanning the range from low to extremely high. Only by solving the differential equation for each star could he learn the star's mass, and see whether it is less than 1.4 Suns.

For stars with low or extremely high central density, which Chandrasekhar had studied on the steamer, he had found the solution to the differential equation and the resulting stellar structures in Eddington's

---

gravity, and they were functions of distance from the star's center. The differential equation was a relation between these functions and the rate that they change as one moves outward through the star. By "solve the differential equation" is meant "compute the functions themselves from this differential equation."

book; but for stars with intermediate densities Eddington's book was of no help and, despite great effort, Chandrasekhar was not able to deduce the solution using mathematical formulas. The mathematics was too complicated. There was no recourse but to solve his differential equation numerically, on a computer.

Now, the computers of 1934 were very different from those of the 1990s. They were more like the simplest of pocket calculators: They could only multiply two numbers at a time, and the user had to enter those numbers by hand, then turn a crank. The crank set into motion a complicated morass of gears and wheels which performed the multiplication and gave the answer.

Such computers were precious machines; it was very hard to gain access to one. But Arthur Eddington owned one, a "Braunschweiger" about the size of an early 1990s desk-top personal computer; so Chandrasekhar, who by now had become well acquainted with the great man, went to Eddington and asked to borrow it. Eddington at the time was embroiled in a controversy with Milne over white dwarfs and was eager to see the full details of white-dwarf structure worked out, so he let Chandrasekhar cart the Braunschweiger off to the rooms in Trinity College where Chandrasekhar was living.

The calculations were long and tedious. Each evening after dinner Eddington, who was a fellow of Trinity College, would ascend the stairs to Chandrasekhar's rooms to see how they were coming and to give him encouragement.

At last, after many days, Chandrasekhar finished. He had met Ambartsumian's challenge. For each of ten representative white-dwarf stars, he had computed the internal structure, and then from it the star's total mass and its circumference. All the masses were less than 1.4 Suns, as he had firmly expected. Moreover, when he plotted the stars' masses and circumferences on a diagram and "connected the dots," he obtained a single, smooth curve (right half of Figure 4.3; see also Box 4.2), and the measured masses and circumferences of Sirius B and other known white dwarfs agreed with that curve moderately well. (With improved, modern astronomical observations, the fit has become much better; note the new, 1990 values of the mass and circumference of Sirius B in Figure 4.3.) Proud of his results and anticipating that the world's astronomers would finally accept his claim that white dwarfs cannot be heavier than 1.4 Suns, Chandrasekhar was very happy.

Especially gratifying would be the opportunity to present these results to the Royal Astronomical Society in London. Chandrasekhar was scheduled for a presentation on Friday, 11 January 1935. Protocol dictated that the details of the meeting's program be kept secret until the meeting started, but Miss Kay Williams, the assistant secretary of the Society and a friend of Chandrasekhar's, was in the habit of sending him programs secretly in advance. On Thursday evening when the program arrived in the mail, he was surprised to discover that immediately following his own talk there would be a talk by Eddington on the subject of "Relativistic Degeneracy." Chandrasekhar was a little annoyed. For the past few months Eddington had been coming to see him at least once a week about his work and had been reading drafts of the

---

Box 4.2

An Explanation of the Masses and Circumferences of White-Dwarf Stars

To understand qualitatively why white dwarfs have the masses and circumferences shown in Figure 4.3, examine the drawing below. It shows the average pressure and gravity inside a white dwarf (plotted upward) as functions of the star's circumference (plotted rightward) or density (plotted leftward). If one compresses the star, so its density increases and its circumference decreases (leftward motion in the drawing), then the star's pressure goes up in the manner of the solid curve, with a sharper rise at low densities where the resistance to compression is 5/3, and a slower rise at high densities where it is 4/3. This same compression of the star causes the star's surface to move in toward its center, thereby increasing the strength of the star's internal gravity in the manner of the dashed lines. The rate of gravity's increase is analogous to a 4/3 resistance: There is a 4/3 percent increase in gravity's strength for each 1 percent compression. The drawing shows several dashed gravity lines, one for each value of the star's mass, because the greater the star's mass, the stronger its gravity.

Inside each star, for example a 1.2-solar-mass star, gravity and pressure must balance each other. The star, therefore, must reside at the intersection of the dashed gravity line marked "1.2 solar masses" and the solid pressure curve; this intersection determines the star's circumference (marked on bottom of graph). If the circumference were larger, then the star's dashed gravity line would be above its solid pressure curve, gravity would overwhelm pressure, and the star would implode. If the circumference were smaller, pressure would overwhelm gravity, and the star would explode.

(continued next page)

(Box 4.2 continued)

The intersections of the several dashed lines with the solid curve correspond to the masses and circumferences of equilibrium white dwarfs, as shown in the right half of Figure 4.3. For a star of small mass (lowest dashed line), the circumference at the intersection is large. For a star of higher mass (higher dashed lines), the circumference is smaller. For a star with mass above 1.4 Suns, there is no intersection whatsoever; the dashed gravity line is always above the solid pressure curve, so gravity always overwhelms pressure, no matter what the star's circumference may be, forcing the star to implode.

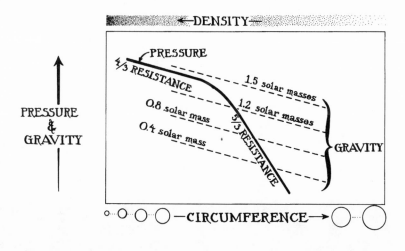

articles he was writing, but never once had Eddington mentioned doing any research of his own on the same subject!

Suppressing his annoyance, Chandrasekhar went down to dinner. Eddington was there, dining at high table, but protocol dictated that, just because you knew so eminent a man, and just because he had been expressing an interest in your work, you did not thereby have a right to go bother him about such a matter as this. So Chandrasekhar sat down elsewhere and held his tongue.

After dinner Eddington himself sought Chandrasekhar out and said, "I've asked Smart to give you half an hour tomorrow instead of the customary fifteen minutes." Chandrasekhar thanked him and waited for him to say something about his own talk, but Eddington just excused himself and left. Chandrasekhar's annoyance acquired an anxious twinge.

The Battle

The next morning Chandrasekhar took the train down to London and a taxi to Burlington House, the home of the Royal Astronomical Society. While he and a friend, Bill McCrae, were waiting for the meeting to start, Eddington came walking by, and McCrae, having just read the program, asked, "Well, Professor Eddington, what are we to understand by 'Relativistic Degeneracy'?" Eddington, in reply, turned to Chandrasekhar and said, "That's a surprise for you," and walked off leaving Chandrasekhar even more anxious.

Left: Arthur Stanley Eddington in 1932. *Right:* Subrahmanyan Chandrasekhar in 1934. [Left: courtesy UPI/Bettmann; right: courtesy S. Chandrasekhar.]

At last the meeting started. Time dragged by as the Society president made various announcements, and various astronomers gave miscellaneous talks. At last it was Chandrasekhar's turn. Suppressing his anxiety, he gave an impeccable presentation, emphasizing particularly his maximum mass for white dwarfs.

After polite applause from the fellows of the Society, the president invited Eddington to speak.

Eddington began gently, by reviewing the history of white-dwarf research. Then, gathering steam, he described the disturbing implications of Chandrasekhar's maximum-mass result:

In Chandrasekhar's diagram of the mass of a star plotted vertically and its circumference plotted horizontally (Figure 4.4), there is only one set of masses and circumferences for which gravity can be counterbalanced by nonthermal pressure (pressure that remains after the star turns cold): that of white dwarfs. In the region to the left of Chandrasekhar's white-dwarf curve (shaded region; stars with smaller circumferences), the star's nonthermal degeneracy pressure completely overwhelms gravity. The degeneracy pressure will drive any star in the shaded region to explode. In the region to the right of the white-dwarf curve (white region; stars with larger circumferences), gravity completely overwhelms the star's degeneracy pressure. Any *cold* star which finds itself in this region will immediately implode under gravity's squeeze.

The Sun can live in the white region only because it is now very hot; its thermal (heat-induced) pressure manages to counterbalance its gravity. However, when the Sun ultimately cools down, its thermal pressure will disappear and it no longer will be able to support itself. Gravity will force it to shrink smaller and smaller, squeezing the Sun's electrons into smaller and smaller cells, until at last they protest with enough degeneracy pressure (nonthermal pressure) to halt the shrinkage. During this shrinkage "death," the Sun's mass will remain nearly constant, but its circumference will decrease, so it will move leftward on a horizontal line in Figure 4.4, finally stopping on the white-dwarf curve—its grave. There, as a white dwarf, the Sun will continue to reside forever, gradually cooling and becoming a black dwarf—a cold, dark, solid object about the size of the Earth but a million times heavier and denser.

This ultimate fate of the Sun seemed quite satisfactory to Eddington. Not so the ultimate fate of a star more massive than Chandrasekhar's white-dwarf limit of 1.4 solar masses—for example, Sirius, the

2.3-solar-mass companion of Sirius B. If Chandrasekhar were right, such a star could never die the gentle death that awaits the Sun. When the radiation it emits into space has carried away enough heat for the star to begin to cool, its thermal pressure will decline, and gravity's squeeze will make it shrink smaller and smaller. For so massive a star as Sirius, the shrinkage cannot be halted by nonthermal degeneracy pressure. This is clear from Figure 4.4, where the shaded region does not extend high enough to intercept Sirius's shrinking track. Eddington found this prediction disturbing.

"The star has to go on radiating and radiating and contracting and contracting," Eddington told his audience, "until, I suppose, it gets down to a few kilometers radius, when gravity becomes strong enough to hold in the radiation, and the star can at last find peace." (In the words of the 1990s, it must form a black hole.) "Dr. Chandrasekhar had got this result before, but he has rubbed it in in his last paper; and, when discussing it with him, I felt driven to the conclusion that this was almost a *reductio ad absurdum* of the relativistic degeneracy formula. Various accidents may intervene to save the star, but I want more protection than that. I think there should be a law of Nature to prevent a star from behaving in this absurd way!"

Then Eddington argued that Chandrasekhar's mathematical proof of his result could not be trusted because it was based on an inadequately sophisticated meshing of special relativity with quantum mechanics. "I do not regard the offspring of such a union as born in lawful wedlock," Eddington said. "I feel satisfied myself that [if the meshing is made correctly], the relativity corrections are compensated, so that we come back to the 'ordinary' formula" (that is, to a $5\!/\!3$ resistance, which would permit white dwarfs to be arbitrarily massive and thereby would enable pressure to halt the contraction of Sirius at the hypothetical dotted curve in Figure 4.4). Eddington then sketched how *he* thought special relativity and quantum mechanics should be meshed, a rather different kind of mesh than Chandrasekhar, Stoner, and Anderson had used, and a mesh, Eddington claimed, that would save all stars from the black-hole fate.

Chandrasekhar was shocked. He had never expected such an attack on his work. Why had Eddington not discussed it with him in advance? And as for Eddington's argument, to Chandrasekhar it looked specious—almost certainly wrong.

Now, Arthur Eddington was *the* great man of British astronomy. His discoveries were almost legendary. He was largely responsible for as-

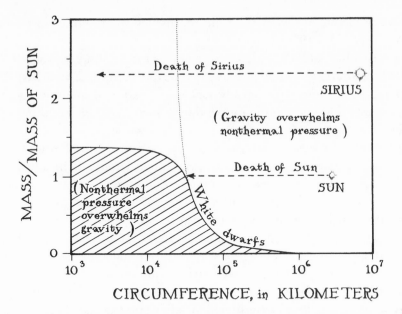

4.4 When a normal star such as the Sun or Sirius (not Sirius B) starts to cool off, it must shrink, moving leftward in this diagram of mass versus circumference. The shrinkage of the Sun will stop when it reaches the edge of the shaded region (the white-dwarf curve). There degeneracy pressure balances gravity's squeeze. The shrinkage of Sirius, by contrast, cannot be so stopped because it never reaches the edge of the shaded region. See Box 4.2 for a different depiction of these conclusions. If, as Eddington claimed, white-dwarf matter's resistance to compression were always 5/3, that is, if relativity did not reduce it to 4/3 at high densities, then the graph of mass versus circumference would have the form of the faint dotted curve, and the shrinkage of Sirius would stop there.

tronomers' understanding of normal stars like the Sun and Sirius, their interiors, their atmospheres, and the light that they emit; so, naturally, the fellows of the Society, and astronomers throughout the world, listened with great respect. Clearly, if Eddington thought Chandrasekhar's analysis incorrect, then it must be incorrect.

After the meeting, one fellow after another came up to Chandrasekhar to offer condolences. "I feel it in my bones that Eddington is right," Milne told him.

The next day, Chandrasekhar began appealing to his physicist friends for help. To Leon Rosenfeld in Copenhagen, he wrote, "If Eddington is right, my last four months' work all goes in the fire. Could Eddington be right? I should very much like to know Bohr's opinion." (Niels Bohr was one of the fathers of quantum mechanics and the most respected physicist of the 1930s.) Rosenfeld replied two days later, with assurances that he and Bohr both were convinced that Eddington was wrong and Chandrasekhar right. "I may say that your letter was *some* surprise for me," he wrote, "for nobody had ever dreamt of questioning the equations [that you used to derive the ⁴/₃ resistance] and Eddington's remark as reported in your letter is utterly obscure. So I think you had better cheer up and not let you scare [*sic*] so much by high priests." In a follow-up letter on the same day, Rosenfeld wrote, "Bohr and I are absolutely unable to see any meaning in Eddington's statements."

But for astronomers, the matter was not so clear at first. They had no expertise in these issues of quantum mechanics and relativity, so Eddington's authority held sway amongst them for several years. Moreover, Eddington stuck to his guns. He was so blinded by his opposition to black holes that his judgment was totally clouded. He so deeply wanted there to "be a law of Nature to prevent a star from behaving in this absurd way" that he continued to believe for the rest of his life that there *is* such a law—when, in fact, there is none.

By the late 1930s, astronomers, having talked to their physicist colleagues, understood Eddington's error, but their respect for his enormous earlier achievements prevented them from saying so in public. In a lecture at an astronomy conference in Paris in 1939, Eddington once again attacked Chandrasekhar's conclusions. As Eddington attacked, Chandrasekhar passed a note to Henry Norris Russell (a famous astronomer from Princeton University in America), who was presiding. Chandrasekhar's note asked for permission to reply. Russell passed back a note of his own saying, "I prefer you don't," even though earlier in the day Russell had told Chandrasekhar privately, "Out there we don't believe in Eddington."

With the world's leading astronomers having finally—at least behind Eddington's back—accepted Chandrasekhar's maximum mass for white dwarfs, were they then ready to admit that black holes might exist in the real Universe? Not at all. If nature provided no law against them of the sort that Eddington had sought, then nature would surely find another way out: Presumably, every massive star would eject enough matter into interstellar space, as it ages or during its death throes, to reduce its mass below 1.4 Suns and thereby enter a safe, white-dwarf grave. This was the view to which most astronomers turned when Eddington lost his battle, and they adhered to it through the 1940s and 1950s, and into the early 1960s.

As for Chandrasekhar, he was badly burned by the controversy with Eddington. As he recalled some forty years later, "I felt that astronomers without exception thought that I was wrong. They considered me as a sort of Don Quixote trying to kill Eddington. As you can imagine, it was a very discouraging experience for me—to find myself in a controversy with the leading figure of astronomy and to have my work completely and totally discredited by the astronomical community. I had to make up my mind as to what to do. Should I go on the rest of my life fighting? After all I was in my middle twenties at that time. I foresaw for myself some thirty to forty years of scientific work, and I simply did not think it was productive to constantly harp on something which was done. It was much better for me to change my field of interest and go into something else."

So in 1939 Chandrasekhar turned his back on white dwarfs and stellar death, and did not return to them for a quarter century (Chapter 7).

And what of Eddington? Why did he treat Chandrasekhar so badly? To Eddington, the treatment may not have seemed bad at all. Rough-and-tumble, freewheeling intellectual conflict was a way of life for him. Treating young Chandrasekhar in this manner may have been, in some sense, a measure of respect, a sign that he was accepting Chandrasekhar as a member of the astronomical establishment. In fact, from their first confrontation in 1935 until Eddington's death in 1944, Eddington displayed warm personal affection for Chandrasekhar, and Chandrasekhar, though burned by the controversy, reciprocated.

5

Implosion
Is Compulsory

*in which even the nuclear force,
supposedly the strongest of all forces,
cannot resist the crush of gravity*

Zwicky

In the 1930s and 1940s, many of Fritz Zwicky's colleagues regarded him as an irritating buffoon. Future generations of astronomers would look back on him as a creative genius.

"By the time I knew Fritz in 1933, he was thoroughly convinced that he had the inside track to ultimate knowledge, and that everyone else was wrong," says William Fowler, then a student at Caltech (the California Institute of Technology) where Zwicky taught and did research. Jesse Greenstein, a Caltech colleague of Zwicky's from the late 1940s onward, recalls Zwicky as "a self-proclaimed genius. . . . There's no doubt that he had a mind which was quite extraordinary. But he was also, although he didn't admit it, untutored and not self-controlled. . . . He taught a course in physics for which admission was at his pleasure. If he thought that a person was sufficiently devoted to his ideas, that person could be admitted. . . . He was very much alone [among the Caltech physics faculty, and was] not popular with the establishment. . . . His publications often included violent attacks on other people."

Zwicky—a stocky, cocky man, always ready for a fight—did not hesitate to proclaim his inside track to ultimate knowledge, or to tout the revelations it brought. In lecture after lecture during the 1930s, and article after published article, he trumpeted the concept of a *neutron star*—a concept that he, Zwicky, had invented to explain the origins of the most energetic phenomena seen by astronomers: supernovae, and cosmic rays. He even went on the air in a nationally broadcast radio show to popularize his neutron stars. But under close scrutiny, his articles and lectures were unconvincing. They contained little substantiation for his ideas.

It was rumored that Robert Millikan (the man who had built Caltech into a powerhouse among science institutions), when asked in the midst of all this hoopla why he kept Zwicky at Caltech, replied that it just might turn out that some of Zwicky's far-out ideas were right. Millikan, unlike some others in the science establishment, must have seen hints of Zwicky's intuitive genius—a genius that became widely

Fritz Zwicky among a gathering of scientists at Caltech in 1931. Also in the photograph are Richard Tolman (who will be an important figure later in this chapter), Robert Millikan, and Albert Einstein. [Courtesy of the Archives, California Institute of Technology.]

Zwicky   Millikan Einstein   Tolman

recognized only thirty-five years later, when observational astronomers discovered real neutron stars in the sky and verified some of Zwicky's extravagant claims about them.

Among Zwicky's claims, the most relevant to this book is the role of neutron stars as stellar corpses. As we shall see, a normal star that is too massive to die a white-dwarf death may die a neutron-star death instead. If *all* massive stars were to die that way, then the Universe would be saved from the most outrageous of hypothesized stellar corpses: black holes. With light stars becoming white dwarfs when they die, and heavy stars becoming neutron stars, there would be no way for nature to make a black hole. Einstein and Eddington, and most physicists and astronomers of their era, would heave a sigh of relief.

Zwicky had been lured to Caltech in 1925 by Millikan. Millikan expected him to do theoretical research on the quantum mechanical structures of atoms and crystals, but more and more during the late 1920s and early 1930s, Zwicky was drawn to astrophysics. It was hard not to be entranced by the astronomical Universe when one worked in Pasadena, the home not only of Caltech but also of the Mount Wilson Observatory, which had the world's largest telescope, a reflector 2.5 meters (100 inches) in diameter.

In 1931 Zwicky latched on to Walter Baade, a new arrival at Mount Wilson from Hamburg and Göttingen and a superb observational astronomer. Baade and Zwicky shared a common cultural background: Baade was German, Zwicky was Swiss, and both spoke German as their native language. They also shared respect for each other's brilliance. But there the commonality ended. Baade's temperament was different from Zwicky's. He was reserved, proud, hard to get to know, universally well informed—and tolerant of his colleagues' peculiarities. Zwicky would test Baade's tolerance during the coming years until finally, during World War II, he and Zwicky would split violently. "Zwicky called Baade a Nazi, which he wasn't, and Baade said he was afraid that Zwicky would kill him. They became a dangerous pair to put in the same room," recalls Jesse Greenstein.

During 1932 and 1933, Baade and Zwicky were often seen in Pasadena, animatedly conversing in German about stars called "novae," which suddenly flare up and shine 10,000 times more brightly than before; and then, after about a month, slowly dim down to normalcy. Baade, with his encyclopedic knowledge of astronomy, was aware of tentative evidence that, in addition to these "ordinary" novae, there might also be unusual, rare, superluminous novae. Astronomers at first had not suspected that those novae were superluminous, since they appeared through telescopes to be roughly the same brightness as ordinary novae. However, they resided in peculiar nebulas (shining "clouds"); and in the 1920s, observations at Mount Wilson and elsewhere began to convince astronomers that those nebulas were not simply clouds of gas in our own Milky Way galaxy, as had been thought, but rather were galaxies in their own right—giant assemblages of nearly 10^{12} (a trillion) stars, far outside our own galaxy. The rare novae seen in these galaxies, being so much farther away than our own galaxy's ordinary novae, would have to be intrinsically far more luminous than ordinary novae in order to have a similar brightness as seen from Earth.

Baade collected from the published literature all the observational data he could find about each of the six superluminous novae that astronomers had seen since the turn of the century. These data he combined with all the observational information he could get about the distances to the galaxies in which they lay, and from this combination he computed how much light the superluminous novae put out. His conclusion was startling: During flare-up these superluminous novae were typically 10^8 (100 million) times more luminous than our

The galaxy NGC 4725 in the constellation Coma Berenices. *Left:* As photographed on 10 May 1940, before a supernova outburst. *Right:* On 2 January 1941 during the supernova outburst. The white line points to the supernova, in the outer reaches of the galaxy. This galaxy is now known to be 30 million light-years from Earth and to contain 3×10^{11} (a third of a trillion) stars. [Courtesy California Institute of Technology.]

Sun! (Today we know, thanks largely to work in 1952 by Baade himself, that the distances were underestimated in the 1930s by nearly a factor of 10, and that correspondingly[1] the superluminous novae are nearly 10^{10}—10 billion—times more luminous than our Sun.)

Zwicky, a lover of extremes, was fascinated by these superluminous novae. He and Baade discussed them endlessly and coined for them the name *supernovae*. Each supernova, they presumed (correctly), was produced by the explosion of a normal star. And the explosion was so hot, they suspected (this time incorrectly), that it radiated far more energy as ultraviolet light and X-rays than as ordinary light. Since the ultraviolet light and X-rays could not penetrate the Earth's atmosphere, it was impossible to measure just how much energy they contained. However, one could try to estimate their energy from the spectrum of the observed light and the laws of physics that govern the hot gas in the exploding supernova.

By combining Baade's knowledge of the observations and of ordinary novae with Zwicky's understanding of theoretical physics, Baade and Zwicky concluded (incorrectly) that the ultraviolet radiation and X-rays from a supernova must carry at least 10,000 and perhaps 10 million times more energy even than the visible light. Zwicky, with his love of extremes, quickly assumed that the larger factor, 10 million, was correct and quoted it with enthusiasm.

This (incorrect) factor of 10 million meant that during the several days that the supernova was at its brightest, it put out an enormous amount of energy: roughly a hundred times more energy than our Sun will radiate in heat and light during its entire 10-billion-year lifetime. This is about as much energy as one would obtain if one could convert a tenth of the mass of our Sun into pure, luminous energy!

(Thanks to decades of subsequent observational studies of supernovae—many of them by Zwicky himself—we now know that the Baade–Zwicky estimate of a supernova's energy was not far off the mark. However, we also know that their calculation of this energy was badly flawed: Almost all the outpouring energy is carried by particles called neutrinos and not by X-ray and ultraviolet radiations as they thought. Baade and Zwicky got the right answer purely by luck.)

What could be the origin of this enormous supernova energy? To explain it, Zwicky invented the neutron star.

---

1. The amount of light received at Earth is inversely proportional to the *square* of the distance to the supernova, so a factor 10 error in distance meant a factor 100 error in Baade's estimate of the total light output.

Zwicky was interested in all branches of physics and astronomy, and he fancied himself a philosopher. He tried to link together all phenomena he encountered in what he later called a "morphological fashion." In 1932, the most popular of all topics in physics or astronomy was *nuclear physics,* the study of the nuclei of atoms. From there, Zwicky extracted the key ingredient for his neutron-star idea: the concept of a *neutron.*

Since the neutron will be so important in this chapter and the next, I shall digress briefly from Zwicky and his neutron stars to describe the discovery of the neutron and the relationship of neutrons to the structure of atoms.

After formulating the "new" laws of quantum mechanics in 1926 (Chapter 4), physicists spent the next five years using those quantum mechanical laws to explore the realm of the small. They unraveled the mysteries of atoms (Box 5.1) and of materials such as molecules, metals, crystals, and white-dwarf matter, which are made from atoms. Then, in 1931, physicists turned their attention inward to the cores of atoms and the atomic nuclei that reside there.

The nature of the atomic nucleus was a great mystery. Most physicists thought it was made from a handful of electrons and twice as many protons, bound together in some as yet ill-understood way. However, Ernest Rutherford in Cambridge, England, had a different hypothesis: protons and neutrons. Now, protons were already known to exist. They had been studied in physics experiments for decades, and were known to be about 2000 times heavier than electrons and to have positive electric charges. Neutrons were unknown. Rutherford had to postulate the neutron's existence in order to get the laws of quantum mechanics to explain the nucleus successfully. A successful explanation required three things: (1) Each neutron must have about the same mass as a proton but have no electric charge, (2) each nucleus must contain about the same number of neutrons as protons, and (3) all the neutrons and protons must be held together tightly in their tiny nucleus by a new type of force, neither electrical nor gravitational—a force called, naturally, the *nuclear force.* (It is now also called the *strong force.*) The neutrons and protons would protest their confinement in the nucleus by claustrophobic, erratic, high-speed motions; these motions would produce degeneracy pressure; and that pressure would counterbalance the nuclear force, holding the nucleus steady at its size of about 10^{-13} centimeter.

Box 5.1

The Internal Structures of Atoms

An atom consists of a cloud of electrons surrounding a central, massive nucleus. The electron cloud is roughly 10^{-8} centimeter in size (about a millionth the diameter of a human hair), and the nucleus at its core is 100,000 times smaller, roughly 10^{-13} centimeter; see the diagram below. If the electron cloud were enlarged to the size of the Earth, then the nucleus would become the size of a football field. Despite its tiny size, the nucleus is several thousand times heavier than the tenuous electron cloud.

The negatively charged electrons are held in their cloud by the electrical pull of the positively charged nucleus, but they do not fall into the nucleus for the same reason as a white-dwarf star does not implode: A quantum mechanical law called the Pauli exclusion principle forbids more than two electrons to occupy the same region of space at the same time (two can do so if they have opposite "spins," a subtlety ignored in Chapter 4). The cloud's electrons therefore get paired together in cells called "orbitals." Each pair of electrons, in protest against being confined to its small cell, undergoes erratic, high-speed "claustrophobic" motions, like those of electrons in a white-dwarf star (Chapter 4). These motions give rise to "electron degeneracy pressure," which counteracts the electrical pull of the nucleus. Thus, one can think of the atom as a tiny white-dwarf star, with an electric force rather than a gravitational force pulling the electrons inward, and with electron degeneracy pressure pushing them outward.

The right-hand diagram below sketches the structure of the atomic nucleus, as discussed in the text; it is a tiny cluster of protons and neutrons, held together by the nuclear force.

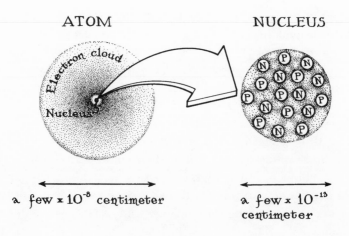

ATOM   NUCLEUS

a few $\times 10^{-8}$ centimeter   a few $\times 10^{-13}$ centimeter

In 1931 and early 1932, experimental physicists competed vigorously with each other to test this description of the nucleus. The method was to try to knock some of Rutherford's postulated neutrons out of atomic nuclei by bombarding the nuclei with high-energy radiation. The competition was won in February 1932 by a member of Rutherford's own experimental team, James Chadwick. Chadwick's bombardment succeeded, neutrons emerged in profusion, and they had just the properties that Rutherford had postulated. The discovery was announced with fanfare by newspapers around the world, and naturally it caught Zwicky's attention.

The neutron arrived on the scene in the same year as Baade and Zwicky were struggling to understand supernovae. This neutron was just what they needed, it seemed to Zwicky. Perhaps, he reasoned, the core of a normal star, with densities of, say, 100 grams per cubic centimeter, could be made to implode until it reached a density like that of an atomic nucleus, 10^{14} (100 trillion) grams per cubic centimeter; and perhaps the matter in that shrunken stellar core would then transform itself into a "gas" of neutrons—a "neutron star" Zwicky called it. If so, Zwicky computed (correctly in this case), the shrunken core's intense gravity would bind it together so tightly that not only would its circumference have been reduced, but so would its mass. The stellar core's mass would now be 10 percent lower than before the implosion. Where would that 10 percent of the core's mass have gone? Into explosive energy, Zwicky reasoned (correctly again; see Figure 5.1 and Box 5.2).

If, as Zwicky believed (correctly), the mass of the shrunken stellar core is about the same as the mass of the Sun, then the 10 percent of it that is converted to explosive energy, when the core becomes a neutron star, would produce 10^{46} joules, which is close to the energy that Zwicky thought was needed to power a supernova. The explosive energy might heat the outer layers of the star to an enormous temperature and blast them off into interstellar space (Figure 5.1), and as the star exploded, its high temperature might make it shine brightly in just the manner of the supernovae that he and Baade had identified.

Zwicky did not know what might initiate the implosion of the star's core and convert it into a neutron star, nor did he know how the core might behave as it imploded, and therefore he could not estimate how long the implosion would take (is it a slow contraction or a high-speed implosion?). (When the full details were ultimately worked out in the 1960s and later, the core turned out to implode violently; its own

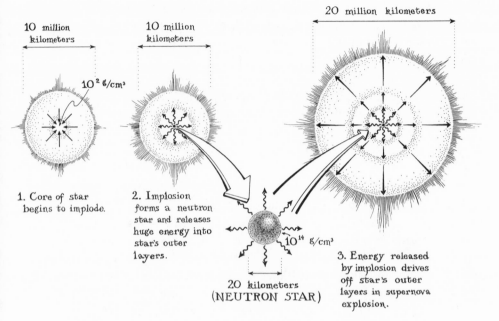

10 million
kilometers

10 million
kilometers

20 million kilometers

10^2 g/cm³

10^{14} g/cm³

1. Core of star
begins to implode.

2. Implosion
forms a neutron
star and releases
huge energy into
star's outer
layers.

20 kilometers
(NEUTRON STAR)

3. Energy released
by implosion drives
off star's outer
layers in supernova
explosion.

5.1 Fritz Zwicky's hypothesis for triggering supernova explosions: The supernova's explosive energy comes from the implosion of a star's normal-density core to form a neutron star.

Box 5.2
The Equivalence of Mass and Energy

Mass, according to Einstein's special relativity laws, is just a very compact form of energy. It is possible, though *how* is a nontrivial issue, to convert any mass, including that of a person, into explosive energy. The amount of energy that comes from such conversion is enormous. It is given by Einstein's famous formula $E = Mc^2$, where E is the explosive energy, M is the mass that gets converted to energy, and $c = 2.99792 \times 10^8$ meters per second is the speed of light. From the 75-kilogram mass of a typical person this formula predicts an explosive energy of 7×10^{18} joules, which is thirty times larger than the energy of the most powerful hydrogen bomb that has ever been exploded.

The conversion of mass into heat or into the kinetic energy of an explosion underlies Zwicky's explanation for supernovae (Figure 5.1), the nuclear burning that keeps the Sun hot (later in this chapter), and nuclear explosions (next chapter).

intense gravity drives it to implode from about the size of the Earth to 100 kilometers circumference in less than 10 seconds.) Zwicky also did not understand in detail how the energy from the core's shrinkage might create a supernova explosion, or why the debris of the explosion would shine very brightly for a few days and remain quite bright for a few months, rather than a few seconds or hours or years. However, he knew—or he thought he knew—that the energy released by forming a neutron star was the right amount, and that was enough for him.

Zwicky was not content with just explaining supernovae; he wanted to explain everything in the Universe. Among all the unexplained things, the one getting the most attention at Caltech in 1932–1933 was *cosmic rays*—high-speed particles that bombard the Earth from space. Caltech's Robert Millikan was the world leader in the study of cosmic rays and had given them their name, and Caltech's Carl Anderson had discovered that some of the cosmic-ray particles were made of *antimatter*.[2] Zwicky, with his love of extremes, managed to convince himself (correctly it turns out) that most of the cosmic rays were coming from outside our solar system, and (incorrectly) that most were from outside our Milky Way galaxy—indeed, from the most distant reaches of the Universe—and he then convinced himself (roughly correctly) that the total energy carried by all the Universe's cosmic rays was about the same as the total energy released by supernovae throughout the Universe. The conclusion was obvious to Zwicky (and perhaps correct[3]): Cosmic rays are made in supernova explosions.

It was late 1933 by the time Zwicky had convinced himself of these connections between supernovae, neutrons, and cosmic rays. Since Baade's encyclopedic knowledge of observational astronomy had been a crucial foundation for these connections, and since many of Zwicky's calculations and much of his reasoning had been carried out in verbal give-and-take with Baade, Zwicky and Baade agreed to present their work together at a meeting of the American Physical Society at Stanford University, an easy day's drive up the coast from Pasadena. The abstract of their talk, published in the 15 January 1934 issue of the

---

2. Antimatter gets its name from the fact that when a particle of matter meets a particle of antimatter, they annihilate each other.

3. It turns out that cosmic rays are made in many different ways. It is not yet known which way produces the most cosmic rays, but a strong possibility is the acceleration of particles to high speeds by shock waves in gas-cloud remnants of supernova explosions, long after the explosions are finished. If this is the case, then in an indirect sense Zwicky was correct.

Physical Review, is shown in Figure 5.2. It is one of the most prescient documents in the history of physics and astronomy.

It asserts unequivocally the existence of supernovae as a distinct class of astronomical objects—although adequate data to prove firmly that they are different from ordinary novae would be produced by Baade and Zwicky only four years later, in 1938. It introduces for the first time the name "supernovae" for these objects. It estimates, correctly, the total energy released in a supernova. It suggests that cosmic rays are produced by supernovae—a hypothesis still thought plausible in 1993, but not firmly established (see Footnote 3). It invents the concept of a star made out of neutrons—a concept that would not become widely accepted as theoretically viable until 1939 and would not be verified observationally until 1968. It coins the name *neutron star* for this concept. And it suggests "with all reserve" (a phrase presumably inserted by the cautious Baade) that supernovae are produced by the transformation of ordinary stars into neutron stars—a suggestion that would be shown theoretically viable only in the early 1960s and would be confirmed by observation only in the late 1960s with the discovery of *pulsars* (spinning, magnetized neutron stars) inside the exploding gas of ancient supernovae.

5.2   Abstract of the talk on supernovae, neutron stars, and cosmic rays given by Walter Baade and Fritz Zwicky at Stanford University in December 1933.

JANUARY 15, 1934   PHYSICAL REVIEW   VOLUME 45

Proceedings
of the
American Physical Society

MINUTES OF THE STANFORD MEETING, DECEMBER 15–16, 1933

38. Supernovae and Cosmic Rays. W. BAADE, *Mt. Wilson Observatory,* AND F. ZWICKY, *California Institute of Technology.*—Supernovae flare up in every stellar system (nebula) once in several centuries. The lifetime of a supernova is about twenty days and its absolute brightness at maximum may be as high as $M_{vis} = -14^M$. The visible radiation L_v of a supernova is about 10^8 times the radiation of our sun, that is, $L_v = 3.78 \times 10^{41}$ ergs/sec. Calculations indicate that the total radiation, visible and invisible, is of the order $L_r = 10^7 L_v = 3.78 \times 10^{48}$ ergs/sec. The supernova therefore emits during its life a total energy $E_r \geq 10^6 L_r = 3.78 \times 10^{53}$ ergs. If supernovae initially are quite ordinary stars of mass $M < 10^{34}$ g, E_r/c^2 is of the same order as M itself. In the *supernova* process *mass in bulk is annihilated.* In addition the hypothesis suggests itself that *cosmic rays are produced by supernovae.* Assuming that in every nebula one supernova occurs every thousand years, the intensity of the cosmic rays to be observed on the earth should be of the order $\sigma = 2 \times 10^{-3}$ erg/cm^2 sec. The observational values are about $\sigma = 3 \times 10^{-3}$ erg/cm^2 sec. (Millikan, Regener). With all reserve we advance the view that supernovae represent the transitions from ordinary stars into *neutron stars,* which in their final stages consist of extremely closely packed neutrons.

Astronomers in the 1930s responded enthusiastically to the Baade–Zwicky concept of a supernova, but treated Zwicky's neutron-star and cosmic-ray ideas with some disdain. "Too speculative" was the general consensus. "Based on unreliable calculations," one might add, quite correctly. Nothing in Zwicky's writings or talks provided more than meager hints of substantiation for the ideas. In fact, it is clear to me from a detailed study of Zwicky's writings of the era that he did not understand the laws of physics well enough to be able to substantiate his ideas. I shall return to this later in the chapter.

Some concepts in science are so obvious in retrospect that it is amazing nobody noticed them sooner. Such was the case with the connection between neutron stars and black holes. Zwicky could have begun to make that connection in 1933, but he didn't; it would get made in a tentative way six years later and definitively a quarter century later. The tortured route that finally rubbed physicists' noses in the connection will occupy much of the rest of this chapter.

To appreciate the story of how physicists came to recognize the neutron-star/black-hole connection, it will help to know something about the connection in advance. Thus, the following digression:

What are the fates of stars when they die? Chapter 4 revealed a partial answer, an answer embodied in the right-hand portion of Figure 5.3 (which is the same as Figure 4.4). That answer depends on whether the star is less massive or more massive than 1.4 Suns (Chandrasekhar's *limiting mass*).

If the star is less massive than the Chandrasekhar limit, for example if the star is the Sun itself, then at the end of its life it follows the path labeled "death of Sun" in Figure 5.3. As it radiates light into space, it gradually cools, losing its thermal (heat-induced) pressure. With its pressure reduced, it no longer can withstand the inward pull of its own gravity; its gravity forces it to shrink. As it shrinks, it moves leftward in Figure 5.3 toward smaller circumferences, while staying always at the same height in the figure because its mass is unchanging. (Notice that the figure plots mass up and circumference to the right.) And as it shrinks, the star squeezes the electrons in its interior into smaller and smaller cells, until finally the electrons protest with such strong degeneracy pressure that the star can shrink no more. The degeneracy pressure counteracts the inward pull of the star's gravity, forcing the star to settle down into a white-dwarf grave on the boundary curve (white-dwarf curve) between the white region of Figure 5.3 and the shaded region. If the star were to shrink even more (that is, move leftward from the white-dwarf curve into the shaded region), its electron degeneracy pressure would grow stronger and make the star expand back to the white-dwarf curve. If the star were to expand into the white region, its electron degeneracy pressure would weaken, permitting gravity to shrink it back to the white-dwarf curve. Thus, the star has no choice but to remain forever on the white-dwarf curve, where gravity and pressure balance perfectly, gradually cooling and turning into a black dwarf—a cold, dark, solid body about the size of the Earth but with the mass of the Sun.

If the star is more massive than Chandrasekhar's 1.4-solar-mass limit, for example if it is the star Sirius, then at the end of its life it will follow the path labeled "death of Sirius." As it emits radiation and cools and shrinks, moving leftward on this path to a smaller and smaller circumference, its electrons get squeezed into smaller and smaller cells; they protest with a rising degeneracy pressure, but they protest in vain. Because of its large mass, the star's gravity is strong enough to squelch all electron protest. The electrons can never produce enough degeneracy pressure to counterbalance the star's gravity[4]; the star must, in Arthur Eddington's words, "go on radiating and radiating and contracting and contracting, until, I suppose, it gets down to a few kilometers radius, when gravity becomes strong enough to hold in the radiation, and the star can at last find peace."

Or that would be its fate, if not for neutron stars. If Zwicky was right that neutron stars can exist, then they must be rather analogous to white-dwarf stars, but with their internal pressure produced by neutrons instead of electrons. This means that there must be a neutron-star curve in Figure 5.3, analogous to the white-dwarf curve, but at circumferences (marked on the horizontal axis) of roughly a hundred kilometers, instead of tens of thousands of kilometers. On this neutron-star curve neutron pressure would balance gravity perfectly, so neutron stars could reside there forever.

Suppose that the neutron-star curve extends upward in Figure 5.3 to large masses; that is, suppose it has the shape labeled *B* in the figure. Then Sirius, when it dies, *cannot* create a black hole. Rather, Sirius will shrink until it hits the neutron-star curve, and then it can shrink no more. If it tries to shrink farther (that is, move to the left of the neutron-star curve into the shaded region), the neutrons inside it will

---

4. The reason was explained in Box 4.2.

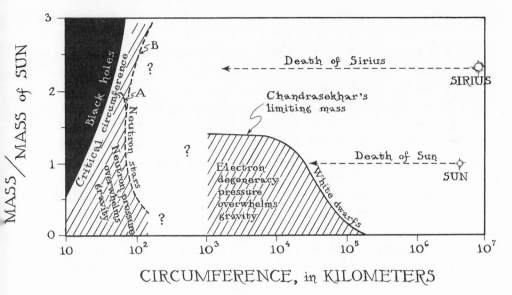

CIRCUMFERENCE, in KILOMETERS

5.3   The ultimate fate of a star more massive than the Chandrasekhar limit of 1.4 Suns depends on how massive neutron stars can be. If they can be arbitrarily massive (curve *B*), then a star such as Sirius, when it dies, can only implode to form a neutron star; it cannot form a black hole. If there is an upper mass limit for neutron stars (as on curve *A*), then a massive dying star can become neither a white dwarf nor a neutron star; and unless there is some other graveyard available, it will die a black-hole death.

protest against being squeezed; they will produce a large pressure (partly due to degeneracy, that is, "claustrophobia," and partly due to the nuclear force); and the pressure will be large enough to overwhelm gravity and drive the star back outward. If the star tries to reexpand into the white region, the neutrons' pressure will decline enough for gravity to take over and squeeze it back inward. Thus, Sirius will have no choice but to settle down onto the neutron-star curve and remain there forever, gradually cooling and becoming a solid, cold, black neutron star.

Suppose, instead, that the neutron-star curve does not extend upward in Figure 5.3 to large masses, but bends over in the manner of the hypothetical curve marked *A*. This will mean that there is a maximum mass that any neutron star can have, analogous to the Chandrasekhar limit of 1.4 Suns for white dwarfs. As for white dwarfs, so also for neutron stars, the existence of a maximum mass would herald a momentous fact: In a star more massive than the maximum, gravity will completely overwhelm the neutron pressure. Therefore, when so massive a star dies, it must either eject enough mass to bring it below the maximum, or else it will shrink inexorably, under gravity's pull, right past the neutron-star curve, and then—*if* there are no other possible stellar graveyards, nothing but white dwarfs, neutron stars, and black holes—it will continue shrinking until it forms a black hole.

Thus, the central question, the question that holds the key to the ultimate fate of massive stars, is this: *How massive can a neutron star be?* If it can be very massive, more massive than any normal star, then black holes can never form in the real Universe. If there is a maximum possible mass for neutron stars, and that maximum is not too large, then black holes *will* form—unless there is yet another stellar graveyard, unsuspected in the 1930s.

This line of reasoning is so obvious in retrospect that it seems amazing that Zwicky did not pursue it, Chandrasekhar did not pursue it, Eddington did not pursue it. Had Zwicky tried to pursue it, however, he would not have got far; he understood too little nuclear physics and too little relativity to be able to discover whether the laws of physics place a mass limit on neutron stars or not. At Caltech there were, however, two others who *did* understand the physics well enough to deduce neutron-star masses: Richard Chace Tolman, a chemist turned physicist who had written a classic textbook called *Relativity, Thermodynamics, and Cosmology;* and J. Robert Oppenheimer, who would later lead the American effort to develop the atomic bomb.

Tolman and Oppenheimer, however, paid no attention at all to Zwicky's neutron stars. They paid no attention, that is, until 1938, when the idea of a neutron star was published (under the slightly different name of *neutron core*) by somebody else, somebody whom, unlike Zwicky, they respected: Lev Davidovich Landau, in Moscow.

Landau

Landau's publication on neutron cores was actually a cry for help: Stalin's purges were in full swing in the U.S.S.R., and Landau was in danger. Landau hoped that by making a big splash in the newspapers with his neutron-core idea, he might protect himself from arrest and death. But of this, Tolman and Oppenheimer knew nothing.

Landau was in danger because of his past contacts with Western scientists:

Soon after the Russian revolution, science had been targeted for special attention by the new Communist leadership. Lenin himself had pushed a resolution through the Eighth Congress of the Bolshevik party in 1919 exempting scientists from requirements for ideological purity: "The Problem of industrial and economic development demands the immediate and widespread use of experts in science and technology whom we have inherited from capitalism, despite the fact that they inevitably are contaminated with bourgeois ideas and customs." Of special concern to the leaders of Soviet science was the sorry state of Soviet theoretical physics, so, with the blessing of the Communist party and the government, the most brilliant and promising young theorists in the U.S.S.R. were brought to Leningrad (Saint Petersburg) for a few years of graduate study, and then, after completing the equivalent of a Ph.D., were sent to Western Europe for one or two years of postdoctoral study.

Why postdoctoral study? Because by the 1920s physics had become so complex that Ph.D.-level training was not sufficient for its mastery. To promote additional training worldwide, a system of postdoctoral fellowships had been set up, funded largely by the Rockefeller Foundation (profits from capitalists' oil ventures). Anyone, even ardent Russian Marxists, could compete for these fellowships. The winners were called "postdoctoral fellows" or simply "postdocs."

Why *Western Europe* for postdoctoral study? Because in the 1920s Western Europe was the mecca of theoretical physics; it was the home of almost every outstanding theoretical physicist in the world. Soviet leaders, in their desperation to transfuse theoretical physics from Western Europe to the U.S.S.R., had no choice but to send their young theorists there for training, despite the dangers of ideological contamination.

Of all the young Soviet theorists who traveled the route to Leningrad, then to Western Europe, and then back to the U.S.S.R., Lev Davidovich Landau would have by far the greatest influence on physics. Born in 1908 into a well-to-do Jewish family (his father was a petroleum engineer in Baku on the Caspian Sea), he entered Leningrad University at age sixteen and finished his undergraduate studies by age nineteen. After just two years of graduate study at the Leningrad Physicotechnical Institute, he completed the equivalent of a Ph.D. and went off to Western Europe, where he spent eighteen months of 1929–30 making the rounds of the great theoretical physics centers in Switzerland, Germany, Denmark, England, Belgium, and Holland.

 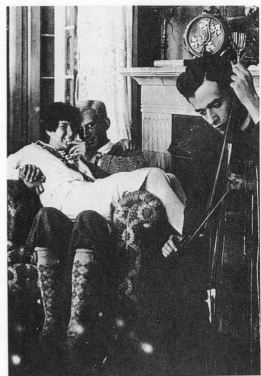

Left: Lev Landau, as a student in Leningrad in the mid-1920s. *Right:* Landau, with fellow physics students George Gamow and Yevgenia Kanegiesser, horsing around in the midst of their studies in Leningrad, ca. 1927. In reality, Landau never played any musical instrument. [Left: courtesy AIP Emilio Segrè Visual Archives, Margarethe Bohr Collection; right: courtesy Library of Congress.]

A fellow postdoctoral student in Zurich, German-born Rudolph Peierls, later wrote, "I vividly remember the great impression Landau made on all of us when he appeared in Wolfgang Pauli's department in Zurich in 1929. . . . It did not take long to discover the depth of his understanding of modern physics, and his skill in solving basic problems. He rarely read a paper on theoretical physics in detail, but looked at it long enough to see whether the problem was interesting, and if so, what was the author's approach. He then set to work to do the calculation himself, and if the answer agreed with the author's he approved of the paper." Peierls and Landau became the best of friends.

Tall, skinny, intensely critical of others as well as himself, Landau despaired that he had been born a few years too late. The golden age of physics, he thought, had been 1925–27 when de Broglie, Schrödinger, Heisenberg, Bohr, and others were creating the new quantum mechanics; if born earlier, he, Landau, could have been a participant. "All the nice girls have been snapped up and married, and all the nice problems have been solved. I don't really like any of those that are left," he said in a moment of despair in Berlin in 1929. But, in fact, explorations of the *consequences* of the laws of quantum mechanics and relativity were only beginning, and those consequences would hold wonderful surprises: the structure of the atomic nucleus, nuclear energy, black holes and their evaporation, superfluidity, superconductivity, transistors, lasers, and magnetic resonance imaging, to name only a few. And Landau, despite his pessimism, would become a central figure in the quest to discover these consequences.

Upon his return to Leningrad in 1931, Landau, who was an ardent Marxist and patriot, resolved to focus his career on transfusing modern theoretical physics into the Soviet Union. He succeeded enormously, as we shall see in later chapters.

Soon after Landau's return, Stalin's iron curtain descended, making further travel to the West almost impossible. As George Gamow, a Leningrad classmate of Landau's, later recalled: "Russian science now had become one of the weapons for fighting the capitalistic world. Just as Hitler was dividing science and the arts into Jewish and Aryan camps, Stalin created the notion of capitalistic and proletarian science. It [was becoming] . . . a crime for Russian scientists to 'fraternize' with scientists of the capitalistic countries."

The political climate went from bad to horrid. In 1936 Stalin, having already killed 6 or 7 million peasants and kulaks (landowners) in his forced collectivization of agriculture, embarked on a several-year-long purge of the country's political and intellectual leadership, a purge now called the Great Terror. The purge included execution of almost all members of Lenin's original Politburo, and execution or forced disappearance, never to be seen again, of the top commanders of the Soviet army, fifty out of seventy-one members of the Central Committee of the Communist party, most of the ambassadors to foreign countries, and the prime ministers and chief officials of the non-Russian Soviet Republics. At lower levels roughly 7 million people were arrested and imprisoned and 2.5 million died—half of them intellectuals, including a large number of scientists and some entire research teams. Soviet biology, genetics, and agricultural sciences were destroyed.

In late 1937 Landau, by now a leader of theoretical physics research in Moscow, felt the heat of the purge nearing himself. In panic he searched for protection. One possible protection might be the focus of public attention on him as an eminent scientist, so he searched among his scientific ideas for one that might make a big splash in West and East alike. His choice was an idea that he had been mulling over since the early 1930s: the idea that "normal" stars like the Sun might possess neutron stars at their centers—*neutron cores* Landau called them.

The reasoning behind Landau's idea was this: The Sun and other normal stars support themselves against the crush of their own gravity by means of thermal (heat-induced) pressure. As the Sun radiates heat and light into space, it must cool, contract, and die in about 30 million years' time—unless it has some way to replenish the heat that it loses. Since there was compelling geological evidence, in the 1920s and 1930s, that the Earth had been kept at roughly constant temperature for 1 billion years or longer, the Sun *must* be replenishing its heat somehow. Arthur Eddington and others had suggested (correctly) in the 1920s that the new heat might come from nuclear reactions, in which one kind of atomic nucleus is transmuted into another—what is now called *nuclear burning* or *nuclear fusion;* see Box 5.3. However, the details of this nuclear burning had not been worked out sufficiently, by 1937, for physicists to know whether it could do the job. Landau's neutron core provided an attractive alternative.

Just as Zwicky could imagine powering a supernova by the energy released when a normal star implodes to form a neutron star, so Landau could imagine powering the Sun and other normal stars by the energy released when its atoms, one by one, get captured onto a neutron core (Figure 5.4).

5.4   Lev Landau's speculation as to the origin of the energy that keeps a normal star hot.

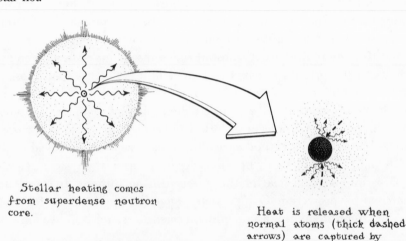

Stellar heating comes from superdense neutron core.

Heat is released when normal atoms (thick dashed arrows) are captured by neutron core.

Box 5.3
Nuclear Burning (Fusion) Contrasted with Ordinary Burning

Ordinary burning is a *chemical reaction.* In chemical reactions, atoms get combined into molecules, where they share their electron clouds with each other; the electron clouds hold the molecules together. Nuclear burning is a *nuclear reaction.* In nuclear burning, atomic nuclei get fused together (*nuclear fusion*) to form more massive atomic nuclei; the nuclear force holds the more massive nuclei together.

The following diagram shows an example of ordinary burning: the burning of hydrogen to produce water (an explosively powerful form of burning that is used to power some rockets that lift payloads into space). Two hydrogen atoms combine with an oxygen atom to form a water molecule. In the water molecule, the hydrogen and oxygen atoms share their electron clouds with each other, but do not share their atomic nuclei.

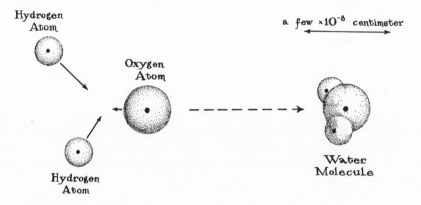

The following diagram shows an example of nuclear burning: the fusion of a deuterium ("heavy hydrogen") nucleus and an ordinary hydrogen nucleus to form a helium-3 nucleus. This is one of the fusion reactions that is now known to power the Sun and other stars, and that powers hydrogen bombs (Chapter 6). The deuterium nucleus contains one neutron and one proton, bound together by the nuclear force; the hydrogen nucleus consists of a single proton; the helium-3 nucleus created by the fusion contains one neutron and two protons.

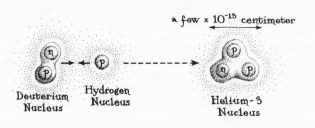

Capturing an atom onto a neutron core was much like dropping a rock onto a cement slab from a great height: Gravity pulls the rock down, accelerating it to high speed, and when it hits the slab, its huge kinetic energy (energy of motion) can shatter it into a thousand pieces. Similarly, gravity above a neutron core should accelerate infalling atoms to very high speeds, Landau reasoned. When such an atom plummets into the core, its shattering stop converts its huge kinetic energy (an amount equivalent to 10 percent of its mass) into heat. In this scenario, the ultimate source of the Sun's heat is the intense gravity of its neutron core; and, as for Zwicky's supernovae, the core's gravity is 10 percent efficient at converting the mass of infalling atoms into heat.

The burning of nuclear fuel (Box 5.3), in contrast to capturing atoms onto a neutron core (Figure 5.4), can convert only a few tenths of 1 percent of the fuel's mass into heat. In other words, Eddington's heat source (nuclear energy) was roughly 30 times less powerful than Landau's heat source (gravitational energy).[5]

Landau had actually developed a more primitive version of his neutron-core idea in 1931. However, the neutron had not yet been discovered then and atomic nuclei had been an enigma, so the capture of atoms onto the core in his 1931 model had released energy by a totally speculative process, one based on an (incorrect) suspicion that the laws of quantum mechanics might fail in atomic nuclei. Now that the neutron had been known for five years and the properties of atomic nuclei were beginning to be understood, Landau could make his idea much more precise and convincing. By presenting it to the world with a big splash of publicity, he might deflect the heat of Stalin's purge.

In late 1937, Landau wrote a manuscript describing his neutron-core idea; to make sure it got maximum public attention, he took a series of unusual steps: He submitted it for publication, in Russian, to *Doklady Akademii Nauk* (*Reports of the Academy of Sciences of the U.S.S.R.*, published in Moscow), and in parallel he mailed an English version to

5. This may seem surprising to people who think of the nuclear force as far more powerful than the gravitational force. The nuclear force is, indeed, far more powerful when one has only a few atoms or atomic nuclei at one's disposal. However, when one has several solar masses' worth of atoms (10^{57} atoms) or more, then the gravitational force of all the atoms put together can become overwhelmingly more powerful than their nuclear force. This simple fact in the end guarantees, as we shall see later in this chapter, that when a massive star dies its huge gravity will overwhelm the repulsion of its atomic nuclei and will crunch them to form a black hole.

the same famous Western physicist as Chandrasekhar had appealed to, when Eddington attacked him (Chapter 4), Niels Bohr in Copenhagen. (Bohr, as an honorary member of the Academy of Sciences of the U.S.S.R., was more or less acceptable to Soviet authorities even during the Great Terror.) With his manuscript, Landau sent Bohr the following letter:

> 5 November 1937, Moscow
>
> Dear Mr. Bohr!
>
> I enclose an article about stellar energy, which I have written. If it makes physical sense to you, I ask that you submit it to *Nature*. If it is not too much trouble for you, I would be very glad to learn your opinion of this work.
>
> With deepest thanks.
>
> Yours, L. Landau

(*Nature* is a British scientific magazine that publishes, quickly, announcements of discoveries in all fields of science and that has one of the highest worldwide circulations among serious scientific journals.)

Landau had friends in high places—high enough to arrange that, as soon as word was received back that Bohr had approved his article and had submitted it to *Nature*, a telegram would be sent to Bohr by the editorial staff of *Izvestia*. (*Izvestia* was one of the two most influential newspapers in the U.S.S.R., a newspaper run by and in behalf of the Soviet government.) The telegram went out on 16 November 1937 saying:

> Inform us, please, of your opinion of the work of Professor Landau. Telegraph to us, please, your brief conclusion.
>
> Editorial Staff, *Izvestia*

Bohr, evidently a bit puzzled and worried by the request, replied from Copenhagen that same day:

> The new idea of Professor Landau about neutron cores of massive stars is of the highest level of excellence and promise. I will be happy to send a short evaluation of it and of various other researches by Landau. Inform me please, more exactly, for what purpose my opinion is needed.
>
> Bohr

The *Izvestia* staff responded that they wanted to publish Bohr's evaluation in their newspaper. They did just that on 23 November, in an article that described Landau's idea and praised it highly:

> This work of Professor Landau's has aroused great interest among Soviet physicists, and his bold idea gives new life to one of the most important processes in astrophysics. There is every reason to think that Landau's new hypothesis will turn out to be correct and will lead to solutions to a whole series of unsolved problems in astrophysics. . . . Niels Bohr has given an extremely complementary evaluation of the work of this Soviet scientist [Landau], saying that "The new idea of L. Landau is excellent and very promising."

This campaign was not enough to save Landau. Early in the morning of 28 April 1938, the knock came on the door of his apartment, and he was taken away in an official black limousine as his wife-to-be Cora watched in shock from the apartment door. The fate that had befallen so many others was now also Landau's.

The limousine took Landau to one of Moscow's most notorious political prisons, the Butyrskaya. There he was told that his activities as a German spy had been discovered, and he was to pay the price for them. That the charges were ludicrous (Landau, a Jew and an ardent Marxist spying for Nazi Germany?) was irrelevant. The charges almost always were ludicrous. In Stalin's Russia one rarely knew the real reason one had been imprisoned—though in Landau's case, there are indications in recently revealed KGB files: In conversations with colleagues, he had criticized the Communist party and the Soviet government for their manner of organizing scientific research, and for the massive arrests of 1936–37 that ushered in the Great Terror. Such criticism was regarded as an "anti-Soviet activity" and could easily land one in prison.

Landau was lucky. His imprisonment lasted but one year, and he survived it—just barely. He was released in April 1939 after Pyotr Kapitsa, the most famous Soviet experimental physicist of the 1930s, appealed directly to Molotov and Stalin to let him go on grounds that Landau and only Landau, of all Soviet theoretical physicists, had the ability to solve the mystery of how superfluidity comes about.[6] (Superfluidity had been discovered in Kapitsa's laboratory, and indepen-

---

6. Superfluidity is a complete absence of viscosity (internal friction) that occurs in some fluids when they are cooled to a few degrees above absolute zero temperature—that is, cooled to about minus 270 degrees Celsius.

dently by J. F. Allen and A. D. Misener in Cambridge, England, and if it could be explained by a Soviet scientist, this would demonstrate doubly to the world the power of Soviet science.)

Landau emerged from prison emaciated and extremely ill. In due course, he recovered physically and mentally, solved the mystery of superfluidity using the laws of quantum mechanics, and received the Nobel Prize for his solution. But his spirit was broken. Never again could he withstand even mild psychological pressure from the political authorities.

Oppenheimer

In California, Robert Oppenheimer was in the habit of reading with care every scientific article published by Landau. Thus, Landau's article on neutron cores in the 19 February 1938 issue of *Nature* caught his immediate attention. Coming from Fritz Zwicky, the idea of a neutron star as the energizer for supernovae was—in Oppenheimer's view—a far-out, flaky speculation. Coming from Lev Landau, a neutron core as the energizer for a normal star was worthy of serious thought. Might the Sun actually possess such a core? Oppenheimer vowed to find out.

Oppenheimer's style of research was completely different from any encountered thus far in this book. Whereas Baade and Zwicky worked together as co-equal colleagues whose talents and knowledge complemented each other, and Chandrasekhar and Einstein each worked very much alone, Oppenheimer worked enthusiastically amidst a large entourage of students. Whereas Einstein had suffered when required to teach, Oppenheimer thrived on teaching.

Like Landau, Oppenheimer had gone to the mecca of theoretical physics, Western Europe, to get educated; and like Landau, Oppenheimer, upon returning home, had launched a transfusion of theoretical physics from Europe to his native land.

By the time of his return to America, Oppenheimer had acquired so tremendous a reputation that he received offers of faculty jobs from ten American universities including Harvard and Caltech, and from two in Europe. Among the offers was one from the University of California at Berkeley, which had no theoretical physics at all. "I visited Berkeley," Oppenheimer recalled later, "and I thought I'd like to go there because it was a desert." At Berkeley he could create something entirely his own. However, fearing the consequences of intellectual isolation, Op-

penheimer accepted both the Berkeley offer and the Caltech offer. He
would spend the autumn and winter in Berkeley, and the spring at
Caltech. "I kept the connection with Caltech. . . . it was a place where I
would be checked if I got too far off base and where I would learn of
things that might not be adequately reflected in the published literature."

At first Oppenheimer, as a teacher, was too fast, too impatient, too
overbearing with his students. He didn't realize how little they knew;
he couldn't bring himself down to their level. His first lecture at Caltech in the spring of 1930 was a tour de force—powerful, elegant,
insightful. When the lecture was over and the room had emptied,
Richard Tolman, the chemist-turned-physicist who by now was a close
friend, remained behind to bring him down to earth: "Well, Robert,"
he said; "that was beautiful but I didn't understand a damned word."

However, Oppenheimer learned quickly. Within a year, graduate
students and postdocs began flocking to Berkeley from all over America
to learn physics from him, and within several years he had made
Berkeley a more attractive place even than Europe for American theoretical physics postdocs.

One of Oppenheimer's postdocs, Robert Serber, later described what
it was like to work with him: "Oppie (as he was known to his Berkeley
students) was quick, impatient, and had a sharp tongue, and in the
earliest days of his teaching he was reputed to have terrorized the
students. But after five years of experience he had mellowed (if his
earlier students were to be believed). His course [on quantum mechan-
ics] was an inspirational as well as an educational achievement. He
transmitted to his students a feeling of the beauty of the logical structure of physics and an excitement about the development of physics.
Almost everyone listened to the course more than once, and Oppie
occasionally had difficulty in dissuading students from coming a third
or fourth time. . . .

"Oppie's way of working with his research students was also original. His group consisted of 8 or 10 graduate students and about a half
dozen postdoctoral fellows. He met the group once a day in his office. A
little before the appointed time, the members straggled in and disposed
themselves on the tables and about the walls. Oppie came in and
discussed with one after another the status of the student's research
problem while the others listened and offered comments. All were
exposed to a broad range of topics. Oppenheimer was interested in
everything; one subject after another was introduced and coexisted

with all the others. In an afternoon they might discuss electrodynamics, cosmic rays, astrophysics and nuclear physics."

Each spring Oppenheimer piled books and papers into his convertible and several students into the rumble seat, and drove down to Pasadena. "We thought nothing of giving up our houses or apartments in Berkeley," said Serber, "confident that we could find a garden cottage in Pasadena for twenty-five dollars a month."

For each problem that interested him, Oppenheimer would select a student or postdoc to work out the details. For Landau's problem, the question of whether a neutron core could keep the Sun hot, he selected Serber.

Robert Serber *(left)* and Robert Oppenheimer *(right)* discussing physics, ca. 1942. [Courtesy U.S. Information Agency.]

Oppenheimer and Serber quickly realized that, if the Sun has a neutron core at its center, and if the core's mass is a large fraction of the Sun's mass, then the core's intense gravity will hold the Sun's outer layers in a tight grip, making the Sun's circumference far smaller than it actually is. Thus, Landau's neutron-core idea could work only if neutron cores can be far less massive than the Sun.

"How *small* can the mass of a neutron core be?" Oppenheimer and Serber were thus driven to ask themselves. "What is the *minimum* possible mass for a neutron core?" Notice that this is the *opposite* question to the one that is crucial for the existence of black holes; to learn whether black holes can form, one needs to know the *maximum* possible mass for a neutron star (Figure 5.3 above). Oppenheimer as yet had no inkling of the importance of the maximum-mass question, but he now knew that the minimum neutron-core mass was central to Landau's idea.

In his article Landau, also aware of the importance of the minimum neutron-core mass, had used the laws of physics to estimate it. With care Oppenheimer and Serber scrutinized Landau's estimate. Yes, they found, Landau had properly taken account of the attractive forces of gravity inside and near the core. And yes, he had properly taken account of the degeneracy pressure of the core's neutrons (the pressure produced by the neutrons' claustrophobic motions when they get squeezed into tiny cells). But no, he had not taken proper account of the nuclear force that neutrons exert on each other. That force was not yet fully understood. However, enough was understood for Oppenheimer and Serber to conclude that probably, not absolutely definitely, but probably, no neutron core can ever be lighter than $\frac{1}{10}$ of a solar mass. If nature ever succeeded in creating a neutron core lighter than this, its gravity would be too weak to hold it together; its pressure would make it explode.

At first sight this did not rule out the Sun's possessing a neutron core; after all, a $\frac{1}{10}$-solar-mass core, which was allowed by Oppenheimer and Serber's estimates, might be small enough to hide inside the Sun without affecting its surface properties very much (without affecting the things we see). But further calculations, balancing the pull of the core's gravity against the pressure of surrounding gas, showed that the core's effects could not be hidden: Around the core there would be a shell of white-dwarf–type matter weighing nearly a full solar mass, and with only a tiny amount of normal gas outside that shell, the Sun could not look at all like we see it. Thus, the Sun could not possess a neutron core, and the energy to keep the Sun hot must come from somewhere else.

Where else? At the same time as Oppenheimer and Serber in Berkeley were doing these calculations, Hans Bethe at Cornell University in Ithaca, New York, and Charles Critchfield at George Washington University in Washington, D.C., were using the newly developed laws of nuclear physics to demonstrate in detail that nuclear burning (the fusion of atomic nuclei; Box 5.3) can keep the Sun and other stars hot. Eddington had been right and Landau had been wrong—at least for the Sun and most other stars. (As of the early 1990s, it appears that a few giant stars might, in fact, use Landau's mechanism.)

Oppenheimer and Serber had no idea that Landau's paper was a desperate attempt to avoid prison and possible death, so on 1 September 1938, as Landau languished in Butyrskaya Prison, they submitted their critique of him to the *Physical Review*. Since Landau was a great enough physicist to take the heat, they said quite frankly: "An estimate of Landau . . . led to the value 0.001 solar masses for the limiting [minimum] mass [of a neutron core]. This figure appears to be wrong. . . . [Nuclear forces] of the often assumed spin exchange type preclude the existence of a [neutron] core for stars with masses comparable to that of the Sun."

Landau's neutron cores and Zwicky's neutron stars are really the same thing. A neutron core is nothing but a neutron star that happens, somehow, to find itself inside a normal star. To Oppenheimer this must have been clear, and now that he had begun to think about neutron stars, he was drawn inexorably to the issue that Zwicky should have tackled but could not: What, precisely, is the fate of massive stars when they exhaust the nuclear fuel that, according to Bethe and Critchfield, keeps them hot? Which corpses will they create: white dwarfs? neutron stars? black holes? others?

Chandrasekhar's calculations had shown unequivocally that stars less massive than 1.4 Suns must become white dwarfs. Zwicky was speculating loudly that at least some stars more massive than 1.4 Suns will implode to form neutron stars, and in the process generate supernovae. Might Zwicky be right? And will all massive stars die this way, thus saving the Universe from black holes?

One of Oppenheimer's great strengths as a theorist was an unerring ability to look at a complicated problem and strip away the complications until he found the central issue that controlled it. Several years later, this talent would contribute to Oppenheimer's brilliance as the leader of the American atomic bomb project. Now, in his struggle to understand stellar death, it told him to ignore all the complications that Zwicky was trumpeting about—the details of the stellar implosion, the transformation of normal matter into neutron matter, the release of enormous energy and its possible powering of supernovae and cosmic rays. All this was irrelevant to the issue of the star's *final fate*. The only relevant thing was the maximum mass that a neutron star can have. If neutron stars can be arbitrarily massive (curve *B* in Figure 5.3 above), then black holes can never form. If there is a maximum possible neutron-star mass (curve *A* in Figure 5.3), then a star heavier than that maximum, when it dies, might form a black hole.

Having posed this maximum-mass question with stark clarity, Oppenheimer went about solving it, methodically and unequivocally— and, as was his standard practice, in collaboration with a student, in this case a young man named George Volkoff. The tale of Oppenheimer and Volkoff's quest to learn the masses of neutron stars, and the central contributions of Oppenheimer's Caltech friend Richard Tolman, is told in Box 5.4. It is a tale that illustrates Oppenheimer's mode of research and several of the strategies by which physicists operate, when they understand clearly *some* of the laws that govern the phenomenon they are studying, but *not all*. In this case Oppenheimer understood the laws of quantum mechanics and general relativity, but neither he nor anyone else understood the nuclear force very well.

Despite their poor knowledge of the nuclear force, Oppenheimer and Volkoff were able to show unequivocally (Box 5.4) that *there is a maximum mass for neutron stars, and it lies between about half a solar mass and several solar masses.*

In the 1990s, after fifty years of additional study, we know that Oppenheimer and Volkoff were correct; neutron stars do, indeed, have a maximum allowed mass, and it is now known to lie between 1.5 and 3 solar masses, roughly the same ballpark as their estimate. Moreover, since 1967 hundreds of neutron stars have been observed by astronomers, and the masses of several have been measured with high accuracy. The measured masses are all close to 1.4 Suns; why, we do not know.

Box 5.4
The Tale of Oppenheimer, Volkoff, and Tolman:
A Quest for Neutron-Star Masses

When embarking on a complicated analysis, it is helpful to get one's bearings by beginning with a rough, "order-of-magnitude" calculation, a calculation accurate only to within a factor of, say, 10. In keeping with this rule of thumb, Oppenheimer began his assault on the issue of whether neutron stars can have a maximum mass by a crude calculation, just a few pages long. The result was intriguing: He found a maximum mass of 6 Suns for any neutron star. If a detailed calculation gave the same result, then Oppenheimer could conclude that black holes might form when stars heavier than 6 Suns die.

A "detailed calculation" meant selecting a mass for a hypothetical neutron star, then asking whether, for that mass, neutron pressure inside the star can balance gravity. If the balance can be achieved, then neutron stars can have that mass. It would be necessary to choose one mass after another, and for each ask about the balance between pressure and gravity. This enterprise is harder than it might sound, because pressure and gravity must balance each other *everywhere* inside the star. However, it was an enterprise that had been pursued once before, by Chandrasekhar, in his analysis of white dwarfs (the analysis performed using Arthur Eddington's Braunschweiger calculator, with Eddington looking over Chandrasekhar's shoulder; Chapter 4).

Oppenheimer could pattern his neutron-star calculations after Chandrasekhar's white-dwarf calculations, but only after making two crucial changes: First, in a white dwarf the pressure is produced by electrons, and in a neutron star by neutrons, so the *equation of state* (the relation between pressure and density) will be different. Second, in a white dwarf, gravity is weak enough that it can be described equally well by Newton's laws or by Einstein's general relativity; the two descriptions will give almost precisely the same predictions, so Chandrasekhar chose the simpler description, Newton's. By contrast, in a neutron star, with its much smaller circumference, gravity is so strong that using Newton's laws might cause serious errors, so Oppenheimer would have to describe gravity by Einstein's general relativistic laws.* Aside from these two changes—a new equation of state (neutron pressure instead of electron) and a new description of gravity (Einstein's instead of Newton's)—Oppenheimer's calculation would be the same as Chandrasekhar's.

Having gotten this far, Oppenheimer was ready to turn the details of the calculation over to a student. He chose George Volkoff, a young man

*See the discussion in the last section of Chapter 1 ("The Nature of Physical Law") of the relationship between different descriptions of the laws of physics and their domains of validity.

(continued next page)

(Box 5.4 continued)

from Toronto, who had emigrated from Russia in 1924.

Oppenheimer explained the problem to Volkoff and told him that the mathematical description of gravity that he would need was in a textbook that Richard Tolman had written, *Relativity, Thermodynamics, and Cosmology*. The equation of state for the neutron pressure, however, was a more difficult issue, since the pressure would be influenced by the nuclear force (with which neutrons push and pull on each other). Although the nuclear force was becoming well understood at the densities inside atomic nuclei, it was very poorly understood at the higher densities that neutrons would face deep inside a massive neutron star. Physicists did not even know whether the nuclear force was attractive at these densities or repulsive (whether neutrons pulled on each other or pushed), and thus there was no way to know whether the nuclear force reduced the neutrons' pressure or increased it. But Oppenheimer had a strategy to deal with these unknowns.

Pretend, at first, that the nuclear force doesn't exist, Oppenheimer suggested to Volkoff. Then all the pressure will be of a sort that is well understood; it will be neutron degeneracy pressure (pressure produced by the neutrons' "claustrophobic" motions). Balance this neutron degeneracy pressure against gravity, and from the balance, calculate the structures and masses that neutron stars would have in a universe without any nuclear force. Then, afterward, try to estimate how the stars' structures and masses will change if, in our real Universe, the nuclear force behaves in this, that, or some other way.

With such well-posed instructions it was hard to miss. It took only a few days for Volkoff, guided by daily discussions with Oppenheimer and by Tolman's book, to derive the general relativistic description of gravity inside a neutron star. And it took only a few days for him to translate the well-known equation of state for degenerate electron pressure into one for degenerate neutron pressure. By balancing the pressure against the gravity, Volkoff obtained a complicated differential equation whose solution would tell him the star's internal structure. Then he was stymied. Try as he might, Volkoff could not solve his differential equation to get a formula for the star's structure; so, like Chandrasekhar with white dwarfs, he was forced to solve his equation numerically. Just as Chandrasekhar had spent many days in 1934 punching the buttons of Eddington's Braunschweiger calculator to compute the analogous white-dwarf structure, so Volkoff labored through much of November and December 1938, punching the buttons of a Marchant calculator.

While Volkoff punched buttons in Berkeley, Richard Tolman in Pasadena was taking a different tack: He strongly preferred to express the stellar structure in terms of formulas instead of just numbers off a calcula-

(continued next page)

(Box 5.4 continued)

tor. A single formula can embody all the information contained in many many tables of numbers. If he could get the right formula, it would contain simultaneously the structures of stars of 1 solar mass, 2 solar masses, 5 solar masses—any mass at all. But even with his brilliant mathematical skills, Tolman was unable to solve Volkoff's equation in terms of formulas.

"On the other hand," Tolman presumably argued to himself, "we know that the correct equation of state is not really the one Volkoff is using. Volkoff has ignored the nuclear force; and since we don't know the details of that force at high densities, we don't know the correct equation of state. So let me ask a different question from Volkoff. Let me ask how the masses of neutron stars depend on the equation of state. Let me pretend that the equation of state is very 'stiff,' that is, that it gives exceptionally high pressures, and let me ask what the neutron-star masses would be in that case. And then let me pretend the equation of state is very 'soft,' that is, that it gives exceptionally low pressures, and ask what then would be the neutron-star masses. In each case, I will adjust the hypothetical equation of state into a form for which I can solve Volkoff's differential equation in formulas. Though the equation of state I use will almost certainly not be the right one, my calculation will still give me a general idea of what the neutron-star masses might be if nature happens to choose a stiff equation of state, and what they might be if nature chooses a soft one."

On 19 October, Tolman sent a long letter to Oppenheimer describing some of the stellar-structure formulas and neutron-star masses he had derived for several hypothetical equations of state. A week or so later, Oppenheimer drove down to Pasadena to spend a few days talking with Tolman about the project. On 9 November, Tolman wrote Oppenheimer another long letter, with more formulas. In the meantime, Volkoff was punching away on his Marchant buttons. In early December, Volkoff finished. He had numerical models for neutron stars with masses 0.3, 0.6, and 0.7 solar mass; and he had found that, *if there were no nuclear force in our Universe, then neutron stars would always be less massive than 0.7 solar mass.*

What a surprise! Oppenheimer's crude estimate, before Volkoff started computing, had been a maximum mass of 6 solar masses. To protect massive stars against forming black holes, the careful calculation would have had to push that maximum mass up to a hundred Suns or more. Instead, it pushed the mass down—way down, to 0.7 solar mass.

Tolman came up to Berkeley to learn more of the details. Fifty years later Volkoff recalled the scene with pleasure: "I remember being greatly overawed by having to explain to Oppenheimer and Tolman what I had done. We were sitting out on the lawn of the old faculty club at Berkeley.

(continued next page)

(Box 5.4 continued)

Amidst the nice green grass and tall trees, here were these two venerated gentlemen and here was I, a graduate student just completing my Ph.D., explaining my calculations."

Now that they knew the masses of neutron stars in an idealized universe with no nuclear force, Oppenheimer and Volkoff were ready to estimate the influence of the nuclear force. Here the formulas that Tolman had worked out so carefully for various hypothetical equations of state were helpful. From Tolman's formulas one could see roughly how the star's structure would change if the nuclear force was repulsive and thereby made the equation of state more "stiff" than the one Volkoff had used, and the change if it was attractive and thereby made the equation of state more "soft." Within the range of believable nuclear forces, those changes were not great. There must still be a maximum mass for neutron stars, Tolman, Oppenheimer, and Volkoff concluded, and it must lie somewhere between about a half solar mass and several solar masses.

O ppenheimer and Volkoff's conclusion cannot have been pleasing to people like Eddington and Einstein, who found black holes anathema. If Chandrasekhar was to be believed (as, in 1938, most astronomers were coming to understand he should), and if Oppenheimer and Volkoff were to be believed (and it was hard to refute them), then neither the white-dwarf graveyard nor the neutron-star graveyard could inter massive stars. Was there any conceivable way at all, then, for massive stars to avoid a black-hole death? Yes; two ways.

First, all massive stars might eject so much matter as they age (for example, by blowing strong winds off their surfaces or by nuclear explosions) that they reduce themselves below 1.4 solar masses and enter the white-dwarf graveyard, or (if one believed Zwicky's mechanism for supernovae, which few people did) they might eject so much matter in supernova explosions that they reduce themselves below about 1 solar mass during the explosion and wind up in the neutron-star graveyard. Most astronomers, through the 1940s and 1950s, and into the early 1960s—if they thought at all about the issue—espoused this view.

Second, besides the white-dwarf, neutron-star, and black-hole graveyards, there might be a fourth graveyard for massive stars, a graveyard unknown in the 1930s. For example, one could imagine a graveyard in Figure 5.3 at circumferences intermediate between neutron stars and white dwarfs—a few hundred or a thousand kilometers. The shrinkage of a massive star might be halted in such a graveyard before the star ever gets small enough to form either a neutron star or a black hole.

If World War II and then the cold war had not intervened, Oppenheimer and his students, or others, would likely have explored such a possibility in the 1940s and would have showed firmly that there is no such fourth graveyard.

However, World War II *did* intervene, and it absorbed the energies of almost all the world's theoretical physicists; then after the war, crash programs to develop hydrogen bombs delayed further the return of physicists to normalcy (see the next chapter).

Finally, in the mid-1950s, two physicists emerged from their respective hydrogen bomb efforts and took up where Oppenheimer and his students had left off. They were John Archibald Wheeler at Princeton University in the United States and Yakov Borisovich Zel'dovich at the Institute of Applied Mathematics in Moscow—two superb physicists, who will be major figures in the rest of this book.

Wheeler

In March 1956, Wheeler devoted several days to studying the articles by Chandrasekhar, Landau, and Oppenheimer and Volkoff. Here, he recognized, was a mystery worth probing. Could it really be true that stars more massive than about 1.4 Suns have no choice, when they die, but to form black holes? "Of all the implications of general relativity for the structure and evolution of the Universe, this question of the fate of great masses of matter is one of the most challenging," Wheeler wrote soon thereafter; and he set out to complete the exploration of stellar graveyards that Chandrasekhar and Oppenheimer and Volkoff had begun.

To make his task very precise, Wheeler formulated a careful characterization of the kind of matter from which cold, dead stars should be made: He called it *matter at the endpoint of thermonuclear evolution*, since the word *thermonuclear* had become popular for the fusion reactions that power nuclear burning in stars and also power the hydrogen bomb. Such matter would be absolutely cold, and it would have burned its nuclear fuel completely; there would be no way, by any kind of nuclear reaction, to extract any more energy from the matter's nuclei.

For this reason, the nickname *cold, dead matter* will be used in this book instead of "matter at the endpoint of thermonuclear evolution."

Wheeler set himself the goal to understand *all* objects that can be made from cold, dead matter. These would include small objects like balls of iron, heavier objects such as cold, dead planets made of iron, and still heavier objects: white dwarfs, neutron stars, and whatever other kinds of cold, dead objects the laws of physics allow. Wheeler wanted a *comprehensive* catalog of cold, dead things.

John Archibald Wheeler, ca. 1954. [Photo by Blackstone-Shelburne, New York City; courtesy J. A. Wheeler.]

Wheeler worked in much the same mode as had Oppenheimer, with an entourage of students and postdocs. From among them he selected B. Kent Harrison, a serious-minded Mormon from Utah, to work out the details of the equation of state for cold, dead matter. This equation of state would describe the details of how the pressure of such matter rises as one gradually compresses it to higher and higher density—or, equivalently, how its resistance to compression changes as its density increases.

Wheeler was superbly prepared to give Harrison guidance in computing the equation of state for cold, dead matter, since he was among the world's greatest experts on the laws of physics that govern the structure of matter: the laws of quantum mechanics and nuclear physics. During the preceding twenty years, he had developed powerful mathematical models to describe how atomic nuclei behave; with Niels Bohr he had developed the laws of nuclear fission (the splitting apart of heavy atomic nuclei such as uranium and plutonium, the principle underlying the atomic bomb); and he had been the leader of a team that designed the American hydrogen bomb (Chapter 6). Drawing on this expertise, Wheeler guided Harrison through the intricacies of the analysis.

The result of their analysis, the equation of state for cold, dead matter, is depicted and discussed in Box 5.5. At the densities of white dwarfs, it was the same equation of state as Chandrasekhar had used in his white-dwarf studies (Chapter 4); at neutron-star densities, it was the same as Oppenheimer and Volkoff had used (Box 5.4); at densities below white dwarfs and between white dwarfs and neutron stars, it was completely new.

With this equation of state for cold, dead matter in hand, John Wheeler asked Masami Wakano, a postdoc from Japan, to do with it what Volkoff had done for neutron stars and Chandrasekhar for white dwarfs: Combine the equation of state with the general relativistic equation describing the balance of gravity and pressure inside a star, and from that combination deduce a differential equation describing the star's structure; then solve the differential equation numerically. The numerical calculations would produce the details of the internal structures of all cold, dead stars and, most important, the stars' masses.

The calculations for the structure of a single star (the distribution of density, pressure, and gravity inside the star) had required Chandrasekhar and Volkoff many days of effort as they punched the buttons of

Box 5.5

The Harrison–Wheeler Equation of State for Cold, Dead Matter

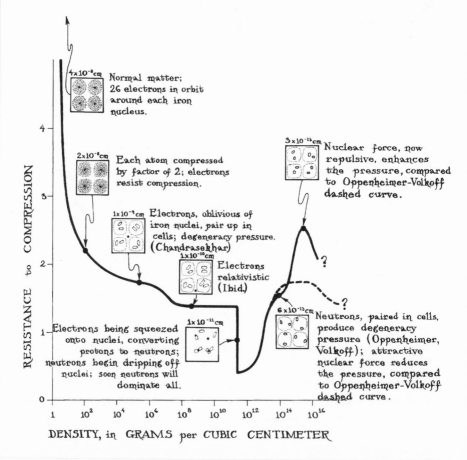

The drawing above depicts the Harrison–Wheeler equation of state. Plotted horizontally is the matter's density. Plotted vertically is its resistance to compression (or adiabatic index, as physicists like to call it)—the percentage increase in pressure that accompanies a 1 percent increase in density. The boxes attached to the curve show what is happening to the matter, microscopically, as it is compressed from low densities to high. The size of each box, in centimeters, is written along the box's top.

At normal densities (left edge of the figure), cold, dead matter is composed of iron. If the matter's atomic nuclei were heavier than iron, energy could be released by splitting them apart to make iron (nuclear fission, as in an atomic bomb). If its nuclei were lighter than iron, energy could be released by joining them together to make iron (nuclear fusion, as in a

(continued next page)

(Box 5.5 continued)

hydrogen bomb). Once in the form of iron, the matter can release no more nuclear energy by any means whatsoever. The nuclear force holds neutrons and protons together more tightly when they form iron nuclei than when they form any other kind of atomic nucleus.

As the iron is squeezed from its normal density of 7.6 grams per cubic centimeter up toward 100, then 1000 grams per cubic centimeter, the iron resists by the same means as a rock resists compression: The electrons of each atom protest with "claustrophobic" (degeneracy-like) motions against being squeezed by the electrons of adjacent atoms. The resistance at first is huge not because the repulsive forces are especially strong, but rather because the starting pressure, at low density, is very low. (Recall that the resistance is the percentage increase in pressure that accompanies a 1 percent increase in density. When the pressure is low, a strong increase in pressure represents a huge percentage increase and thus a huge "resistance." Later, at higher densities where the pressure has grown large, a strong pressure increase represents a much more modest percentage increase and thus a more modest resistance.)

At first, as the cold matter is compressed, the electrons congregate tightly around their iron nuclei, forming electron clouds made of electron orbitals. (There are actually two electrons, not one, in each orbital—a subtlety overlooked in Chapter 4 but discussed briefly in Box 5.1.) As the compression proceeds, each orbital and its two electrons are gradually confined into a smaller and smaller cell of space; the claustrophobic electrons protest this confinement by becoming more wave-like and developing higher-speed, erratic, claustrophobic motions ("degeneracy motions"; see Chapter 4). When the density has reached 10^5 (100,000) grams per cubic centimeter, the electrons' degeneracy motions and the degeneracy pressure they produce have become so large that they completely overwhelm the electric forces with which the nuclei pull on the electrons. The electrons no longer congregate around the iron nuclei; they completely ignore the nuclei. The cold, dead matter, which began as a lump of iron, has now become the kind of stuff of which white dwarfs are made, and the equation of state has become the one that Chandrasekhar, Anderson, and Stoner computed in the early 1930s (Figure 4.3): a resistance of 5/3, and then a smooth switch to 4/3 at a density of about 10^7 grams per cubic centimeter when the erratic speeds of the electrons near the speed of light.

The transition from white-dwarf matter to neutron-star matter begins at a density of 4×10^{11} grams per cubic centimeter, according to the Harrison–Wheeler calculations. The calculations show several phases to the transition: In the first phase, the electrons begin to be squeezed into the atomic nuclei, and the nuclei's protons swallow them to form neutrons. The matter, having thereby lost some of its pressure-sustaining

(continued next page)

(Box 5.5 continued)

electrons, suddenly becomes much less resistant to compression; this causes the sharp cliff in the equation of state (see diagram above). As this first phase proceeds and the resistance plunges, the atomic nuclei become more and more bloated with neutrons, thereby triggering the second phase: Neutrons begin to drip out of (get squeezed out of) the nuclei and into the space between them, alongside the few remaining electrons. These dripped-out neutrons, like the electrons, protest the continuing squeeze with a degeneracy pressure of their own. This neutron degeneracy pressure terminates the over-the-cliff plunge in the equation of state; the resistance to compression recovers and starts rising. In the third phase, at densities between about 10^{12} and 4×10^{12} grams per cubic centimeter, each neutron-bloated nucleus completely disintegrates, that is, breaks up into individual neutrons, forming the neutron gas studied by Oppenheimer and Volkoff, plus a tiny smattering of electrons and protons. From there on upward in density, the equation of state takes on the Oppenheimer–Volkoff neutron-star form (dashed curve in the diagram when nuclear forces are ignored; solid curve using the best 1990s understanding of the influence of nuclear forces).

their calculators in Cambridge and Berkeley in the 1930s. Wakano in Princeton in the 1950s, by contrast, had at his disposal one of the world's first digital computers, the MANIAC—a room full of vacuum tubes and wires that had been constructed at the Princeton Institute for Advanced Study for use in the design of the hydrogen bomb. With the MANIAC, Wakano could crunch out the structure of each star in less than an hour.

The results of Wakano's calculations are shown in Figure 5.5. *This figure is the firm and final catalog of cold, dead objects; it answers all the questions we raised, early in this chapter, in our discussion of Figure 5.3.*

In Figure 5.5, the circumference of a star is plotted rightward and its mass upward. Any star with circumference and mass in the white region of the figure has a stronger internal gravity than its pressure, so its gravity makes the star shrink leftward in the diagram. Any star in the shaded region has a stronger pressure than gravity, so its pressure makes the star expand rightward in the diagram. Only along the boundary of white and shaded do gravity and pressure balance each other perfectly; thus, the boundary curve is the curve of cold, dead stars that are in pressure/gravity equilibrium.

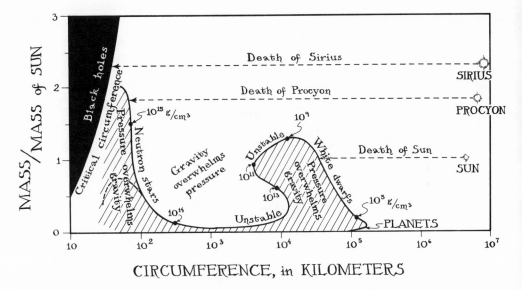

CIRCUMFERENCE, in KILOMETERS

5.5 The circumferences (plotted horizontally), masses (plotted vertically), and central densities (labeled on curve) for cold, dead stars, as computed by Masami Wakano under the direction of John Wheeler, using the equation of state of Box 5.5. At central densities above those of an atomic nucleus (above 2×10^{14} grams per cubic centimeter), the solid curve is a modern, 1990s, one that takes proper account of the nuclear force, and the dashed curve is that of Oppenheimer and Volkoff without nuclear forces.

As one moves along this *equilibrium curve,* one is tracing out dead "stars" of higher and higher densities. At the lowest densities (along the bottom edge of the figure and largely hidden from view), these "stars" are not stars at all; rather, they are cold planets made of iron. (When Jupiter ultimately exhausts its internal supply of radioactive heat and cools off, although it is made mostly of hydrogen rather than iron, it will nevertheless lie near the rightmost point on the equilibrium curve.) At higher densities than the planets are Chandrasekhar's white dwarfs.

When one reaches the topmost point on the white-dwarf part of the curve (the white dwarf with Chandrasekhar's maximum mass of 1.4 Suns[7]) and then moves on to still higher densities, one meets cold, dead

7. Actually, the maximum white-dwarf mass in Figure 5.5 (Wakano's calculation) is 1.2 Suns, which is slightly less than the 1.4 Suns that Chandrasekhar calculated. The difference is due to a different chemical composition: Wakano's stars were made of "cold, dead matter" (mostly iron), which has 46 percent as many electrons as nucleons (neutrons and protons). Chandrasekhar's stars were made of elements such as helium, carbon, nitrogen, and oxygen, which have 50 percent as many electrons as nucleons. In fact, most white dwarfs in our Universe are more nearly like Chandrasekhar's than like Wakano's. That is why, in this book, I consistently quote Chandrasekhar's value for the maximum mass: 1.4 Suns.

Unstable Inhabitants of the Gap between White Dwarfs and Neutron Stars

Along the equilibrium curve in Figure 5.5, all the stars between the white dwarfs and the neutron stars are unstable. An example is the star with central density 10^{13} grams per cubic centimeter, whose mass and circumference are those of the point in Figure 5.5 marked 10^{13}. At the 10^{13} point this star is in equilibrium; its gravity and pressure balance each other perfectly. However, the star is as unstable as a pencil standing on its tip.

If some tiny random force (for example, the fall of interstellar gas onto the star) squeezes the star ever so slightly, that is, reduces its circumference so it moves leftward a bit in Figure 5.5 into the white region, then the star's gravity will begin to overwhelm its pressure and will pull the star into an implosion; as the star implodes, it will move strongly leftward through Figure 5.5 until it crosses the neutron-star curve into the shaded region; there its neutron pressure will skyrocket, halt the implosion, and push the star's surface back outward until the star settles down into a neutron-star grave, on the neutron-star curve.

By contrast, if, when the star is at the 10^{13} point, instead of being squeezed inward by a tiny random force, its surface gets pushed outward a bit (for example, by a random increase in the erratic motions of some of its neutrons), then it will enter the shaded region where pressure overwhelms gravity; the star's pressure will then make its surface explode on outward across the white-dwarf curve and into the white region of the figure; and there its gravity will take over and pull it back inward to the white-dwarf curve and a white-dwarf grave.

This instability (squeeze the 10^{13} star a tiny bit and it will implode to become a neutron star; expand it a tiny bit and it will explode to become a white dwarf) means that no real star can ever live for long at the 10^{13} point—or at any other point along the portion of the equilibrium curve marked "unstable."

stars that cannot exist in nature because they are unstable against implosion or explosion (Box 5.6). As one moves from white-dwarf densities toward neutron-star densities, the masses of these unstable equilibrium stars decrease until they reach a minimum of about 0.1 solar mass at a circumference of 1000 kilometers and a central density of 3×10^{13} grams per cubic centimeter. This is the first of the neutron stars; it is the "neutron core" that Oppenheimer and Serber studied and showed cannot possibly be as light as the 0.001 solar mass that Landau wanted for a core inside the Sun.

Moving on along the equilibrium curve, we meet the entire family of neutron stars, with masses ranging from 0.1 to about 2 Suns. The maximum neutron-star mass of about 2 Suns is somewhat uncertain even in the 1990s because the behavior of the nuclear force at very high densities is still not well understood. The maximum could be as low as 1.5 Suns but not much lower, or as high as 3 Suns but not much higher.

At the (approximately) 2-solar-mass peak of the equilibrium curve, the neutron stars end. As one moves further along the curve to still higher densities, the equilibrium stars become unstable in the same manner as those between white dwarfs and neutron stars (Box 5.6). Because of this instability, these "stars," like those between white dwarfs and neutron stars, cannot exist in nature. Were they to form, they would immediately implode to become black holes or explode to become neutron stars.

Figure 5.5 is absolutely firm and unequivocal: There is *no* third family of stable, massive, cold, dead objects between the white dwarfs and the neutron stars. Therefore, when stars such as Sirius, which are more massive than about 2 Suns, exhaust their nuclear fuel, either they must eject all of their excess mass or they will implode inward past white-dwarf densities, past neutron-star densities, and into the critical circumference—where today, in the 1990s, we are completely certain they must form black holes. Implosion is compulsory. For stars of sufficiently large mass, neither the degeneracy pressure of electrons nor the nuclear force between neutrons can stop the implosion. Gravity overwhelms even the nuclear force.

There remains, however, a way out, a way to save all stars, even the most massive, from the black-hole fate: Perhaps *all* massive stars eject enough mass late in their lives (in winds or explosions), or during their deaths, to bring them below about 2 Suns so they can end up in the neutron-star or white-dwarf graveyard. During the 1940s, 1950s, and early 1960s, astronomers tended to espouse this view, when they thought at all about the issue of the final fates of stars. (By and large, however, they didn't think about the issue. There were no observational data pushing them to think about it; and the observational data that they were gathering on other kinds of objects—normal stars, nebulas, galaxies—were so rich, challenging, and rewarding as to absorb the astronomers' full attention.)

Today, in the 1990s, we know that heavy stars *do* eject enormous amounts of mass as they age and die; they eject so much, in fact, that

most stars born with masses as large as 8 Suns lose enough to wind up
in the white-dwarf graveyard, and most born between about 8 and 20
Suns lose enough to wind up in the neutron-star graveyard. Thus,
nature seems *almost* to protect herself against black holes. But not
quite: The preponderance of the observational data suggest (but do not
yet firmly prove) that most stars born heavier than about 20 Suns
remain so heavy when they die that their pressure provides no protection against gravity. When they exhaust their nuclear fuel and begin to
cool, gravity overwhelms their pressure and they implode to form black
holes. We shall meet some of the observational data suggesting this in
Chapter 8.

There is much to be learned about the nature of science and scientists
from the neutron-star and neutron-core studies of the 1930s.

The objects that Oppenheimer and Volkoff studied were Zwicky's
neutron stars and not Landau's neutron cores, since they had no surrounding envelope of stellar matter. Nevertheless, Oppenheimer had so
little respect for Zwicky that he declined to use Zwicky's name for
them, and insisted on using Landau's instead. Thus, his article with
Volkoff describing their results, which was published in the 15 February 1939 issue of the *Physical Review,* carries the title "On Massive
Neutron Cores." And to make sure that nobody would mistake the
origin of his ideas about these stars, Oppenheimer sprinkled the article
with references to Landau. Not once did he cite Zwicky's plethora of
prior neutron-star publications.

Zwicky, for his part, watched with growing consternation in 1938 as
Tolman, Oppenheimer, and Volkoff pursued their studies of neutron-
star structure. How could they do this? he fumed. Neutron stars were
his babies, not theirs; they had no business working on neutron stars—
and, besides, although Tolman would talk to him occasionally, Oppenheimer was not consulting him at all!

In the plethora of papers that Zwicky had written about neutron
stars, however, there was only talk and speculation, no real details. He
had been so busy getting under way a major (and highly successful)
observational search for supernovae and giving lectures and writing
papers about the idea of a neutron star and its role in supernovae that
he had never gotten around to trying to fill in the details. But now his
competitive spirit demanded action. Early in 1938 he did his best to

develop a detailed mathematical theory of neutron stars and tie it to his supernova observations. His best effort was published in the 15 April 1939 issue of the *Physical Review* under the title "On the Theory and Observation of Highly Collapsed Stars." His paper is two and a half times longer than that of Oppenheimer and Volkoff; it contains not a single reference to the two-months-earlier Oppenheimer–Volkoff article, though it does refer to a subsidiary and minor article by Volkoff alone; and it contains nothing memorable. Indeed, much of it is simply wrong. By contrast, the Oppenheimer–Volkoff paper is a tour de force, elegant, rich in insights, correct in all details.

Despite this, Zwicky is venerated today, more than half a century later, for inventing the concept of a neutron star, for recognizing, correctly, that neutron stars are created in supernova explosions and energize them, for proving observationally, with Baade, that supernovae are indeed a unique class of astronomical objects, for initiating and carrying through a definitive, decades-long observational study of supernovae—and for a variety of other insights unrelated to neutron stars or supernovae.

How is it that a man with so meager an understanding of the laws of physics could have been so prescient? My own opinion is that he embodied a remarkable combination of character traits: enough understanding of theoretical physics to get things right qualitatively, if not quantitatively; so intense a curiosity as to keep up with everything happening in all of physics and astronomy; an ability to discern, intuitively, in a way that few others could, connections between disparate phenomena; and, of not least importance, such great faith in his own inside track to truth that he had no fear whatsoever of making a fool of himself by his speculations. He knew he was right (though he often was not), and no mountain of evidence could convince him to the contrary.

Landau, like Zwicky, had great self-confidence and little fear of appearing a fool. For example, he did not hesitate to publish his 1931 idea that stars are energized by superdense stellar cores in which the laws of quantum mechanics fail. In mastery of theoretical physics, Landau totally outclassed Zwicky; he was among the top ten theorists of the twentieth century. Yet his speculations were wrong and Zwicky's were right. The Sun is *not* energized by neutron cores; supernovae *are* energized by neutron stars. Was Landau, by contrast with Zwicky, simply unlucky? Perhaps partly. But there is another factor: Zwicky was immersed in the atmosphere of Mount Wilson, then the world's greatest center for astronomical observations. And he collaborated with one of the world's greatest observational astronomers, Walter Baade, who was a master of the observational data. And at Caltech he could and did talk almost daily with the world's greatest cosmic-ray observers. By contrast, Landau had almost no direct contact with observational astronomy, and his articles show it. Without such contact, he could not develop an acute sense for what things are like out there, far beyond the Earth. Landau's greatest triumph was his masterful use of the laws of quantum mechanics to explain the phenomenon of superfluidity, and in this research, he interacted extensively with the experimenter, Pyotr Kapitsa, who was probing superfluidity's details.

For Einstein, by contrast with Zwicky and Landau, close contact between observation and theory was of little importance; he discovered his general relativistic laws of gravity with almost no observational input. But that was a rare exception. A rich interplay between observation and theory is essential to progress in most branches of physics and astronomy.

And what of Oppenheimer, a man whose mastery of physics was comparable to Landau's? His article, with Volkoff, on the structure of neutron stars is one of the great astrophysics articles of all time. But, as great and beautiful as it is, it "merely" filled in the details of the neutron-star concept. The concept was, indeed, Zwicky's baby—as were supernovae and the powering of supernovae by the implosion of a stellar core to form a neutron star. Why was Oppenheimer, with so much going for him, far less innovative than Zwicky? Primarily, I think, because he declined—perhaps even feared—to speculate. Isidore I. Rabi, a close friend and admirer of Oppenheimer, has described this in a much deeper way:

"[I]t seems to me that in some respects Oppenheimer was over-educated in those fields which lie outside the scientific tradition, such as his interest in religion, in the Hindu religion in particular, which resulted in a feeling for the mystery of the Universe that surrounded him almost like a fog. He saw physics clearly, looking toward what had already been done, but at the border he tended to feel that there was much more of the mysterious and novel than there actually was. He was insufficiently confident of the power of the intellectual tools he already possessed and did not drive his thought to the very end because he felt instinctively that new ideas and new methods were necessary to go further than he and his students had already gone."

6

Implosion to What?

*in which all the armaments
of theoretical physics
cannot ward off the conclusion:
implosion produces black holes*

The confrontation was inevitable. These two intellectual giants, J. Robert Oppenheimer and John Archibald Wheeler, had such different views of the Universe and of the human condition that time after time they found themselves on opposite sides of deep issues: national security, nuclear weapons policy—and now black holes.

The scene was a lecture hall at the University of Brussels in Belgium. Oppenheimer and Wheeler, neighbors in Princeton, New Jersey, had journeyed there along with thirty-one other leading physicists and astronomers from around the world for a full week of discussions on the structure and evolution of the Universe.

It was Tuesday, 10 June 1958. Wheeler had just finished presenting, to the assembled savants, the results of his recent calculations with Kent Harrison and Masami Wakano—the calculations that had identified, unequivocally, the masses and circumferences of all possible cold, dead stars (Chapter 5). He had filled in the missing gaps in the Chandrasekhar and Oppenheimer–Volkoff calculations, and had confirmed their conclusions: Implosion is compulsory when a star more massive than about 2 Suns dies, and the implosion cannot produce a white dwarf, or a neutron star, or any other kind of cold, dead star, unless the dying star ejects enough mass to pull itself below the maximum-mass limit of about 2 Suns.

"Of all the implications of general relativity for the structure and evolution of the Universe, this question of the fate of great masses of matter is one of the most challenging," Wheeler asserted. On this his audience could agree. Wheeler then, in a near replay of Arthur Eddington's attack on Chandrasekhar twenty-four years earlier (Chapter 4), described Oppenheimer's view that massive stars must die by imploding to form black holes, and then he opposed it: Such implosion "does not give an acceptable answer," Wheeler asserted. Why not? For essentially the same reason as Eddington had rejected it; in Eddington's words, "there should be a law of Nature to prevent a star from behaving in this absurd way." But there was a deep difference between Eddington and Wheeler: Whereas Eddington's 1935 speculative mechanism to save the Universe from black holes was immediately branded as wrong by such experts as Niels Bohr, Wheeler's 1958 speculative mechanism could not at the time be proved or disproved—and fifteen years later it would turn out to be partially right (Chapter 12).

Wheeler's speculation was this. Since (in his view) implosion to a black hole must be rejected as physically implausible, "there seems no escape from the conclusion that the nucleons [neutrons and protons] at the center of an imploding star must necessarily dissolve away into radiation, and that this radiation must escape from the star fast enough to reduce its mass [below about 2 Suns]" and permit it to wind up in the neutron-star graveyard. Wheeler readily acknowledged that such a conversion of nucleons into escaping radiation was outside the bounds of the known laws of physics. However, such conversion might result from the as yet ill-understood "marriage" of the laws of general relativity with the laws of quantum mechanics (Chapters 12–14). This, to Wheeler, was the most enticing aspect of "the problem of great masses": The absurdity of implosion to form a black hole forced him to contemplate an entirely new physical process. (See Figure 6.1.)

Oppenheimer was not impressed. When Wheeler finished speaking, he was the first to take the floor. Maintaining a politeness that he had not displayed as a younger man, he affirmed his own view: "I do not know whether non-rotating masses much heavier than the sun really occur in the course of stellar evolution; but if they do, I believe their implosion can be described in the framework of general relativity [without asserting new laws of physics]. Would not the simplest as-

OPPENHEIMER-SNYDER VIEW:

Massive star
exhausts nuclear fuel

Star implodes

Star cuts itself
off from rest of the
Universe.

WHEELER'S 1958 VIEW:

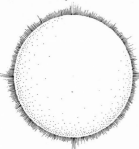

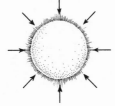

Massive star
exhausts nuclear fuel

Star's implosion squeezes its
center to extreme density. Quantum
laws plus spacetime curvature convert
nucleons at center into radiation. Radiation
flies away, reducing star's mass and carrying off
information about marriage of quantum physics with
general relativity.

Star, with reduced
mass, settles down
into a pressure/
gravity equi-
librium (neutron
star).

6.1 Contrast of Oppenheimer's view of the fates of large masses *(upper sequence)* with Wheeler's 1958 view *(lower sequence).*

sumption be that such masses undergo continued gravitational contraction and ultimately cut themselves off more and more from the rest of the Universe [that is, form black holes]?" (See Figure 6.1.)

Wheeler was equally polite, but held his ground. "It is very difficult to believe 'gravitational cutoff' is a satisfactory answer," he asserted.

Oppenheimer's confidence in black holes grew out of detailed calculations he had done nineteen years earlier:

Black-Hole Birth:
A First Glimpse

In the winter of 1938–39, having just completed his computation with George Volkoff of the masses and circumferences of neutron stars (Chapter 5), Oppenheimer was firmly convinced that massive stars, when they die, must implode. The next challenge was obvious: use the laws of physics to compute the details of the implosion. What would the implosion look like as seen by people in orbit around the star? What would it look like as seen by people riding on the star's surface? What would be the final state of the imploded star, thousands of years after the implosion?

This computation would not be easy. Its mathematical manipulations would be the most challenging that Oppenheimer and his students had yet tackled: The imploding star would change its properties rapidly as time passes, whereas the Oppenheimer–Volkoff neutron stars had been static, unchanging. Spacetime curvature would become enormous inside the imploding star, whereas it had been much more modest in neutron stars. To deal with these complexities would require a very special student. The choice was obvious: Hartland Snyder.

Snyder was different from Oppenheimer's other students. The others came from middle-class families; Snyder was working class. Berkeley rumor had it that he was a truck driver in Utah before turning physicist. As Robert Serber recalls, "Hartland pooh-poohed a lot of things that were standard for Oppie's students, like appreciating Bach and Mozart and going to string quartets and liking fine food and liberal politics."

The Caltech nuclear physicists were a more rowdy bunch than Oppenheimer's entourage; on Oppenheimer's annual spring trek to Pasadena, Hartland fit right in. Says Caltech's William Fowler, "Oppie was extremely cultured; knew literature, art, music, Sanskrit. But Hartland—he was like the rest of us bums. He loved the Kellogg Lab parties, where Tommy Lauritsen played the piano and Charlie Lauritsen [leader of the lab] played the fiddle and we sang college songs and drinking songs. Of all of Oppie's students, Hartland was the most independent."

Hartland was also different mentally. "Hartland had more talent for difficult mathematics than the rest of us," recalls Serber. "He was very good at improving the cruder calculations that the rest of us did." It was this talent that made him a natural for the implosion calculation.

Before embarking on the full, complicated calculation, Oppenheimer insisted (as always) on making a first, quick survey of the problem. How much could be learned with only a little effort? The key to this first survey was Schwarzschild's geometry for the curved spacetime outside a star (Chapter 3).

Schwarzschild had discovered his spacetime geometry as a solution to Einstein's general relativistic field equation. It was the solution for the exterior of a static star, one that neither implodes nor explodes nor pulsates. However, in 1923 George Birkhoff, a mathematician at Harvard, had proved a remarkable mathematical theorem: Schwarzschild's geometry describes the exterior of any star that is spherical, including not only static stars but also imploding, exploding, and pulsating ones.

For their quick calculation, then, Oppenheimer and Snyder simply assumed that a spherical star, upon exhausting its nuclear fuel, would implode indefinitely, and without probing what happens inside the star, they computed what the imploding star would look like to somebody far away. With ease they inferred that, since the spacetime geometry outside the imploding star is the same as outside any static star, the imploding star would look very much like a sequence of static stars, each one more compact than the previous one.

Now, the external appearance of such static stars had been studied two decades earlier, around 1920. Figure 6.2 reproduces the *embedding diagrams* that we used in Chapter 3 to discuss that appearance. Recall

6.2 (Same as Figure 3.4.) General relativity's predictions for the curvature of space and the redshift of light from a sequence of three highly compact, static (non-imploding) stars that all have the same mass but have different circumferences.

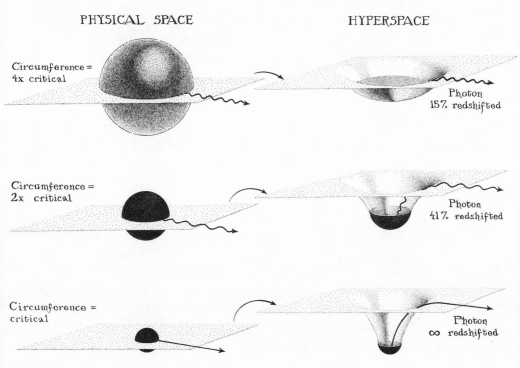

that each embedding diagram depicts the curvature of space inside and near a star. To make the depiction comprehensible, the diagram displays the curvature of only two of the three dimensions of space: the two dimensions on a sheet that lies precisely in the star's equatorial "plane" (left half of the figure). The curvature of space on this sheet is visualized by imagining that we pull the sheet out of the star and out of the physical space in which we and the star live, and move it into a flat (uncurved), fictitious *hyperspace*. In the uncurved hyperspace, the sheet can maintain its curved geometry only by bending downward like a bowl (right half of the figure).

The figure shows a sequence of three static stars that mimic the implosion that Oppenheimer and Snyder were preparing to analyze. Each star has the same mass, but they have different circumferences. The first is four times bigger around than the *critical circumference* (four times bigger than the circumference at which the star's gravity would become so strong that it forms a black hole). The second is twice the critical circumference, and the third is precisely at the critical circumference. The embedding diagrams show that the closer the star is to its critical circumference, the more extreme is the curvature of space around the star. However, the curvature does not become infinitely extreme. The bowl-like geometry is smooth everywhere with no sharp cusps or points or creases, even when the star is at its critical circumference; that is, the spacetime curvature is not infinite, and, correspondingly, since *tidal gravitational forces* (the kinds of forces that stretch one from head to foot and produce the tides on the Earth) are the physical manifestation of spacetime curvature, tidal gravity is not infinite at the critical circumference.

In Chapter 3 we also discussed the fate of light emitted from the surfaces of static stars. We learned that because time flows more slowly at the stellar surface than far away *(gravitational time dilation)*, light waves emitted from the star's surface and received far away will have a lengthened period of oscillation and correspondingly a lengthened wavelength and a redder color. The light's wavelength gets shifted toward the red end of the spectrum as the light climbs out of the star's intense gravitational field *(gravitational redshift)*. When the static star is four times larger than its critical circumference, the light's wavelength is lengthened by 15 percent (see the photon of light in the upper right part of the figure); when the star is at twice its critical circumference, the redshift is 41 percent (middle right); and when the star is precisely at its critical circumference, the light's wavelength is infinitely redshifted, which means that the light has no energy left at all and therefore has ceased to exist.

Oppenheimer and Snyder, in their quick calculation, inferred two things from this sequence of static stars: First, an imploding star, like these static stars, would probably develop strong spacetime curvature as it nears its critical circumference, but not infinite curvature and therefore not infinite tidal gravitational forces. Second, as the star implodes, light from its surface should get more and more redshifted, and when it reaches the critical circumference, the redshift should become infinite, making the star become completely invisible. In Oppenheimer's words, the star should "cut itself off" visually from our external Universe.

Was there any way, Oppenheimer and Snyder asked themselves, that the star's internal properties—ignored in this quick calculation— could save the star from this cutoff fate? For example, might the implosion be forced to go so slowly that never, even after an infinite time, would the critical circumference actually be reached?

Oppenheimer and Snyder would have liked to answer these questions by calculating the details of a realistic stellar implosion, as depicted in the left half of Figure 6.3. Any real star will spin, as does the Earth, at least a little bit. Centrifugal forces due to that spin will force the star's equator to bulge out at least a little bit, as does Earth's equator. Thus, the star cannot be precisely spherical. As it implodes, the star must spin faster and faster like a figure skater pulling in his arms; and its faster spin will cause centrifugal forces inside the star to grow, making the equatorial bulge more pronounced—sufficiently pronounced, perhaps, that it even halts the implosion, with the outward centrifugal forces then fully balancing gravity's pull. Any real star has high density and pressure in its center, and lower density and pressure in its outer layers; as it implodes, high-density lumps will develop here and there like blueberries in a blueberry muffin. Moreover, the star's gaseous matter, as it implodes, will form shock waves—analogues of breaking ocean waves—and these shocks may eject matter and mass from some parts of the star's surface just as an ocean wave can eject droplets of water into the air. Finally, radiation (electromagnetic waves, gravitational waves, neutrinos) will pour out of the star, carrying away mass.

All these effects Oppenheimer and Snyder would have liked to include in their calculations, but to do so was a formidable task, far beyond the capabilities of any physicist or computing machine in 1939. It would not become feasible until the advent of supercomputers in the

1980s. Thus, to make any progress at all, it was necessary to build an idealized model of the imploding star and then compute the predictions of the laws of physics for that model.

Such idealizations were Oppenheimer's forte: When confronted with a horrendously complex situation such as this one, he could discern almost unerringly which phenomena were of crucial importance and which were peripheral.

For the imploding star, one feature was crucial above all others, Oppenheimer believed: gravity as described by Einstein's general relativistic laws. It, and only it, must not be compromised when formulating a calculation that could be done. By contrast, the star's spin and its nonspherical shape could be ignored; they might be crucially important for *some* imploding stars, but for stars that spin slowly, they probably would have no strong effect. Oppenheimer could not really prove this mathematically, but intuitively it seemed clear, and indeed it has turned out to be true. Similarly, his intuition said, the outpouring of radiation was an unimportant detail, as were shock waves and density lumps. Moreover, since (as Oppenheimer and Volkoff had shown) gravity could overwhelm all pressure in massive, dead stars, it seemed safe to pretend (incorrectly, of course) that the imploding star has no internal pressure whatsoever—neither thermal pressure, nor pressure arising from the electrons' or neutrons' claustrophobic degeneracy motions, nor pressure arising from the nuclear force. A real star, with its real pressure, might implode in a different manner from an idealized, pressureless star; but the differences of implosion should be only modest, not great, Oppenheimer's intuition insisted.

Thus it was that Oppenheimer suggested to Snyder an idealized computational problem: Study, using the precise laws of general relativity, the implosion of a star that is idealized as precisely spherical, nonspinning, and nonradiating, a star with uniform density (the same near its surface as at its center) and with no internal pressure whatsoever; see Figure 6.3.

Even with all these idealizations—idealizations that would generate skepticism in other physicists for thirty years to come—the calculation was exceedingly difficult. Fortunately, Richard Tolman was available in Pasadena for help. Leaning heavily on Tolman and Oppenheimer for advice, Snyder worked out the equations governing the entire implosion—and in a tour de force, he managed to solve them. He now had the full details of the implosion, expressed in formulas! By scrutinizing those formulas, first from one direction and then another, physi-

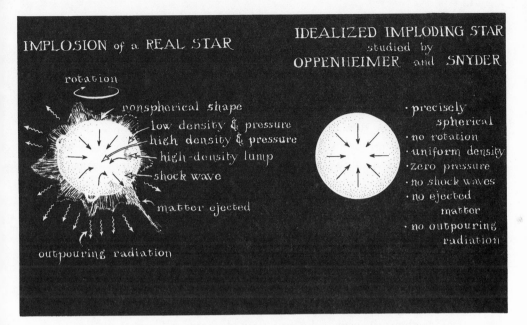

6.3 *Left:* Physical phenomena in a realistic, imploding star. *Right:* The idealizations that Oppenheimer and Snyder made in order to compute stellar implosion.

cists could read off whatever aspect of the implosion they wished—how it looks from outside the star, how it looks from inside, how it looks on the star's surface, and so forth.

Especially intriguing is the appearance of the imploding star as observed from a *static, external reference frame,* that is, as seen by observers outside the star who remain always at the same fixed circumference instead of riding inward with the star's imploding matter. The star, as seen in a static, external frame, begins its implosion in just the way one would expect. Like a rock dropped from a rooftop, the star's surface falls downward (shrinks inward) slowly at first, then more and more rapidly. Had Newton's laws of gravity been correct, this acceleration of the implosion would continue inexorably until the star, lacking any internal pressure, is crushed to a point at high speed. Not so according to Oppenheimer and Snyder's relativistic formulas. Instead, as the star nears its critical circumference, its shrinkage slows to a crawl. The smaller the star gets, the more slowly it implodes, until it becomes *frozen* precisely at the critical circumference. No matter how long a time one waits, if one is at rest outside the star (that is, at rest in the static, external reference frame), one will never be able to see the star implode through the critical circumference. That is the unequivocal message of Oppenheimer and Snyder's formulas.

Is this freezing of the implosion caused by some unexpected, general relativistic force inside the star? No, not at all, Oppenheimer and Snyder realized. Rather, it is caused by gravitational time dilation (the slowing of the flow of time) near the critical circumference. Time on the imploding star's surface, as seen by static external observers, must flow more and more slowly when the star approaches the critical circumference, and correspondingly everything occurring on or inside the star including its implosion must appear to go into slow motion and then gradually freeze.

As peculiar as this might seem, even more peculiar was another prediction made by Oppenheimer and Snyder's formulas: Although, as seen by static external observers, the implosion freezes at the critical circumference, it *does not freeze at all* as viewed by observers riding inward on the star's surface. If the star weighs a few solar masses and begins about the size of the Sun, then as observed from its own surface, it implodes to the critical circumference in about an hour's time, and then keeps right on imploding past criticality and on in to smaller circumferences.

By 1939, when Oppenheimer and Snyder discovered these things, physicists had become accustomed to the fact that time is relative; the flow of time is different as measured in different reference frames that move in different ways through the Universe. But never before had anyone encountered such an extreme difference between reference frames. That *the implosion freezes forever as measured in the static, external frame but continues rapidly on past the freezing point as measured in the frame of the star's surface* was extremely hard to comprehend. Nobody who studied Oppenheimer and Snyder's mathematics felt comfortable with such an extreme warpage of time. Yet there it was in their formulas. One might wave one's arms with heuristic explanations, but no explanation seemed very satisfying. It would not be fully understood until the late 1950s (near the end of this chapter).

By looking at Oppenheimer and Snyder's formulas from the viewpoint of an observer on the star's surface, one can deduce the details of the implosion even after the star sinks within its critical circumference; that is, one can discover that the star gets crunched to infinite density and zero volume, and one can deduce the details of the spacetime curvature at the crunch. However, in their article describing their calculations, Oppenheimer and Snyder avoided any discussion of the crunch whatsoever. Presumably Oppenheimer was prevented from discussing it by his own innate scientific conservatism, his unwillingness to speculate (see the last two paragraphs of Chapter 5).

If reading the star's final crunch off their formulas was too much for Oppenheimer and Snyder to face, even the details outside and at the critical circumference were too bizarre for most physicists in 1939. At Caltech, for example, Tolman was a believer; after all, the predictions were unequivocal consequences of general relativity. But nobody else at Caltech was very convinced. General relativity had been tested experimentally only in the solar system, where gravity is so weak that Newton's laws give almost the same predictions as general relativity. By contrast, the bizarre Oppenheimer–Snyder predictions relied on ultra-strong gravity. General relativity might well fail before gravity ever became so strong, most physicists thought; and even if it did not fail, Oppenheimer and Snyder might be misinterpreting what their mathematics was trying to say; and even if they were not misinterpreting their mathematics, their calculation was so idealized, so devoid of spin, lumps, shocks, and radiation, that it should not be taken seriously.

Such skepticism held sway throughout the United States and Western Europe, but not in the U.S.S.R. There Lev Landau, still recuperating from his year in prison, kept a "Golden List" of the most important physics research articles published anywhere in the world. Upon reading the Oppenheimer–Snyder paper, Landau entered it in his List, and he proclaimed to his friends and associates that these latest Oppenheimer revelations had to be right, even though they were extremely difficult for the human mind to comprehend. So great was Landau's influence that his view took hold among leading Soviet theoretical physicists from that day forward.

Nuclear Interlude

W̲ere Oppenheimer and Snyder right, or were they wrong? The answer would likely have been learned definitively during the 1940s had World War II and then crash programs to develop the hydrogen bomb not intervened. But the war and the bomb did intervene, and research on impractical, esoteric issues like black holes became frozen in time as physicists turned their full energies to weapons design.

Only in the late 1950s did the weapons efforts wind down enough to bring stellar implosion back into physicists' consciousness. Only then did the skeptics launch their first serious attack on the Oppenheimer–Snyder predictions. Carrying the banner of the skeptics at first, but not for long, was John Archibald Wheeler. From the outset, a leader of the believers was Wheeler's Soviet counterpart, Yakov Borisovich Zel'-dovich.

The characters of Wheeler and Zel'dovich were shaped in the fire of nuclear weapons projects during the nearly two decades that black-hole research was frozen in time, the decades of the 1940s and 1950s. From their weapons work, Wheeler and Zel'dovich emerged with crucial tools for analyzing black holes: powerful computational techniques, a deep understanding of the laws of physics, and interactive research styles in which they would continually stimulate younger colleagues. They also emerged carrying difficult baggage—a set of complex rela-tionships with some of their key colleagues: Wheeler with Oppen-heimer; Zel'dovich with Landau and with Andrei Sakharov.

John Wheeler, fresh out of graduate school in 1933, and the winner of a Rockefeller-financed National Research Council postdoctoral fellow-ship, had a choice of where and with whom to do his postdoctoral study. He could have chosen Berkeley and Oppenheimer, as did most NRC theoretical physics postdocs in those days; instead he chose New York University and Gregory Breit. "In personality they [Oppen-heimer and Breit] were utterly different," Wheeler says. "Oppen-heimer saw things in black and white and was a quick decider. Breit worked in shades of grey. Attracted to issues that require long reflec-tion, I chose Breit."

From New York University in 1933, Wheeler moved on to Copenha-gen to study with Niels Bohr, then to an assistant professorship at the University of North Carolina, followed by one at Princeton University, in New Jersey. In 1939, while Oppenheimer and students in California were probing neutron stars and black holes, Wheeler and Bohr at Princeton (where Bohr was visiting) were developing the theory of *nuclear fission:* the breakup of heavy atomic nuclei such as uranium into smaller pieces, when the nuclei are bombarded by neutrons (Box 6.1). Fission had just been discovered quite unexpectedly by Otto Hahn and Fritz Strassman in Germany, and its implications were ominous: By a chain reaction of fissions a weapon of unprecedented power might be made. But Bohr and Wheeler did not concern themselves with chain reactions or weapons; they just wanted to understand how fission comes

Box 6.1
Fusion, Fission, and Chain Reactions

The *fusion* of very light nuclei to form medium-sized nuclei releases huge amounts of energy. A simple example from Box 5.3 is the fusion of a deuterium nucleus ("heavy hydrogen," with one proton and one neutron) and an ordinary hydrogen nucleus (a single proton) to form a helium-3 nucleus (two protons and one neutron):

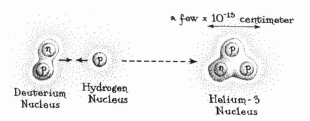

Such fusion reactions keep the Sun hot and power the hydrogen bomb (the "superbomb" as it was called in the 1940s and 1950s).

The *fission* (splitting apart) of a very heavy nucleus to form two medium-sized nuclei releases a large amount of energy—far more than comes from chemical reactions (since the nuclear force which governs nuclei is far stronger than the electromagnetic force which governs chemically reacting atoms), but much less energy than comes from the fusion of light nuclei. A few very heavy nuclei undergo fission naturally, without any outside help. More interesting for this chapter are fission reactions in which a neutron hits a very heavy nucleus such as uranium-235 (a uranium nucleus with 235 protons and neutrons) and splits it roughly in half:

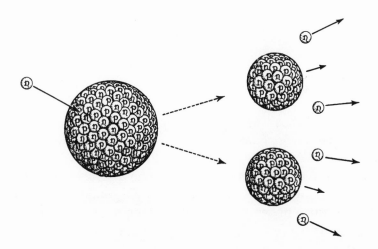

(continued next page)

(Box 6.1 continued)

There are two special, heavy nuclei, uranium-235 and plutonium-239, with the property that their fission produces not only two medium-sized nuclei, but also a handful of neutrons (as in the drawing above). These neutrons make possible a *chain reaction:* If one concentrates enough uranium-235 or plutonium-239 into a small enough package, then the neutrons released from one fission will hit other uranium or plutonium nuclei and fission them, producing more neutrons that fission more nuclei, producing still more neutrons that fission still more nuclei, and so on. The result of this chain reaction, if uncontrolled, is a huge explosion (an atomic bomb blast); if controlled in a reactor, the result can be highly efficient electric power.

about. What is the underlying mechanism? How do the laws of physics produce it?

Bohr and Wheeler were remarkably successful. They discovered how the laws of physics produce fission, and they predicted which nuclei would be the most effective at sustaining chain reactions: uranium-235 (which would become the fuel for the bomb to destroy Hiroshima) and plutonium-239 (a type of nucleus that does not exist in nature but that the American physicists would soon learn how to make in nuclear reactors and would use to fuel the bomb to destroy Nagasaki). However, Bohr and Wheeler were not thinking of bombs in 1939; they only wanted to understand.

The Bohr–Wheeler article explaining nuclear fission was published in the same issue of the *Physical Review* as the Oppenheimer–Snyder article describing the implosion of a star. The publication date was 1 September 1939, the very day that Hitler's troops invaded Poland, triggering World War II.

Yakov Borisovich Zel'dovich was born into a Jewish family in Minsk in 1914; later that year his family moved to Saint Petersburg (renamed Leningrad in the 1920s, then restored to Saint Petersburg in the 1990s). Zel'dovich completed high school at age fifteen and then, instead of entering university, went to work as a laboratory assistant at the Physicotechnical Institute in Leningrad. There he taught himself so much physics and chemistry and did such impressive research that, without any formal university training, he was awarded a Ph.D. in 1934, at age twenty.

In 1939, while Wheeler and Bohr were developing the theory of nuclear fission, Zel'dovich and a close friend, Yuli Borisovich Khariton, were developing the theory of chain reactions produced by nuclear fission: Their research was triggered by an intriguing (incorrect) suggestion from French physicist Francis Perrin that volcanic eruptions might be powered by natural, underground nuclear explosions, which result from a chain reaction of fissions of atomic nuclei. However, nobody including Perrin had worked out the details of such a chain reaction. Zel'dovich and Khariton—already among the world's best experts on chemical explosions—leaped on the problem. Within a few months they had shown (as, in parallel, did others in the West) that such an explosion cannot occur in nature, because naturally occurring uranium consists mostly of uranium-238 and not enough uranium-235. However, they concluded, if one were to artificially separate out uranium-235 and concentrate it, then one could make a chain-reaction explosion. (The Americans would soon embark on such separation to make the fuel for their Hiroshima bomb.) The curtain of secrecy had not yet descended around nuclear research, so Zel'dovich and Khariton published their calculations in the most prestigious of Soviet physics journals, the *Journal of Experimental and Theoretical Physics,* for all the world to see.

During the six years of World War II, physicists of the warring nations developed sonar, mine sweepers, rockets, radar, and, most fatefully, the atomic bomb. Oppenheimer led the "Manhattan Project" at Los Alamos, New Mexico, to design and build the American bombs. Wheeler was the lead scientist in the design and construction of the world's first production-scale nuclear reactors, in Hanford, Washington, which made the plutonium-239 for the Nagasaki bomb.

After the bombs' decimation of Hiroshima and Nagasaki and the deaths of several hundred thousand people, Oppenheimer was in anguish: "If atomic bombs are to be added to the arsenals of a warring world, or to the arsenals of nations preparing for war, then the time will come when mankind will curse the name of Los Alamos and Hiroshima." "In some sort of crude sense which no vulgarity, no humor, no overstatement can quite extinguish, the physicists have known sin; and this is a knowledge which they cannot lose."

But Wheeler had the opposite kind of regret: "As I look back on [1939 and my fission theory work with Bohr], I feel a great sadness. How did it come about that I looked on fission first as a physicist [simply curious to know how fission works], and only secondarily as a citizen [intent on defending my country]? Why did I not look at it first as a citizen and only secondarily as a physicist? A simple survey of the records shows that between twenty and twenty-five million people perished in World War II and more of them in the later years than in the earlier years. Every month by which the war was shortened would have meant a saving of the order of half a million to a million lives. Among those granted life would have been my brother Joe, killed in October 1944 in the Battle for Italy. What a difference it would have made if the critical date [of the atomic bomb's first use in the war] had been not August 6, 1945, but August 6, 1943."

In the U.S.S.R., physicists abandoned all nuclear research in June 1941, when Germany attacked Russia, since other physics would produce quicker payoffs for national defense. As the German army marched on and surrounded Leningrad, Zel'dovich and his friend Khariton were evacuated to Kazan, where they worked intensely on the theory of the explosion of ordinary types of bombs, trying to improve the bombs' explosive power. Then, in 1943, they were summoned to Moscow. It had become clear, they were told, that both the Americans and the Germans were mounting efforts to construct an atomic bomb. They were to be part of a small, elite, Soviet bomb development effort under the leadership of Igor V. Kurchatov.

By two years later, when the Americans bombed Hiroshima and Nagasaki, Kurchatov's team had developed a thorough theoretical understanding of nuclear reactors for making plutonium-239, and had developed several possible bomb designs—and Khariton and Zel'-dovich had become the lead theorists on the project.

When Stalin learned of the American atomic bomb explosions, he angrily berated Kurchatov for the Soviet team's slowness. Kurchatov defended his team: Amidst the war's devastation, and with its limited resources, the team could not move more rapidly. Stalin told him angrily that if a child doesn't cry, its mother can't know what it needs. Ask for anything you need, he commanded, nothing will be refused; and he then demanded a no-holds-barred, crash project to construct the bomb, a project under the ultimate authority of Lavrenty Pavlovich Beria, the fearsome head of the secret police.

The magnitude of the effort that Beria mounted is hard to imagine. He commandeered the forced labor of millions of Soviet citizens from Stalin's prison camps. These *zeks*, as they were colloquially called, constructed uranium mines, uranium purification factories, nuclear reactors, theoretical research centers, weapons test centers, and self-contained, small cities to support these facilities. The facilities, scattered across the face of the nation, were surrounded by levels of security unheard of in the Americans' Manhattan Project. Zel'dovich and Khariton were moved to one of these facilities, in "a far away place" whose location, though almost certainly well known to Western authorities by the late 1950s, was forbidden to be revealed by Soviet citizens until 1990.[1] The facility was known simply as *Obyekt* ("the Installation"); Khariton became its director, and Zel'dovich the leader of one of its key bomb design teams. Under Beria's authority, Kurchatov set up several teams of physicists to pursue, in parallel and completely independently, each aspect of the bomb project; redundancy brings security. The teams at the Installation fed design problems to the other teams, including a small one led by Lev Landau at the Institute of Physical Problems in Moscow.

While this massive effort was rolling inexorably forward, Soviet spies were acquiring, through Klaus Fuchs (a British physicist who had worked on the American bomb project), the design of the Americans' plutonium-based bomb. It differed somewhat from the design that Zel'dovich and his colleagues had produced, so Kurchatov, Khariton, and company faced a tough decision: They were under excruciating pressure from Stalin and Beria for results, and they feared the consequences of an unsuccessful bomb test in an era when failure often meant execution; they knew that the American design had worked at Alamogordo and Nagasaki, but they could not be completely sure of their own design; and they possessed enough plutonium for only one bomb. The decision was clear but painful: They put their own design on hold[2] and converted their crash program over to the American design.

At last, on 29 August 1949—after four years of crash effort, untold misery, untold deaths of slave-labor zeks, and the beginning of an accumulation of waste from nuclear reactors near Cheliabinsk that would explode ten years later, contaminating hundreds of square miles of countryside—the crash program reached fruition. The first Soviet atomic bomb was exploded near Semipalatinsk in Soviet Asia, in a test

1. It is near the town of Arzimas, between Cheliabinsk and the Ural Mountains.
2. After their successful test of a bomb based on the American design, the Soviets returned to their own design, constructed a bomb based on it, and tested it successfully in 1951.

witnessed by the Supreme Command of the Soviet army and government leaders.

On 3 September 1949 an American WB-29 weather reconnaissance plane, on a routine flight from Japan to Alaska, discovered products of nuclear fission from the Soviet test. The data were given to a committee of experts, including Oppenheimer, for evaluation. The verdict was unequivocal. The Russians had tested an atomic bomb!

Amidst the panic that ensued (backyard bomb shelters; atomic bomb drills for schoolchildren; McCarthy's "witch hunts" to root out spies, Communists, and their fellow travelers from government, army, media, and universities), a profound debate occurred amongst physicists and politicians. Edward Teller, one of the most innovative of the American atomic bomb design physicists, advocated a crash program to design and build the "superbomb" (or "hydrogen bomb")—a weapon based on the fusion of hydrogen nuclei to form helium. The hydrogen bomb, if it could be built, would be awesome. There seemed no limit to its power. Did one want a bomb ten times more powerful than Hiroshima? a hundred times more powerful? a thousand? a million? If the bomb could be made to work at all, it could be made as powerful as one wished.

John Wheeler backed Teller: A crash program for the "super" was essential to counter the Soviet threat, he believed. Robert Oppenheimer and his General Advisory Committee to the U.S. Atomic Energy Commission were opposed. It was not at all obvious whether a superbomb as then conceived could ever be made to work, Oppenheimer and his committee argued. Moreover, even if it did work, any super that was vastly more powerful than an ordinary atomic bomb would likely be too heavy for delivery by airplane or rocket. And then there were the moral issues, which Oppenheimer and his committee addressed as follows. "We base our recommendations [against a crash program] on our belief that the extreme dangers to mankind inherent in the proposal wholly outweigh any military advantage that could come from this development. Let it be clearly realized that this is a super weapon; it is in a totally different category from an atomic bomb. The reason for developing such super bombs would be to have the capacity to devastate a vast area with a single bomb. Its use would involve a decision to slaughter a vast number of civilians. We are alarmed as to the possible global effects of the radioactivity generated by the explosion of a few super bombs of conceivable magnitude. If super bombs will work at all, there is no inherent limit in the destructive power that may be attained with them. Therefore, a super bomb might become a weapon of genocide."

To Edward Teller and John Wheeler these arguments made no sense at all. The Russians surely would push forward with the hydrogen bomb; if America did not push forward as well, the free world could be put in enormous danger, they believed.

The Teller–Wheeler view prevailed. On 10 March 1950, President Truman ordered a crash program to develop the super.

The Americans' 1949 design for the super appears in retrospect to have been a prescription for failure, just as Oppenheimer's committee had suspected. However, since it was not certain to fail, and since nothing better was known, it was pursued intensely until March 1951, when Teller and Stanislaw Ulam invented a radically new design, one that showed bright promise.

The Teller–Ulam invention at first was just an idea for a design. As Hans Bethe has said, "Nine out of ten of Teller's ideas are useless. He needs men with more judgement, even if they be less gifted, to select the tenth idea, which often is a stroke of genius." To test whether this idea was a stroke of genius or a deceptive dud required turning it into a concrete and detailed bomb design, then carrying out extensive computations on the biggest available computers to see whether the design might work, and then, if the calculations predicted success, constructing and testing an actual bomb.

Two teams were set up to carry out the calculations: One at Los Alamos, the other at Princeton University. John Wheeler led the Princeton team. Wheeler's team worked night and day for several months to develop a full bomb design based on the Teller–Ulam idea, and to test by computer calculations whether it would work. As Wheeler recalls, "We did an immense amount of calculation. We were using the computer facilities of New York, Philadelphia, and Washington—in fact, a very large fraction of the computer capacity of the United States. Larry Wilets, John Toll, Ken Ford, Louis Henyey, Carl Hausman, Dick l'Olivier, and others worked three six-hour stretches each day to get things out."

When the calculations made it clear that the Teller–Ulam idea probably would work, a meeting was called, at the Institute for Advanced Study in Princeton (where Oppenheimer was the director), to present the idea to Oppenheimer's General Advisory Committee and its parent U.S. Atomic Energy Commission. Teller described the idea, and then Wheeler described his team's specific design and its predicted explosion. Wheeler recalls, "While I was starting to give my talk, Ken Ford rushed up to the window from outside, lifted it up, and passed in this big chart. I unrolled it and put it on the wall; it showed the progress of the thermonuclear combustion [as we had computed it.] . . . The Committee had no option but to conclude that this thing made sense. . . . Our calculation turned Oppie around on the project."

A portion of John Wheeler's hydrogen bomb design team at Princeton University in 1952. *Front row, left to right:* Margaret Fellows, Margaret Murray, Dorothea Ruffel, Audrey Ojala, Christene Shack, Roberta Casey. *Second row:* Walter Aron, William Clendenin, Solomon Bochner, John Toll, John Wheeler, Kenneth Ford. *Third and fourth rows:* David Layzer, Lawrence Wilets, David Carter, Edward Frieman, Jay Berger, John McIntosh, Ralph Pennington, unidentified, Robert Goerss. [Photo by Howard Schrader; courtesy Lawrence Wilets and John A Wheeler.]

Oppenheimer has described his own reaction: "The program we had in 1949 [the 'prescription for failure'] was a tortured thing that you could well argue did not make a great deal of technical sense. It was therefore possible to argue also that you did not want it even if you could have it. The program in 1952 [the new design based on the Teller–Ulam idea] was technically so sweet that you could not argue about that. The issues became purely the military, the political and the humane problems of what you were going to do about it once you had it." Suppressing his deep misgivings about the ethical issues, Oppenheimer, together with the other members of his committee, closed ranks with Teller, Wheeler, and the super's proponents, and the project moved forward at an accelerated pace to construct and test the bomb. It worked as predicted by Wheeler's team and by parallel calculations at Los Alamos.

Wheeler's team's extensive design calculations were ultimately written up as the secret *Project Matterhorn Division B Report 31* or PMB-31. "I'm told," says Wheeler, "that for at least ten years PMB-31 was the bible for design of thermonuclear devices" (hydrogen bombs).

In 1949–50, while America was in a state of panic, and Oppenheimer, Teller, and others were debating whether America should mount a crash program to develop the super, the Soviet Union was already in the midst of a crash superbomb project of its own.

In spring 1948, fifteen months before the first Soviet atomic bomb test, Zel'dovich and his team at the Installation had carried out theoretical calculations on a superbomb design similar to the Americans' "prescription for failure."[3] In June 1948, a second superbomb team was established in Moscow under the leadership of Igor Tamm, one of the most eminent of Soviet theoretical physicists. Its members were Vitaly Ginzburg (of whom we shall hear much in Chapters 8 and 10), Andrei Sakharov (who would become a dissident in the 1970s, and then a hero and Soviet saint in the late 1980s and 1990s), Semyon Belen'ky, and Yuri Romanov. Tamm's team was charged with the task of checking and refining the Zel'dovich team's design calculations.

3. Sakharov has speculated that this design was directly inspired by information acquired from the Americans through espionage, perhaps via the spy Klaus Fuchs. Zel'dovich by contrast has asserted that neither Fuchs nor any other spy produced any significant information about the superbomb that his design team did not already know; the principal value of the Soviet superbomb espionage was to convince Soviet political authorities that their physicists knew what they were doing.

The Tamm team's attitude toward this task is epitomized by a statement of Belen'ky's at the time: "Our job is to lick Zel'dovich's anus." Zel'dovich, with his paradoxical combination of a forceful, demanding personality and extreme political timidity, was *not* among the most popular of Soviet physicists. But he *was* among the most brilliant. Landau, who as a leader of a small subsidiary design team occasionally received orders from Zel'dovich's team to analyze this, that, or another facet of the bomb design, sometimes referred to him behind his back as "that bitch, Zel'dovich." Zel'dovich, by contrast, revered Landau as a great judge of the correctness of physics ideas, and as his greatest teacher—though Zel'dovich had never taken a formal course from him.

It required only a few months for Sakharov and Ginzburg, in Tamm's team, to come up with a far better design for a superbomb than the "prescription for failure" that Zel'dovich and the Americans were pursuing. Sakharov proposed constructing the bomb as a *layered cake* of alternating shells of a heavy fission fuel (uranium) and light fusion fuel, and Ginzburg proposed for the fusion fuel lithium deuteride (LiD). In the bomb's intense blast, the LiD's lithium nuclei would fission into two tritium nuclei, and these tritiums, together with the LiD's deuterium, would then fuse to form helium nuclei, releasing enormous amounts of energy. The heavy uranium would strengthen the explosion by preventing its energy from leaking out too quickly, by helping compress the fusion fuel, and by adding fission energy to the fusion. When Sakharov presented these ideas, Zel'dovich grasped their promise immediately. Sakharov's layered cake and Ginzburg's LiD quickly became the focus of the Soviet superbomb effort.

To push the superbomb forward more rapidly, Sakharov, Tamm, Belen'ky, and Romanov were ordered transferred from Moscow to the Installation. But not Ginzburg. The reason seems obvious: Three years earlier, Ginzburg had married Nina Ivanovna, a vivacious, brilliant woman, who in the early 1940s had been thrown into prison on a trumped-up charge of plotting to kill Stalin. She and her fellow plotters supposedly were planning to shoot Stalin from a window in the room where she lived, as he passed by on Arbat Street below. When a troika of judges met to decide her fate, it was pointed out that her room did not have any windows at all looking out on Arbat Street, so in an unusual exhibition of mercy, her life was spared; she was merely sentenced to prison and then to exile, not death. Her imprisonment and exile presumably were enough to taint Ginzburg, the inventor of the

LiD fuel for the bomb, and lock him out of the Installation. Ginzburg, preferring basic physics research over bomb design, was pleased, and the world of science reaped the rewards: While Zel'dovich, Sakharov, and Wheeler concentrated on bombs, Ginzburg solved the mystery of how cosmic rays propagate through our galaxy, and with Landau he used the laws of quantum mechanics to explain the origin of superconductivity.

In 1949, as the Soviet atomic bomb project reached fruition, Stalin ordered that the full resources of the Soviet state be switched over, without pause, to a superbomb effort. The slave labor of zeks, the theoretical research facilities, the manufacturing facilities, the test facilities, the multiple teams of physicists on each aspect of the design and construction, all must be focused on trying to beat the Americans to the hydrogen bomb. Of this the Americans, in the midst of their debate over whether to mount a crash effort on the super, knew nothing. However, the Americans had superior technology and a large head start.

On 1 November 1952, the Americans exploded a hydrogen bomb–type device code-named *Mike.* Mike was designed to test the 1951 Teller–Ulam invention and was based on the design computations of Wheeler's team and the parallel team at Los Alamos. It used liquid deuterium as its principal fuel. To liquify the deuterium and pipe it into the explosion region required an enormous, factory-like apparatus. Thus, this was not the kind of bomb that one could deliver on any airplane or rocket. Nevertheless, it totally destroyed the island of Elugelab in the Eniwetok Atoll in the Pacific Ocean; it was 800 times more powerful than the bomb that killed over 100,000 people in Hiroshima.

On 5 March 1953, amidst somber music, Radio Moscow announced that Joseph Stalin had died. There was rejoicing in America, and grief in the U.S.S.R. Andrei Sakharov wrote to his wife, Klava, "I am under the influence of a great man's death. I am thinking of his humanity."

On 12 August 1953, at Semipalatinsk, the Soviets exploded their first hydrogen bomb. Dubbed *Joe-4* by the Americans, it used Sakharov's layered-cake design and Ginzburg's LiD fusion fuel, and it was small enough to deliver in an airplane. However, the fuel in Joe-4 was *not* ignited by the Teller–Ulam method, and as a result Joe-4 was rather less powerful than the Americans' Mike: "only" about 30 Hiroshimas, compared to Mike's 800.

In fact, in the language of the American bomb design physicists,

Joe-4 was not a hydrogen bomb at all; it was a *boosted atomic bomb*, that is, an atomic bomb whose power is boosted by the inclusion of some fusion fuel. Such boosted atomic bombs were already part of the American arsenal, and the Americans refused to regard them as hydrogen bombs because their layered-cake design did not enable them to ignite an *arbitrarily large* amount of fusion fuel. There was no way by this design to make, for example, a "doomsday weapon" thousands of times more powerful than Hiroshima.

But 30 Hiroshimas was not to be sneezed at, nor was deliverability. Joe-4 was an awesome weapon indeed, and Wheeler and other Americans heaved a sigh of relief that, thanks to their own, true superbomb, the new Soviet leader, Georgi Malenkov, could not threaten America with it.

On 1 March 1954, the Americans exploded their first LiD-fueled, deliverable superbomb. It was code named *Bravo* and like Mike, it relied on design calculations by the Wheeler and Los Alamos teams and used the Teller–Ulam invention. The explosive energy was 1300 Hiroshimas.

In March 1954, Sakharov and Zel'dovich jointly invented (independently of the Americans) the Teller–Ulam idea, and within a few months Soviet resources were focused on implementing it in a real superbomb, one that could have as large a destructive power as anyone might wish. It took just eighteen months to fully design and construct the bomb. On 23 November 1955, it was detonated, with an explosive energy of 300 Hiroshimas.

As Oppenheimer's General Advisory Committee had suspected, in their opposition to the crash program for the super, these enormously powerful bombs—and the behemoth 5000-Hiroshima weapon exploded later by the Soviets in an attempt to intimidate John Kennedy—have not been very attractive to the military establishments of either the United States or the U.S.S.R. The weapons currently in Russian and American arsenals are around 30 Hiroshimas, not thousands. Although they are true hydrogen bombs, they are no more powerful than a large atomic bomb. The military neither needed nor wanted a "doomsday" device. The sole use of such a device would be psychological intimidation of the adversary—but intimidation can be a serious matter in a world with leaders like Joseph Stalin.

On 2 July 1953, Lewis Strauss, a member of the Atomic Energy Commission who had fought bitterly with Oppenheimer over the crash

Box 6.2
Why Did Soviet Physicists Build the Bomb for Stalin?

Why did Zel'dovich, Sakharov, and other great Soviet physicists work so hard to build atomic bombs and hydrogen bombs for Joseph Stalin? Stalin was responsible for the deaths of millions of Soviet citizens: 6 million or 7 million peasants and kulaks in forced collectivization in the early 1930s, 2.5 million from the top strata of the military, government, and society in the Great Terror of 1937–39, 10 million from all strata of society in the prisons and labor camps of the 1930s through 1950s. How could any physicist, in good conscience, put the *ultimate* weapon into the hands of such an *evil* man?

Those who ask such questions forget or don't know the conditions— physical and psychological—that pervaded the Soviet Union in the late 1940s and early 1950s:

1. The Soviet Union had just barely emerged from the bloodiest, most devastating war in its history—a war in which Germany, the aggressor, had killed 27 million Soviet people and had laid waste to their homeland—when Winston Churchill fired an early salvo of the cold war: In a 5 March 1946 speech in Fulton, Missouri, Churchill warned the West about a Soviet threat and coined the phrase "iron curtain" to describe the boundaries that Stalin had established around his empire. Stalin's propaganda machinery milked Churchill's speech for all it could, creating a deep fear among Soviet citizens that the British and Americans might attack. The Americans, the subsequent propaganda claimed,* were planning a nuclear war against the Soviet Union, with hundreds of atomic bombs, carried by airplanes, and targeted on hundreds of Soviet cities. Most Soviet physicists believed the propaganda and accepted the absolute necessity that the U.S.S.R. create nuclear weapons to protect against a repeat of Hitler's devastation.

2. The machinery of Stalin's state was so effective at controlling information and at brainwashing even the leading scientists that few of them understood the evil of the man. Stalin was revered by most Soviet physicists (even Sakharov), as by most Soviet citizens, as the *Great Leader*—a harsh but benevolent dictator who had masterminded the victory over Germany and would protect his people against a hostile world. The Soviet physicists were frightfully aware that evil pervaded lower levels of the government: The flimsiest of denunciations by somebody one hardly knew could send one to

---

*Beginning in 1945, American strategic planning did, indeed, include an option—if the U.S.S.R initiated a conventional war—for a massive nuclear attack on Soviet cities and on military and industrial targets; see Brown (1978).

(continued next page)

(Box 6.2 continued)

prison, and often to death. (In the late 1960s, Zel'dovich recalled for me what it was like: "Life is so wonderful now," he said; "the knocks no longer come in the middle of the night, and one's friends no longer disappear, never to be heard from again.") But the source of this evil, most physicists believed, could not be the Great Leader; it must be others below him. (Landau knew better; he had learned much in prison. But, psychologically devastated by his imprisonment, he rarely spoke of Stalin's guilt, and when he did, his friends did not believe.)

3. Though one lived a life of fear, information was so tightly controlled that one could not deduce the enormity of the toll that Stalin had taken. That toll would only become known in Gorbachev's epoch of glasnost, the late 1980s.

4. Many Soviet physicists were "fatalists." They didn't think about these issues at all. Life was so hard that one merely struggled to keep going, doing one's job as best one could, whatever it might be. Besides, the technical challenge of figuring out how to make a bomb that works was fascinating, and there was some joy to be had in the camaraderie of the design team and the prestige and substantial salary that one's work brought.

program for the super, became the Commission's chairman. As one of his first acts in power, he ordered removal of all classified material from Oppenheimer's Princeton office. Strauss and many others in Washington were deeply suspicious of Oppenheimer's loyalty. How could a man loyal to America oppose the super effort, as he had before Wheeler's team demonstrated that the Teller–Ulam invention would work? William Borden, who had been chief counsel of Congress's Joint Committee on Atomic Energy during the super debate, sent a letter to J. Edgar Hoover saying, in part: "The purpose of this letter is to state my own exhaustively considered opinion, based upon years of study of the available classified evidence, that more probably than not J. Robert Oppenheimer is an agent of the Soviet Union." Oppenheimer's security clearance was canceled, and in April and May of 1954, simultaneous with the first American tests of deliverable hydrogen bombs, the Atomic Energy Commission conducted hearings to determine whether or not Oppenheimer was really a security risk.

Wheeler was in Washington on other business at the time of the hearings. He was not involved in any way. However, Teller, a close personal friend, went to Wheeler's hotel room the night before he was to testify, and paced the floor for hours. If Teller said what he really thought, it would severely damage Oppenheimer. But how could he not say it? Wheeler had no doubts; in his view, Teller's integrity would force him to testify fully.

Wheeler was right. The next day Teller, espousing a viewpoint that Wheeler understood, said: "In a great number of cases I have seen Dr. Oppenheimer act . . . in a way which for me was exceedingly hard to understand. I thoroughly disagreed with him in numerous issues and his actions frankly appeared to me confused and complicated. To this extent I feel that I would like to see the vital interests of the country in hands which I understand better, and therefore trust more. . . . I believe, and that is merely a question of belief and there is no expertness, no real information behind it, that Dr. Oppenheimer's character is such that he would not knowingly and willingly do anything that is designed to endanger the safety of this country. To the extent, therefore, that your question is directed toward intent, I would say I do not see any reason to deny clearance. If it is a question of wisdom and judgment, as demonstrated by actions since 1945, then I would say one would be wiser not to grant clearance."

Almost all the other physicists who testified were unequivocal in their support of Oppenheimer—and were aghast at Teller's testimony. Despite this, and despite the absence of credible evidence that Oppenheimer was "an agent of the Soviet Union," the climate of the times prevailed: Oppenheimer was declared a security risk and was denied restoration of his security clearance.

To most American physicists, Oppenheimer became an instant martyr and Teller an instant villain. Teller would be ostracized by the physics community for the rest of his life. But to Wheeler, it was Teller who was the martyr: Teller had "had the courage to express his honest judgment, putting his country's security ahead of solidarity of the community of physicists," Wheeler believed. Such testimony, in Wheeler's view, "deserved consideration," not ostracism. Andrei Sakharov, thirty-five years later, came to agree.[4]

4. Just for the record, I strongly disagree with Wheeler (though he is one of my closest friends and my mentor) and with Sakharov. For thoughtful and knowledgeable insights into the Teller–Oppenheimer controversy and the pros and cons of the American debate over whether to build the superbomb, I recommend reading Bethe (1982) and York (1976). For Sakharov's view, see Sakharov (1990); for a critique of Sakharov's view, see Bethe (1990). For a transcript of the Oppenheimer hearings, see USAEC (1954).

Black-Hole Birth:
Deeper Understanding

Not only did Wheeler and Oppenheimer differ profoundly on issues of national security, they also differed profoundly in their approach to theoretical physics. Where Oppenheimer hewed narrowly to the predictions of well-established physical law, Wheeler was driven by a deep yearning to know what lies beyond well-established law. He was continually reaching, mentally, toward the domain where known laws break down and new laws come into play. He tried to leapfrog his way into the twenty-first century, to catch a glimpse of what the laws of physics might be like beyond twentieth-century frontiers.

Of all the places that such a glimpse might be had, none looked more promising to Wheeler, from the 1950s onward, than the interface between general relativity (the domain of the large) and quantum mechanics (the domain of the small). General relativity and quantum mechanics did not mesh with each other in a logically consistent way. They were like the rows and columns of a crossword puzzle early in one's attempts to solve it. One has a tentative set of words written along the rows and a tentative set written down the columns,

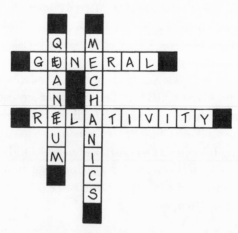

and one discovers a logical inconsistency at some of the intersections of rows and columns: Where the row word GENERAL demands an E, the column word QUANTUM demands a U; where the row word RELA-TIVITY demands an E, the column word QUANTUM demands a T. Looking at the row and column, it is obvious that one or the other or both must be changed to get consistency. Similarly, looking at the laws of general relativity and the laws of quantum mechanics, it was obvious that one or the other or both must be changed to make them mesh logically. If such a mesh could be achieved, the resulting union of general relativity and quantum mechanics would produce a powerful new set of laws that physicists were calling *quantum gravity.* However, physicists' understanding of how to marry general relativity with quantum mechanics was so primitive in the 1950s that, despite great effort, nobody was making much progress.

Progress was also slow on trying to understand the fundamental building blocks of atomic nuclei—the neutron, the proton, the electron, and the plethora of other *elementary particles* that were being created in particle accelerators.

Wheeler had a dream of leaping over these impasses and catching a simultaneous glimpse of the nature of quantum gravity and the nature of elementary particles. Such a glimpse, he thought, might come from seeking out those places in theoretical physics where paradoxes abound. From resolving a paradox comes deep understanding. The deeper the paradox, the more likely that the understanding would probe beyond twentieth-century frontiers.

It was in this spirit that, soon after emerging from the superbomb effort, Wheeler, with Harrison and Wakano, filled in the missing gaps in our knowledge of cold, dead stars (Chapter 5); and it was in this spirit that Wheeler contemplated the resulting "fate of great masses." Here was a deep paradox of just the sort Wheeler was seeking: No cold, dead star can be more massive than about 2 Suns; and yet the heavens seem to abound in hot stars far more massive than that—stars which some day must cool and die. Oppenheimer, in his straightforward way, had asked the well-established laws of physics what happens to such stars, and had got (with Snyder) an answer that seemed outrageous to Wheeler. This reinforced Wheeler's conviction that here, in the fates of great masses, he might catch a glimpse of physics beyond twentieth-century frontiers. Wheeler was right, as we shall see in Chapters 12 and 13.

Wheeler had fire in his belly—a deep, unremitting need to know the fate of great masses and learn whether their fate might unlock the mysteries of quantum gravity and elementary particles. Oppenheimer, by contrast, seemed not to care much in 1958. He believed his own calculations with Snyder but showed no need to push further, no drive for deeper understanding. Perhaps he was tired from the intense battles of the preceding two decades—weapons design battles, political battles, personal battles. Perhaps he was overawed by the mysteries of the unknown. In any event, he would never again contribute answers. The torch was being passed to a new generation. Oppenheimer's legacy would become Wheeler's foundation; and in the U.S.S.R., Landau's legacy would become Zel'dovich's foundation.

In his 1958 Brussels confrontation with Oppenheimer, Wheeler asserted that the Oppenheimer–Snyder calculations could not be trusted. Why? Because of their severe idealizations (Figure 6.3 above). Most especially, Oppenheimer had pretended from the outset that the imploding star has no pressure whatsoever. Without pressure, it was impossible for the imploding material to form shock waves (the analogue of breaking ocean waves, with their froth and foam). Without pressure and shock waves, there was no way the imploding material could heat up. Without heat and pressure, there was no way for nuclear reactions to be triggered and no way to emit radiation. Without outpouring radiation, and without the outward ejection of material by nuclear reactions, pressure, or shock waves, there was no way for the star to lose mass. With mass loss forbidden from the outset, there was no way the massive star could ever reduce itself below 2 Suns and become a cold, dead, neutron star. No wonder Oppenheimer's imploding star had formed a black hole, Wheeler reasoned; his idealizations prevented it from doing anything else!

In 1939, when Oppenheimer and Snyder did their work, it had been hopeless to compute the details of implosion with realistic pressure (thermal pressure, degeneracy pressure, and pressure produced by the nuclear force) and with nuclear reactions, shock waves, heat, radiation, and mass ejection. However, the nuclear weapons design efforts of the intervening twenty years provided precisely the necessary tools. Pressure, nuclear reactions, shock waves, heat, radiation, and mass ejection are all central features of a hydrogen bomb; without them, the bomb won't explode. To design a bomb, one had to incorporate all these things into one's computer calculations. Wheeler's team, of course, had done so. Thus, it would have been natural for Wheeler's team now to rewrite their computer programs so that, instead of simulating the explosion of a hydrogen bomb, they simulated the implosion of a massive star.

It would have been natural, that is, if the team still existed. However, the team was now disbanded; they had written their PMB-31 report and had dispersed to teach, do physics research, and become administrators at a variety of universities and government laboratories.

America's bomb design expertise was now concentrated at Los Alamos, and at a new government laboratory in Livermore, California. At Livermore in the late 1950s, Stirling Colgate became fascinated by the problem of stellar implosion. With encouragement from Edward Teller, and in collaboration with Richard White and later Michael May, Colgate set out to simulate such an implosion on a computer. The Colgate–White–May simulations kept some of Oppenheimer's idealizations: They insisted from the outset that the imploding star be spherical and not rotate. Without this restriction, their computations would have been enormously more difficult. However, their simulations took account of all the things that worried Wheeler—pressure, nuclear reactions, shock waves, heat, radiation, mass ejection—and did so by relying heavily on bomb design expertise and computer codes. To perfect the simulations required several years of effort, but by the early 1960s they were working well.

One day in the early 1960s, John Wheeler rushed into a relativity class at Princeton University that I, as a graduate student, was taking from him. He was slightly late, but beaming with pleasure. He had just returned from a visit to Livermore, where he had seen the results of the most recent Colgate, White, and May simulations. With excitement in his voice, he drew diagram after diagram on the blackboard, explaining what his Livermore friends had learned:

When the imploding star had a small mass, it triggered a supernova explosion and formed a neutron star in just the manner that Fritz Zwicky had speculated thirty years earlier. When the mass of the star was much larger than the 2-Suns maximum for a neutron star, the implosion—despite its pressure, nuclear reactions, shock waves, heat, and radiation—produced a black hole. And the black hole's birth was remarkably similar to the highly idealized one computed nearly twenty-five years earlier by Oppenheimer and Snyder. As seen from outside, the implosion slowed and became frozen at the critical circumference, but as seen by someone on the star's surface, the implosion did not freeze at all. The star's surface shrank right through the critical circumference and on inward, without hesitation.

Wheeler, in fact, had already come to expect this. Other insights (to be described below) had already transformed him from a critic of Oppenheimer's black holes to an enthusiastic supporter. But here, for the first time, was a concrete proof from a realistic computer simulation: Implosion must produce black holes.

Was Oppenheimer pleased by Wheeler's conversion? He showed little interest and little pleasure. At a December 1963 international conference in Dallas, Texas, on the occasion of the discovery of quasars (Chapter 9), Wheeler gave a long lecture on stellar implosion. In his lecture, he described with enthusiasm the 1939 calculations of Oppenheimer and Snyder. Oppenheimer attended the conference, but during Wheeler's lecture he sat on a bench in the hallway chatting with friends about other things. Thirty years later, Wheeler recalls the scene with sadness in his eyes and voice.

In the late 1950s, Zel'dovich began to get bored with weapons design work. Most of the really interesting problems had been solved. In search of new challenges, he forayed, part time, into the theory of elementary particles and then into astrophysics, while keeping command of his bomb design team at the Installation and of another team that did subsidiary bomb calculations at the Institute of Applied Mathematics, in Moscow.

In his bomb design work, Zel'dovich would pummel his teams with ideas, and the team members would do calculations to see whether the ideas worked. "Zel'dovich's sparks and his team's gasoline" was the way Ginzburg described it. As he moved into astrophysics, Zel'dovich retained this style.

Stellar implosion was among the astrophysical problems that caught Zel'dovich's fancy. It was obvious to him, as to Wheeler, Colgate, May, and White in America, that the tools of hydrogen bomb design were ideally suited to the mathematical simulation of imploding stars.

To puzzle out the details of realistic stellar implosion, Zel'dovich collared several young colleagues: Dmitri Nadezhin and Vladimir Imshennik at the Institute of Applied Mathematics, and Mikhail Podurets at the Installation. In a series of intense discussions, he gave them his vision of how the implosion could be simulated on a computer, including all the key effects that were so important for the hydrogen bomb: pressure, nuclear reactions, shock waves, heat, radiation, mass ejection.

Stimulated by these discussions, Imshennik and Nadezhin simulated the implosion of stars with small mass—and verified, independently of Colgate and White in America, Zwicky's conjectures about supernovae. In parallel, Podurets simulated the implosion of a massive star. Podurets's results, published almost simultaneously with those from May and White in America, were nearly identical to the Americans'. There could be no doubt. Implosion produces black holes, and does so in just the way that Oppenheimer and Snyder had claimed.

The adaptation of bomb design codes to simulate stellar implosion is just one of many intimate connections between nuclear weapons and astrophysics. These connections were obvious to Sakharov in 1948. Upon being ordered to join Tamm's bomb design team, he embarked on a study of astrophysics to prepare himself. My own nose was rubbed into the connections unexpectedly in 1969.

I never really wanted to know what the Teller–Ulam/Sakharov–Zel'dovich idea was. The superbomb, one that by virtue of their idea could "be arbitrarily powerful," seemed obscene to me, and I didn't want even to speculate about how it worked. But my quest to understand the possible roles of neutron stars in the Universe forced the Teller–Ulam idea onto my consciousness.

Zel'dovich, several years earlier, had pointed out that gas from interstellar space or a nearby star, falling onto a neutron star, should heat up and shine brightly: It should become so hot, in fact, that it radiates mostly high-energy X-rays rather than less energetic light. The infalling gas controls the rate of outflow of X-rays, Zel'dovich argued, and conversely, the outflowing X-rays control the rate of infall of gas. Thereby, the two, gas and X-rays, working together, produce a steady, *self-regulated flow.* If the gas falls in at too high a rate, then it will produce lots of X-rays, and the outpouring X-rays will strike the infalling gas, producing an outward pressure that slows the gas's fall (Figure 6.4a). On the other hand, if the gas falls in at too low a rate, then it produces so few X-rays that they are powerless to slow the infalling gas, so the infall rate increases. There is just one unique rate of gas infall, not too high and not too low, at which the X-rays and gas are in mutual equilibrium.

This picture of the flow of gas and X-rays disturbed me. I knew full well that if, on Earth, one tries to hold a dense fluid such as liquid mercury up by means of a less dense fluid such as water below it, tongues of mercury quickly eat their way down into the water, the mercury goes whooshing down, and the water goes whooshing up (Figure 6.4b). This phenomenon is called the *Rayleigh–Taylor instability.* In Zel'dovich's picture, the X-rays were like the low-density water and the infalling gas was like the high-density mercury. Wouldn't tongues of gas eat their way into the X-rays, and wouldn't the gas then fall freely down those tongues, destroying Zel'dovich's self-regulated flow (Figure 6.4c)? A detailed calculation with the laws of physics could tell me whether this happens, but such a calculation would be very complex and time consuming; so, rather than calculate, I asked Zel'dovich one afternoon in 1969, when we were discussing physics in his apartment in Moscow.

Zel'dovich looked a bit uncomfortable when I raised the question, but his answer was firm: "No, Kip, that doesn't happen. There are no tongues into the X-rays. The gas flow is stable." "How do you know, Yakov Borisovich?" I asked. Amazingly, I could not get an answer. It seemed clear that Zel'dovich or somebody had done a detailed calculation or experiment showing that X-rays can push hard on gas without Rayleigh–Taylor tongues destroying the push, but Zel'dovich could not point me to any such calculation or experiment in the published literature, nor would he describe for me the detailed physics that goes on. How uncharacteristic of him!

A few months later I was hiking in the high Sierras in California with Stirling Colgate. (Colgate is one of the best American experts on the flows of fluids and radiation, he was deeply involved in the late stages of the American superbomb effort, and he was one of the three Livermore physicists who had simulated a star's implosion on a com-

6.4   (a) Gas falling onto a neutron star is slowed by the pressure of outpouring X-rays. (b) Liquid mercury trying to fall in the Earth's gravitational field is held back by water beneath it; a Rayleigh–Taylor instability results. (c) Is it possible that there is also a Rayleigh–Taylor instability for the infalling gas held back by a neutron star's X-rays?

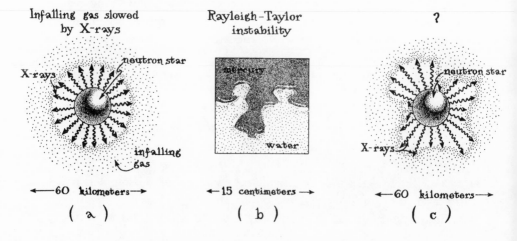

puter.) As we hiked, I posed to Colgate the same question I had asked of Zel'dovich, and he gave me the same answer: The flow is stable; the gas cannot escape the force of the X-rays by developing tongues. "How do you know, Stirling?" I asked. "It has been shown," he replied. "Where can I find the calculations or experiments?" I asked. "I don't know . . ."

"That's very peculiar," I told Stirling. "Zel'dovich told me precisely the same thing—the flow is stable. But he, like you, would not point me to any proofs." "Oh! That's fascinating. So Zel'dovich really knew," said Stirling.

And then I knew as well. I hadn't wanted to know. But the conclusion was unavoidable. The Teller–Ulam idea must be the use of X-rays, emitted in the first microsecond of the fission (atomic bomb) trigger, to heat, help compress, and ignite the superbomb's fusion fuel (Figure 6.5). That this is, indeed, part of the Teller–Ulam idea was confirmed in the 1980s in several unclassified American publications; otherwise I would not mention it here.

6.5 Schematic diagram showing one aspect of the Teller–Ulam/Sakharov–Zel'-dovich idea for the design of a hydrogen bomb: A fission-powered explosion (atomic bomb trigger) produces intense X-rays that somehow are focused onto the fusion fuel (lithium deuteride, LiD). The X-rays presumably heat the fusion fuel and help compress it long enough for fusion reactions to occur. The technology for focusing the X-rays and other practical problems are so formidable that by knowing this piece of the Teller–Ulam "secret," one is only an infinitesimal distance along the way toward building a working superbomb.

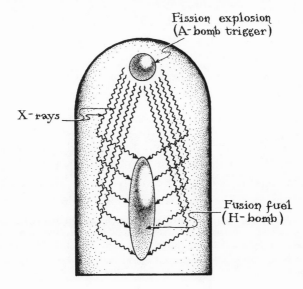

What converted Wheeler from a skeptic of black holes to a believer and advocate? Computer simulations of imploding stars were only the final validation of his conversion. Far more important was the destruction of a mental block. This mental block pervaded the world's community of theoretical physicists from the 1920s through the 1950s. It was fostered in part by the same *Schwarzschild singularity* that was then being used for a black hole. It was also fostered by the mysterious, seemingly paradoxical conclusion, from Oppenheimer and Snyder's idealized calculations, that an imploding star becomes frozen forever at the critical circumference ("Schwarzschild singularity") from the viewpoint of a static, external observer, but it implodes quickly through the freezing point and on inward from the viewpoint of an observer on the star's surface.

In Moscow, Landau and his colleagues, while believing Oppenheimer and Snyder's calculations, had severe trouble reconciling these two viewpoints. "You cannot appreciate how difficult it was for the human mind to understand how both viewpoints can be true simultaneously," Landau's closest friend, Evgeny Lifshitz, told me some years later.

Then one day in 1958, the same year as Wheeler was attacking Oppenheimer and Snyder's conclusions, there arrived in Moscow an issue of the *Physical Review* with an article by David Finkelstein, an unknown postdoc at a little known American university, the Stevens Institute of Technology in Hoboken, New Jersey. Landau and Lifshitz read the article. It was a revelation. Suddenly everything was clear.[5]

Finkelstein visited England that year and lectured at Kings College in London. Roger Penrose (who later would revolutionize our understanding of what goes on *inside* black holes; see Chapter 13) took the train down to London to hear Finkelstein's lecture, and returned to Cambridge enthusiastic.

In Princeton, Wheeler was intrigued at first, but was not fully convinced. He would become convinced only gradually, over the next several years. He was slower than Landau or Penrose, I believe, because he was looking deeper. He was fixated on his vision that quantum gravity must make nucleons (neutrons and protons) in an imploding star dissolve away into radiation and escape the implosion, and it

5. Finkelstein's insight had actually been found earlier, in other contexts by other physicists including Arthur Eddington; but they had not understood its significance and it was quickly forgotten.

seemed impossible to reconcile this vision with Finkelstein's insight. Nevertheless, as we shall see later, in a certain deep sense both Wheeler's vision and Finkelstein's insight were correct.

So just what was Finkelstein's insight? Finkelstein discovered, quite by chance and in just two lines of mathematics, a new reference frame in which to describe Schwarzschild's spacetime geometry. Finkelstein was not motivated by the implosion of stars and he did not make the connection between his new reference frame and stellar implosion. However, to others the implication of his new reference frame was clear. It gave them a totally new perspective on stellar implosion.

David Finkelstein, ca. 1958. [Photo by Herbert S. Sonnenfeld; courtesy David Finkelstein.]

The geometry of spacetime outside an imploding star is that of Schwarzschild, and thus the star's implosion could be described using Finkelstein's new reference frame. Now, Finkelstein's new frame was quite different from the reference frames we have met previously (Chapters 1 and 2). Most of those frames (imaginary laboratories) were small, and all portions of each frame (top, bottom, sides, middle) were at rest with respect to each other. By contrast, Finkelstein's reference frame was large enough to cover simultaneously the regions of space-time far from the imploding star, the regions near it, and all regions in between. More important, the various parts of Finkelstein's frame were in motion with respect to each other: The parts far from the star were static, that is, not imploding, while the parts near the star were falling inward along with the imploding star's surface. Correspondingly, Finkelstein's frame could be used to describe the star's implosion simultaneously from the viewpoint of faraway static observers and from the viewpoint of observers who ride inward with the imploding star. The resulting description reconciled beautifully the freezing of the implosion as observed from far away with the continued implosion as observed from the star's surface.

In 1962, two members of Wheeler's Princeton research group, David Beckedorff and Charles Misner, constructed a set of embedding diagrams to illustrate this reconciliation, and in 1967 I converted their embedding diagrams into the following fanciful analogy for an article in *Scientific American.*

Once upon a time, six ants lived on a large rubber membrane (Figure 6.6). These ants, being highly intelligent, had learned to communicate using signal balls that roll with a constant speed (the "speed of light") along the membrane's surface. Regrettably, the ants had not calculated the membrane's strength.

One day five of the ants happened to gather near the center of the membrane, and their weight made it begin to collapse. They were trapped; they could not crawl out fast enough to escape. The sixth ant—an astronomer ant—was a safe distance away with her signal-ball telescope. As the membrane collapsed, the trapped ants dispatched signal balls to the astronomer ant so she could follow their fate.

The membrane did two things as it collapsed: First, its surface contracted inward, dragging surrounding objects toward the center of the collapse in much the same manner as an imploding star's gravity pulls objects toward its center. Second, the membrane sagged and became curved into a bowl-like shape analogous to the curved shape of space around an imploding star (compare with Figure 6.2).

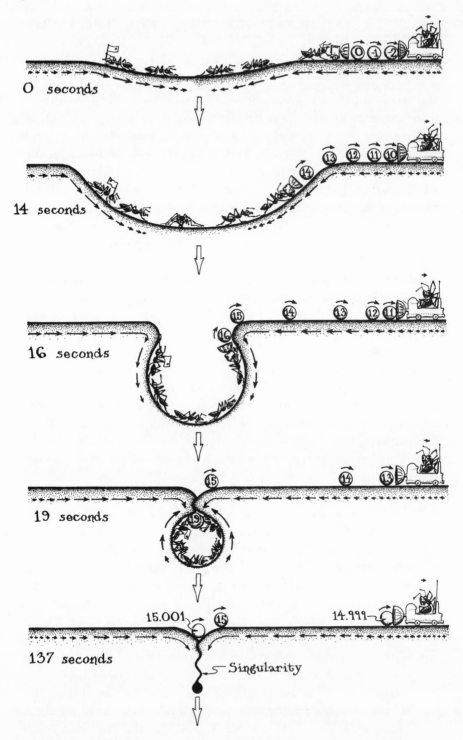

6.6 Collapsing rubber membrane populated by ants provides a fanciful analogue of the gravitational implosion of a star to form a black hole. [Adapted from Thorne (1967).]

The membrane's surface contracted faster and faster as the collapse proceeded. As a result, the signal balls, which were uniformly spaced in time when dispatched by the trapped ants, were received by the astronomer ant at more and more widely spaced time intervals. (This is analogous to the reddening of light from an imploding star.) Ball number 15 was dispatched 15 seconds after the collapse began, at the precise moment when the trapped ants were being sucked through the membrane's critical circumference. Ball 15 stayed forever at the critical circumference because the membrane there was contracting with precisely the speed of the ball's motion (speed of light). Just 0.001 second before reaching the critical circumference, the trapped ants dispatched ball number 14.999 (shown only in the last diagram). This ball, barely outracing the contracting membrane, did not reach the astronomer ant until 137 seconds after the collapse began. Ball number 15.001, sent out 0.001 second after the critical circumference, got inexorably sucked into the highly curved region and was crushed along with the five trapped ants.

But the astronomer ant could never learn about the crushing. She would never receive signal ball number 15, or any signal balls emitted after it; and those just before 15 would take so long to escape that to her the collapse would appear to slow and freeze right at the critical circumference.

This analogy is remarkably faithful in reproducing the behavior of an imploding star:

1. The shape of the membrane is precisely that of the curved space around the star—as embodied in an embedding diagram.
2. The motions of the signal balls on the membrane are precisely the same as the motions of photons of light in the imploding star's curved space. In particular, the signal balls move with the speed of light as measured locally by any ant at rest on the membrane; yet balls emitted just before number 15 take a very long time to escape, so long that to the astronomer ant the collapse seems to freeze. Similarly, photons emitted from the star's surface move with the speed of light as measured locally by anyone; yet the photons emitted just before the star shrinks inside its critical circumference (its horizon) take a very long time to escape, so long that to external observers the implosion must appear to freeze.
3. The trapped ants do not see any freezing whatsoever at the criti-

cal circumference. They are sucked through the critical circumference without hesitation, and crushed. Similarly, anyone on the surface of an imploding star will not see the implosion freeze. He will experience implosion with no hesitation, and get crushed by tidal gravity (Chapter 13).

This, translated into embedding diagrams, was the insight that came from Finkelstein's new reference frame. With this way of thinking about the implosion, there was no more mystery. An imploding star really does shrink through the critical circumference without hesitation. That it appears to freeze as seen from far away is an illusion.

The embedding diagrams of the parable of the ants capture only some of the insight that came from Finkelstein's new reference frame, not all. Further insight is embodied in Figure 6.7, which is a *spacetime diagram* for the imploding star.

Until now, the only spacetime diagrams we have met were in the flat spacetime of special relativity; for example, Figure 1.3. In Figure 1.3, we drew our diagrams from two different viewpoints: that of an inertial reference frame at rest in the city of Pasadena (with the downward pull of gravity ignored), Figure 1.3c; and that of an inertial frame attached to your high-speed sports car as you zoom down Pasadena's Colorado Boulevard, Figure 1.3b. In each diagram we plotted our chosen frame's space horizontally, and its time vertically.

In Figure 6.7, the chosen reference frame is that of Finkelstein. Accordingly, we plot horizontally two of the three dimensions of space, as measured in Finkelstein's frame ("Finkelstein's space"), and we plot vertically time as measured in his frame ("Finkelstein's time"). Since, far from the star, Finkelstein's frame is static (not imploding), Finkelstein's time there is that experienced by a static observer. And since, near the star, Finkelstein's frame falls inward with the imploding stellar surface, Finkelstein's time there is that experienced by an infalling observer.

Two horizontal slices are shown in the diagram. Each depicts two of the dimensions of space at a specific moment of time, but with the space's curvature removed so the space looks flat. More specifically, circumferences around the star's center are faithfully represented on these horizontal slices, but radii (distances from the center) are not. To represent both radii and circumferences faithfully, we would have to use embedding diagrams like those of Figure 6.2 or those of the parable of the ants, Figure 6.6. The space curvature would then show clearly: Circumferences would be less than 2π times radii. By drawing the horizontal slices flat, we are artificially removing their curvature. This incorrect flattening of the space is a price we pay to make the diagram legible. The payoff we gain is our ability to see space and time together on a single, legible diagram.

At the earliest time shown in the diagram (bottom horizontal slice), the star, with one spatial dimension absent, is the interior of a large circle; if the missing dimension were restored, the star would be the interior of a large sphere. At a later time (second slice), the star has shrunk; it is now the interior of a smaller circle. At a still later time, the star passes through its critical circumference, and still later it shrinks to zero circumference, creating there a *singularity* in which, according to general relativity, the star is crunched out of existence. We shall not discuss the details of this singularity until Chapter 13, but it is crucial to know that it is a completely different thing from the "Schwarzschild singularity" of which physicists spoke from the 1920s through the 1950s. The "Schwarzschild singularity" was their ill-conceived name for the critical circumference or for a black hole; this "singularity" is the object that resides at the black hole's center.

The black hole itself is the region of spacetime that is shown black in the diagram, that is, the region inside the critical circumference and to the future of the imploding star's surface. The hole's surface (its *horizon*) is at the critical circumference.

Also shown in the diagram are the world lines (trajectories through spacetime) of some particles attached to the star's surface. As one's eye travels upward in the diagram (that is, as time passes), one sees these world lines move in closer and closer to the center of the star (to the central axis of the diagram). This motion exhibits the star's shrinkage with time.

Of greatest interest are the world lines of four photons (four particles of light). These photons are the analogues of the signal balls in the parable of the ants. Photon A is emitted outward from the star's surface at the moment when the star begins to implode (bottom slice). It travels outward with ease, to larger and larger circumferences, as time passes (as one's eye travels upward in the diagram). Photon B, emitted shortly before the star reaches its critical circumference, requires a long time to escape; it is the analogue of signal ball number 14.999 in the parable of the ants. Photon C, emitted precisely at the critical circumference, remains always there, just like signal ball number 15. And

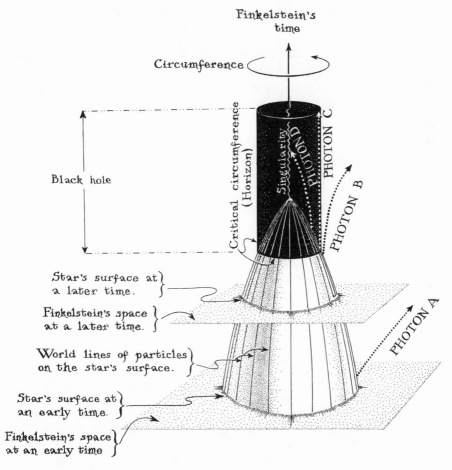

6.7 A spacetime diagram depicting the implosion of a star to form a black hole. Plotted upward is time as measured in Finkelstein's reference frame. Plotted horizontally are two of the three dimensions of that frame's space. Horizontal slices are two-dimensional "snapshots" of the imploding star and the black hole it creates at specific moments of Finkelstein's time, but with the curvature of space suppressed.

photon D, emitted from inside the critical circumference (inside the black hole), never escapes; it gets pulled into the singularity by the hole's intense gravity, just like signal ball 15.001.

It is interesting to contrast this modern understanding of the propagation of light from an imploding star with eighteenth-century predictions for light emitted from a star smaller than its critical circumference.

Recall (Chapter 3) that in the late eighteenth century John Michell in England and Pierre Simon Laplace in France used Newton's laws of gravity and Newton's corpuscular description of light to predict the existence of black holes. These "Newtonian black holes" were actually static stars with circumferences so small (less than the critical circumference) that gravity prevented light from escaping from the stars' vicinities.

The left half of Figure 6.8 (a space diagram, not a spacetime diagram) depicts such a star inside its critical circumference, and depicts the spatial trajectory of a photon (light corpuscle) emitted from the star's surface nearly vertically (radially). The outflying photon, like a thrown rock, is slowed by the pull of the star's gravity, it draws to a halt, and it then falls back into the star.

The right half of the figure depicts in a spacetime diagram the motions of two such photons. Plotted upward is Newton's universal time; plotted outward, his absolute space. With the passage of time, the circular star sweeps out the vertical cylinder; at any moment of time (horizontal slice through the diagram) the star is described by the same circle as in the left picture. As time passes, photon A flies out and then falls back into the star, and photon B, emitted a little later, does the same.

6.8   The predictions from Newton's laws of physics for the motion of light corpuscles (photons) emitted by a star that is inside its critical circumference. *Left:* a spatial diagram (similar to Figure 3.1). *Right:* a spacetime diagram.

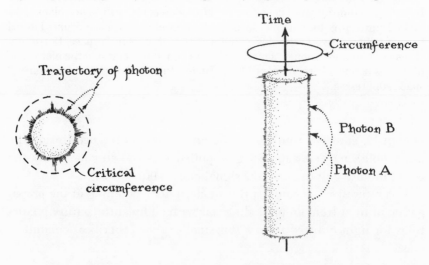

It is instructive to compare this (incorrect) Newtonian version of a star inside its critical circumference and the photons it emits with the (correct) relativistic version, Figure 6.7. The comparison shows two profound differences between the predictions of Newton's laws and those of Einstein:

1. Newton's laws (Figure 6.8) permit a star smaller than the critical circumference to live a happy, non-imploding life, with its gravitational squeeze forever counterbalanced by its internal pressure. Einstein's laws (Figure 6.7) insist that when any star is smaller than its critical circumference, its gravitational squeeze will be so strong that no internal pressure can possibly counterbalance it. The star has no choice but to implode.

2. Newton's laws (Figure 6.8) predict that photons emitted from the star's surface at first will fly out to larger circumferences, even in some cases to circumferences larger than critical, and then will be pulled back in. Einstein's laws (Figure 6.7) demand that any photon emitted from inside the critical circumference move always toward smaller and smaller circumferences. The only reason that such a photon can escape the star's surface is that the star itself is shrinking faster than the outward-directed photon moves inward (Figure 6.7).

Although Finkelstein's insight and the bomb code simulations fully convinced Wheeler that the implosion of a massive star must produce a black hole, the fate of the imploding stellar matter continued to disturb him in the 1960s, just as it had disturbed him in Brussels in his 1958 confrontation with Oppenheimer. General relativity insisted that the star's matter will be crunched out of existence in the singularity at the hole's center (Chapter 13), but such a prediction seemed physically unacceptable. To Wheeler it seemed clear that the laws of general relativity must fail at the hole's center and be replaced by new laws of *quantum gravity*, and these new laws must halt the crunch. Perhaps, Wheeler speculated, building on views he had expounded in Brussels, the new laws would convert the imploding matter into radiation that quantum mechanically "tunnels" its way out of the hole and escapes into interstellar space. To test this speculation would require understanding in depth the marriage of quantum mechanics and general relativity. Therein lay the beauty of the speculation. It was a testbed to assist in discovering the new laws of quantum gravity.

As Wheeler's student in the early 1960s, I thought that his speculation of matter being converted into radiation at the singularity and
then tunneling its way out of the hole was outrageous. How could
Wheeler believe such a thing? The new laws of quantum gravity would
surely be important in the singularity at the hole's center, as Wheeler
asserted. But not near the critical circumference. The critical circumference was in the "domain of the large," where general relativity must
be highly accurate; and the general relativistic laws were unequivocal—nothing can escape out of the critical circumference. Gravity
holds everything in. Thus, there can be no "quantum mechanical tunneling" (whatever that was) to let radiation out; I was firmly convinced
of it.

In 1964 and 1965 Wheeler and I wrote a technical book, together
with Kent Harrison and Masami Wakano, about cold, dead stars and
stellar implosion. I was shocked when Wheeler insisted on including in
the last chapter his speculation that radiation might tunnel its way out
of the hole and escape into interstellar space. In a last-minute struggle
to convince Wheeler to delete his speculation from the book, I called on
David Sharp, one of Wheeler's postdocs, for help. David and I argued
vigorously with Wheeler in a three-way telephone call, and Wheeler
finally capitulated.

Wheeler was right; David and I were wrong. Ten years later, Zel'-
dovich and Stephen Hawking would use a newly developed partial
marriage of general relativity and quantum mechanics to prove, mathematically, that radiation *can* tunnel its way out of a black hole—
though very, very slowly (Chapter 12). In other words, black holes can
evaporate, though they do it so slowly that a hole formed by the
implosion of a star will require far longer than the age of our Universe
to disappear.

The names that we give to things are important. The agents of movie
stars, who change their clients' names from Norma Jean Baker to
Marilyn Monroe and from Béla Blasko to Béla Lugosi, know this well.
So do physicists. In the movie industry a name helps set the tone, the
frame of mind with which the viewer regards the star—glamour for
Marilyn Monroe, horror for Béla Lugosi. In physics a name helps set
the frame of mind with which we view a physical concept. A good
name will conjure up a mental image that emphasizes the concept's
most important properties, and thereby it will help trigger, in a sub-

conscious, intuitive sort of a way, good research. A bad name can produce mental blocks that hinder research.

Perhaps nothing was more influential in preventing physicists, between 1939 and 1958, from understanding the implosion of a star than the name they used for the critical circumference: "Schwarzschild singularity." The word "singularity" conjured up an image of a region where gravity becomes infinitely strong, causing the laws of physics as we know them to break down—an image that we now understand is correct for the object at the center of a black hole, but not for the critical circumference. This image made it difficult for physicists to accept the Oppenheimer–Snyder conclusion that a person who rides through the Schwarzschild singularity (the critical circumference) on an imploding star will feel *no* infinite gravity and see *no* breakdown of physical law.

How truly *nonsingular* the Schwarzschild singularity (critical circumference) is did not become fully clear until David Finkelstein discovered his new reference frame and used it to show that the Schwarzschild singularity is nothing but a location into which things can fall but out of which nothing can come—and a location, therefore, into which we on the outside can never see. An imploding star continues to exist after it sinks through the Schwarzschild singularity, Finkelstein's reference frame showed, just as the Sun continues to exist after it sinks below the horizon on Earth. But just as we, sitting on Earth, cannot see the Sun beyond our horizon, so observers far from an imploding star cannot see the star after it implodes through the Schwarzschild singularity. This analogy motivated Wolfgang Rindler, a physicist at Cornell University in the 1950s, to give the Schwarzschild singularity (critical circumference) a new name, a name that has since stuck: He called it the *horizon*.

There remained the issue of what to call the object created by the stellar implosion. From 1958 to 1968 different names were used in East and West: Soviet physicists used a name that emphasized a distant astronomer's vision of the implosion. Recall that because of the enormous difficulty light has escaping gravity's grip, as seen from afar the implosion seems to take forever; the star's surface seems never quite to reach the critical circumference, and the horizon never quite forms. It looks to astronomers (or would if their telescopes were powerful enough to see the imploding star) as though the star becomes frozen just outside the critical circumference. For this reason, Soviet physicists called the object produced by implosion a *frozen star*—and this name helped set the tone and frame of mind for their implosion research in the 1960s.

In the West, by contrast, the emphasis was on the viewpoint of the person who rides inward on the imploding star's surface, through the horizon and into the true singularity; and, accordingly, the object thereby created was called a *collapsed star*. This name helped focus physicists' minds on the issue that became of greatest concern to John Wheeler: the nature of the singularity in which quantum physics and spacetime curvature would be married.

Neither name was satisfactory. Neither paid particular attention to the horizon which surrounds the collapsed star and which is responsible for the optical illusion of stellar "freezing." During the 1960s, physicists' calculations gradually revealed the enormous importance of the horizon, and gradually John Wheeler—the person who, more than anyone else, worries about using optimal names—became more and more dissatisfied.

It is Wheeler's habit to meditate about the names we call things when relaxing in the bathtub or lying in bed at night. He sometimes will search for months in this way for just the right name for something. Such was his search for a replacement for "frozen star"/"collapsed star." Finally, in late 1967, he found the perfect name.

In typical Wheeler style, he did not go to his colleagues and say, "I've got a great new name for these things; let's call them da-de-da-de-da." Rather, he simply started to use the name as though no other name had ever existed, as though everyone had already agreed that this was the right name. He tried it out at a conference on pulsars in New York City in the late fall of 1967, and he then firmly adopted it in a lecture in December 1967 to the American Association for the Advancement of Science, entitled "Our Universe, the Known and the Unknown." Those of us not there encountered it first in the written version of his lecture: "[B]y reason of its faster and faster infall [the surface of the imploding star] moves away from the [distant] observer more and more rapidly. The light is shifted to the red. It becomes dimmer millisecond by millisecond, and in less than a second is too dark to see . . . [The star,] like the Cheshire cat, fades from view. One leaves behind only its grin, the other, only its gravitational attraction. Gravitational attraction, yes; light, no. No more than light do any particles emerge. Moreover, light and particles incident from outside

... [and] going down the black hole only add to its mass and increase its gravitational attraction."

Black hole was Wheeler's new name. Within months it was adopted enthusiastically by relativity physicists, astrophysicists, and the general public, in East as well as West—with one exception: In France, where the phrase *trou noir* (black hole) has obscene connotations, there was resistance for several years.

7

The Golden Age

*in which black holes are found
to spin and pulsate,
store energy and release it,
and have no hair*

T he year was 1975; the place, the University of Chicago on the south side of the city, near the shore of Lake Michigan. There, in a corner office overlooking 56th Street, Subrahmanyan Chandrasekhar was immersed in developing a full mathematical description of black holes. The black holes he was analyzing were radically different beasts from those of the early 1960s, when physicists had begun to embrace the concept of a black hole. The intervening decade had been a *golden age* of black-hole research, an era that revolutionized our understanding of general relativity's predictions.

In 1964, at the beginning of the golden age, black holes were thought to be just what their name suggests: holes in space, down which things can fall, out of which nothing can emerge. But during the golden age, one calculation after another, by more than a hundred physicists using Einstein's general relativity equations, had changed that picture. Now, as Chandrasekhar sat in his Chicago office, calculating, black holes were regarded not as mere quiescent holes in space, but rather as dynamical objects: A black hole should be able to spin, and as it spins it should create a tornado-like swirling motion in the curved spacetime around itself. Stored in that swirl should be enormous energies, energies that nature might tap and use to power cosmic explosions. When stars or planets or smaller holes fall into a big hole, they should set the big hole pulsating. The horizon of the big hole should pulsate in and out, just as the surface of the Earth pulsates up and down after an earthquake, and those pulsations should produce gravitational waves—ripples in the curvature of spacetime that propagate out through the Universe, carrying a symphonic description of the hole.

Perhaps the greatest surprise to emerge from the golden age was general relativity's insistence that all the properties of a black hole are precisely predictable from just three numbers: the hole's mass, its rate of spin, and its electric charge. From those three numbers, if one is sufficiently clever at mathematics, one should be able to compute, for example, the shape of the hole's horizon, the strength of its gravitational pull, the details of the swirl of spacetime around it, and its frequencies of pulsation. Many of these properties were known by 1975, but not all. To compute and thereby learn all the remaining black-hole properties was a difficult challenge, precisely the kind of challenge that Chandrasekhar loved. He took it up, in 1975, as his personal quest.

For nearly forty years, the pain of his 1930s battles with Eddington had smoldered inside Chandrasekhar, impeding him from a return to research on the black-hole fates of massive stars. In those forty years he had laid many of the foundations for modern astrophysics—foundations for the theories of stars and their pulsations, of galaxies, of interstellar gas clouds, and much more. But throughout it all, the fascination of the fates of massive stars had attracted him. Finally, in the golden age, he had overcome his pain and returned.

He returned to a family of researchers who were almost all students and postdocs. The golden age was dominated by youth, and Chandrasekhar, young at heart but middle-aged and conservative in demeanor, was welcomed into their midst. On extended visits to Caltech and Cambridge, he could often be seen in cafeterias, surrounded by brightly and informally bedecked graduate students but himself attired in a conservative dark gray suit—"Chandrasekhar gray" his youthful friends called its color.

The golden age was brief. Caltech graduate student Bill Press had given the golden age its name, and in the summer of 1975, just as Chandrasekhar was embarking on his quest to compute the properties of black holes, Press organized its funeral: a four-day conference at

Top: Subrahmanyan Chandrasekhar at Caltech's student cafeteria ("the Greasy") with graduate students Saul Teukolsky *(left)* and Alan Lightman *(right),* in autumn 1971. *Bottom:* The participants in the conference/funeral for the golden age of black-hole research, Princeton University, summer 1975. *Front row, left to right:* Jacobus Petterson, Philip Yasskin, Bill Press, Larry Smarr, Beverly Berger, Georgia Witt, Bob Wald. *Second and third rows, left to right:* Philip Marcus, Peter D'Eath, Paul Schechter, Saul Teukolsky, Jim Nestor, Paul Wiita, Michael Schull, Bernard Carr, Clifford Will, Tom Chester, Bill Unruh, Steve Christensen. [Top: courtesy Sándor J. Kovács; bottom: courtesy Saul Teukolsky.]

Princeton University to which only researchers under the age of thirty were invited.[1] At the conference, Press and many of his young colleagues agreed that now was the time to move on to other research topics. The broad outlines of black holes as spinning, pulsating, dynamical objects were now in place, and the rapid pace of theoretical discoveries was beginning to slow. All that was left, it seemed, was to fill in the details. Chandrasekhar and a few others could do that handily, while his young (but now aging) friends sought new challenges elsewhere. Chandrasekhar was not pleased.

The Mentors:
Wheeler, Zel'dovich, Sciama

Who were these youths who revolutionized our understanding of black holes? Most of them were students, postdocs, and intellectual "grandchildren" of three remarkable master teachers: John Archibald Wheeler in Princeton, New Jersey, U.S.A.; Yakov Borisovich Zel'-dovich in Moscow, Russia, U.S.S.R.; and Dennis Sciama in Cambridge, England, U.K. Through their intellectual progeny, Wheeler, Zel'-dovich, and Sciama put their personal stamps on our modern understanding of black holes.

Each of these mentors had his own style. In fact, styles more different are hard to find. Wheeler was a charismatic, inspirational visionary. Zel'dovich was the hard-driving player/coach of a tightly knit team. Sciama was a self-sacrificing catalyst. We shall meet each of them in turn in the following pages.

How well I recall my first meeting with Wheeler. It was September 1962, two years before the advent of the golden age. Wheeler was a recent convert to the concept of a black hole, and I, at twenty-two years of age, had just graduated from Caltech and come to Princeton to pursue graduate study toward a Ph.D. My dream was to work on relativity research under Wheeler's guidance, so I knocked on his office door that first time with trepidation.

---

1. As Saul Teukolsky, a compatriot of Bill Press's, recalls it, "This conference was Bill's response to what he considered a provocation. There was another conference going on, to which none of us had been invited. But all the gray eminences were attending, so Bill decided to have a conference only for young people."

Professor Wheeler greeted me with a warm smile, ushered me into his office, and began immediately (as though I were an esteemed colleague, not a total novice) to discuss the mysteries of stellar implosion. The mood and content of that stirring private discussion are captured in Wheeler's writings of that era: "There have been few occasions in the history of physics when one could surmise more surely than one does now [in the study of stellar implosion] that he confronts a new phenomenon, with a mysterious nature of its own, waiting to be unravelled. . . . Whatever the outcome [of future studies], one feels that one has at last [in stellar implosion] a phenomenon where general relativity dramatically comes into its own, and where its fiery marriage with quantum physics will be consummated." I emerged, an hour later, a convert.

Wheeler gave inspiration to an entourage of five to ten Princeton students and postdocs—inspiration, but not detailed guidance. He presumed that we were brilliant enough to develop the details for ourselves. To each of us he suggested a first research problem—some issue that might yield a bit of new insight about stellar implosion, or black holes, or the "fiery marriage" of general relativity with quantum physics. If that first problem turned out to be too hard, he would gently nudge us in some easier direction. If it turned out easy, he would prod us to extract from it all the insight we possibly could, then write a technical article on the insight, and then move on to a more challenging problem. We soon learned to keep several problems going at once—one problem so hard that it must be visited and revisited time after time over many months or years before it cracked, hopefully with a big payoff; and other problems much easier, with quicker payoffs. Through it all, Wheeler gave just barely enough advice to keep us from totally floundering, never so much that we felt he had solved our problem for us.

My first problem was a lulu: Take a bar magnet with a magnetic field threading through it and emerging from its two ends. The field consists of field lines, which children are taught to make visible using iron filings on a piece of paper with the magnet below it (Figure 7.1a). Adjacent field lines repel each other. (Their repulsion is felt when one pushes the north poles of two magnets toward each other.) Each magnet's field lines are held together, despite their mutual repulsion, by the magnet's iron. Remove the iron, and their repulsion will make the field lines explode (Figure 7.1b). All this was familiar to me from my undergraduate studies. Wheeler reminded me of it in a long, private

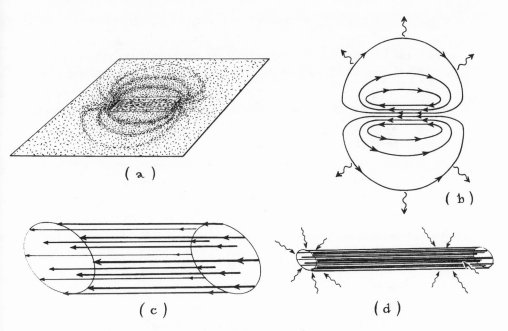

7.1 (a) The magnetic field lines around a bar magnet, made visible by iron filings on a piece of paper with the magnet below it. (b) The same field lines, with the paper and the magnet removed. Pressure between adjacent field lines makes them explode in the directions of the wavy arrows. (c) An infinitely long, cylindrical bundle of magnetic field lines whose field is so intense that its energy creates enough spacetime curvature (gravity) to hold the bundle together, despite the repulsion between field lines. (d) Wheeler's conjecture that when the bundle of field lines in (c) is squeezed slightly, its gravity would become so strong as to compress the bundle into implosion (wiggly lines).

discussion in his Princeton office. He then described a recent discovery by his friend Professor Mael Melvin at Florida State University in Tallahassee.

Melvin had shown, using Einstein's field equation, that not only can magnetic field lines be held together against explosion by the iron in a bar magnet, they can also be held together by gravity without the aid of any magnet. The reason is simple: The magnetic field has energy, and its energy gravitates. [To see why the energy gravitates, recall that energy and mass are "equivalent" (Box 5.2): It is possible to convert mass of any sort (uranium, hydrogen, or whatever) into energy; and conversely, it is possible to convert energy of any sort (magnetic energy, explosive energy, or whatever) into mass. Thus, in a deep sense, mass and energy are merely different names for the same thing, and this means that, since all forms of mass produce gravity, so must all forms of energy. The Einstein field equation, when examined carefully, insists on it.] Now, if we have an enormously intense magnetic field—a field far more intense than ever encountered on Earth—then the field's intense energy will produce intense gravity, and that gravity will compress the field; it will hold the field lines together despite the pressure between them (Figure 7.1c). This was Melvin's discovery.

Wheeler's intuition told him that such "gravitationally bundled" field lines might be as unstable as a pencil standing on its tip: Push the pencil slightly, and gravity will make it fall. Compress the magnetic field lines slightly, and gravity might overwhelm their pressure, pulling them into implosion (Figure 7.1d). Implosion to what? Perhaps to form an infinitely long, cylindrical black hole; perhaps to form a *naked singularity* (a singularity without an enshrouding horizon).

It did not matter to Wheeler that magnetic fields in the real Universe are too weak for gravity to hold them them together against explosion. Wheeler's quest was *not* to understand the Universe as it exists, but rather to understand the fundamental laws that govern the Universe. By posing idealized problems which push those laws to the extreme, he expected to gain new insights into the laws. In this spirit, he offered me my first gravitational research problem: Use the Einstein field equation to try to deduce whether Melvin's bundle of magnetic field lines will implode, and if so, to what.

For many months I struggled with this problem. The scene of the daytime struggle was the attic of Palmer Physical Laboratory in Princeton, where I shared a huge office with other physics students and we shared our problems with each other, in a camaraderie of verbal give-and-take. The nighttime struggle was in the tiny apartment, in a converted World War II army barracks, where I lived with my wife, Linda (an artist and mathematics student), our baby daughter, Kares, and our huge collie dog, Prince. Each day I carried the problem back and forth with me between army barracks and laboratory attic. Every few days I collared Wheeler for advice. I beat at the problem with pencil and paper; I beat at it with numerical calculations on a computer; I beat at it in long arguments at the blackboard with my fellow students; and gradually the truth became clear. Einstein's equation, pummeled, manipulated, and distorted by my beatings, finally told me that Wheeler's guess was wrong. No matter how hard one might squeeze it, Melvin's cylindrical bundle of magnetic field lines will always spring back. Gravity can never overcome the field's repulsive pressure. There is no implosion.

This was the best possible result, Wheeler explained to me enthusiastically: When a calculation confirms one's expectations, one merely firms up a bit one's intuitive understanding of the laws of physics. But when a calculation contradicts expectations, one is on the way toward new insight.

The contrast between a spherical star and Melvin's cylindrical bundle of magnetic field lines was extreme, Wheeler and I realized: When a spherical star is very compact, gravity inside it overwhelms any and all internal pressure that the star can muster. *The implosion of massive, spherical stars is compulsory* (Chapter 5). By contrast, regardless of how hard one squeezes a cylindrical bundle of magnetic field lines, regardless of how compact one makes the bundle's circular cross section (Figure 7.1d), the bundle's pressure will always overcome gravity and push the field lines back outward. *The implosion of cylindrical, magnetic field lines is forbidden;* it can never occur.

Why do spherical stars and a cylindrical magnetic field behave so differently? Wheeler encouraged me to probe this question from every possible direction; the answer might bring deep insight into the laws of physics. But he did not tell me how to probe. I was becoming an independent researcher; it would be best, he believed, for me to develop my own research strategy without further guidance from him. Independence breeds strength.

From 1963 to 1972, through most of the golden age, I struggled to understand the contrast between spherical stars and cylindrical magnetic fields, but only in fits and starts. The question was deep and difficult, and there were other, easier issues to study with most of my effort: the pulsations of stars, the gravitational waves that stars should emit when they pulsate, the effects of spacetime curvature on huge clusters of stars and on their implosion. Amidst those studies, once or twice a year I would pull from my desk drawer the stacks of manila folders containing my magnetic field calculations. Gradually I augmented those calculations with computations of other idealized infinitely long, cylindrical objects: cylindrical "stars" made of hot gas, cylindrical clouds of dust that implode, or that spin and implode simultaneously. Although these objects do not exist in the real Universe, my calculations about them, done in fits and starts, gradually brought understanding.

By 1972 the truth was evident: Only if an object is compressed in *all three* of its spatial directions, north–south, east–west, and up–down (for example, if it is compressed spherically), can gravity become so strong that it overwhelms all forms of internal pressure. If, instead, the object is compressed in only two spatial directions (for example, if it is compressed cylindrically into a long thin thread), gravity grows strong, but not nearly strong enough to win the battle with pressure. Very modest pressure, whether due to hot gas, electron degeneracy, or magnetic field lines, can easily overwhelm gravity and make the cylindrical object explode. And if the object is compressed in only a single direction, into a very thin pancake, pressure will overwhelm gravity even more easily.

My calculations showed this clearly and unequivocally in the case of spheres, infinitely long cylinders, and infinitely extended pancakes. For such objects, the calculations were manageable. Much harder to compute—indeed, far beyond my talents—were nonspherical objects of finite size. But physical intuition emerging from my calculations and from calculations by my youthful comrades told me what to expect. That expectation I formulated as a *hoop conjecture:*

Take any kind of object you might wish—a star, a cluster of stars, a bundle of magnetic field lines, or whatever. Measure the object's mass, for example, by measuring the strength of its gravitational pull on orbiting planets. Compute from that mass the object's critical circumference (18.5 kilometers times the object's mass in units of the mass of the Sun). If the object were spherical (which it is not) and were to implode or be squeezed, it would form a black hole when it gets compressed inside this critical circumference. What happens if the object is not spherical? The hoop conjecture purports to give an answer (Figure 7.2).

Construct a hoop with circumference equal to the critical circumference of your object. Then try to place the object at the center of the hoop, and try to rotate the hoop completely around the object. If you succeed, then the object must already have created a black-hole horizon around itself. If you fail, then the object is not yet compact enough to create a black hole.

In other words, the hoop conjecture claims that, if an object (a star, a star cluster, or whatever) gets compressed in a highly nonspherical manner, then the object will form a black hole around itself when, and only when, its circumference in all directions has become less than the critical circumference.

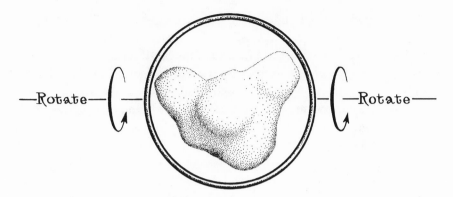

7.2 According to the hoop conjecture, an imploding object forms a black hole when, and only when, a hoop with the critical circumference can be placed around the object and rotated.

I proposed this hoop conjecture in 1972. Since then, I and others have tried hard to learn whether it is correct or not. The answer is buried in Einstein's field equation, but to extract the answer has proved exceedingly difficult. In the meantime, circumstantial evidence in favor of the hoop conjecture has continued to mount. Most recently, in 1991, Stuart Shapiro and Saul Teukolsky at Cornell University have simulated, on a supercomputer, the implosion of a highly nonspherical star and have seen black holes form around the imploded star precisely when the hoop conjecture predicts it. If a hoop can be slipped over the imploded star and rotated, a black hole forms; if it cannot, there is no black hole. But only a few such stars were simulated and with special nonspherical shapes. We therefore still do not know for certain, nearly a quarter century after I proposed it, whether the hoop conjecture is correct, but it looks promising.

Igor Dmitrievich Novikov in many ways was my Soviet counterpart, just as Yakov Borisovich Zel'dovich was Wheeler's. In 1962, when I was first meeting Wheeler and embarking on my career under his mentorship, Novikov was first meeting Zel'dovich and becoming a member of his research team.

Whereas I had had a simple and supportive early life—born and reared in a large, tightly knit Mormon family[2] in Logan, Utah—Igor

---

2. In the late 1980s, at my mother's suggestion, the entire family requested excommunication from the Mormon Church in response to the Church's suppression of the rights of women.

Novikov had had it rough. In 1937, when Igor was two, his father, a high official in the Railway Ministry in Moscow, was entrapped by Stalin's Great Terror, arrested, and (less lucky than Landau) executed. His mother's life was spared; she was sent to prison and then exile, and Igor was reared by an aunt. (Such Stalin-era family tragedies were frightfully common among my Russian friends and colleagues.)

In the early 1960s, while I was studying physics as an undergraduate at Caltech, Igor was studying it as a graduate student at Moscow University.

In 1962, when I was preparing to go to Princeton for graduate study and do general relativity research with John Wheeler, one of my Caltech professors warned me against this course: General relativity has little relevance for the real Universe, he warned; one should look elsewhere for interesting physics challenges. (This was the era of widespread skepticism about black holes and lack of interest in them.) At this same time, in Moscow, Igor was completing his kandidat degree (Ph.D.) with a specialty in general relativity, and his wife, Nora, also a physicist, was being warned by friends that relativity was a backwater with no relevance to the real Universe. Her husband, for the sake of his career, should leave it.

While I was ignoring these warnings and pushing onward to Princeton, Nora, worried by the warnings, seized an opportunity at a physics conference in Estonia to get advice from the famous physicist Yakov Borisovich Zel'dovich. She sought Zel'dovich out and asked whether he thought general relativity was of any importance. Zel'dovich, in his dynamic, forceful way, replied that relativity was going to become extremely important for astrophysics research. Nora then described an idea on which her husband was working, the idea that the implosion of a star to form a black hole might be similar to the big-bang origin of our Universe, but with time turned around and run backward.[3] As Nora spoke, Zel'dovich became more and more excited. He himself had developed the same idea and was exploring it.

A few days later, Zel'dovich barged into an office that Igor Novikov shared with many other students at Moscow University's Shternberg Astronomical Institute, and began grilling Novikov about his research. Though their ideas were similar, their research methods were completely different. Novikov, already a great expert in relativity, had used

---

3. This idea, while correct, has not yet produced any big payoffs, so I shall not discuss it in this book.

an elegant mathematical calculation to demonstrate the similarity between the big bang and stellar implosion. Zel'dovich, who knew hardly any relativity, had demonstrated it using deep physical insight and crude calculations. Here was an ideal match of talents, Zel'dovich realized. He was just then emerging from his life as an inventor and designer of nuclear weapons and was beginning to build a new team of researchers, a team to work on his newfound love: astrophysics. Novikov, as a master of general relativity, would be an ideal member of the team.

When Novikov, happy at Moscow University, hesitated to sign up, Zel'dovich exerted pressure. He went to Mstislav Keldysh, the director of the Institute of Applied Mathematics where Zel'dovich's team was being assembled; Keldysh telephoned Ivan Petrovsky, the *Rektor* (president) of Moscow University, and Petrovsky sent for Novikov. With trepidation Novikov entered Petrovsky's office, high in the central tower of the University, a place to which Novikov had never imagined venturing. Petrovsky was unequivocal: "Maybe you *now* don't want to leave the University to work with Zel'dovich, but you *will* want to." Novikov signed up, and despite some difficult times, never regretted it.

Zel'dovich's style as a mentor for young astrophysicists was the one he had developed while working with his nuclear weapons design team: "Zel'dovich's sparks [ideas] and his team's gasoline"—unless, perchance, some other member of the team could compete in inventing ideas (as Novikov usually did, when relativity was involved). Then Zel'dovich would enthusiastically take up his young colleague's idea and knock it about with the team in a vigorous thrust and parry, bringing the idea quickly to maturity and making it the joint property of himself and its inventor.

Novikov has described Zel'dovich's style vividly. Calling his mentor by first name plus abbreviated patronymic (a form of Russian address that is simultaneously respectful and intimate), Novikov says: "Yakov Boris'ch would often awaken me by telephone at five or six in the morning. 'I have a new idea! a new idea! Come to my apartment! Let's talk!' I would go, and we would talk for a long, long time. Yakov Boris'ch thought we all could work as long as he. He would work with his team from six in the morning to, say, ten, on one subject. Then a new subject until lunch. After lunch we would take a small walk or exercise or a short nap. Then coffee and more interaction until five or six. In the evening we were freed to calculate, think, or write, in preparation for the next day."

Coddled in his weapons design days, Zel'dovich continued to demand that the world adjust to him: follow his schedule, start work when he started, nap when he napped. (In 1968, John Wheeler, Andrei Sakharov, and I spent an afternoon discussing physics with him in a hotel room in the deep south of the Soviet Union. After several hours of intense discussion, Zel'dovich abruptly announced that it was time to nap. He then laid down and slept for twenty minutes, while Wheeler, Sakharov, and I relaxed and read quietly in our respective corners of the room, waiting for him to awaken.)

Impatient with perfectionists like me, who insist on getting all the details of a calculation right, Zel'dovich cared only about the main concepts. Like Oppenheimer, he could scatter irrelevant details to the winds and zero in, almost unerringly, on the central issues. A few arrows and curves on the blackboard, an equation not longer than half a line, a few sentences of vivid prose, with these he would bring his team to the heart of a research problem.

He was quick to judge an idea or a physicist's worth, and slow to change his judgments. He could retain faith in a wrong snap judgment for years, thereby blinding himself to an important truth, as when he rejected the idea that tiny black holes can evaporate (Chapter 12). But when (as was usually the case) his snap judgments were right, they enabled him to move forward across the frontiers of knowledge at a tremendous pace, faster than anyone I have ever met.

The contrast between Zel'dovich and Wheeler was stark: Zel'dovich whipped his team into shape with a firm hand, a constant barrage of his own ideas, and joint exploitation of his team's ideas. Wheeler offered his fledglings a philosophical ambience, a sense that there were exciting ideas all around, ready for the plucking; but he rarely pressed an idea, in concrete form, onto a student, and he absolutely never joined his students in exploiting their ideas. Wheeler's paramount goal was the education of his fledglings, even if that slowed the pace of discovery. Zel'dovich—still infused with the spirit of the race for the superbomb—sought the fastest pace possible, whatever the expense.

Zel'dovich was on the telephone at ungodly hours of the morning, demanding attention, demanding interaction, demanding progress. Wheeler seemed to us, his fledglings, the busiest man in the world; far too busy with his own projects to demand our attention. Yet he was always available at our request, to give advice, wisdom, encouragement.

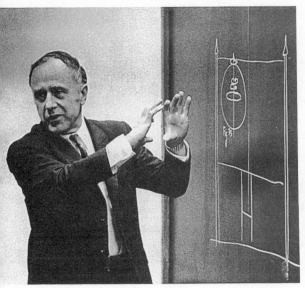

Top left: John Archibald Wheeler, ca. 1970. *Top right:* Igor Dmitrievich Novikov and Yakov Borisovich Zel'dovich in 1962. *Bottom:* Dennis Sciama in 1955. [Top left: courtesy Joseph Henry Laboratories, Princeton University; top right: courtesy S. Chandrasekhar; bottom: courtesy Dennis W. Sciama.]

Dennis Sciama, the third great mentor of the era, had yet another style. He devoted the 1960s and early 1970s almost exclusively to providing an optimal environment for his Cambridge University students to grow in. Because he relegated his own personal research and career to second place, after those of his students, he was never promoted to the august position of "Professor" at Cambridge (a position much higher than being a professor in America). It was his students, far more than he, who reaped the rewards and the kudos. By the end of the 1970s two of his former students, Stephen Hawking and Martin Rees, were Cambridge Professors.

Sciama was a catalyst; he kept his students closely in touch with the most important new developments in physics, worldwide. Whenever an interesting discovery was published, he would assign a student to read and report on it to the others. Whenever an interesting lecture was scheduled in London, he would take or send his entourage of students down on the train to hear it. He had exquisitely good sense about what ideas were interesting, what issues were worth pursuing, what one should read in order to get started on any research project, and whom one should go to for technical advice.

Sciama was driven by a desperate desire to know how the Universe is made. He himself described this drive as a sort of metaphysical angst. The Universe seemed so crazy, bizarre, and fantastic that the only way to deal with it was to try to understand it, and the best way to understand it was through his students. By having his students solve the most challenging problems, he could move more quickly from issue to issue than if he paused to try to solve them himself.

Black Holes Have No Hair

Among the discoveries of the golden age, one of the greatest was that "a black hole has no hair." (The meaning of this phrase will become clear gradually in the coming pages.) Some discoveries in science are made quickly, by individuals; others emerge slowly, as a result of diverse contributions from many researchers. The hairlessness of black holes was of the second sort. It grew out of research by the intellectual progeny of all three great mentors, Zel'dovich, Wheeler, and Sciama, and out of research by many others. In the following pages, we shall watch as this myriad of researchers struggles step by step, bit by bit, to formulate the concept of a black hole's hairlessness, prove it, and grasp its implications.

The first hints that "a black hole has no hair" came in 1964, from Vitaly Lazarevich Ginzburg, the man who had invented the LiD fuel for the Soviet hydrogen bomb, and whose wife's alleged complicity in a plot to kill Stalin had freed him from further bomb design work (Chapter 6). Astronomers at Caltech had just discovered *quasars*, enigmatic, explosive objects in the most distant reaches of the Universe, and Ginzburg was trying to understand how quasars might be powered (Chapter 9). One possibility, Ginzburg thought, might be the implosion of a magnetized, supermassive star to form a black hole. The magnetic field lines of such a star would have the shape shown in the upper part of Figure 7.3a—the same shape as the Earth's magnetic field lines. As the star implodes, its field lines might become strongly compressed and then explode violently, releasing huge energy, Ginzburg speculated; and this might help to explain quasars.

Left: Vitaly Lazarevich Ginzburg (ca. 1962), the person who produced the first evidence for the "no-hair conjecture." *Right:* Werner Israel (in 1964), the person who devised the first rigorous proof that the "no-hair conjecture" is correct. [Left: courtesy Vitaly Ginzburg; right: courtesy Werner Israel.]

To test this speculation by computing the full details of the star's implosion would have been exceedingly difficult, so Ginzburg did the second best thing. Like Oppenheimer in his first crude exploration of what happens when a star implodes (Chapter 6), Ginzburg examined a sequence of static stars, each one more compact than the previous one, and all with the same number of magnetic field lines threading through their interiors. This sequence of static stars should mimic a single imploding star, Ginzburg reasoned. Ginzburg derived a formula that described the shapes of the magnetic field lines for each of the stars in his sequence—and found a great surprise. When a star was nearly at its critical circumference and beginning to form a black hole around itself, its gravity sucked its magnetic field lines down onto its surface, plastering them there tightly. When the black hole was formed, the plastered-down field lines were all inside its horizon. No field lines remained, sticking out of the hole (Figure 7.3a). This did not bode well for Ginzburg's idea of how to power quasars, but it did suggest an intriguing possibility: When a magnetized star implodes to form a black hole, the hole might well be born with no magnetic field whatsoever.

7.3 Some examples of the "no-hair conjecture": (a) When a magnetized star implodes, the hole it forms has no magnetic field. (b) When a square star implodes, the hole it forms is round, not square. (c) When a star with a mountain on its surface implodes, the hole it forms has no mountain.

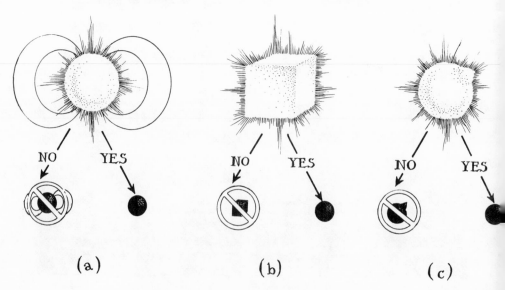

At about the time that Ginzburg was making this discovery, only a few kilometers away in Moscow Zel'dovich's team—with Igor Novikov and Andrei Doroshkevich taking the lead—began to ask themselves, "Since a round star produces a round hole when it implodes, will a deformed star produce a deformed hole?" As an extreme example, will a square star produce a square hole? (Figure 7.3b). To compute the implosion of a hypothetical square star would be exceedingly difficult, so Doroshkevich, Novikov, and Zel'dovich focused on an easier example: When a nearly spherical star implodes with a tiny mountain sticking out of its surface, will the hole it forms have a mountain-like protrusion on its horizon? By asking about nearly spherical stars with tiny mountains, the Zel'dovich team could simplify their calculations greatly; they could use mathematical techniques called *perturbation methods* that John Wheeler and a postdoc, Tullio Regge, had pioneered a few years earlier. These perturbation methods, which are explained a bit in Box 7.1, were carefully designed for the study of any small "perturbation" (any small disturbance) of an otherwise spherical situation. The gravitational distortion due to a tiny mountain on the Zel'dovich team's star was just such a perturbation.

Doroshkevich, Novikov, and Zel'dovich simplified their calculation still further by the same trick that Oppenheimer and Ginzburg used: Instead of simulating the full, dynamical implosion of a mountain-endowed star, they examined only a sequence of static, mountainous stars, each one more compact than the one before. With this trick, and with perturbation techniques, and with intensive give-and-take amongst themselves, Doroshkevich, Novikov, and Zel'dovich quickly discovered a remarkable result: When a static, mountain-endowed star is small enough to form a black hole around itself, the hole's horizon must be precisely round, with no protrusion (Figure 7.3c).

Similarly, it was tempting to conjecture that if an imploding square star were to form a black hole, its horizon would also be round, not square (Figure 7.3b). If this conjecture was correct, then a black hole should bear no evidence whatsoever of whether the star that created it was square, or round, or mountain-endowed, and also (according to Ginzburg) no evidence of whether the star was magnetized or free of magnetism.

Seven years later, as this conjecture was gradually turning out to be correct, John Wheeler invented a pithy phrase to describe it: *A black hole has no hair*—the hair being anything that might stick out of the hole to reveal the details of the star from which it was formed.

Box 7.1

An Explanation of Perturbation Methods, for Readers Who Like Algebra

In algebra one learns to compute the square of a sum of two numbers, a and b, from the formula

$$(a + b)^2 = a^2 + 2ab + b^2.$$

Suppose that a is a huge number, for example 1000, and that b is very small by comparison, for example 3. Then the third term in this formula, b^2, will be very small compared to the other two and thus can be thrown away without making much error:

$$(1000 + 3)^2 = 1000^2 + 2 \times 1000 \times 3 + 3^2 = 1,006,009$$
$$\simeq 1000^2 + 2 \times 1000 \times 3 = 1,006,000.$$

Perturbation methods are based on this approximation. The $a = 1000$ is like a precisely spherical star, $b = 3$ is like the star's tiny mountain, and $(a + b)^2$ is like the spacetime curvature produced by the star and mountain together. In computing that curvature, perturbation methods keep only effects that are linear in the mountain's properties (effects like $2ab = 6000$, which is linear in $b = 3$); these methods throw away all other effects of the mountain (effects like $b^2 = 9$). So long as the mountain remains small compared to the star, perturbation methods are highly accurate. However, if the mountain were to grow as big as the rest of the star (as it would need to do to make the star square rather than round), then perturbation methods would produce serious errors—errors like those in the above formulas with $a = 1000$ and $b = 1000$:

$$(1000 + 1000)^2 = 1000^2 + 2 \times 1000 \times 1000 + 1000^2 = 4,000,000$$
$$\neq 1000^2 + 2 \times 1000 \times 1000 = 3,000,000.$$

These two results differ significantly.

It is hard for most of Wheeler's colleagues to believe that this conservative, highly proper man was aware of his phrase's prurient interpretation. But I suspect otherwise; I have seen his impish streak, in
private, on rare occasion.[4] Wheeler's phrase quickly took hold, despite
resistance from Simon Pasternak, the editor-in-chief of the *Physical
Review*, the journal in which most Western black-hole research is published. When Werner Israel tried to use the phrase in a technical paper
around 1971, Pasternak fired off a peremptory note that under no
circumstances would he allow such obscenities in his journal. But Pasternak could not hold back for long the flood of "no-hair" papers. In
France and the U.S.S.R., where the French- and Russian-language
translations of Wheeler's phrase were also regarded as unsavory, the
resistance lasted longer. By the late 1970s, however, Wheeler's phrase
was being used and published by physicists worldwide, in all languages, without even a flicker of a childish grin.

It was the winter of 1964–65 by the time Ginzburg, and Doroshkevich, Novikov, and Zel'dovich, had invented their *no-hair conjecture*
and mustered their evidence for it. Once every three years, experts on
general relativity gathered somewhere in the world for a one-week
scientific conference to exchange ideas and show each other the results
of their researches. The fourth such conference would be held in London in June.

Nobody on Zel'dovich's team had ever traveled beyond the borders
of the Communist bloc of nations. Zel'dovich himself would surely not
be allowed to go; his contact with weapons research was much too
recent. Novikov, however, was too young to have been involved in the
hydrogen bomb project, his knowledge of general relativity was the
best of anyone on the team (which is why Zel'dovich had recruited him
onto the team in the first place), he was now the team's captain (Zel'-
dovich was the coach), and his English was passable though far from
fluent. He was the logical choice.

This was a good period in East–West relations. Stalin's death a
dozen years earlier had triggered a gradual resumption of correspondence and visits between Soviet scientists and their Western colleagues

---

4. I have seen it unleashed in public only once. In 1971, on the occasion of his sixtieth
birthday, Wheeler happened to be at an elegant banquet in a castle in Copenhagen—a banquet
in honor of an international conference, not in honor of him. To celebrate his birthday,
Wheeler set off a string of firecrackers behind his banquet chair, creating chaos amongst the
nearby diners.

(though not nearly so free a correspondence or visits as in the 1920s and early 1930s before Stalin's iron curtain descended). As a matter of course, the Soviet Union was now sending a small delegation of scientists to every major international conference; such delegations were important not only for maintaining the strength of Soviet science, but also for demonstrating the Soviets' strength to Western scientists. Since the time of the tsars, Russian bureaucrats have had an inferiority complex with respect to the West; it is very important for them to be able to hold their heads up in Western public view and show with pride what their nation can do.

Thus it was that Zel'dovich, having arranged an invitation from London for Novikov to give one of the major lectures of the Relativity Conference, found it easy to convince the bureaucrats to include his young colleague in the Soviet delegation. Novikov had many impressive things to report; he would create a very positive impression of the strength of Soviet physics.

In London, Novikov presented a one-hour lecture to an audience of three hundred of the world's leading relativity physicists. His lecture was a tour de force. The results on the gravitational implosion of a mountain-endowed star were but one small part of the lecture; the remainder was a series of equally major contributions to our understanding of relativistic gravity, neutron stars, stellar implosion, black holes, the nature of quasars, gravitational radiation, and the origin of the Universe. As I sat there in London listening to Novikov, I was stunned by the breadth and power of the Zel'dovich team's research. I had never before seen anything like it.

After Novikov's lecture, I joined the enthusiastic crowd around him and discovered, much to my pleasure, that my Russian was slightly better than his English and that I was needed to help with translating the discussion. As the crowd thinned, Novikov and I went off together to continue our discussion privately. Thus began one of my finest friendships.

It was not possible for me or anyone else to absorb fully in London the details of the Zel'dovich team's no-hair analysis. The details were too complex. We had to await a written version of the work, one in which the details were spelled out with care.

The written version arrived in Princeton in September 1965, in Russian. Once again I was thankful for the many boring hours I had spent in Russian class as an undergraduate. The written analysis contained two pieces. The first piece, clearly the work of Doroshkevich and Novikov, was a mathematical proof that, when a static star with a tiny mountain is made more and more compact, there are just two possible outcomes. Either the star creates a precisely spherical hole around itself, or else the mountain produces such enormous spacetime curvature, as the star nears its critical circumference, that the mountain's effects are no longer a "small perturbation"; the method of calculation then fails, and the outcome of the implosion is unknown. The second piece of the analysis was what I soon learned to identify as a "typical Zel'dovich" argument: If the mountain initially is tiny, it is *intuitively obvious* that the mountain *cannot* produce enormous curvature as the star nears its critical circumference. We must discard that possibility. The other possibility must be the truth: The star must produce a precisely spherical hole.

What was intuitively obvious to Zel'dovich (and would ultimately turn out to be true) was far from obvious to most Western physicists. Controversy began to swirl.

The power of a controversial research result is enormous. It attracts physicists like picnics attract ants. Thus it was with the Zel'dovich team's no-hair evidence. The physicists, like ants, came one by one at first, but then in droves.

The first was Werner Israel, born in Berlin, reared in South Africa, trained in the laws of relativity in Ireland, and now struggling to start a relativity research group in Edmonton, Canada. In a mathematical tour de force, Israel improved on the first, Doroshkevich–Novikov, part of the Soviet proof: He treated not just tiny mountains, as had the Soviets, but mountains of any size and shape. In fact, his calculations worked correctly for *any* implosion, no matter how nonspherical, even a square one, and they allowed the implosion to be dynamical, not just an idealized sequence of static stars. Equally remarkable was Israel's conclusion, which was similar to the Doroshkevich–Novikov conclusion, but far stronger: *A highly nonspherical implosion can have only two outcomes: either it produces no black hole at all, or else it produces a black hole that is precisely spherical.* For this conclusion to be true, however, the imploding body had to have two special properties: It must be completely devoid of any electric charge, and it must not spin at all. The reasons will become clear below.

Israel first presented his analysis and results on 8 February 1967, at a lecture at Kings College in London. The title of the lecture was a little enigmatic, but Dennis Sciama in Cambridge urged his students to journey down to London and hear it. As George Ellis, one of the students, recalls, "It was a very, very interesting lecture. Israel proved a theorem that came totally out of the blue; it was totally unexpected; nothing remotely like it had ever been done before." When Israel brought his lecture to a close, Charles Misner (a former student of Wheeler's) rose to his feet and offered a speculation: What happens if the imploding star spins and has electric charge? Might there again be just two possibilities: no hole at all, or a hole with a unique form, determined entirely by the imploding star's mass, spin, and charge? The answer would ultimately turn out to be yes, but not until after Zel'dovich's intuitive insight had been tested.

Zel'dovich, Doroshkevich, and Novikov, you will recall, had studied not highly deformed stars, but rather nearly spherical stars, with small mountains. Their analysis and Zel'dovich's claims triggered a plethora of questions:

If an imploding star has a tiny mountain on its surface, what is the implosion's outcome? Does the mountain produce enormous spacetime curvature, as the star nears its critical circumference (the outcome rejected by Zel'dovich's intuition)? Or does the mountain's influence disappear, leaving behind a perfectly spherical black hole (the outcome Zel'dovich favored)? And if a perfectly spherical hole is formed, how does the hole manage to rid itself of the mountain's gravitational influence? *What makes the hole become spherical?*

As one of Wheeler's students, I pondered these questions. However, I pondered them not as a challenge for myself, but rather as a challenge for my own students. It was now 1968; I had completed my Ph.D. at Princeton and had returned to my alma mater, Caltech, first as a postdoc and now as a professor; and I was beginning to build around myself an entourage of students similar to Wheeler's at Princeton.

Richard Price, a rough-bearded, two-hundred-pound, physically powerful young man from Brooklyn with a black belt in karate, had already worked with me on several small research projects, including one using the kind of mathematical methods needed to answer these questions: perturbations methods. He was now mature enough to tackle a more challenging project. The test of Zel'dovich's intuition looked ideal, but for one thing. It was a hot topic; others elsewhere were struggling with it; the ants were beginning to attack the picnic in droves. Price would have to move fast.

He didn't. Others beat him to the answers. He got there third, after Novikov and after Israel, but he got there more firmly, more completely, with deeper insight.

Price's insight was immortalized by Jack Smith, a humorous columnist for the *Los Angeles Times*. In the 27 August 1970 issue of the *Times*, Smith described a visit the previous day to Caltech: "After luncheon at the Faculty Club I walked alone around the campus. I could feel the deep thought in the air. Even in summer it stirs the olive trees. I looked in a window. A blackboard was covered with equations, thick as leaves on a walk, and three sentences in English: *Price's Theorem: Whatever can be radiated is radiated. Schutz's Observation: Whatever is radiated can be radiated. Things can be radiated if and only if they are radiated.* I walked on, wondering how it will affect Caltech this fall when they let girls in as freshmen for the first time. I don't think they'll do the place a bit of harm . . . I have a hunch they'll radiate."

This quote requires some explanation. "Schutz's observation" was facetious, but Price's theorem, "Whatever can be radiated is radiated," was a serious confirmation of a 1969 speculation by Roger Penrose.

Price's theorem is illustrated by the implosion of a mountain-endowed star. Figure 7.4 depicts the implosion. The left half of this figure is a spacetime diagram of the type introduced in Figure 6.7 of Chapter 6; the right side is a sequence of snapshots of the star's and horizon's shape as time passes, with the earliest times at the bottom and the latest at the top.

As the star implodes (bottom two snapshots in Figure 7.4), its mountain grows larger, producing a growing, mountain-shaped distortion in the star's spacetime curvature. Then, as the star sinks inside its critical circumference and creates a black hole horizon around itself (middle snapshot), the distorted spacetime curvature deforms the horizon, giving it a mountain-like protrusion. The horizon's protrusion, however, cannot live long. The stellar mountain that generated it is now inside the hole, so the horizon can no longer feel the mountain's influence. The horizon is no longer being forced, by the mountain, to keep its protrusion. The horizon ejects the protrusion in the only way it can: It converts the protrusion into ripples of spacetime curvature (gravitational waves—Chapter 10) that propagate away in all directions (top two snapshots). Some of the ripples go down the hole, others fly out into the surrounding Universe, and as they fly away, the ripples leave the hole with a perfectly spherical shape.

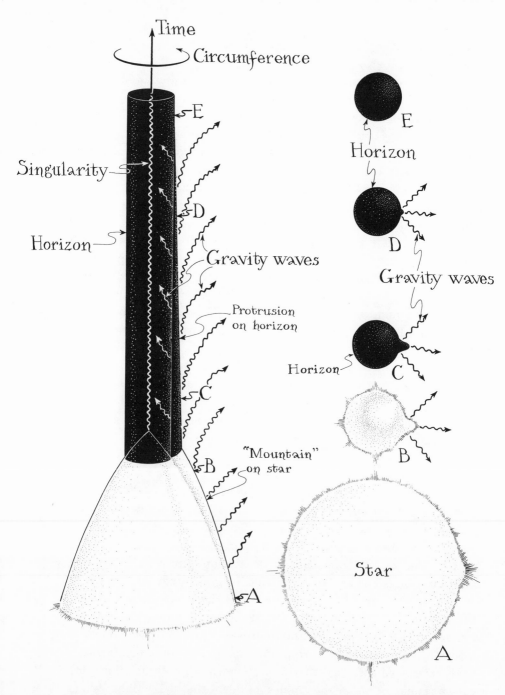

7.4 Spacetime diagram *(left)* and a sequence of snapshots *(right)* showing the implosion of a mountain-endowed star to form a black hole.

A familiar analogue is the plucking of a violin string. So long as one's finger holds the string in a deformed shape, it remains deformed; so long as the mountain is protruding out of the hole, it keeps the newborn horizon deformed. When one removes one's finger from the string, the string vibrates, sending sound waves out into the room; the sound waves carry away the energy of the string's deformation, and the string settles down into an absolutely straight shape. Similarly, when the mountain sinks inside the hole, it can no longer keep the horizon deformed, so the horizon vibrates, sending off gravitational waves; the waves carry away the energy of the horizon's deformation, and the horizon settles down into an absolutely spherical shape.

How does this mountain-endowed implosion relate to Price's theorem? According to the laws of physics, the horizon's mountain-like protrusion *can* be converted into gravitational radiation (ripples of curvature). Price's theorem tells us, then, that the protrusion *must* be converted into gravitational waves, and that this radiation must carry the protrusion completely away. *This is the mechanism that makes the hole hairless.*

Price's theorem tells us not only how a deformed hole loses its deformation, but also how a magnetized hole loses its magnetic field (Figure 7.5). (The mechanism, in this case, was already clear before

7.5 A sequence of snapshots showing the implosion of a magnetized star (a) to form a black hole (b). The hole at first inherits the magnetic field from the star. However, the hole has no power to hold on to the field. The field slips off it (c), is converted into electromagnetic radiation, and flies away (d).

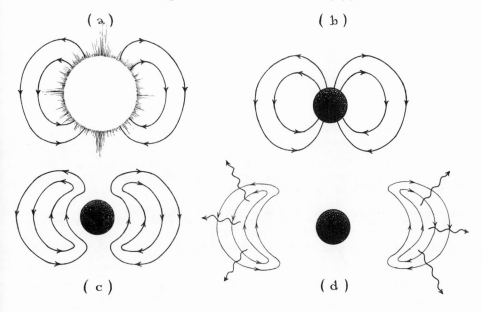

(a)   (b)

(c)   (d)

Price's theorem from a computer simulation by Werner Israel and two of his Canadian students, Vicente de la Cruz and Ted Chase.) The magnetized hole is created by the implosion of a magnetized star. Before the horizon engulfs the imploding star (Figure 7.5a), the magnetic field is firmly anchored in the star's interior; electric currents inside the star prevent the field from escaping. After the star is swallowed by the horizon (Figure 7.5b), the field can no longer feel the star's electric currents; they no longer anchor it. The field now threads the horizon, rather than the star, but the horizon is a worthless anchor. The laws of physics permit the field to turn itself into electromagnetic radiation (ripples of magnetic and electric force), and Price's theorem therefore demands that it do so (Figure 7.5c). The electromagnetic radiation flies away, partly down the hole and partly away from it, leaving the hole unmagnetized (Figure 7.5d).

If, as we have seen, mountains can be radiated away and magnetic fields can be radiated away, then what is left? What *cannot* be turned into radiation? The answer is simple: Among the laws of physics there is a special set of laws called *conservation laws*. According to these conservation laws, there are certain quantities that can never oscillate or vibrate in a radiative manner, and that therefore can never be converted into radiation and be ejected from a black hole's vicinity. These conserved quantities are the gravitational pull due to the hole's mass, the swirl of space due to the hole's spin (discussed below), and *radially* pointing electric field lines, that is, electric fields that point directly outward (discussed below) due to the hole's electric charge.[5]

Thus, according to Price's theorem, the influences of the hole's mass, spin, and charge are the only things that can remain behind when all the radiation has cleared away. All of the hole's other features will be gone, carried away by the radiation. This means that no measurement one might ever make of the properties of the final hole can possibly reveal any features of the star that imploded to form it, except the star's mass, spin, and charge. From the hole's properties one cannot even discern (according to calculations by James Hartle and Jacob Beken-

---

5. In the late 1980s it became clear that the laws of quantum mechanics can give rise to additional conserved quantities, associated with "quantum fields" (a type of field discussed in Chapter 12); and since these quantities, like a hole's mass, spin, and electric charge, cannot be radiated, they will remain as "quantum hair" when a black hole is born. Although this quantum hair might strongly influence the final fate of a microscopic, evaporating black hole (Chapter 12), it is of no consequence for the macroscopic holes (holes weighing more than the Sun) of this and the next few chapters, since quantum mechanics is generally unimportant on macroscopic scales.

stein, both Wheeler students) whether the star that formed the hole was made of matter or antimatter, of protons and electrons, or of neutrinos and antineutrinos. In Wheeler's words, made more precise, a black hole has *almost* no hair; its only "hair" is its mass, its spin, and its electric charge.

The firm, ultimate proof that a black hole has no hair (except its mass, spin, and electric charge) was actually not Price's. Price's analysis was restricted to imploding stars that are very nearly spherical, and that spin, if at all, only very slowly. The perturbation methods he used required this restriction. To learn the ultimate fate of a highly deformed, rapidly spinning, imploding star required a set of mathematical techniques very different from perturbation methods.

Dennis Sciama's students at Cambridge University were masters of the required techniques, but the techniques were difficult; extremely so. It took fifteen years for Sciama's students and their intellectual descendants, using those techniques, to produce a firm and complete proof that black holes have no hair—that even if a hole spins fast and is strongly deformed by its spin, the hole's final properties (after all radiation has flown away) are uniquely fixed by the hole's mass, spin, and charge. The lion's share of the credit for the proof goes to two of Sciama's students, Brandon Carter and Stephen Hawking, and to Werner Israel; but major contributions came also from David Robinson, Gary Bunting, and Pavel Mazur.

In Chapter 3, I commented on the great difference between the laws of physics in our real Universe and the society of ants in T. H. White's epic novel *The Once and Future King.* White's ants were governed by the motto "Everything not forbidden is compulsory," but the laws of physics violate that motto flagrantly. Many things allowed by physical law are so highly improbable that they never occur. Price's theorem is a remarkable exception. It is one of the few situations I have ever encountered in physics where the ants' motto holds sway: If physical law does not forbid a black hole to eject something as radiation, then ejection is compulsory.

Equally unusual are the implications of a black hole's resulting "hairless" state. Normally we physicists build simplified theoretical or computer models to try to understand the complicated Universe around us. As an aid to understanding weather, atmospheric physicists build computer models of the Earth's circulating atmosphere. As an aid to understanding earthquakes, geophysicists build simple theoretical models of slipping rocks. As an aid to understanding stellar implosion, Oppenheimer and Snyder in 1939 built a simple theoretical model: an imploding cloud of matter that was perfectly spherical, perfectly homogeneous, and completely devoid of pressure. And as we physicists build all these models, we are intensely aware of their limitations. They are but pale images of the complexity that abounds "out there," in the "real" Universe.

Not so for a black hole—or, at least, not so once the radiation has flown away, carrying off all the hole's "hair." Then the hole is so exceedingly simple that we can describe it by precise, simple mathematical formulas. We need no idealizations at all. Nowhere else in the macroscopic world (that is, on scales larger than a subatomic particle) is this true. Nowhere else is our mathematics expected to be so precise. Nowhere else are we freed from the limitations of idealized models.

Why are black holes so different from all other objects in the macroscopic Universe? Why are they, and they alone, so elegantly simple? If I knew the answer, it would probably tell me something very deep about the nature of physical laws. But I don't know. Perhaps the next generation of physicists will figure it out.

Black Holes Spin and Pulsate

What are the properties of the hairless holes, which are described so perfectly by the mathematics of general relativity?

If a black hole is idealized as having absolutely no electric charge and no spin, then it is precisely the spherical hole that we met in previous chapters. It is described, mathematically, by Karl Schwarzschild's 1916 solution to Einstein's field equation (Chapters 3 and 6).

When electric charge is dropped into such a hole, then the hole acquires just one new feature: electric field lines, which stick out of it radially like quills out of a hedgehog. If the charge is positive, then these electric field lines push protons away from the hole and attract electrons; if it is negative, then the field lines push electrons away and attract protons. Such a charge-endowed hole is described mathematically, with perfect precision, by a solution to Einstein's field equation found by the German and Dutch physicists Hans Reissner in 1916 and Gunnar Nordström in 1918. However, nobody understood the physical meaning of Reissner's and Nordström's solution until 1960, when two

Box 7.2

The Organization of Soviet and Western Science: Contrasts and Consequences

As I and my young physicist colleagues struggled to develop the hoop conjecture and to prove that black holes have no hair and to discover how they lose their hair, we also were discovering how very differently physics was organized in the U.S.S.R. than in Britain and America, and what profound effects those differences have. The lessons we learned may have some value in planning for the future, especially in the former Soviet Union, where all state institutions—scientific as well as governmental and economic—are now (1993) struggling to reorganize along Western lines. The Western model is not completely perfect, and the Soviet system was not uniformly bad!

In America and Britain there is a constant flow of young talent through a research group such as Wheeler's or Sciama's. Undergraduates may join the group for their last, senior year, but they then are sent away for graduate study. Graduate students join it for three to five years, and then are sent elsewhere for postdoctoral study. Postdocs join it for two or three years and then are sent away and expected either to start a research group of their own elsewhere (as I did at Caltech) or to join a small, struggling group elsewhere. Almost nobody in Britain or America, no matter how talented, is allowed to stay on in the nest of his or her mentor.

In the U.S.S.R., by contrast, outstanding young physicists (such as Novikov) usually remained in the nest of their mentor for ten, twenty, and sometimes even thirty or forty years. A great Soviet mentor like Zel'dovich or Landau usually worked in an Institute of the Academy of Sciences, rather than in a university, so his teaching load was small or nonexistent; by keeping his best former students, he built around himself a permanent team of full-time researchers, which became tightly knit and extremely powerful, and which might even stay with him until the end of his career.

Some of my Soviet friends attributed this difference to the failings of the British/American system: Almost all great British or American physicists work at universities, where research is often subservient to teaching and where there are inadequate numbers of permanent positions available to permit building up a strong, lasting group of researchers. As a result, there have been *no* theoretical physics research groups in Britain or America that can pretend to be the equal of Landau's group in the 1930s through 1950s, or of Zel'dovich's group in the 1960s and 1970s. The West, in this sense, had no hope of competing with the Soviet Union.

Some of my American friends attributed the difference to the failings of the Soviet system: It was very difficult, logistically, to move from institute

(continued next page)

(*Box 7.2 continued*)

to institute and city to city in the U.S.S.R., so young physicists were forced to remain with their mentors; they had no opportunity to get out and start independent groups of their own. The result, the critics asserted, was a feudal system. The mentor was like a lord and his team like serfs, indentured for most of their careers. The lord and serfs were interdependent in a complex way, but there was no question who was boss. If the lord was a master craftsman like Zel'dovich or Landau, the lord/serf team could be richly productive. If the lord was authoritarian and not so outstanding (as was commonly the case), the result could be tragic: a waste of human talent and a miserable life for the serfs.

In the Soviet system, each great mentor such as Zel'dovich produced just one research team, albeit a tremendously powerful one, one unequaled anywhere in the West. By contrast, great American or British mentors like Wheeler and Sciama produce as their progeny many smaller and weaker research groups, scattered throughout the land, but those groups can have a large cumulative impact on physics. The American and British mentors have a constant influx of new, young people to help keep their minds and ideas fresh. In those rare cases where Soviet mentors wanted to start over afresh, they had to break their ties with their old team in a manner which could be highly traumatic.

This, in fact, was destined to happen to Zel'dovich: He began building his astrophysics team in 1961; by 1964 it was superior to any other theoretical astrophysics team anywhere in the world; then in 1978, soon after the golden age ended, came a traumatic, explosive split in which almost everybody in Zel'dovich's team went one way and he went another, psychologically wounded but free from encumbrances, free to begin building afresh. Sadly, his rebuilding would not be successful. Never again would he surround himself with a team so talented and powerful as that which he, with Novikov's assistance, had led. But Novikov, now an independent researcher, would come into his own in the 1980s as the talented leader of a reconstructed team.

---

of Wheeler's students, John Graves and Dieter Brill, discovered that it describes a charged black hole.

We can depict the curvature of space around a charged black hole, and the hole's electric field lines, using an embedding diagram (left half of Figure 7.6). This diagram is essentially the same as the one in the lower right of Figure 3.4, but with the star (black portion of Figure 3.4) removed because the star is inside the black hole and thus no longer has contact with the external Universe. Stated more carefully,

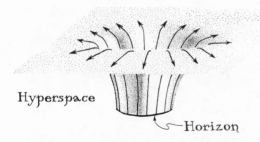 

7.6 Electric field lines emerging from the horizon of an electrically charged black hole. *Left:* Embedding diagram. *Right:* View of the embedding diagram from above.

this diagram depicts the equatorial "plane"—a two-dimensional piece of the hole's space—outside the black hole, embedded in a flat, three-dimensional hyperspace. (For a discussion of the meaning of such diagrams, see Figure 3.3 and the accompanying text.) The equatorial "plane" is cut off at the hole's horizon, so we are seeing only the hole's exterior, not its interior. The horizon, which in reality is the surface of a sphere, looks like a circle in the diagram because we are seeing only its equator. The diagram shows the hole's electric field lines sticking radially out of the horizon. If we look down on the diagram from above (right side of Figure 7.6), then we do not see the curvature of space, but we do see the electric field lines more clearly.

The effects of spin on a black hole were not understood until the late 1960s. The understanding came largely from Brandon Carter, one of Dennis Sciama's students at Cambridge University.

When Carter joined Sciama's group in autumn 1964, Sciama immediately suggested, as his first research problem, a study of the implosion of realistic, spinning stars. Sciama explained that all previous calculations of implosion had dealt with idealized, nonspinning stars, but that the time and tools now seemed right for an assault on the effects of spin. A New Zealander mathematician named Roy Kerr had just published a paper giving a solution of Einstein's field equation that describes the spacetime curvature outside a spinning star. This was the first solution for a spinning star that anyone had ever found. Unfortunately, Sciama explained, it was a very special solution; it surely could

Left: Roy Kerr ca. 1975. *Right:* Brandon Carter lecturing about black holes at a summer school in the French Alps in June 1972. [Left: courtesy Roy Kerr; right: photo by Kip Thorne.]

not describe *all* spinning stars. Spinning stars have lots of "hair" (lots of properties such as complicated shapes and complicated internal motions of their gas), and Kerr's solution did not have much "hair" at all: The shapes of its spacetime curvature were very smooth, very simple; too simple to correspond to typical spinning stars. Nevertheless, Kerr's solution of Einstein's field equation was a place to start.

Few research problems have the immediate payoff that this one did: Within a year Carter had shown mathematically that Kerr's solution describes not a spinning star, but rather a spinning black hole. (This discovery was also made, independently, by Roger Penrose in London, and by Robert Boyer in Liverpool and Richard Lindquist, a former student of Wheeler's who was working at Wesleyan University in Middletown, Connecticut.) By the mid-1970s, Carter and others had gone on to show that Kerr's solution describes not just one special type of spinning black hole, but rather every spinning black hole that can possibly exist.

The physical properties of a spinning black hole are embodied in the mathematics of Kerr's solution, and Carter, by plumbing that mathematics, discovered just what those properties should be. One of the most interesting is a tornado-like swirl that the hole creates in the space around itself.

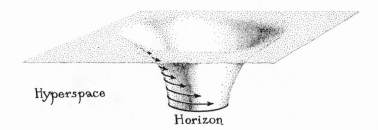

Hyperspace

Horizon

7.7   An embedding diagram showing the "tornado-like swirl" of space created by the spin of a black hole.

This swirl is depicted in the embedding diagram of Figure 7.7. The trumpet-horn-shaped surface is the hole's equatorial sheet (a two-dimensional piece of the hole's space), as embedded in a flat, three-dimensional hyperspace. The hole's spin grabs hold of its surrounding space (the trumpet-horn surface) and forces it to rotate in a tornado-like manner, with speeds proportional to the lengths of the arrows on the diagram. Far from a tornado's core the air rotates slowly, and, similarly, far from the hole's horizon space rotates slowly. Near the tornado's core the air rotates fast, and, similarly, near the horizon space rotates fast. At the horizon, space is locked tightly onto the horizon: It rotates at precisely the same rate as the horizon spins.

This swirl of space has an inexorable influence on the motions of particles that fall into the hole. Figure 7.8 shows the trajectories of two such particles, as viewed in the reference frame of a static, external observer—that is, in the frame of an observer who does not fall through the horizon and into the hole.

The first particle (Figure 7.8a) is dropped gently into the hole. If the hole were not spinning, this particle, like the surface of an imploding star, would move radially inward faster and faster at first; but then, as observed by the static, external observer, it would slow its infall and become frozen right at the horizon. (Recall the "frozen stars" of Chapter 6.) The hole's spin changes this in a very simple way: The spin makes space swirl, and the swirl of space makes the particle, as it nears the horizon, rotate in lockstep with the horizon itself. The particle thereby becomes frozen onto the spinning horizon and, as seen by the static, external observer, it circles around and around with the horizon forever. (Similarly, when a spinning star implodes to form a spinning hole, as seen by a static, external observer the star's surface "freezes" onto the spinning horizon, circling around and around with it forever.)

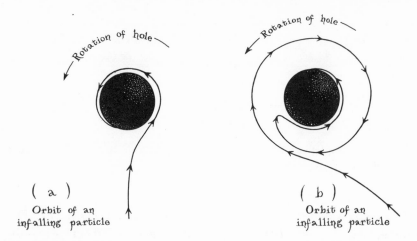

7.8 The trajectories in space of two particles that are thrown toward a black hole. (The trajectories are those that would be measured in a static, external reference frame.) Despite their very different initial motions, both particles are dragged, by the swirl of space, into precisely the same lockstep rotation with the hole as they near the horizon.

Though external observers see the particle of Figure 7.8a freeze onto the spinning horizon and stay there forever, the particle itself sees something quite different. As the particle nears the horizon, gravitational time dilation forces the particle's time to flow more and more slowly, compared with the time of a static, external reference frame. When an infinite amount of external time has passed, the particle has experienced only a finite and very small amount of time. In that finite time, the particle has reached the hole's horizon, and in the next few moments of its time, it plunges right on through the horizon and down toward the hole's center. This enormous difference between the particle's infall as seen by the particle and as seen by external observers is completely analogous to the difference between a stellar implosion as seen on the star's surface (rapid plunge through the horizon) and as seen by external observers (freezing of the implosion; last part of Chapter 6).

The second particle (Figure 7.8b) is thrown toward the hole on an inspiraling trajectory that rotates oppositely to the hole's spin. However, as the particle spirals closer and closer to the horizon, the swirl of space grabs hold of it and reverses its rotational motion. Like the first particle, it is forced into lockstep rotation with the horizon, as seen by external observers.

Besides creating a swirl in space, the spin of a black hole also distorts the hole's horizon, in much the same way as the spin of the Earth distorts the Earth's surface. Centrifugal forces push the spinning Earth's equator outward a distance of 22 kilometers relative to its poles. Similarly, centrifugal forces make a black hole's horizon bulge out at its equator in the manner depicted in Figure 7.9. If the hole does not spin, its horizon is spherical (left half of figure). If the hole spins rapidly, its horizon bulges out strongly (right half of figure).

If the hole were to spin extremely rapidly, centrifugal forces would tear its horizon apart much like they fling water out of a bucket when the bucket spins extremely rapidly. Thus, there is some maximum spin rate at which the hole can survive. The hole on the right half of Figure 7.9 is spinning at 58 percent of this maximum.

Is it possible to spin a hole up beyond its maximum allowed rate, and thereby destroy the horizon and catch a glimpse of what is inside? Unfortunately not. In 1986, a decade after the golden age, Werner Israel showed that, if one tries to make the hole spin faster than its maximum by any method at all, one will always fail. For example, if one tries to speed up a maximally spinning hole by throwing fast-spinning matter into it, centrifugal forces will prevent the fast-spinning matter from reaching the horizon and entering the hole. More to the point, perhaps, any tiny random interaction of a maximally spinning hole with the surrounding Universe (for example, the gravitational pull of distant stars) acts to slow the spin a bit. The laws of

7.9 The shapes of the horizons of two black holes, one *(left)* not spinning, and the other *(right)* spinning with a spin rate 58 percent of the maximum. The effect of the spin on the horizon shape was discovered in 1973 by Larry Smarr, a student at Stanford University who was inspired by Wheeler.

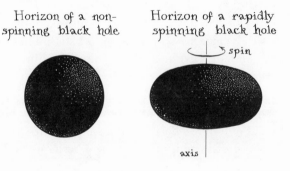

physics, it seems, don't want to let anyone outside the hole peek into its interior and discover the quantum gravity secrets locked up in the hole's central singularity (Chapter 13).

For a hole with the mass of the Sun, the maximum spin rate is one revolution each 0.000062 second (62 microseconds). Since the hole's circumference is about 18.5 kilometers, this corresponds to a spin speed of about (18.5 kilometers)/(0.000062 second), which is about the speed of light, 299,792 kilometers per second (not entirely a coincidence!). A hole whose mass is 1 million Suns has a 1 million times larger circumference than a 1-solar-mass hole, so its maximum spin rate (the rate which makes it spin at about the speed of light) is 1 million times smaller, one revolution each 62 seconds.

In 1969, Roger Penrose (about whom we shall learn much in Chapter 13) made a marvelous discovery. By manipulating the equations of Kerr's solution to the Einstein field equation, he discovered that a spinning black hole stores *rotational energy* in the swirl of space around itself, and because the swirl of space and the swirl's energy are *outside* the hole's horizon and not inside, this energy can actually be extracted and used to power things. Penrose's discovery was marvelous because the hole's rotational energy is huge. If the hole spins at its maximum possible rate, its efficiency at storing and releasing energy is 48 times higher than the efficiency of all the Sun's nuclear fuel. If it were to burn all its nuclear fuel over its entire lifetime (actually, it will not burn *all*), the Sun would only be able to convert a fraction 0.006 of its mass into heat and light. If one were to extract all of a fast-spinning hole's rotational energy (thereby halting its spin), one would get out 48 × 0.006 = 29 percent of the hole's mass as usable energy.

Amazingly, physicists had to search for seven years before they discovered a practical method by which nature might extract a hole's spin energy and put it to use. Their search led the physicists through one crazy method after another, all of which would work in principle but none of which showed much practical promise, before they finally discovered nature's cleverness. In Chapter 9 I shall describe this search and discovery, and its payoff: a black-hole "machine" for powering quasars and gigantic jets.

If, as we have seen, electric charge produces electric field lines that stick radially out of a hole's horizon, and spin produces a swirl in space around the hole, a distortion of the horizon's shape, and a storage of energy, then what happens when a hole has both charge and spin? Unfortunately, the answer is not terribly interesting; it contains little new. The hole's charge produces the usual electric field lines. The hole's spin creates the usual swirl of the hole's space, it stores the usual rotational energy, and it makes the horizon's equator bulge out in the usual manner. The only things new are a few rather uninteresting magnetic field lines, created by the swirl of space as it flows through the electric field. (These field lines are *not* a new form of "hair" on the hole; they are merely a manifestation of the interaction of the old, standard forms of hair: the interaction of the spin-induced swirl with the charge-induced electric field.) All the properties of a spinning, charged black hole are embodied in an elegant solution to the Einstein field equation derived in 1965 by Ted Newman at the University of Pittsburgh and a bevy of his students: Eugene Couch, K. Chinnapared, Albert Exton, A. Prakash, and Robert Torrence.

Not only can black holes spin; they can also pulsate. Their pulsations, however, were not discovered mathematically until nearly a decade after their spin; the discovery was impeded by a powerful mental block.

For three years (1969–71) John Wheeler's progeny "watched" black holes pulsate, and didn't know what they were seeing. The progeny were Richard Price (my student, and thus Wheeler's intellectual grandson), C. V. Vishveshwara and Lester Edelstein (students of Charles Misner's at the University of Maryland, and thus also Wheeler's intellectual grandsons), and Frank Zerilli (Wheeler's own student at Princeton). Vishveshwara, Edelstein, Price, and Zerilli watched black holes pulsate in computer simulations and in pencil-and-paper calculations. What they thought they were seeing was gravitational radiation (ripples of spacetime curvature) bouncing around in the vicinity of a hole, trapped there by the hole's own spacetime curvature. The trapping was not complete; the ripples would gradually leak out of the hole's vicinity, and fly away. This was sort of cute, but not terribly interesting.

In autumn 1971, Bill Press, a new graduate student in my group, realized that the ripples of spacetime curvature bouncing around near a hole could be thought of as pulsations of the black hole itself. After all, as seen from outside its horizon, the hole consists of nothing but space-time curvature. The ripples of curvature were thus nothing more nor less than pulsations of the hole's curvature, and therefore pulsations of the hole itself.

This change of viewpoint had a huge impact. If we think of black holes as able to pulsate, then it is natural to ask whether there are any similarities between their pulsations and the pulsations ("ringing") of a bell, or the pulsations of a star. Before Press's insight, such questions weren't asked. Afterwards, such questions were obvious.

A bell and a star have natural frequencies at which they like to pulsate. (The bell's natural frequencies produce its pure ringing tone.) Are there similarly natural frequencies at which a black hole likes to pulsate? Yes, Press discovered, using computer simulations. This discovery triggered Chandrasekhar, together with Steven Detweiler (an intellectual great-grandson of Wheeler's), to embark on a project of cataloging all of a black hole's natural frequencies of pulsation. We shall return to those frequencies, the bell-like tones of a black hole, in Chapter 10.

When a rapidly spinning automobile wheel is slightly out of alignment, it can begin to vibrate, and its vibrations can begin to extract energy from the spin and use that energy to grow stronger and stronger. The vibrations can grow so strong, in fact, that in extreme cases they can even tear the wheel off the car. Physicists describe this by the phrase "the wheel's vibrations are unstable." Bill Press was aware of this and of an analogous behavior of spinning stars, so it was natural for him to ask, when he discovered that black holes can pulsate, "If a black hole spins rapidly, will its pulsations be unstable? Will they extract energy from the hole's spin and use that energy to grow stronger and stronger, and can the pulsations grow so strong that they tear the hole apart?" Chandrasekhar (who was not yet deeply immersed in black-hole research) thought yes. I thought no. In November 1971, we made a bet.

The tools did not yet exist for resolving the bet. What kinds of tools were needed? Since the pulsations would begin weak and only gradually grow strong (if they grew at all), they could be regarded as small "perturbations" of the hole's spacetime curvature—just as the vibrations of a ringing wine glass are small perturbations of the glass's shape. This meant that the hole's pulsations could be analyzed using the perturbation methods whose spirit was described in Box 7.1 above. However, the specific perturbation methods which Price, Press, Vishveshwara, Chandrasekhar, and others were using in the autumn of 1971 would work only for perturbations of *non*spinning, or very slowly spinning, black holes. What they needed were entirely new perturbation methods, methods for perturbations of rapidly spinning holes.

The effort to devise such perturbation methods became a hot topic in 1971 and 1972. My students, Misner's students, Wheeler's students, and Chandrasekhar with his student John Friedman all worked on it, as did others. The competition was stiff. The winner was Saul Teukolsky, a student of mine from South Africa.

Teukolsky recalls vividly the scene when the equations of his method fell into place. "Sometimes when you play with mathematics, your mind starts picking out patterns," he says. "I was sitting at the kitchen table in our apartment in Pasadena one May evening in 1972, playing with the mathematics; and my wife Roz was making crepes in a Teflon pan, which was supposed not to stick. The crepes kept sticking. Everytime she poured the batter in she would bang the pan on the countertop. She was cursing and banging, and I was yelling at her to be quiet because I was getting excited; the mathematical terms were start-

A party at Mama Kovács's home in New York City, December 1972. *Left to right:* Kip Thorne, Margaret Press, Bill Press, Roselyn Teukolsky, and Saul Teukolsky. [Courtesy Sándor J. Kovács.]

ing to cancel each other in my formulas. Everything was canceling! The equations were falling into place! As I sat there staring at my amazingly simple equations, I was filled with this feeling of how dumb I had been; I could have done it six months earlier; all I had to do was collect the right terms together."

Using Teukolsky's equations, one could analyze all sorts of problems: the natural frequencies of black-hole pulsations, the stability of a hole's pulsations, the gravitational radiation produced when a neutron star gets swallowed by a black hole, and more. Such analyses, and extensions of Teukolsky's methods, were immediately undertaken by a small army of researchers: Alexi Starobinsky (a student of Zel'dovich's), Bob Wald (a student of Wheeler's), Jeff Cohen (a student of Dieter Brill's, who was a student of Wheeler's), and many others. Teukolsky himself, with Bill Press, commanded the most important problem: the stability of black-hole pulsations.

Their conclusion, derived by a mixture of computer calculations and calculations with formulas, was disappointing: No matter how fast a black hole spins, its pulsations are stable.[6] The hole's pulsations *do* extract rotational energy from the hole, but they also radiate energy away as gravitational waves; and the rate at which they radiate energy is always greater than the rate they extract it from the hole's spin. Their pulsational energy thus always dies out. It never grows, and the hole therefore cannot be destroyed by its pulsations.

Chandrasekhar, dissatisfied with this Press–Teukolsky conclusion because of its crucial reliance on computer calculations, refused to concede our bet. Only when the entire proof could be done directly with formulas would he be fully convinced. Fifteen years later Bernard Whiting, a former postdoc of Hawking's (and thus an intellectual grandson of Sciama's), gave such a proof, and Chandrasekhar threw in the towel.[7]

Chandrasekhar is even more of a perfectionist than I. He and Zel'dovich are at opposite ends of the perfectionist spectrum. So in 1975, when the youths of the golden age declared the golden age finished

---

6. A significant, mathematical piece of the proof of stability was provided, independently, by Steven Detweiler and James Ipser at Chicago, and a missing piece of the proof was supplied a year later by James Hartle and Dan Wilkins at the University of California at Santa Barbara.

7. Chandrasekhar was supposed to give me a subscription to *Playboy* as my reward, but my feminist mother and sisters made me feel so guilty that I requested instead a subscription to *The Listener.*

and exited from black-hole research en masse, Chandrasekhar was annoyed. These youths had carried Teukolsky's perturbation methods far enough to prove that black holes are probably stable, but they had not brought the methods into a form where other physicists could automatically compute *all* details of *any* desired black-hole perturbation—be it a pulsation, the gravitational waves from an infalling neutron star, a black-hole bomb, or whatever. This incompleteness was rankling.

Thus Chandrasekhar, in 1975 at age sixty-five, turned the full force of his mathematical prowess onto Teukolsky's equations. With unfailing energy and mathematical insight, he drove forward, through the complex mathematics, organizing it into a form that has been characterized as "rococo: splendorous, joyful, and immensely ornate." Finally in 1983, at age seventy-three, he completed his task and published a treatise entitled *The Mathematical Theory of Black Holes*—a treatise that will be a mathematical handbook for black-hole researchers for decades to come, a handbook from which they can extract methods for solving any black-hole perturbation problem that catches their fancy.

8

The Search

*in which a method to search
for black holes in the sky
is proposed and pursued
and succeeds (probably)*

The Method

Imagine yourself as J. Robert Oppenheimer. It is 1939; you have just convinced yourself that massive stars, when they die, must form black holes (Chapters 5 and 6). Do you now sit down with astronomers and plan a search of the sky for evidence that black holes truly exist? No, not at all. If you are Oppenheimer, then your interests are in fundamental physics; you may offer your ideas to astronomers, but your own attention is now fixed on the atomic nucleus—and on the outbreak of World War II, which soon will embroil you in the development of the atomic bomb. And what of the astronomers; do they take up your idea? No, not at all. There is a conservatism abroad in the astronomical community, except for that "wild man" Zwicky, pushing his neutron stars (Chapter 5). The worldview that rejected Chandrasekhar's maximum mass for a white-dwarf star (Chapter 4) still holds sway.

Imagine yourself as John Archibald Wheeler. It is 1962; you are beginning to be convinced, after mighty resistance, that some massive stars must create black holes when they die (Chapters 6 and 7). Do you now sit down with astronomers and plan a search for them? No, not at all. If you are Wheeler, then your interest is riveted on the fiery marriage of general relativity with quantum mechanics, a marriage that may take place at the center of a black hole (Chapter 13). You are preaching to physicists that the endpoint of stellar implosion is a great crisis, from which deep new understanding may emerge. You are not preaching to astronomers that they should search for black holes, or even neutron stars. Of searches for black holes you say nothing; of the more promising idea to search for a neutron star, you echo in your writings the conservative view of the astronomical community: "Such an object will have a diameter of the order of 30 kilometers. . . . it will cool rapidly. . . . There is about as little hope of seeing such a faint object as there is of seeing a planet belonging to another star" (in other words, no hope at all).

Imagine yourself as Yakov Borisovich Zel'dovich. It is 1964; Mikhail Podurets, a member of your old hydrogen bomb design team, has just finished his computer simulations of stellar implosion including the effects of pressure, shock waves, heat, radiation, and mass ejection (Chapter 6). The simulations produced a black hole (or, rather, a computer's version of one). You are now fully convinced that some massive stars, when they die, must form black holes. Do you next sit down with astronomers and plan a search for them? Yes, by all means. If you are Zel'dovich, then you have little sympathy for Wheeler's obsession with the endpoint of stellar implosion. The endpoint will be hidden by the hole's horizon; it will be invisible. By contrast, the horizon itself and the hole's influence on its surroundings might well be observable; you just need to be clever enough to figure out how. Understanding the observable part of the Universe is your obsession, if you are Zel'dovich; how could you possibly resist the challenge of searching for black holes?

Where should your search begin? Clearly, you should begin in our own Milky Way galaxy—our disk-shaped assemblage of 10^{12} stars. The other big galaxy nearest to our own, Andromeda, is 2 million light-years away, 20 times farther than the size of the Milky Way; see Figure 8.1. Thus, any star or gas cloud or other object in Andromeda will appear 20 times smaller and 400 times dimmer than a similar one in the Milky Way. Therefore, if black holes are hard to detect in the Milky Way, they will be 400 times harder to detect in Andromeda—and enormously harder still in the 1 billion or so large galaxies beyond Andromeda.

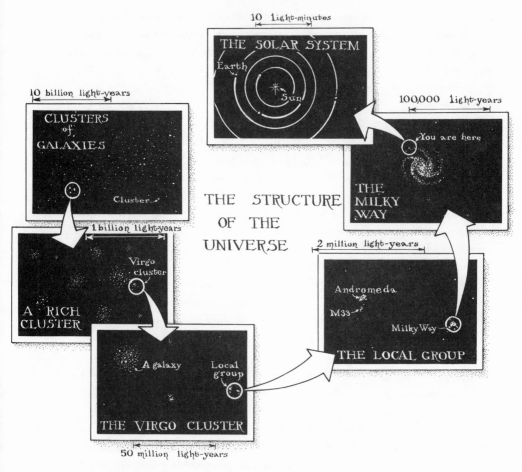

8.1 A sketch of the structure of our Universe.

If searching nearby is so important, then why not search in our own solar system, the realm stretching from the Sun out to the planet Pluto? Might there be a black hole here, among the planets, unnoticed because of its darkness? No, clearly not. The gravitational pull of such a hole would be greater than that of the Sun; it would totally disrupt the orbits of the planets; no such disruption is seen. The nearest hole, therefore, must be far beyond the orbit of Pluto.

How far beyond Pluto? You can make a rough estimate. If black holes are formed by the deaths of massive stars, then the nearest hole is not likely to be much closer than the closest massive star, Sirius, at 8 light-years from Earth; and it almost certainly won't be closer than the closest of all stars (aside from the Sun), Alpha Centauri, at 4 light-years distance.

How could an astronomer possibly detect a black hole at such a great distance? Could an astronomer just watch the sky for a moving, dark object which blots out the light from stars behind it? No. With its circumference of roughly 50 kilometers and its distance of at least 4 light-years, the hole's dark disk will subtend an angle no larger than 10^{-7} arc second. That is roughly the thickness of a human hair as seen from the distance of the Moon, and 10 million times smaller than the resolution of the world's best telescopes. The moving dark object would be invisibly tiny.

If one could not see the hole's dark disk as the hole goes in front of a star, might one see the hole's gravity act like a lens to magnify the star's light (Figure 8.2)? Might the star appear dim at first, then brighten as the hole moves between Earth and the star, then dim again as the hole moves on? No, this method of search also will fail. The reason it will fail depends on whether the star and the hole are orbiting around each other and thus are close together, or are separated by

8.2   A black hole's gravity should act like a lens to change the apparent size and shape of a star as seen from Earth. In this figure the hole is precisely on the line between the star and the Earth, so light rays from the star can reach the Earth equally well by going over the top of the hole, or under the bottom, or around the front, or around the back. All the light rays reaching Earth move outward from the star on a diverging cone; as they pass the hole they get bent down toward Earth; they then arrive at Earth on a converging cone. The resulting image of the star on the Earth's sky is a thin ring. This ring has far larger surface area, and hence far larger total brightness, than the star's image would have if the black hole were absent. The ring is too small to be resolved by a telescope, but the star's total brightness can be increased by a factor of 10 or 100 or more.

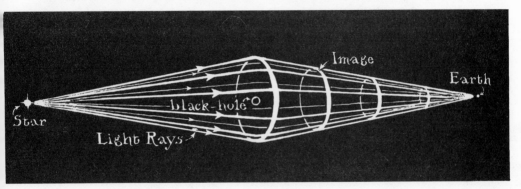

typical interstellar distances. If they are close together, then the tiny hole will be like a hand-held magnifying glass placed upright on a windowsill on the eighty-ninth floor of the Empire State Building, and then viewed from several kilometers distance. Of course, the tiny magnifying glass has no power to magnify the building's appearance, and similarly the hole has no effect on the star's appearance.

If the star and the hole are far apart as in Figure 8.2, however, the strength of the focusing can be large, an increase of 10 or 100 or more in stellar brightness. But interstellar distances are so vast that the necessary Earth–hole–star lineup would be an exceedingly rare event, so rare that to search for one would be hopeless. Moreover, even if such a lensing were observed, the light rays from star to Earth would pass the hole at so large a distance (Figure 8.2) that there would be room for an entire star to sit at the hole's location and act as the lens. An astronomer on Earth thus could not know whether the lens was a black hole or merely an ordinary, but dim, star.

Zel'dovich must have gone through a chain of reasoning much like this as he sought a method to observe black holes. His chain led finally to a method with some promise (Figure 8.3): Suppose that a black hole and a star are in orbit around each other (they form a *binary system*). When astronomers train their telescopes on this binary, they will see light from only the star; the hole will be invisible. However, the star's light will give evidence of the hole's presence: As the star moves around the hole in its orbit, it will travel first toward the Earth and then away. When it is traveling toward us, the Doppler effect should shift the star's light toward the blue, and when moving away, toward the red. Astronomers can measure such shifts with high precision, since the star's light, when sent through a spectrograph (a sophisticated form of prism), exhibits sharp spectral lines, and a slight change in the wavelength (color) of such a line stands out clearly. From a measurement of the shift in wavelength, astronomers can infer the velocity of the star toward or away from Earth, and by monitoring the shift as time passes, they can infer how the star's velocity changes with time. The magnitude of those changes might typically be somewhere between 10 and 100 kilometers per second, and the accuracy of the measurements is typically 0.1 kilometer per second.

What does one learn from such high-precision measurements of the star's velocity? One learns something about the mass of the hole: The more massive is the hole, the stronger is its gravitational pull on the star, and thus the stronger must be the centrifugal forces by which the star resists getting pulled into the hole. To acquire strong centrifugal forces, the star must move rapidly in its orbit. Thus, large orbital velocity goes hand in hand with large black-hole mass.

To search for a black hole, then, astronomers should look for a star whose spectra show a telltale periodic shift from red to blue to red to blue. Such a shift is an unequivocal sign that the star has a companion. The astronomers should measure the star's spectra to infer the velocity of the star around its companion, and from that velocity they should infer the companion's mass. If the companion is very massive and no

8.3 Zel'dovich's proposed method of searching for a black hole. (a) The hole and a star are in orbit around each other. If the hole is heavier than the star, then its orbit is smaller than the star's as shown (that is, the hole moves only a little while the star moves a lot). If the hole were lighter than the star, then its orbit would be the larger one (that is, the star would move only a little while the hole moves a lot). When the star is moving away from Earth, as shown, its light is shifted toward the red (toward longer wavelength). (b) The light, upon entering a telescope on Earth, is sent through a spectrograph to form a spectrum. Here are shown two spectra, the top recorded when the star is moving away from Earth, the bottom a half orbit later when the star is moving toward Earth. The wavelengths of the sharp lines in the spectra are shifted relative to each other. (c) By measuring a sequence of such spectra, astronomers can determine how the velocity of the star toward and away from the Earth changes with time, and from that changing velocity, they can determine the mass of the object around which the star orbits. If the mass is larger than about 2 Suns and no light is seen from the object, then the object might be a black hole.

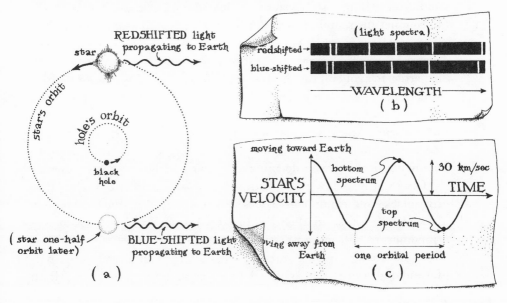

light is seen from it at all, then the companion might well be a black hole. This was Zel'dovich's proposal.

Although this method was vastly superior to any previous one, it nevertheless is fraught with many pitfalls, of which I shall discuss just two: First, the weighing of the dark companion is not straightforward. The star's measured velocity depends not only on the companion's mass, but also on the mass of the star itself, and on the inclination of the binary's orbital plane to our line of sight. While the star's mass and the inclination may be inferred from careful observations, one cannot do so with ease or with good accuracy. As a result, one can readily make large errors (say, a factor of 2 or 3) in one's estimate of the mass of the dark companion. Second, black holes are not the only kind of dark companions that a star might have. For example, a neutron-star companion would also be dark. To be certain the companion is not a neutron star, one needs to be very confident that it is much heavier than the maximum allowed for a neutron star, about 2 solar masses. Two neutron stars in a tight orbit around each other could also be dark and could weigh as much as 4 Suns. The dark companion might be such a system; or it might be two cold white dwarfs in a tight orbit with total mass as much as 3 Suns. And there are other kinds of stars that, while not completely dark, can be rather massive and abnormally dim. One must look very carefully at the measured spectra to be certain there is no sign of tiny amounts of light from such stars.

Astronomers had worked hard over the preceding decades to observe and catalog binary star systems, so it was not necessary for Zel'dovich to conduct his search directly in the sky; he could search the astronomers' catalogs instead. However, he had neither the time nor the patience to comb through the catalogs himself, nor did he have the expertise to avoid all the pitfalls. Therefore, as was his custom in such a situation, he commandeered the time and the talents of someone else—in this case, Oktay Guseinov, an astronomy graduate student who already knew much about binary stars. Together, Guseinov and Zel'dovich found five promising black-hole candidates among the many hundreds of well-documented binary systems in the catalogs.

Over the next few years, astronomers paid little attention to these five black-hole candidates. I was rather annoyed at the astronomers' lack of interest, so in 1968 I enlisted Virginia Trimble, a Caltech astronomer, to help me revise and extend the Zel'dovich–Guseinov list. Trimble, though only months past her Ph.D., had already acquired a formidable knowledge of the lore of astronomy. She knew all the pitfalls we might encounter—those described above and many more—and she could gauge them accurately. By searching through the catalogs ourselves, and by collating all the published data we could find on the most promising binaries, we came up with a new list of eight black-hole candidates. Unfortunately, in all eight cases, Trimble could invent a semi-reasonable non–black-hole explanation for why the companion was so dark. Today, a quarter century later, none of our candidates has survived. It now seems likely that none of them is truly a black hole.

Zel'dovich knew, when he conceived it, that this binary star method of search was a gamble, by no means assured of success. Fortunately, his brainstorming on how to search for black holes produced a second idea—an idea conceived simultaneously and independently, in 1964, by Edwin Salpeter, an astrophysicist at Cornell University in Ithaca, New York.

Suppose that a black hole is traveling through a cloud of gas—or, equivalently, as seen by the hole a gas cloud is traveling past it (Figure 8.4). Then streams of gas, accelerated to near the speed of light by the hole's gravity, will fly around opposite sides of the hole and come crashing together at the hole's rear. The crash, in the form of a *shock front* (a sudden, large increase in density), will convert the gas's huge energy of infall into heat, causing it to radiate strongly. In effect, then,

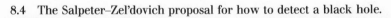

8.4   The Salpeter–Zel'dovich proposal for how to detect a black hole.

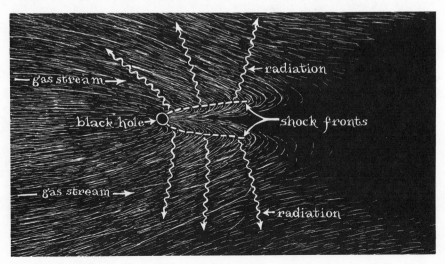

the black hole will serve as a machine for converting some of the mass of infalling gas into heat and then radiation. This "machine" could be highly efficient, Zel'dovich and Salpeter deduced—far more efficient, for example, than the burning of nuclear fuel.

Zel'dovich and his team mulled over this idea for two years, looking at it first from this direction and then that, searching for ways to make it more promising. However, it was but one of dozens of ideas about black holes, neutron stars, supernovae, and the origin of the Universe that they were pursuing, and it got only a little attention. Then, one day in 1966, in an intense discussion, Zel'dovich and Novikov together realized they could combine the binary star idea with the infalling gas idea (Figure 8.5).

Strong winds of gas (mostly hydrogen and helium) blow off the surfaces of some stars. (The Sun emits such a wind, though only a weak one.) Suppose that a black hole and a wind-emitting star are in orbit around each other. The hole will capture some of the wind's gas, heat it in a shock front, and force it to radiate. At the one-meter-square black-

8.5 The Zel'dovich–Novikov proposal of how to search for a black hole. A wind, blowing off the surface of a companion star, is captured by the hole's gravity. The wind's streams of gas swing around the hole in opposite directions and collide in a sharp shock front, where they are heated to millions of degrees temperature and emit X-rays. Optical telescopes should see the star orbiting around a heavy, dark companion. X-ray telescopes should see X-rays from the companion.

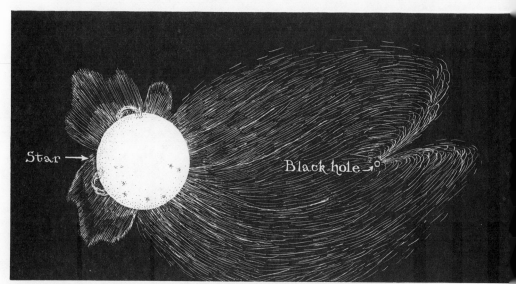

board in Zel'dovich's Moscow apartment, he and Novikov estimated the temperature of the shocked gas: several million degrees.

Gas at such a temperature does not emit much light. It emits X-rays instead. Thus, Zel'dovich and Novikov realized, among those black holes which orbit around stellar companions, a few (though not most) might shine brightly with X-rays.

To search for black holes, then, one could use a combination of optical telescopes and X-ray telescopes. The black-hole candidates would be binaries in which one object is an optically bright but X-ray-dark star, and the other is an optically dark but X-ray-bright object (the black hole). Since a neutron star could also capture gas from a companion, heat it in shock fronts, and produce X-rays, the weighing of the optically dark but X-ray-bright object would be crucial. One must be sure it is heavier than 2 Suns and thus not a neutron star.

There was but one problem with this search strategy. In 1966, X-ray telescopes were extremely primitive.

The Search

The trouble with X-rays, if you are an astronomer, is that they cannot penetrate the Earth's atmosphere. (If you are a human, that is a virtue, since X-rays cause cancer and mutations.)

Fortunately, experimental physicists with vision, led by Herbert Friedman of the U.S. Naval Research Laboratory (NRL), had been working since the 1940s to lay the groundwork for space-based X-ray astronomy. Friedman and his colleagues had begun, soon after World War II, by flying instruments to study the Sun on captured German V-2 rockets. Friedman has described their first flight, on 28 June 1946, which carried in the rocket's nose a spectrograph for studying the Sun's far ultraviolet radiation. (Far ultraviolet rays, like X-rays, cannot penetrate the Earth's atmosphere.) After soaring above the atmosphere briefly and collecting data, "the rocket returned to Earth, nose down, in streamlined flight and buried itself in an enormous crater some 80 feet in diameter and 30 feet deep. Several weeks of digging recovered just a small heap of unidentifiable debris; it was as if the rocket had vaporized on impact."

From this inauspicious beginning, the inventiveness, persistence, and hard work of Friedman and others brought ultraviolet and X-ray astronomy step by step to fruition. By 1949 Friedman and his colleagues were flying Geiger counters on V-2 rockets to study X-rays from the Sun. By the late 1950s, now flying their counters on American-made Aerobee rockets, Friedman and colleagues were studying ultraviolet radiation not only from the Sun, but also from stars. X-rays, however, were another matter. Each second the Sun dumped 1 million X-rays onto a square centimeter of their Geiger counter, so detecting the Sun with X-rays was relatively easy. Theoretical estimates, however, suggested that the brightest X-ray stars would be 1 billion times fainter than the Sun. To detect so faint a star would require an X-ray detector 10 million times more sensitive than those that Friedman was flying in 1958. Such an improvement was a tall order, but not impossible.

By 1962, the detectors had been improved 10,000-fold. With just another factor of a thousand to go, other research groups, impressed by Friedman's progress, were beginning to compete with him. One, a team led by Riccardo Giacconi, would become a formidable competitor.

In a peculiar way, Zel'dovich may have shared responsibility for Giacconi's success. In 1961, the Soviet Union unexpectedly abrogated a mutual Soviet/American three-year moratorium on the testing of nuclear weapons, and tested the most powerful bomb ever exploded by humans—a bomb designed by Zel'dovich's and Sakharov's teams at the Installation (Chapter 6). In panic, the Americans prepared new bomb tests of their own. These would be the first American tests in the era of Earth-orbiting spacecraft. For the first time it would be possible to measure, from space, the X-rays, gamma rays, and high-energy particles emerging from nuclear explosions. Such measurements would be crucial for monitoring future Soviet bomb tests. To make such measurements on the impending series of American tests, however, would require a crash program. The task of organizing and leading it went to Giacconi, a twenty-eight-year-old experimental physicist at American Science and Engineering (a private Cambridge, Massachusetts, company), who had recently begun to design and fly X-ray detectors like Friedman's. The U.S. Air Force gave Giacconi all the money he needed, but little time. In less than a year, he augmented his six-person X-ray astronomy team by seventy new people, designed, built, and tested a variety of weapons-blast monitoring instruments, and flew them with a 95 percent success rate in twenty-four rockets and six satellites. This experience molded the core members of his group into a loyal, dedicated, and highly skilled team, ideally primed to beat all competitors in the creation of X-ray astronomy.

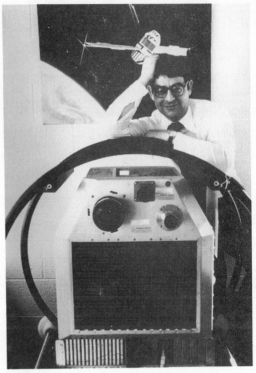

Left: Herbert Friedman, with payload from an Aerobee rocket, in 1968. *Right:* Riccardo Giacconi with the Uhuru X-ray detector, ca. 1970. [Left: courtesy U.S. Naval Research Laboratory; right: courtesy R. Giacconi.]

Giacconi's seasoned team took its first astronomical step with a search for X-rays from the Moon, using a detector patterned after Friedman's, and like Friedman, flying it on an Aerobee rocket. Their rocket, launched from White Sands, New Mexico, at one minute before midnight on 18 June 1962, climbed quickly to an altitude of 230 kilometers, then fell back to Earth. For 350 seconds it was high enough above the Earth's atmosphere to detect the Moon's X-rays. The data, telemetered back to the ground, were puzzling; the X-rays were far stronger than expected. When examined more closely, the data were even more surprising. The X-rays seemed to be coming not from the Moon, but from the constellation Scorpius (Figure 8.6b). For two months, Giacconi and his team members (Herbert Gursky, Frank Paolini, and Bruno Rossi) sought errors in their data and apparatus. When none could be found, they announced their discovery: The first X-ray star ever detected, *5000 times brighter than theoretical astrophysicists had predicted.* Ten months later, Friedman's team confirmed the discovery, and the star was given the name Sco X-1 (*1* for "the brightest," *X* for "X-ray source," *Sco* for "in the constellation Scorpius").

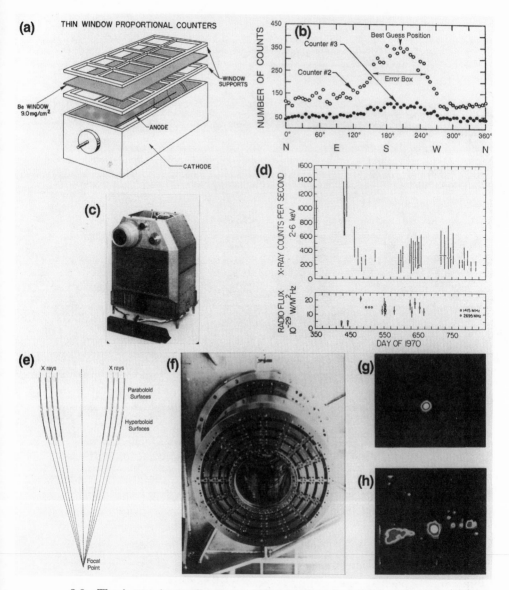

8.6 The improving technology and performance of X-ray astronomy's tools, 1962–1978. (a) Schematic design of the Geiger counter used by Giacconi's team in their 1962 discovery of the first X-ray star. (b) The data from that Geiger counter, showing that the star was not at the location of the Moon; note the very poor angular resolution (large error box), 90 degrees. (c) The 1970 Uhuru X-ray detector: A vastly improved Geiger counter sits inside the box, and in front of the counter one sees venetian-blind slats that prevent the counter from detecting an X-ray unless it arrives nearly perpendicular to the counter's window. (d) Uhuru's measurements of X-rays from the black-hole candidate Cygnus X-1. (e) Schematic diagram and (f) photograph of the mirrors that focus X-rays in the 1978 X-ray telescope Einstein. (g, h) Photographs made by the Einstein telescope of two black-hole candidates, Cygnus X-1 and SS-433. [Individual drawings and pictures courtesy R. Giacconi.]

How had the theorists gone wrong? How had they underestimated by a factor of 5000 the strengths of cosmic X-rays? They had presumed, wrongly, that the X-ray sky would be dominated by objects already known in the optical sky—objects like the Moon, planets, and ordinary stars that are poor emitters of X-rays. However, Sco X-1 and other X-ray stars soon to be discovered were not a type of object anyone had ever seen before. They were neutron stars and black holes, capturing gas from normal-star companions and heating it to high temperatures in the manner soon to be proposed by Zel'dovich and Novikov (Figure 8.5 above). To deduce that this was indeed the nature of the observed X-ray stars, however, would require another decade of hand-in-hand hard work by experimenters like Friedman and Giacconi and theorists like Zel'dovich and Novikov.

Giacconi's 1962 detector was exceedingly simple (Figure 8.6a): an electrified chamber of gas, with a thin window in its top face. When an X-ray passed through the window into the chamber, it knocked electrons off some of the gas's atoms; and those electrons were pulled by an electric field onto a wire, where they created an electric current that announced the X-ray's arrival. (Such chambers are sometimes called *Geiger counters* and sometimes *proportional counters.*) The rocket carrying the chamber was spinning at two rotations per second and its nose slowly swung around from pointing up to pointing down. These motions caused the chamber's window to sweep out a wide swath of sky, pointing first in one direction and then another. When pointed toward the constellation Scorpius, the chamber recorded many X-ray counts. When pointed elsewhere, it recorded few. However, because X-rays could enter the chamber from a wide range of directions, the chamber's estimate of the location of Sco X-1 on the sky was highly uncertain. It could report only a best-guess location, and a surrounding 90-degree-wide *error box* indicating how far wrong the best guess was likely to be (see Figure 8.6b).

To discover that Sco X-1 and other X-ray stars soon to be found were in fact neutron stars and black holes in binary systems would require error boxes (uncertainties in position on the sky) a few minutes of arc in size or smaller. That was a very tall order: a 1000-fold improvement in angular accuracy.

The needed improvement, and much more, came step by step over the next sixteen years, with several teams (Friedman's, Giacconi's, and others) competing at each step of the way. A succession of rocket flights by one team after another with continually improving detectors was followed, in December 1970, by the launch of *Uhuru*, the first X-ray satellite (Figure 8.6c). Built by Giacconi's team, Uhuru contained a gas-filled, X-ray counting chamber one hundred times larger than the one they flew on their 1962 rocket. In front of the chamber's window were slats, like venetian blinds, to prevent the chamber from seeing X-rays from any direction except a few degrees around the perpendicular (Figure 8.6d). Uhuru, which discovered and cataloged 339 X-ray stars, was followed by several other similar but special-purpose X-ray satellites, built by American, British, and Dutch scientists. Then in 1978 Giacconi's team flew a grand successor to Uhuru: *Einstein*, the world's first true X-ray *telescope*. Because X-rays penetrate right through any object that they strike perpendicularly, even a mirror, the Einstein telescope used a set of nested mirrors along which the X-rays slide, like a tobogan sliding down an icy slope (Figures 8.6e,f). These mirrors focused the X-rays to make images of the X-ray sky 1 arc second in size—images as accurate as those made by the world's best optical telescopes (Figures 8.6g,h).

From Giacconi's rocket to the Einstein telescope in just sixteen years (1962 to 1978), a 300,000-fold improvement of angular accuracy had been achieved, and in the process our understanding of the Universe had been revolutionized: The X-rays had revealed neutron stars, black-hole candidates, hot diffuse gas that bathes galaxies when they reside in huge clusters, hot gas in the remnants of supernovae and in the coronas (outer atmospheres) of some types of stars, and particles with ultra-high energies in the nuclei of galaxies and quasars.

Of the several black-hole candidates discovered by X-ray detectors and X-ray telescopes, Cygnus X-1 (Cyg X-1 for short) was one of the most believable. In 1974, soon after it became a good candidate, Stephen Hawking and I made a bet; he wagered that it is not a black hole, I that it is.

Carolee Winstein, whom I married a decade after the bet was made, was mortified by the stakes (*Penthouse* magazine for me if I win; *Private Eye* magazine for Stephen if he wins). So were my siblings and mother. But they didn't need to worry that I would actually win the *Penthouse* subscription (or so I thought in the 1980s); our information about the nature of Cyg X-1 was improving only very slowly. By 1990, in my view, we could be only 95 percent confident it was a black hole, still not confident enough for Stephen to concede. Evidently Stephen

Whereas Stephen Hawking has such a large investment in General Relativity and Black Holes and desires an insurance policy, and whereas Kip Thorne likes to live dangerously without an insurance policy,

Therefore be it resolved that Stephen Hawking bets 1 year's subscription to "Penthouse" as against Kip Thorne's wager of a 4-year Subscription to "Private Eye", that Cygnus X 1 does not contain a black hole of mass above the Chandrasekhar limit.

Conceded Stephen Hawking

Kip S. Thorne

June 1990

Witnessed this tenth day of December 1974
Hraniman Anna Zytkow Werner J

Right: The bet between Stephen Hawking and me as to whether Cygnus X-1 is a black hole. *Left:* Hawking lecturing at the University of Southern California in June 1990, just two hours before breaking into my office and signing off on our bet. [Hawking photo courtesy Irene Fertik, University of Southern California.]

read the evidence differently. Late one night in June 1990, while I was in Moscow working on research with Soviet colleagues, Stephen and an entourage of family, nurses, and friends broke into my office at Caltech, found the framed bet, and wrote a concessionary note on it with validation by Stephen's thumbprint.

The evidence that Cyg X-1 contains a black hole is of just the sort that Zel'dovich and Novikov envisioned when they proposed the method of search: Cyg X-1 is a binary made of an optically bright and X-ray-dark star orbiting around an X-ray-bright and optically dark companion, and the companion has been weighed to make sure it is too heavy to be a neutron star and thus is probably a black hole.

The evidence that this is the nature of Cyg X-1 was not developed easily. It required a cooperative, massive, worldwide effort carried out in the 1960s and 1970s by hundreds of experimental physicists, theoretical astrophysicists, and observational astronomers.

The experimental physicists were people like Herbert Friedman, Stuart Bowyer, Edward Byram, and Talbot Chubb, who discovered Cyg

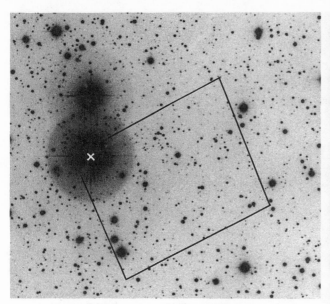

8.7 *Left:* A negative print of a photograph taken with the 5-meter (200-inch) optical telescope at Palomar Mountain by Jerome Kristian in 1971. The black rectangle outlines the error box in which Uhuru's 1971 data say that Cygnus X-1 lies. The white x marks the location of a radio flare, measured by radio telescopes, which coincided with a sudden change in the X-rays from Cyg X-1. The x coincides with the optical star HDE 226868, and thus identifies it as a binary companion of Cyg X-1. In 1978 the X-ray telescope Einstein confirmed this identification; see Figure 8.6g. *Right:* Artist's conception of Cyg X-1 and HDE 226868, based on all the optical and X-ray data. [Left: photo courtesy Dr. Jerome Kristian, Carnegie Observatories; right: painting by Victor J. Kelley, courtesy the National Geographic Society.]

X-1 in a rocket flight in 1964; Harvey Tananbaum, Edwin Kellog, Herbert Gursky, Stephen Murray, Ethan Schrier, and Riccardo Giacconi, who used Uhuru in 1971 to produce a 2-arc-minute-sized error box for the position of Cyg X-1 (Figure 8.7); and many others who discovered and studied violent, chaotic fluctuations of the X-rays and their energies—fluctuations that are what one would expect from hot, turbulent gas around a black hole.

The observational astronomers contributing to the worldwide effort were people like Robert Hjellming, Cam Wade, Luc Braes, and George Miley, who discovered in 1971 a flare of radio waves in Uhuru's Cyg X-1 error box simultaneous with a huge, Uhuru-measured change in Cyg X-1's X-rays, and thereby pinned down the location of Cyg X-1 to within 1 second of arc (Figures 8.6d and 8.7); Louise Webster, Paul Murdin, and Charles Bolton, who discovered with optical telescopes that an optical star, HDE 226868, at the location of the radio flare is orbiting around a massive, optically dark but X-ray-bright companion (Cyg X-1); and a hundred or so other optical astronomers who made painstaking measurements of HDE 226868 and other stars in its vicinity, measurements crucial to avoiding severe pitfalls in estimating the mass of Cyg X-1.

The theoretical astrophysicists contributing to the effort included people like Zel'dovich and Novikov, who proposed the method of search; Bohdan Paczyński, Yoram Avni, and John Bahcall, who developed complex but reliable ways to circumvent the mass-estimate pitfalls; Geoffrey Burbidge and Kevin Prendergast, who realized that the hot, X-ray-emitting gas should form a disk around the hole; and Nikolai Shakura, Rashid Sunyaev, James Pringle, Martin Rees, Jerry Ostriker, and many others, who developed detailed theoretical models of the X-ray-emitting gas and its disk, for comparison with the X-ray observations.

By 1974 this massive effort had led, with roughly 80 percent confidence, to the picture of Cyg X-1 and its companion star HDE 226868 that is shown in an artist's sketch in the right half of Figure 8.7. It was just the kind of picture that Zel'dovich and Novikov had envisioned, but with far greater detail: The black hole at the center of Cyg X-1 has a mass definitely greater than 3 Suns, probably greater than 7 Suns, and most likely about 16; its optically bright but X-ray-dark companion HDE 226868 has a mass probably greater than 20 Suns and most likely about 33, and it is roughly 20 times larger in radius than the Sun; the distance from the star's surface to the hole is about 20 solar radii (14 million kilometers); and the binary is about 6000 light-years from Earth. Cyg X-1 is the second brightest object in the X-ray sky; HDE 226868, while very bright in comparison with most stars seen by a large telescope, is nevertheless far too dim to be seen by the naked eye.

In the nearly two decades since 1974, our confidence in this picture of Cyg X-1 has increased from roughly 80 percent to, say, 95 percent. (These are my personal estimates.) Our confidence is not 100 percent because, despite enormous efforts, no unequivocal signature of a black hole has yet been found in Cyg X-1. No signal, in X-rays or light, cries out at astronomers saying unmistakably, "I come from a black hole." It is still possible to devise other, non–black-hole explanations for all the observations, though those explanations are so contorted that few astronomers take them seriously.

By contrast, some neutron stars, called pulsars, produce an unequivocal "I am a neutron star" cry: Their X-rays, or in some cases radio waves, come in sharp pulses that are very precisely timed. The timing is as precise, in some cases, as the ticking of our best atomic clocks. Those pulses can *only* be explained as due to beams of radiation shining off a neutron star's surface and swinging past Earth as the star rotates—the analogue of a rotating light beacon at a rural airport or in a lighthouse. Why is this the only possible explanation? Such precise timing can come only from the rotation of a massive object with massive inertia and thus massive resistance to erratic forces that would make the timing erratic; of all the massive objects ever conceived by the minds of astrophysicists, only neutron stars and black holes can spin at the enormous rates (hundreds of rotations per second) of some pulsars; and only neutron stars, not black holes, can produce rotating beams, because black holes cannot have "hair." (Any source of such a beam, attached to the hole's horizon, would be an example of the type of "hair" that a black hole cannot hang on to.[1])

An unequivocal black-hole signature, analogous to a pulsar's pulses, has been sought by astronomers in Cyg X-1 for twenty years, to no avail. An example of such a signature (suggested in 1972 by Rashid Sunyaev, a member of Zel'dovich's team) is pulsar-like pulses of radiation produced by a swinging beam that originates in a coherent lump of gas orbiting around the hole. If the lump were close to the hole and held itself together for many orbits until it finally began to plunge into the horizon, then the details of its gradually shifting interval between pulses might provide a clear and unambiguous "I am a black hole" signature. Unfortunately, such a signature has never been seen. There seem to be several reasons: (1) The hot, X-ray-emitting gas moves around the black hole so turbulently and chaotically that coherent lumps may hold themselves together for only one or a few orbits, not many. (2) If a few lumps do manage to hold themselves together for a long time and produce a black-hole signature, the turbulent X-rays from the rest of the turbulent gas evidently bury their signature. (3) If Cyg X-1 is indeed a black hole, then mathematical simulations show that most of the X-rays should come from far outside its horizon—from circumferences roughly 10 times critical or more, where there is much more volume from which X-rays can be emitted than near the horizon. At such large distances from the hole, the gravitational predictions of general relativity and Newton's theory of gravity are approximately

---

1. Chapter 7. The electric field hair of a charged black hole is evenly distributed around the hole's spin axis and thus cannot produce a concentrated beam.

the same, so if there were pulses from orbiting lumps, they would not carry a strongly definitive black-hole signature.

For reasons similar to these, astronomers might *never* find any kind of definitive black-hole signature in any electromagnetic waves produced from the vicinity of a black hole. Fortunately, the prospects are excellent for a completely different kind of black-hole signature: one carried by gravitational radiation. To this we shall return in Chapter 10.

The golden age of theoretical black-hole research (Chapter 7) coincided with the observational search for black holes and the discovery of Cyg X-1 and deciphering of its nature. Thus, one might have expected the youths who dominated the golden age (Penrose, Hawking, Novikov, Carter, Israel, Price, Teukolsky, Press, and others) to play key roles in the black-hole search. Not so, except for Novikov. The talents and knowledge that those youths had developed, and the remarkable discoveries they were making about black-hole spin, pulsation, and hairlessness, were irrelevant to the search and to deciphering Cyg X-1. It might have been different if Cyg X-1 had had an unequivocal black-hole signature. But there was none.

These youths and other theoretical physicists like them are sometimes called *relativists*, because they spend so much time working with the laws of general relativity. The theorists who *did* contribute to the search (Zel'dovich, Paczyński, Sunyaev, Rees, and others) were a very different breed called *astrophysicists*. For the search, these astrophysicists needed to master only a tiny amount of general relativity—just enough to be confident that curved spacetime was quite irrelevant, and that a Newtonian description of gravity would be quite sufficient for modeling an object like Cyg X-1. However, they needed enormous amounts of *other* knowledge, knowledge that is part of the standard tool kit of an astrophysicist. They needed a mastery of extensive astronomical lore about binary star systems, and about the structures and evolutions and spectra of the companion stars of black-hole candidates, and about the reddening of starlight by interstellar dust—a key tool in determining the distance to Cyg X-1. They also needed to be experts on such issues as the flow of hot gas, shock waves formed when streams of hot gas collide, turbulence in the gas, frictional forces in the gas caused by turbulence and by chaotic magnetic fields, violent breaking and reconnection of magnetic field lines, the formation of X-rays in hot gas, the propagation of X-rays through the gas, and much much more. Few people could be masters of all this and, simultaneously, be masters of the intricate mathematics of curved spacetime. Human limitations forced a split in the community of researchers. Either you specialized in the theoretical physics of black holes, in deducing from general relativity the properties that black holes ought to have, or you specialized in the astrophysics of binary systems and hot gas falling onto black holes and radiation produced by the gas. You were either a *relativist* or an *astrophysicist.*

Some of us tried to be both, with only modest success. Zel'dovich, the consummate astrophysicist, had occasional new insights about the fundamentals of black holes. I, as a somewhat talented relativist, tried to build general relativistic models of flowing gas near the black hole in Cyg X-1. But Zel'dovich didn't understand relativity deeply, and I didn't understand the astronomical lore very well. The barrier to cross over was enormous. Of all the researchers I knew in the golden age, only Novikov and Chandrasekhar had one foot firmly planted in astrophysics and the other in relativity.

Experimental physicists like Giacconi, who designed and flew X-ray detectors and satellites, faced a similar barrier. But there was a difference. Relativists were not needed in the search for black holes, whereas experimental physicists were essential. The observational astronomers and the astrophysicists, with their mastery of the tools for understanding binaries, gas flow, and X-ray propagation, could do nothing until the experimental physicists gave them detailed X-ray data. The experimental physicists often tried to decipher what their own data said about the gas flow and the possible black hole producing it, before turning the data over to the astronomers and astrophysicists, but with only modest success. The astronomers and astrophysicists thanked them very kindly, took the data, and then interpreted them in their own, more sophisticated and reliable ways.

This dependence of the astronomers and astrophysicists on the experimental physicists is but one of many interdependencies that were crucial to success in the search for black holes. Success, in fact, was a product of joint, mutually interdependent efforts by six different communities of people. Each community played an essential role. *Relativists* deduced, using the laws of general relativity, that black holes must exist. *Astrophysicists* proposed the method of search and gave crucial guidance at several steps along the way. *Observational astronomers*

identified HDE 226868, the companion of Cyg X-1; they used periodi-
cally shifting spectral lines from it to weigh Cyg X-1; and they made
extensive other observations to firm up their estimate of its weight.
Experimental physicists created the instruments and techniques that
made possible the search for X-ray stars, and they carried out the
search that identified Cyg X-1. *Engineers and managers* at NASA created the rockets and spacecraft that carried the X-ray detectors into
Earth orbit. And, not least in importance, *American taxpayers* provided
the funds, several hundreds of millions of dollars, for the rockets, spacecraft, X-ray detectors and X-ray telescopes, and the salaries of the
engineers, managers, and scientists who worked with them.

Thanks to this remarkable teamwork, we now, in the 1990s, are
almost 100 percent sure that black holes exist not only in Cyg X-1, but
also in a number of other binaries in our galaxy.

9

Serendipity

*in which astronomers are forced to conclude,
without any prior predictions,
that black holes a millionfold heavier than the Sun
inhabit the cores of galaxies (probably)*

Radio Galaxies

If, in 1962 (when theoretical physicists were just beginning to accept
the concept of a black hole), anyone had asserted that the Universe
contains gigantic black holes, millions or billions of times heavier than
the Sun, astronomers would have laughed. Nevertheless, astronomers
unknowingly had been observing such gigantic holes since 1939, using
radio waves. Or so we strongly suspect today.

Radio waves are the opposite extreme to X-rays. X-rays are electromagnetic waves with extremely short wavelengths, typically 10,000
times *shorter* than the wavelength of light (Figure P.2 in the Prologue). Radio waves are also electromagnetic, but they have long wavelengths, typically a few meters from wave crest to wave crest, which is
a million times *longer* than the wavelength of light. X-rays and radio
waves are also opposites in terms of wave/particle duality (Box 4.1)—
the propensity of electromagnetic waves to behave sometimes like a
wave and sometimes like a particle (a photon). X-rays typically behave
like high-energy particles (photons) and thus are most easily detected
with Geiger counters in which the X-ray photons hit atoms, knocking
electrons off them (Chapter 8). Radio waves almost always behave like

waves of electric and magnetic force, and thus are most easily detected with wire or metal antennas in which the waves' oscillating electric force pushes electrons up and down, thereby creating oscillating signals in a radio receiver attached to the antenna.

Cosmic radio waves (radio waves coming from outside the Earth) were discovered serendipitously in 1932 by Karl Jansky, a radio engineer at the Bell Telephone Laboratories in Holmdel, New Jersey. Fresh out of college, Jansky had been assigned the task of identifying the noise that plagued telephone calls to Europe. In those days, telephone calls crossed the Atlantic by radio transmission, so Jansky constructed a special radio antenna, made of a long array of metal pipes, to search for sources of radio static (Figure 9.1a). Most of the static, he soon discovered, came from thunderstorms, but when the storms were gone, there remained a faint, hissing static. By 1935 he had identified the source of the hiss; it was coming, mostly, from the central regions of our Milky Way galaxy. When the central regions were overhead, the hiss was strong; when they sank below the horizon, the hiss weakened but did not entirely disappear.

This was an amazing discovery. Anyone who had ever thought about cosmic radio waves had expected the Sun to be the brightest source of radio waves in the sky, just as it is the brightest source of light. After all, the Sun is a billion (10^9) times closer to us than most other stars in the Milky Way, so its radio waves ought to be roughly $10^9 \times 10^9 = 10^{18}$ times brighter than those from other stars. Since there are only 10^{12} stars in our galaxy, the Sun should be brighter than all the others put together by a factor of roughly $10^{18}/10^{12} = 10^6$ (a million). How could this argument fail? How could the radio waves from the distant central regions of the Milky Way be so much brighter than those from the nearby Sun?

As amazing as this mystery might be, it is even more amazing, in retrospect, that astronomers paid almost no attention to the mystery. In fact, despite extensive publicity by the Bell Telephone Company, only two astronomers seem to have taken any interest at all in Jansky's discovery. It was doomed to near oblivion by the same astronomical conservatism that Chandrasekhar was encountering with his claims that no white dwarf can be heavier than 1.4 Suns (Chapter 4).

The two exceptions to this general lack of interest were a graduate student, Jesse Greenstein, and a lecturer, Fred Whipple, in Harvard University's astronomy department. Greenstein and Whipple, pondering Jansky's discovery, showed that, if the then-current ideas about how cosmic radio waves might be generated were correct, it was *impossible* for our Milky Way galaxy to produce radio waves as strong as Jansky was seeing. Despite this apparent impossibility, Greenstein and Whipple believed Jansky's observations; they were sure the problem lay with astrophysical theory, not with Jansky. But with no hints as to where the theory was going wrong, and since, as Greenstein recalls, "I never met anybody else [in the 1930s] who had any interest in the subject, not one astronomer," they turned their attention elsewhere.

By 1935 (about the time that Zwicky was inventing the concept of a neutron star; Chapter 5), Jansky had learned everything about the galactic hiss that his primitive antenna would allow him to discover. In a quest to learn more, he proposed to Bell Telephone Laboratories the construction of the world's first real radio telescope: a huge metal bowl, 100 feet (30 meters) in diameter, which would reflect incoming radio waves up to a radio antenna and receiver in much the same way that an optical reflecting telescope reflects light from its mirror up to an eyepiece or a photographic plate. The Bell bureaucracy rejected the proposal; there was no profit in it. Jansky, ever the good employee, acquiesced. He abandoned his study of the sky, and in the shadow of the approach of World War II, turned his efforts toward radio-wave communication at shorter wavelengths.

So uninterested were professional scientists in Jansky's discovery that the only person to build a radio telescope during the next decade was Grote Reber, an eccentric bachelor and ham radio operator in Wheaton, Illinois, call number W9GFZ. Having read of Jansky's radio hiss in the magazine *Popular Astronomy*, Reber set out to study its details. Reber had a very poor education in science, but that was unimportant. What mattered was his good training in engineering and his strong practical streak. Using enormous ingenuity and his own modest savings, he designed and constructed with his own hands, in his mother's backyard, the world's first radio telescope, a 30-foot (that is, 9-meter)-diameter dish (Figure 9.1c); and with it, he made radio maps of the sky

9.1 (a) Karl Jansky and the antenna with which he discovered, in 1932, cosmic radio waves from our galaxy. (b) Grote Reber, ca. 1940. (c) The world's first radio telescope, constructed by Reber in his mother's backyard in Wheaton, Illinois. (d) A map of radio waves from the sky constructed by Reber with his backyard radio telescope. [(a) Photo by Bell Telephone Laboratories, courtesy AIP Emilio Segrè Visual Archives; (b) and (c) courtesy Grote Reber; (d) is adapted from Reber (1944).]

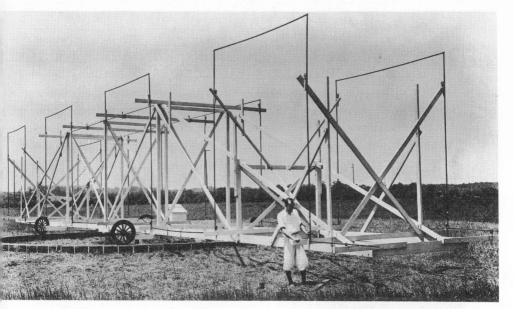

(a)

(b)

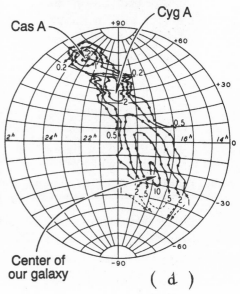

(c)

(d)

(Figure 9.1d). In his maps one can see clearly not only the central region of our Milky Way galaxy, but also two other radio sources, later called Cyg A and Cas A—*A* for the "brightest radio sources," *Cyg* and *Cas* for "in the constellations Cygnus and Cassiopeia." Four decades of detective work would ultimately show, with high probability, that Cyg A and many other radio sources discovered in the ensuing years are powered by gigantic black holes.

The story of this detective work will be the central thread of this chapter. I have chosen to devote a whole chapter to the story for several reasons:

First, this story illustrates a mode of astronomical discovery quite different from that illustrated in Chapter 8. In Chapter 8, Zel'dovich and Novikov proposed a concrete method to search for black holes; experimental physicists, astronomers, and astrophysicists implemented that method; and it paid off. In this chapter, gigantic black holes are already being observed by Reber in 1939, long before anyone ever thought to look for them, but it will take forty years for the mounting observational evidence to force astronomers to the conclusion that black holes are what they are seeing.

Second, Chapter 8 illustrated the powers of astrophysicists and relativists; this chapter shows their limitations. The types of black holes discovered in Chapter 8 were predicted to exist a quarter century before anyone ever went searching for them. They were the Oppenheimer–Snyder holes: a few times heavier than the Sun and created by the implosion of heavy stars. The gigantic black holes of this chapter, by contrast, were never predicted to exist by any theorist. They are thousands or millions of times heavier than any star that any astronomer has ever seen in the sky, so they cannot possibly be created by the implosion of such stars. Any theorist predicting these gigantic holes would have tarnished his or her scientific reputation. The discovery of these holes was serendipity in its purest form.

Third, this chapter's story of discovery will illustrate, even more clearly than Chapter 8, the complex interactions and interdependencies of four communities of scientists: relativists, astrophysicists, astronomers, and experimental physicists.

Fourth, it will turn out, late in this chapter, that the spin and the rotational energy of gigantic black holes play central roles in explaining the observed radio waves. By contrast, a hole's spin was of no importance for the observed properties of the modest-sized holes in Chapter 8.

In 1940, having made his first radio scans of the sky, Reber carefully wrote up a technical description of his telescope, his measurements, and his map, and mailed it to Subrahmanyan Chandrasekhar, who was now the editor of the *Astrophysical Journal* at the University of Chicago's Yerkes Observatory, on the shore of Lake Geneva in Wisconsin. Chandrasekhar circulated Reber's remarkable manuscript among the Yerkes astronomers. Bemused by the manuscript and skeptical of this completely unknown amateur, several of the astronomers drove down to Wheaton, Illinois, to look at his instrument. They returned, impressed. Chandrasekhar approved the paper for publication.

Jesse Greenstein, who had become an astronomer at Yerkes after completing his Harvard graduate studies, made a number of trips down to Wheaton over the next few years and became a close friend of Reber's. Greenstein describes Reber as "the ideal American inventor. If he had not been interested in radio astronomy, he would have made a million dollars."

Enthusiastic about Reber's research, Greenstein tried, after a few years, to move him to the University of Chicago. "The University didn't want to spend a dime on radio astronomy," Greenstein recalls. But Otto Struve, the director of the University's Yerkes Observatory, agreed to a research appointment provided the money to pay Reber and support his research came from Washington. Reber, however, "was an independent cuss," Greenstein says. He refused to explain to the bureaucrats in any detail how the money for new telescopes would be spent. The deal fell through.

In the meantime, World War II had ended, and scientists who had done technical work in the war effort were looking for new challenges. Among them were experimental physicists who had developed radar for tracking enemy aircraft during the war. Since radar is nothing but radio waves that are sent out from a radio-telescope-like transmitter, bounce off an airplane, and return back to the transmitter, these experimental physicists were ideally poised to give life to the new field of radio astronomy—and some of them were eager to do so; the technical challenges were great, and the intellectual payoffs promising. Of the many who tried their hand at it, three teams quickly came to dominate the field: Bernard Lovell's team at Jodrell Bank/Manchester University in England; Martin Ryle's team at Cambridge University in England; and a team put together by J. L. Pawsey and John Bolton in Australia.

In America there was little effort of note; Grote Reber continued his radio astronomy research virtually alone.

Optical astronomers (astronomers who study the sky with light,[1] the only kind of astronomer that existed in those days) paid little attention to the experimental physicists' feverish activity. They would remain uninterested until radio telescopes could measure a source's position on the sky accurately enough to determine which light-emitting object was responsible for the radio waves. This would require a 100-fold improvement in resolution over that achieved by Reber, that is, a 100-fold improvement in the accuracy with which the positions, sizes, and shapes of the radio sources were measured.

Such an improvement was a tall order. An optical telescope, or even a naked human eye, can achieve a high resolution with ease, because the waves it works with (light) have very short wavelengths, less than 10^{-6} meter. By contrast, the human ear cannot distinguish very accurately the direction from which a sound comes because sound waves have wavelengths that are long, roughly a meter. Similarly, radio waves, with their meter-sized wavelengths, give poor resolution—unless the telescope one uses is enormously larger than a meter. Reber's telescope was only modestly larger; hence, its modest resolution. To achieve a 100-fold improvement in resolution would require a telescope 100 times larger, roughly a kilometer in size, and/or the use of shorter wavelength radio waves, for example, a few centimeters instead of one meter.

The experimental physicists actually achieved this 100-fold improvement in 1949, not by brute force, but by cleverness. The key to their cleverness can be understood by analogy with something very simple and familiar. (This is just an analogy; it in fact is a slight cheat, but it gives an impression of the general idea.) We humans can see the three-dimensionality of the world around us using just two eyes, not more. The left eye sees around an object a little bit on the left side, and the right eye sees around it a bit on the right side. If we turn our heads over on their sides we can see around the top of the object a bit and around the bottom of the object a bit; and if we were to move our eyes farther apart (as in effect is done with the pair of cameras that make

---

1. By *light,* I always mean in this book the type of electromagnetic waves that the human eye can see; that is, optical radiation.

3-D movies with exaggerated three-dimensionality), we would see somewhat farther around the object. However, our three-dimensional vision would not be improved enormously by having a huge number of eyes, covering the entire fronts of our faces. We would see things far more brightly with all those extra eyes (we would have a higher *sensitivity*), but we would gain only modestly in three-dimensional *resolution.*

Now a huge, 1-kilometer radio telescope (left half of Figure 9.2) would be somewhat like our face covered with eyes. The telescope would consist of a 1-kilometer-sized bowl covered with metal that reflects and focuses the radio waves up to a wire antenna and radio receiver. If we were to remove the metal everywhere except for a few spots widely scattered over the bowl, it would be like removing most of those extra eyes from our face, and keeping only a few. In both cases, there is a modest loss of resolution, but a large loss of sensitivity. What the experimental physicists wanted most was an improved resolution

9.2 The principle of a radio interferometer. *Left:* In order to achieve good angular resolution, one would like to have a huge, say, 1-kilometer, telescope. However, it would be sufficient if only a few spots (solid) on the radio-wave-reflecting bowl are actually covered with metal and reflect. *Right:* It is not necessary for the radio waves reflected from those spots to be focused to an antenna and radio receiver at the huge bowl's center. Rather, each spot can focus its waves to its own antenna and receiver, and the resulting radio signals can then be carried by wire from all the receivers to a central receiving station, where they are combined in the same manner as they would have been at the huge telescope's receiver. The result is a network of small radio telescopes with linked and combined outputs, a *radio interferometer.*

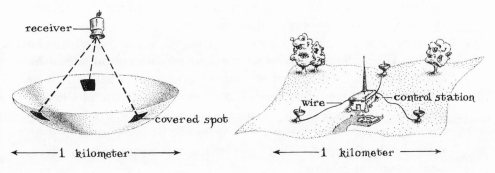

(they wanted to find out where the radio waves were coming from and what the shapes of the radio sources were), not an improved sensitivity (not an ability to see more, dimmer radio sources—at least not for now). Therefore, they needed only a spotty bowl, not a fully covered bowl.

A practical way to make such a spotty bowl was by constructing a network of small radio telescopes connected by wires to a central radio receiving station (right half of Figure 9.2). Each small telescope was like a spot of metal on the big bowl, the wires carrying each small telescope's radio signal were like radio beams reflected from the big bowl's spots, and the central receiving station which combines the signals from the wires was like the big bowl's antenna and receiver, which combine the beams from the bowl's spots. Such networks of small telescopes, the centerpieces of the experimental physicists' efforts, were called *radio interferometers*, because the principle behind their operation was *interferometry*: By "interfering" the outputs of the small telescopes with each other in a manner we shall meet in Box 10.3 of Chapter 10, the central receiving station constructs a radio map or picture of the sky.

Through the late 1940s, the 1950s, and into the 1960s, the three teams of experimental physicists (Jodrell Bank, Cambridge, and Australia) competed with each other in building ever larger and more sophisticated radio interferometers, with ever improving resolutions. The first crucial benchmark, the 100-fold improvement necessary to begin to stir an interest among optical astronomers, came in 1949, when John Bolton, Gordon Stanley, and Bruce Slee of the Australian team produced 10-arc-minute-sized *error boxes* for the positions of a number of radio sources; that is, when they identified 10-arc-minute-sized regions on the sky in which the radio sources must lie. (Ten arc minutes is one-third the diameter of the Sun as seen from Earth, and thus much poorer resolution than the human eye can achieve with light, but it is a remarkably good resolution when working with radio waves.) When the error boxes were examined with optical telescopes, some, including Cyg A, showed nothing bright of special note; finer radio resolution would be needed to reveal which of the plethora of optically dim objects in these error boxes might be the true sources of the radio waves. In three of the error boxes, however, there was an unusually bright optical object: one remnant of an ancient supernova, and two distant galaxies.

As difficult as it may have been for astrophysicists to explain the radio waves that Jansky had discovered emanating from our own galaxy, it was even more difficult to understand how distant galaxies could emit such strong radio signals. That some of the brightest radio sources in the sky might be objects so extremely distant was too incredible for belief (though it ultimately would turn out to be true). Therefore, it seemed a good bet (but those who made the bet would lose) that each error box's radio signals were coming not from the distant galaxy, but rather from one of the plethora of optically dim but nearby stars in the error box. Only better resolution could tell for sure. The experimental physicists pushed forward, and a few optical astronomers began to watch with half an eye, mildly interested.

By summer 1951, Ryle's team at Cambridge had achieved a further 10-fold improvement of resolution, and Graham Smith, a graduate student of Ryle's, used it to produce a 1-arc-minute error box for Cyg A—a box small enough that it could contain only a hundred or so optical objects (objects seen with light). Smith airmailed his best-guess position and its error box to the famous optical astronomer Walter Baade at the Carnegie Institute in Pasadena. (Baade was the man who seventeen years earlier, with Zwicky, had identified supernovae and proposed that neutron stars power them—Chapter 5.) The Carnegie Institute owned the 2.5-meter (100-inch) optical telescope on Mount Wilson, until recently the world's largest; Caltech, down the street in Pasadena, had just finished building the larger 5-meter (200-inch) telescope on Palomar Mountain; and the Carnegie and Caltech astronomers shared their telescopes with each other. At his next scheduled observing session on the Palomar 5-meter (Figure 9.3a), Baade photographed the error box on the sky where Smith said Cyg A lies. (This spot on the sky, like most spots, had never before been examined through a large optical telescope.) When Baade developed the photograph, he could hardly believe his eyes. There, in the error box, was an object unlike any ever before seen. It appeared to be two galaxies colliding with each other (center of Figure 9.3d). (We now know, thanks to observations with infrared telescopes in the 1980s, that the galaxy collision was an optical illusion. Cyg A is actually a single galaxy with a band of dust running across its face. The dust absorbs light in just such a way as to make the single galaxy look like two galaxies in collision.) The whole system, central galaxy plus radio source, would later come to be called a *radio galaxy*.

Astronomers were convinced for two years that the radio waves were

(a)

(b)

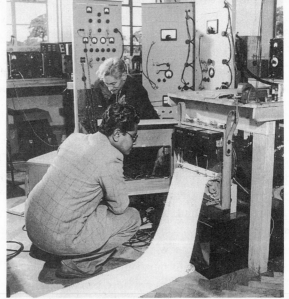

(c)

(d)

being produced by a galactic collision. Then, in 1953, came another surprise. R. C. Jennison and M. K. Das Gupta of Lovell's Jodrell Bank team studied Cyg A using a new interferometer consisting of two telescopes, one fixed to the ground and the other moving around the countryside on a truck so as to cover, one after another, a number of "spots" on the "bowl" of an imaginary 4-kilometer-square telescope (see left half of Figure 9.2). With this new interferometer (Figures 9.3b, c), they discovered that the Cyg A radio waves were not coming from the "colliding galaxies," but rather from two giant, roughly rectangular regions of space, about 200,000 light-years in size and 200,000 light-years apart, on opposite sides of the "colliding galaxies." These radio-emitting regions, or *lobes* as they are called, are shown as rectangles in Figure 9.3d, together with Baade's optical photograph of the "colliding galaxies." Also shown in the figure is a more detailed map of the lobes' radio emission, constructed sixteen years later using more sophisticated interferometers; this map is shown as thin lined contours that exhibit the brightness of the radio emission in the same way as the contours of a topographic map exhibit the height of the land. These contours confirm the 1953 conclusion that the radio waves come from gigantic lobes of gas on either side of the "colliding galaxies." How both of these enormous lobes can be powered by a single, gigantic black hole will become a major issue later in this chapter.

9.3 The discovery that Cyg A is a distant *radio galaxy:* (a) The 5-meter optical telescope used in 1951 by Baade to discover that Cyg A is connected with what appeared to be two colliding galaxies. (b) The radio interferometer at Jodrell Bank used in 1953 by Jennison and Das Gupta to show that the radio waves are coming from two giant lobes outside the colliding galaxies. The interferometer's two antennas (each an array of wires on a wooden framework) are shown here side by side. In the measurements, one was put on a truck and moved around the countryside, while the other remained behind, at rest on the ground. (c) Jennison and Das Gupta, inspecting the radio data in the control room of their interferometer. (d) The two giant lobes of radio emission (rectangles) as revealed in the 1953 measurements, shown together with Baade's optical photograph of the "colliding galaxies." Also shown in (d) is a high-resolution contour map of the lobes' radio emission (thin solid contours), produced in 1969 by Ryle's group at Cambridge. [(a) Courtesy Palomar Observatory/California Institute of Technology; (b) and (c) courtesy Nuffield Radio Astronomy Laboratories, University of Manchester; (d) adapted from Mitton and Ryle (1969), Baade and Minkowski (1954), Jennison and Das Gupta (1953).]

These discoveries were startling enough to generate, at long last, strong interest among optical astronomers. Jesse Greenstein was no longer the only one paying serious attention.

For Greenstein himself, these discoveries were the final straw. Having failed to push into radio work right after the war, Americans were now bystanders in the greatest revolution to hit astronomy since Galileo invented the optical telescope. The rewards of the revolution were being reaped in Britain and Australia, and not in America.

Greenstein was now a professor at Caltech. He had been brought there from Yerkes to build an astronomy program around the new 5-meter optical telescope, so naturally, he now went to Lee DuBridge, the Caltech president, and urged that Caltech build a radio interferometer to be used hand in hand with the 5-meter in exploring distant galaxies. DuBridge, having been director of the American radar effort during the war, was sympathetic, but cautious. To swing DuBridge into action, Greenstein organized an international conference on the future of radio astronomy in Washington, D.C., on 5 and 6 January 1954.

In Washington, after the representatives from the great British and Australian radio observatories had described their remarkable discoveries, Greenstein posed his question: Must the United States continue as a radio astronomy wasteland? The answer was obvious.

With strong backing from the National Science Foundation, American physicists, engineers, and astronomers embarked on a crash program to construct a National Radio Astronomy Observatory in Greenbank, West Virginia; and DuBridge approved Greenstein's proposal for a state-of-the-art Caltech radio inteferometer, to be built in Owens Valley, California, just southeast of Yosemite National Park. Since nobody at Caltech had the expertise to build such an instrument, Greenstein lured John Bolton from Australia to spearhead the effort.

Quasars

By the late 1950s, the Americans were competitive. Radio telescopes at Greenbank were coming into operation, and at Caltech, Tom Mathews, Per Eugen Maltby, and Alan Moffett on the new Owens Valley radio interferometer were working hand in hand with Baade, Greenstein, and others on the Palomar 5-meter optical telescope to discover and study large numbers of radio galaxies.

In 1960 this effort brought another surprise: Tom Mathews at Caltech received word from Henry Palmer that, according to Jodrell Bank measurements, a radio source named 3C48 (the 48th source in the third version of a catalog constructed by Ryle's group at Cambridge) was extremely small, no more than 1 arc second in diameter (1/10,000 of the angular size of the Sun). So tiny a source would be something quite new. However, Palmer and his Jodrell Bank colleagues could not provide a tight error box for the source's location. Mathews, in exquisitely beautiful work with Caltech's new radio interferometer, produced an error box just 5 seconds of arc in size, and gave it to Allan Sandage, an optical astronomer at the Carnegie Institute in Pasadena. On his next observing run on the 5-meter optical telescope, Sandage took a photograph centered on Mathews's error box and found, to his great surprise, not a galaxy, but a single, blue point of light; it looked like a star. "I took a spectrum the next night and it was the weirdest spectrum I'd ever seen," Sandage recalls. The wavelengths of the spectral lines were not at all like those of stars or of any hot gas ever manufactured on Earth; they were unlike anything ever before encountered by astronomers or physicists. Sandage could not make any sense at all out of this weird object.

Over the next two years a half-dozen similar objects were discovered by the same route, each as puzzling as 3C48. All the optical astronomers at Caltech and Carnegie began photographing them, taking spectra, struggling to understand their nature. The answer should have been obvious, but it was not. A mental block held sway. These weird objects looked so much like stars that the astronomers kept trying to interpret them as a type of star in our own galaxy that had never before been seen, but the interpretations were horrendously contorted, not really believable.

The mental block was broken by Maarten Schmidt, a thirty-two-year-old Dutch astronomer who had recently joined the Caltech faculty. For months he had struggled to understand a spectrum he had taken of 3C273, one of the weird objects. On 5 February 1963, as he sat in his Caltech office carefully sketching the spectrum for inclusion in a manuscript he was writing, the answer suddenly hit him. The four brightest lines in the spectrum were the four standard "Balmer lines" produced by hydrogen gas—the most famous of all spectral lines, the first lines that college physics students learned about in their courses on quantum mechanics. However, these four lines did not have their usual wavelengths. Each was shifted to the red by 16 percent. 3C273 must be an object containing a massive amount of hydrogen gas and moving away from the Earth at 16 percent of the speed of light—enormously faster than any star that any astronomer had ever seen.

Schmidt flew out into the hall, ran into Greenstein, and excitedly described his discovery. Greenstein turned, headed back to his office, pulled out his spectrum of 3C48, and stared at it for a while. Balmer lines were not present at any redshift; but lines emitted by magnesium, oxygen, and neon were there staring him in the face, and they had a redshift of 37 percent. 3C48 was, at least in part, a massive amount of gas containing magnesium, oxygen, and neon, and moving away from Earth at 37 percent of the speed of light.

What was producing these high speeds? If, as everyone had thought, these weird objects (which would later be named *quasars*) were some type of star in our own Milky Way galaxy, then they must have been ejected from somewhere, perhaps the Milky Way's central nucleus, with enormous force. This was too incredible to believe, and a close examination of the quasars' spectra made it seem extremely unlikely.

Left: Jesse L. Greenstein with a drawing of the Palomar 5-meter optical telescope, ca. 1955. *Right:* Maarten Schmidt, with an instrument for measuring spectra made by the 5-meter telescope, ca. 1963. [Courtesy the Archives, California Institute of Technology.]

The only reasonable alternative, Greenstein and Schmidt argued (correctly), was that these quasars are very far away in our Universe, and move away from Earth at high speed as a result of the Universe's expansion.

Recall that the expansion of the Universe is like the expansion of the surface of a balloon that is being blown up. If a number of ants are standing on the balloon's surface, each ant will see all the other ants move away from him as a result of the balloon's expansion. The farther away another ant is, the faster the first ant will see it move. Similarly, the farther away a distant object is from Earth, the faster we on Earth will see it move as a result of the Universe's expansion. In other words, the object's speed is proportional to its distance. Therefore, from the speeds of 3C273 and 3C48, Schmidt and Greenstein could infer their distances: 2 billion light-years and 4.5 billion light-years, respectively.

These were enormous distances, nearly the largest distances ever yet recorded. This meant that, in order for 3C273 and 3C48 to be as bright as they appear in the 5-meter telescope, they had to radiate enormous amounts of power: 100 times more power than the most luminous galaxies ever seen.

3C273, in fact, was so bright that, along with many other objects near it on the sky, it had been photographed more than 2000 times since 1895 using modest-sized telescopes. Upon learning of Schmidt's discovery, Harlan Smith of the University of Texas organized a close examination of this treasure trove of photographs, archived largely at Harvard, and discovered that 3C273 had been fluctuating in brightness during the past seventy years. Its light output had changed substantially within periods as short as a month. This means that a large portion of the light from 3C273 must come from a region smaller than the distance light travels in a month, that is, smaller than 1 "light-month." (If the region were larger, then there would be no way that any force, traveling, of course, at a speed less than or equal to that of light, could make the emitting gas all brighten up or dim out simultaneously to within an accuracy of a month.)

The implications were extremely hard to believe. This weird quasar, this 3C273, was shining 100 times more brightly than the brightest galaxies in the Universe; but whereas galaxies produce their light in regions 100,000 light-years in size, 3C273 produces its light in a region at least a million times smaller in diameter and 10^{18} times smaller in volume: just a light-month or less. The light must come from a massive, compact, gaseous object that is heated by an enormously powerful engine. The engine would ultimately turn out to be, with high but not complete confidence, a gigantic black hole, but strong evidence for this was still fifteen years into the future.

If explaining Jansky's radio waves from our own Milky Way galaxy was difficult, and explaining the radio waves from distant radio galaxies was even more difficult, then the explanation for radio waves from these superdistant quasars would have to be superdifficult.

The difficulty, it turned out, was an extreme mental block. Jesse Greenstein, Fred Whipple, and all other astronomers of the 1930s and 1940s had presumed that cosmic radio waves, like light from stars, are emitted by the heat-induced jiggling of atoms, molecules, and electrons. Astronomers of the thirties and forties could not conceive of any other way for nature to create the observed radio waves, even though their calculations showed unequivocally that this way can't work.

Another way, however, had been known to physicists since the early twentieth century: When an electron, traveling at high speed, encounters a magnetic field, the field's magnetic force twists the electron's motion into a spiral. The electron is forced to spiral around and around the magnetic field lines (Figure 9.4), and as it spirals, it emits electro-

9.4 Cosmic radio waves are produced by near-light-speed electrons that spiral around and around in magnetic fields. The magnetic field forces an electron to spiral instead of moving on a straight line, and the electron's spiraling motion produces the radio waves.

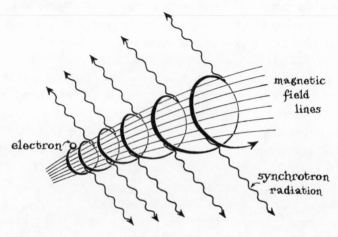

magnetic field lines

electron

synchrotron radiation

magnetic radiation. Physicists in the 1940s began to call this radiation *synchrotron radiation*, because it is produced by spiraling electrons in the particle accelerators called "synchrotrons" that they were then building. Remarkably, in the 1940s, despite physicists' considerable interest in synchrotron radiation, astronomers paid no attention to it. The astronomers' mental block held sway.

In 1950 Karl Otto Kiepenheuer in Chicago and Vitaly Lazarevich Ginzburg in Moscow (the same Ginzburg who had invented the LiD fuel for the Soviet hydrogen bomb, and who had discovered the first hint that black holes have no hair[2]) broke the mental block. Building on seminal ideas of Hans Alfvén and Nicolai Herlofson, Kiepenheuer and Ginzburg proposed (correctly) that Jansky's radio waves from our own galaxy are synchrotron radiation produced by near-light-speed electrons spiraling around magnetic field lines that fill interstellar space (Figure 9.4).

A few years later, when the giant radio-emitting lobes of radio galaxies and then quasars were discovered, it was natural (and correct) to conclude that their radio waves were also produced by electrons spiraling around magnetic field lines. From the physical laws governing such spiraling and the properties of the observed radio waves, Geoffrey Burbidge at the University of California in San Diego computed how much energy the lobes' magnetic field and high-speed electrons must have. His startling answer: In the most extreme cases, the radio-emitting lobes must have about as much magnetic energy and high-speed *(kinetic)* energy as one would get by converting all the mass of 10 million (10^7) Suns into pure energy with 100 percent efficiency.

These energy requirements of quasars and radio galaxies were so staggering that they forced astrophysicists, in 1963, to examine all conceivable sources of power in search of an explanation.

Chemical power (the burning of gasoline, oil, coal, or dynamite), which is the basis of human civilization, was clearly inadequate. The chemical efficiency for converting mass into energy is only 1 part in 100 million (1 part in 10^8). To energize a quasar's radio-emitting gas

2. See Figure 7.3. Ginzburg is best known not for these discoveries, but for yet another: his development, with Lev Landau, of the "Ginzburg–Landau theory" of superconductivity (that is, an explanation for how it is that some metals, when made very cold, lose all their resistance to the flow of electricity). Ginzburg is one of the world's few true "Renaissance physicists," a man who has contributed significantly to almost all branches of theoretical physics.

would therefore require $10^8 \times 10^7 = 10^{15}$ solar masses of chemical fuel—10,000 times more fuel than is contained in our entire Milky Way galaxy. This seemed totally unreasonable.

Nuclear power, the basis of the hydrogen bomb and of the Sun's heat and light, looked only marginal as a way to energize a quasar. Nuclear fuel's efficiency for mass-to-energy conversion is roughly 1 percent (1 part in 10^2), so a quasar would need $10^2 \times 10^7 = 10^9$ (1 billion) solar masses of nuclear fuel to energize its radio-emitting lobes. And this 1 billion solar masses would be adequate only if the nuclear fuel were burned completely and the resulting energy were converted completely into magnetic fields and kinetic energy of high-speed electrons. Complete burning and complete energy conversion seemed highly unlikely. Even with carefully contrived machines, humans rarely achieve better than a few percent conversion of fuel energy into useful energy, and nature without careful designs might well do worse. Thus, 10 billion or 100 billion solar masses of nuclear fuel seemed more reasonable. Now, this is less than the mass of a giant galaxy, but not a lot less, and how nature might achieve the conversion of the fuel's nuclear energy into magnetic and kinetic energy was very unclear. Thus, nuclear fuel *was* a possibility, but not a likely one.

The annihilation of matter with antimatter[3] could give 100 percent conversion of mass to energy, so 10 million solar masses of antimatter annihilating with 10 million solar masses of matter could satisfy a quasar's energy needs. However, there is no evidence that any antimatter exists in our Universe, except tiny bits created artifically by humans in particle accelerators and tiny bits created by nature in collisions between matter particles. Moreover, even if so much matter and antimatter were to annihilate in a quasar, their annihilation energy would go into very high energy gamma rays, and not into magnetic energy and electron kinetic energy. Thus, matter/antimatter annihilation appeared to be a very unsatisfactory way to energize a quasar.

One other possibility remained: *gravity*. The implosion of a normal star to form a neutron star or a black hole might, conceivably, convert 10 percent of the star's mass into magnetic and kinetic energy—though precisely how was unclear. If it managed to do so, then the implosions of $10 \times 10^7 = 10^8$ (100 million) normal stars might provide a quasar's energy, as would the implosion of a single, hypothetical,

---

3. For background, see the entry "antimatter" in the glossary, and Footnote 2 in Chapter 5.

supermassive star 100 million times heavier than the Sun. [The correct idea, that the gigantic black hole produced by the implosion of such a supermassive star might itself be the engine that powers the quasar, did not occur to anybody in 1963. Black holes were but poorly understood. Wheeler had not yet coined the phrase "black hole" (Chapter 6). Salpeter and Zel'dovich had not yet realized that gas falling toward a black hole could heat and radiate with high efficiency (Chapter 8). Penrose had not yet discovered that a black hole can store up to 29 percent of its mass as rotational energy, and release it (Chapter 7). The golden age of black-hole research had not yet begun.]

The idea that the implosion of a star to form a black hole might energize quasars was a radical departure from tradition. This was the first time in history that astronomers and astrophysicists had felt a need to appeal to effects of general relativity to explain an object that was being observed. Previously, relativists had lived in one world and astronomers and astrophysicists in another, hardly communicating. Their insularity was about to end.

To foster dialogue between the relativists and the astronomers and astrophysicists, and to catalyze progress in the study of quasars, a conference of three hundred scientists was held on 16–18 December 1963, in Dallas, Texas. In an after-dinner speech at this First Texas Symposium on Relativistic Astrophysics, Thomas Gold of Cornell University described the situation, only partially with tongue in cheek: "[The mystery of the quasars] allows one to suggest that the relativists with their sophisticated work are not only magnificent cultural ornaments but might actually be useful to science! Everyone is pleased: the relativists who feel they are being appreciated and are experts in a field they hardly knew existed, the astrophysicists for having enlarged their domain, their empire, by the annexation of another subject—general relativity. It is all very pleasing, so let us all hope that it is right. What a shame it would be if we had to go and dismiss all the relativists again."

Lectures went on almost continuously from 8:30 in the morning until 6 in the evening with an hour out for lunch, plus 6 P.M. until typically 2 A.M. for informal discussions and arguments. Slipped in among the lectures was a short, ten-minute presentation by a young New Zealander mathematician, Roy Kerr, who was unknown to the other participants. Kerr had just discovered his solution of the Einstein field equation—the solution which, one decade later, would turn out to describe all properties of spinning black holes, including their storage and release of rotational energy (Chapters 7 and 11); the solution which, as we shall see below, would ultimately become a foundation for explaining the quasars' energy. However, in 1963 Kerr's solution seemed to most scientists only a mathematical curiosity; nobody even knew it described a black hole—though Kerr speculated it might somehow give insight into the implosion of rotating stars.

The astronomers and astrophysicists had come to Dallas to discuss quasars; they were not at all interested in Kerr's esoteric mathematical topic. So, as Kerr got up to speak, many slipped out of the lecture hall and into the foyer to argue with each other about their favorite theories of quasars. Others, less polite, remained seated in the hall and argued in whispers. Many of the rest catnapped in a fruitless effort to remedy their sleep deficits from late-night science. Only a handful of relativists listened, with rapt attention.

This was more than Achilles Papapetrou, one of the world's leading relativists, could stand. As Kerr finished, Papapetrou demanded the floor, stood up, and with deep feeling explained the importance of Kerr's feat. He, Papapetrou, had been trying for thirty years to find such a solution of Einstein's equation, and had failed, as had many other relativists. The astronomers and astrophysicists nodded politely, and then, as the next speaker began to hold forth on a theory of quasars, they refocused their attention, and the meeting picked up pace.

The 1960s marked a turning point in the study of radio sources. Previously the study was totally dominated by observational astronomers—that is, optical astronomers and the radio-observing experimental physicists, who were now being integrated into the astronomical community and called *radio astronomers*. Theoretical astrophysicists, by contrast, had contributed little, because the radio observations were not yet detailed enough to guide their theorizing very much. Their only contributions had been the realization that the radio waves are produced by high-speed electrons spiraling around magnetic field lines in the giant radio-emitting lobes, and their calculation of how much magnetic and kinetic energy this entails.

In the 1960s, as the resolutions of radio telescopes continued to improve and optical observations began to reveal new features of the radio sources (for example, the tiny sizes of the light-emitting cores of quasars), this growing body of information became grist for the minds of astrophysicists. From this rich information, the astrophysicists generated dozens of detailed models to explain radio galaxies and quasars, and then one by one their models were disproved by accumulating observational data. This, at last, was how science was supposed to work!

One key piece of information was the radio astronomers' discovery that radio galaxies emit radio waves not only from their giant double lobes, one on each side of the central galaxy, but also from the core of the central galaxy itself. In 1971, this suggested to Martin Rees, a recent student of Dennis Sciama's in Cambridge, a radically new idea about the powering of the double lobes. Perhaps a single engine in the galaxy's core was responsible for *all* the galaxy's radio waves. Perhaps this engine was directly energizing the core's radio-emitting electrons and magnetic fields, perhaps it was also beaming up power to the giant lobes, to energize their electrons and fields, and perhaps this engine in the cores of radio galaxies was of the same sort (whatever that might be) that powers quasars.

Rees initially suspected that the beams that carry power from the core to the lobes were made of ultra-low-frequency electromagnetic waves. However, theoretical calculations soon made it clear that such electromagnetic beams cannot penetrate through the galaxy's interstellar gas, no matter how hard they try.

As is often the case, Rees's not quite correct idea stimulated a correct one. Malcolm Longair, Martin Ryle, and Peter Scheuer in Cambridge took the idea and modified it in a simple way: They kept Rees's beams, but made them of hot, magnetized gas rather than electromagnetic waves. Rees quickly agreed that this kind of *gas jet* would do the job, and with his student Roger Blandford he computed the properties that the gas jets should have.

A few years later, this prediction, that the radio-emitting lobes are powered by jets of gas emerging from a central engine, was spectacularly confirmed using huge new radio interferometers in Britain, Holland, and America—most notably the American VLA *(very large array)* on the plains of St. Augustin in New Mexico (Figure 9.5). The interferometers saw the jets, and the jets had just the predicted properties. They reached from the galaxy's core to the two lobes, and they could even be seen ramming into gas in the lobes and being slowed to a halt.

The VLA uses the same "spots on the bowl" technique as the radio

9.5 *Top:* The VLA radio interferometer on the plains of St. Augustin in New Mexico. *Bottom:* A picture of the radio emission from the radio galaxy Cygnus A made with the VLA by R. A. Perley, J.W. Dreyer, and J.J. Cowan. The jet that feeds the right-hand radio lobe is quite clear; the jet feeding the left lobe is much fainter. Notice the enormous improvement in resolution of this radio-wave picture compared with Reber's 1944 contour map which did not show the double lobes at all (Figure 9.1d), and with Jennison and Das Gupta's 1953 radio map which barely revealed the existence of the lobes (two rectangles in Figure 9.3d), and with Ryle's 1969 contour map (Figure 9.3d). [Both pictures courtesy NRAO/AUI.]

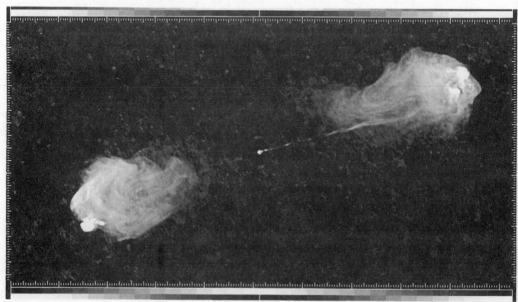

interferometers of the 1940s and 1950s (Figure 9.2), but its bowl is much larger and it uses many more spots (many more linked radio telescopes). It achieves resolutions as good as 1 arc second, about the same as the world's best optical telescopes—a tremendous achievement when one contemplates the crudeness of Jansky's and Reber's original instruments forty years earlier. But the improvements did not stop there. By the early 1980s, pictures of the cores of radio galaxies and quasars, with resolutions 1000 times better than optical telescopes, were being produced by *very long baseline interferometers* (VLBIs) composed of radio telescopes on opposite sides of a continent or the world. (The output of each telescope in a VLBI is recorded on magnetic tape, along with time markings from an atomic clock, and the tapes from all the telescopes are then played into a computer where they are "interfered" with each other to make the pictures.)

These VLBI pictures showed, in the early 1980s, that the jets extend right into the innermost few light-years of the core of a galaxy or quasar—the very region in which resides, in the case of some quasars such as 3C273, a brilliantly luminous, light-emitting object no larger than a light-month in size. Presumably the central engine is inside the light-emitting object, and it is powering not only that object, but also the jets, which then feed the radio lobes.

The jets gave yet another clue to the nature of the central engine. Some jets were absolutely straight over distances of a million light-years or more. If the source of such jets were turning, then, like a rotating water nozzle on a sprinkler, it would produce bent jets. The observed jets' straightness thus meant that the central engine had been firing its jets in precisely the same direction for a very long time. How long? Since the jets' gas cannot move faster than the speed of light, and since some straight jets were longer than a million light-years, the firing direction must have been steady for more than a million years. To achieve such steadiness, the engine's "nozzles," which eject the jets, must be attached to a superbly steady object—a long-lived *gyroscope* of some sort. (Recall that a gyroscope is a rapidly spinning object that holds the direction of its spin axis steadily fixed over a very long time. Such gyroscopes are key components of inertial navigation systems for airplanes and missiles.)

Of the dozens of ideas that had been proposed by the early 1980s to explain the central engine, only one entailed a superb gyroscope with a long life, a size less than a light-month, and an ability to generate powerful jets. That unique idea was a gigantic, spinning black hole.

Gigantic Black Holes

The idea that gigantic black holes might power quasars and radio galaxies was conceived by Edwin Salpeter and Yakov Borisovich Zel'dovich in 1964 (the first year of the golden age—Chapter 7). This idea was an obvious application of the Salpeter–Zel'dovich discovery that gas streams, falling toward a black hole, should collide and radiate (see Figure 8.4).

A more complete and realistic description of the fall of gas streams toward a black hole was devised in 1969 by Donald Lynden-Bell, a British astrophysicist in Cambridge. Lynden-Bell argued, convincingly, that after the gas streams collide, they will join together, and then centrifugal forces will make them spiral around and around the hole many times before falling in; and as they spiral, they will form a disk-shaped object, much like the rings around the planet Saturn—an *accretion disk* Lynden-Bell called it, since the gas is "accreting" onto the hole. (The right half of Figure 8.7 shows an artist's conception of such an accretion disk around the modest-sized hole in Cygnus X-1.) In the accretion disk, adjacent gas streams will rub against each other, and intense friction from that rubbing will heat the disk to high temperatures.

In the 1980s, astrophysicists realized that the brilliant light-emitting object at the center of 3C273, the object 1 light-month or less in size, was probably Lynden-Bell's friction-heated accretion disk.

We normally think of friction as a poor source of heat. Recall the unfortunate Boy Scout who tries to start a fire by rubbing two sticks together! However, the Boy Scout is limited by his meager muscle power, while an accretion disk's friction feeds off gravitational energy. Since the gravitational energy is enormous, far larger than nuclear energy, the friction is easily up to the task of heating the disk and making it shine 100 times more brightly than the most luminous galaxies.

How can a black hole act as a gyroscope? James Bardeen and Jacobus Petterson of Yale University realized the answer in 1975: If the black hole spins rapidly, then it behaves precisely like a gyroscope. Its spin direction remains always firmly fixed and unchanging, and the swirl of space near the hole created by the spin (Figure 7.7) remains always

firmly oriented in the same direction. Bardeen and Petterson showed
by a mathematical calculation that this near-hole swirl of space must
grab the inner part of the accretion disk and hold it firmly in the hole's
equatorial plane—and must do so no matter how the disk is oriented
far from the hole (Figure 9.6). As new gas from interstellar space is
captured into the distant part of the disk, it may change the distant
disk's orientation, but it can never change the disk's orientation near
the hole. The hole's gyroscopic action prevents it. Near the hole the
disk remains always in the hole's equatorial plane.

Without Kerr's solution to the Einstein field equation, this gyroscopic action would have been unknown, and it might have been impossible to explain quasars. With Kerr's solution in hand, astrophysicists in the mid-1970s were arriving at a clear and elegant explanation.
For the first time, the concept of a black hole as a dynamical body,
more than just a "hole in space," was playing a central role in explaining astronomers' observations.

How strong will the swirl of space be near the gigantic hole? In other
words, how fast will gigantic holes spin? James Bardeen deduced the
answer: He showed mathematically that gas accreting into the hole
from its disk should gradually make the hole spin faster and faster. By

9.6   The spin of a black hole produces a swirl of space around the hole, and that
swirl holds the inner part of the accretion disk in the hole's equatorial plane.

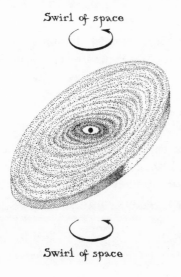

Swirl of space

Swirl of space

the time the hole has swallowed enough inspiraling gas to double its mass, the hole should be spinning at nearly its maximum possible rate—the rate beyond which centrifugal forces prevent any further speedup (Chapter 7). Thus, gigantic holes should typically have near-maximal spins.

How can a black hole and its disk produce two oppositely pointed jets? Amazingly easily, Blandford, Rees, and Lynden-Bell at Cambridge University recognized in the mid-1970s. There are four possible ways to produce jets; any one of them might do the job.

First, Blandford and Rees realized, the disk may be surrounded by a cool gas cloud (Figure 9.7a). A wind blowing off the upper and lower faces of the disk (analogous to the wind that blows off the Sun's surface) may create a bubble of hot gas inside the cool cloud. The hot gas may then punch orifices in the cool cloud's upper and lower faces and flow out of them. Just as a nozzle on a garden hose collimates outflowing water to form a fast, thin stream, so the orifices in the cool cloud should collimate the outflowing hot gas to form thin jets. The directions of the jets will depend on the locations of the orifices. The most likely locations, if the cool cloud spins about the same axis as the black hole, are along the common spin axis, that is, perpendicular to the plane of the inner part of the accretion disk—and the orifices at these locations will produce jets whose direction is anchored to the black hole's gyroscopic spin.

Second, because the disk is so hot, its internal pressure is very high, and this pressure might puff the disk up until it becomes very thick (Figure 9.7b). In this case, Lynden-Bell pointed out, the orbital motion of the disk's gas will produce centrifugal forces that create whirlpool-like funnels in the top and bottom faces of the disk. These funnels are precisely analogous to the vortex that sometimes forms when water swirls down the drainhole of a bathtub. The black hole is like the drainhole, and the disk's gas is like the water. The faces of the vortex-like funnels should be so hot, because of friction in the gas, that they blow a strong wind off themselves, and the funnels might then collimate this wind into jets, Lynden-Bell reasoned. The jets' directions will be the same as the funnels', which in turn are firmly anchored to the hole's gyroscopic spin axis.

Third, Blandford realized, magnetic field lines anchored in the disk and sticking out of it will be forced, by the disk's orbital motion, to spin

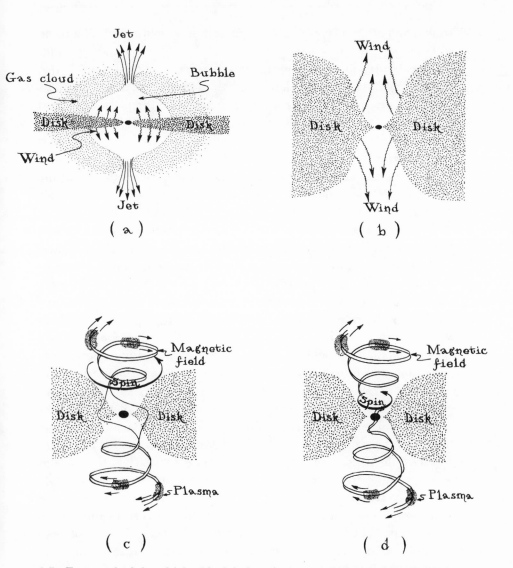

9.7 Four methods by which a black hole or its accretion disk could power twin jets. (a) A wind from the disk blows a bubble in a surrounding, spinning gas cloud; the bubble's hot gas punches orifices through the cloud, along its spin axis; and jets of hot gas shoot out the orifices. (b) The disk is puffed up by the pressure of its great internal heat, and the surface of the puffed, rotating disk forms two funnels that collimate the disk's wind into two jets. (c) Magnetic field lines anchored in the disk are forced to spin by the disk's orbital rotation; as they spin, the field lines fling plasma upward and downward, and the plasma, sliding along the field lines, forms two magnetized jets. (d) Magnetic field lines threading through the black hole are forced to spin by the swirl of the hole's space, and as they spin, the field lines fling plasma upward and downward to form two magnetized jets.

around and around (Figure 9.7c). The spinning field lines will assume
an outward and upward (or outward and downward) spiraling shape.
Electrical forces should anchor hot gas (plasma) onto the spinning field
lines; the plasma can slide along the field lines but not across them. As
the field lines spin, centrifugal forces should fling the plasma outward
along them to form two magnetized jets, one shooting outward and
upward, the other outward and downward. Again the jets' directions
will be firmly anchored to the hole's spin.

The fourth method of producing jets is more interesting than the
others and requires more explanation. In this fourth method, the hole
is threaded by magnetic field lines as shown in Figure 9.7d. As the hole
spins, it drags the field lines around and around, causing them to fling
plasma upward and downward in much the same manner as the third
method, to form two jets. The jets shoot out along the hole's spin axis
and their direction thus is firmly anchored to the hole's gyroscopic spin.
This method was conceived of by Blandford soon after he received his
Ph.D. in Cambridge, together with a Cambridge graduate student,
Roman Znajek, and it thus is called the *Blandford–Znajek process.*

The Blandford–Znajek process is especially interesting, because the
power that goes into the jets comes from the hole's enormous rotational
energy. (This should be obvious since it is the hole's spin that causes
space to swirl, and the swirl of space that causes the magnetic field lines
to rotate, and the field lines' rotation that flings plasma outward.)

How is it possible, in this Blandford–Znajek process, for the hole's
horizon to be threaded by magnetic field lines? Such field lines would
be a form of "hair" that can be converted into electromagnetic radiation and be radiated away, and therefore, according to Price's theorem
(Chapter 7), they *must* be radiated away. In fact, Price's theorem is
correct only if the black hole is sitting alone, far from all other objects.
The hole we are discussing, however, is not alone; it is surrounded by
an accretion disk. If the field lines of Figure 9.7d pop off the hole, the
lines going out the hole's northern hemisphere and those going out its
southern hemisphere will turn out to be continuations of each other,
and the only way these lines can then escape is by pushing their way
out through the accretion disk's hot gas. But the hot gas will not let the
field lines through; it confines them firmly into the region of space
inside the disk's inner face, and since most of that region is occupied by
the hole, most of the confined field lines thread through the hole.

Where do these magnetic field lines come from? From the disk itself.
All gas in the Universe is magnetized, at least a little bit, and the disk's

gas is no exception.[4] As, bit by bit, the disk's gas accretes into the hole, it carries its magnetic field lines with it. Upon nearing the hole, each bit of gas slides down its magnetic field lines and through the horizon, leaving the field lines behind, sticking out of the horizon and threading it in the manner of Figure 9.7d. These threading field lines, firmly confined by the surrounding disk, should then extract the hole's rotational energy by the Blandford–Znajek process.

All four methods of producing jets (orifices in a gas cloud, wind from a funnel, whirling field lines anchored in a disk, and the Blandford–Znajek process) probably operate, to varying degrees, in quasars, in radio galaxies, and in the peculiar cores of some other types of galaxies (cores that are called *active galactic nuclei*).

If quasars and radio galaxies are powered by the same kind of black-hole engine, what makes them look so different? Why does the light of a quasar appear to come from an intensely luminous, star-like object, 1 light-month in size or less, while the light of a radio galaxy comes from a Milky Way–like assemblage of stars, 100,000 light-years in size?

It seems almost certain that quasars are not much different from radio galaxies; their central engines are also surrounded by a 100,000-light-year-sized galaxy of stars. However, in a quasar, the central black hole is fueled at an especially high rate by accreting gas (Figure 9.8), and frictional heating in the disk is correspondingly high. This huge heating makes the disk shine so strongly that its optical brilliance is hundreds or thousands of times greater than that of all the stars in the surrounding galaxy put together. Astronomers, blinded by the brilliance of the disk, cannot see the galaxy's stars, and thus the object looks "quasi-stellar" (that is, star-like; like a tiny, intense point of light) instead of looking like a galaxy.[5]

The innermost region of the disk is so hot that it emits X-rays; a little farther out, the disk is cooler and emits ultraviolet radiation; still farther out it is cooler still and emits optical radiation (light); and in its outermost region it is even cooler and emits infrared radiation. The light-emitting region is typically about a light-year in size, though in

4. The magnetic fields have been built up continually over the life of the Universe by the motions of interstellar and stellar gas, and once generated, the magnetic fields are extremely hard to get rid of. When interstellar gas accumulates into the accretion disk, it carries its magnetic fields with itself.

5. The word "quasar" is shorthand for "quasi-stellar."

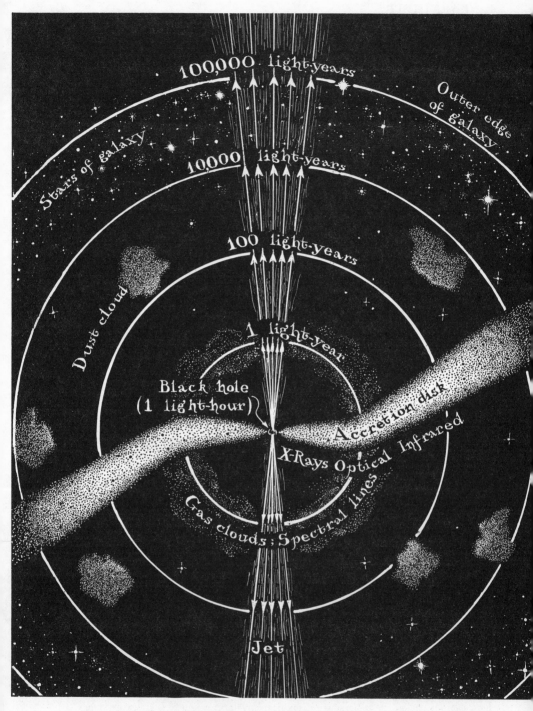

9.8 Our best present understanding of the structures of quasars and radio galaxies. This detailed model, based on all the observational data, has been developed by Sterl Phinney of Caltech and others.

some cases such as 3C273 it can be a light-month or smaller and thus can vary in brightness over periods as short as a month. Much of the X-ray radiation and ultraviolet light pouring out of the innermost region hits and heats gas clouds several light-years from the disk; it is those heated clouds that emit the spectral lines by which the quasars were first discovered. A magnetized wind blowing off the disk, in some quasars but not all, will be strong enough and well enough collimated to produce radio-emitting jets.

In a radio galaxy, by contrast with a quasar, the central accretion disk presumably is rather quiescent. Quiescence means small friction in the disk, and thus small heating and low luminosity, so that the disk shines much less brightly than the rest of the galaxy. Astronomers thus see the galaxy and not the disk through their optical telescopes. However, the disk, the spinning hole, and magnetic fields threading through the hole together produce strong jets, probably in the manner of Figure 9.7d (the Blandford–Znajek process), and those jets shoot out through the galaxy and into intergalactic space, where they feed energy into the galaxy's huge radio-emitting lobes.

These black-hole-based explanations for quasars and radio galaxies are so successful that it is tempting to assert they *must* be right, and a galaxy's jets must be a unique signature crying out to us "I come from a black hole!" However, astrophysicists are a bit cautious. They would like a more ironclad case. It is still possible to explain all the observed properties of radio galaxies and quasars using an alternative, non–black-hole engine: a rapidly spinning, magnetized, supermassive star, one weighing millions or billions of times as much as the Sun—a type of star that has never been seen by astronomers, but that theory suggests might form at the centers of galaxies. Such a supermassive star would behave much like a hole's accretion disk. By contracting to a small size (but a size still larger than its critical circumference), it could release a huge amount of gravitational energy; that energy, by way of friction, could heat the star so it shines brightly like an accretion disk; and magnetic field lines anchored in the star could spin and fling out plasma in jets.

It might be that some radio galaxies or quasars are powered by such supermassive stars. However, the laws of physics insist that such a star should gradually contract to a smaller and smaller size, and then, as it nears its critical circumference, should implode to form a black hole. The star's total lifetime before implosion should be much less than the age of the Universe. This suggests that, although the youngest of radio galaxies and quasars *might* be powered by supermassive stars, older ones are almost certainly powered, instead, by gigantic holes—*almost* certainly, but not *absolutely* certainly. These arguments are not iron-clad.

How common are gigantic black holes? Evidence, gradually accumulated during the 1980s, suggests that such holes inhabit not only the cores of most quasars and radio galaxies, but also the cores of most large, normal (non-radio) galaxies such as the Milky Way and Andromeda, and even the cores of some small galaxies such as Andromeda's dwarf companion, M32. In normal galaxies (the Milky Way, Andromeda, M32) the black hole presumably is surrounded by no accretion disk at all, or by only a tenuous disk that pours out only modest amounts of energy.

The evidence for such a hole in our own Milky Way galaxy (as of 1993) is suggestive, but far from firm. One key bit of evidence comes from the orbital motions of gas clouds near the center of the galaxy. Infrared observations of those clouds, by Charles Townes and colleagues at the University of California at Berkeley, show that they are orbiting around an object which weighs about 3 million times as much as the Sun, and radio observations reveal a very peculiar, though not strong, radio source at the position of the central object—a radio source amazingly small, no larger than our solar system. These are the types of observations one might expect from a quiescent, 3-million-solar-mass black hole with only a tenuous accretion disk; but they are also readily explained in other ways.

The possibility that gigantic black holes might exist and inhabit the cores of galaxies came as a tremendous surprise to astronomers. In retrospect, however, it is easy to understand how such holes might form in a galactic core.

In any galaxy, whenever two stars pass near each other, their gravitational forces swing them around each other and then fling them off in directions different from their original paths. (This same kind of swing and fling changes the orbits of NASA's spacecraft when they encounter planets such as Jupiter.) In the swing and fling, one of the stars typically gets flung inward, toward the galaxy's center, while the other gets flung outward, away from the center. The cumulative effect of many such swings and flings is to drive some of the galaxy's stars deep down into the galaxy's core. Similarly, it turns out, the cumulative effect of friction in the galaxy's interstellar gas is to drive much of the gas down into the galaxy's core.

As more and more gas and stars accumulate in the core, the gravity of the agglomerate they form should become stronger and stronger. Ultimately, the agglomerate's gravity may become so strong as to overwhelm its internal pressure, and the agglomerate may implode to form a gigantic hole. Alternatively, massive stars in the agglomerate may implode to form small holes, and those small holes may collide with each other and with stars and gas to form ever larger and larger holes, until a single gigantic hole dominates the core. Estimates of the time required for such implosions, collisions, and coalescences make it seem plausible (though not compelling) that most galaxies will have grown gigantic black holes in their cores long before now.

If astronomical observations did not strongly suggest that the cores of galaxies are inhabited by gigantic black holes, astrophysicists even today, in the 1990s, would probably not predict it. However, since the observations *do* suggest gigantic holes, astrophysicists easily accommodate themselves to the suggestion. This is indicative of our poor understanding of what really goes on in the cores of galaxies.

What of the future? Need we worry that the gigantic hole in our Milky Way galaxy might swallow the Earth? A few numbers set one's mind at ease. Our galaxy's central hole (if it indeed exists) weighs about 3 million times what the Sun weighs, and thus has a circumference of about 50 million kilometers, or 200 light-seconds—about one-tenth the circumference of the Earth's orbit around the Sun. This is tiny by comparison with the size of the galaxy itself. Our Earth, along with the Sun, is orbiting around the galaxy's center on an orbit with a circumference of 200,000 light-years—about 30 billion times larger than the circumference of the hole. If the hole were ultimately to swallow most of the mass of the galaxy, its circumference would expand only to about 1 light-year, still 200,000 times smaller than the circumference of our orbit.

Of course, in the roughly 10^{18} years (100 million times the Universe's present age) that it will require for our central hole to swallow a large fraction of the mass of our galaxy, the orbit of the Earth and Sun will change substantially. It is not possible to predict the details of those changes, since we do not know well enough the locations and motions of all the other stars that the Sun and Earth may encounter during 10^{18} years. Thus, we cannot predict whether the Earth and Sun will wind up, ultimately, inside the galaxy's central hole, or will be flung out of the galaxy. However, we can be confident that, if the Earth ultimately gets swallowed, its demise is roughly 10^{18} years in the future—so far off that many other catastrophes will almost certainly befall the Earth and humanity in the meantime.

10

Ripples
of Curvature

*in which gravitational waves carry to Earth
encoded symphonies of black holes colliding,
and physicists devise instruments
to monitor the waves
and decipher their symphonies*

Symphonies

In the core of a far-off galaxy, a billion light-years from Earth and a billion years ago, there accumulated a dense agglomerate of gas and hundreds of millions of stars. The agglomerate gradually shrank, as one star after another was flung out and the remaining 100 million stars sank closer to the center. After 100 million years, the agglomerate had shrunk to several light-years in size, and small stars began, occasionally, to collide and coalesce, forming larger stars. The larger stars consumed their fuel and then imploded to form black holes, and pairs of holes, flying close to each other, occasionally were captured into orbit around each other.

Figure 10.1 shows an embedding diagram for one such *black-hole binary*. Each hole creates a deep pit (strong spacetime curvature) in the embedded surface, and as the holes encircle each other, the orbiting pits produce ripples of curvature that propagate outward with the speed of light. The ripples form a spiral in the fabric of spacetime around the binary, much like the spiraling pattern of water from a

10.1 An embedding diagram depicting the curvature of space in the orbital "plane" of a binary system made of two black holes. At the center are two pits that represent the strong spacetime curvature around the two holes. These pits are the same as encountered in previous black-hole embedding diagrams, for example, Figure 7.6. As the holes orbit each other, they create outward propagating ripples of curvature called *gravitational waves*. [Courtesy LIGO Project, California Institute of Technology.]

rapidly rotating lawn sprinkler. Just as each drop of water from the sprinkler flies nearly radially outward, so each bit of curvature flies nearly radially outward; and just as the outward flying drops together form a spiraling stream of water, so all the bits of curvature together form spiraling ridges and valleys in the fabric of spacetime.

Since spacetime curvature is the same thing as gravity, these ripples of curvature are actually waves of gravity, or *gravitational waves*. Einstein's general theory of relativity predicts, unequivocally, that such gravitational waves must be produced whenever two black holes orbit each other—and also whenever two stars orbit each other.

As they depart for outer space, the gravitational waves push back on the holes in much the same way as a bullet kicks back on the gun that fires it. The waves' push drives the holes closer together and up to higher speeds; that is, it makes them slowly spiral inward toward each other. The inspiral gradually releases gravitational energy, with half of the released energy going into the waves and the other half into increasing the holes' orbital speeds.

The holes' inspiral is slow at first, but the closer the holes draw to each other, the faster they move, the more strongly they radiate their ripples of curvature, and the more rapidly they lose energy and spiral inward (Figures 10.2a,b). Ultimately, when each hole is moving at nearly the speed of light, their horizons touch and merge. Where once there were two holes, now there is one—a rapidly spinning, dumbbell-shaped hole (Figure 10.2c). As the horizon spins, its dumbbell shape radiates ripples of curvature, and those ripples push back on the hole, gradually reducing its dumbbell protrusions until they are gone (Figure 10.2d). The spinning hole's horizon is left perfectly smooth and circular in equatorial cross section, with precisely the shape described by Kerr's solution to the Einstein field equation (Chapter 7).

By examining the final, smooth black hole, one cannot in any way discover its past history. One cannot discern whether it was created by the coalescence of two smaller holes, or by the direct implosion of a star made of matter, or by the direct implosion of a star made of antimatter. The black hole has no "hair" from which to decipher its history (Chapter 7).

10.2 Embedding diagrams depicting the curvature of space around a binary system made of two black holes. The diagrams have been embellished by the artist to give a sense of motion. Each successive diagram is at a later moment of time, when the two holes have spiraled closer together. In diagrams (a) and (b), the holes' horizons are the circles at the bottoms of the pits. The horizons merge just before diagram (c), to form a single, dumbbell-shaped horizon. The rotating dumbbell emits gravitational waves, which carry away its deformation, leaving behind a smooth, spinning, Kerr black hole in diagram (d). [Courtesy LIGO Project, California Institute of Technology.]

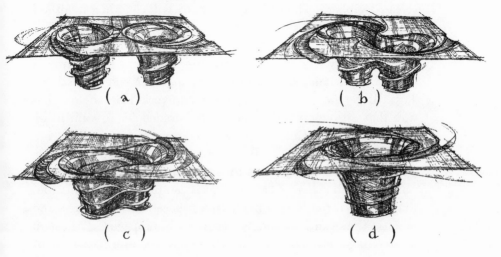

(a)   (b)

(c)   (d)

However, the history is not entirely lost. A record has been kept: It has been encoded in the ripples of spacetime curvature that the coalescing holes emitted. Those curvature ripples are much like the sound waves from a symphony. Just as the symphony is encoded in the sound waves' modulations (larger amplitude here, smaller there; higher frequency wiggles here, lower there), so the coalescence history is encoded in modulations of the curvature ripples. And just as the sound waves carry their encoded symphony from the orchestra that produces it to the audience, so the curvature ripples carry their encoded history from the coalescing holes to the distant Universe.

The curvature ripples travel outward in the fabric of spacetime, through the agglomerate of stars and gas where the two holes were born. The agglomerate absorbs none of the ripples and distorts them not at all; the ripples' encoded history remains perfectly unchanged. On outward the ripples propagate, through the agglomerate's parent galaxy and into intergalactic space, through the cluster of galaxies in which the parent galaxy resides, then onward through one cluster of galaxies after another and into our own cluster, into our own Milky Way galaxy, and into our solar system, through the Earth, and on out toward other, distant galaxies.

If we humans are clever enough, we should be able to monitor the ripples of spacetime curvature as they pass. Our computers can translate them from ripples of curvature to ripples of sound, and we then will hear the holes' symphony: a symphony that gradually rises in pitch and intensity as the holes spiral together, then gyrates in a wild way as they coalesce into one, deformed hole, then slowly fades with steady pitch as the hole's protrusions gradually shrink and disappear.

If we can decipher it, the ripples' symphony will contain a wealth of information:

1. The symphony will contain a signature that says, "I come from a pair of black holes that are spiraling together and coalescing." This will be the kind of absolutely unequivocal black-hole signature that astronomers thus far have searched for in vain using light and X-rays (Chapter 8) and radio waves (Chapter 9). Because the light, X-rays, and radio waves are produced far outside a hole's horizon, and because they are emitted by a type of material (hot, high-speed electrons) that is completely different from that of which the hole is made (pure spacetime curvature), and because they can be strongly distorted by propagating through intervening matter, they can bring us but little information about

the hole, and no definitive signature. The ripples of curvature (gravitational waves), by contrast, are produced very near the coalescing holes' horizons, they are made of the same material (a warpage of the fabric of spacetime) as the holes, they are not distorted at all by propagating through intervening matter—and, as a consequence, they can bring us detailed information about the holes and an unequivocal black-hole signature.

2. The ripples' symphony can tell us just how heavy each of the holes was, how fast they were spinning, the shape of their orbit (circular? elongated?), where the holes are on our sky, and how far they are from Earth.

3. The symphony will contain a partial map of the inspiraling holes' spacetime curvature. For the first time we will be able to test definitively general relativity's black-hole predictions: Does the symphony's map agree with Kerr's solution of the Einstein field equation (Chapter 7)? Does the map show space swirling near the spinning hole, as Kerr's solution demands? Does the amount of swirl agree with Kerr's solution? Does the way the swirl changes as one approaches the horizon agree with Kerr's solution?

4. The symphony will describe the merging of the two holes' horizons and the wild vibrations of the newly merged holes—merging and vibrations of which, today, we have only the vaguest understanding. We understand them only vaguely because they are governed by a feature of Einstein's general relativity laws that we comprehend only poorly: the laws' *nonlinearity* (Box 10.1). By "nonlinearity" is meant the propensity of strong curvature itself to produce more curvature, which in turn produces still more curvature—much like the growth of an avalanche, where a trickle of sliding snow pulls new snow into the flow, which in turn grabs more snow until an entire mountainside of snow is in motion. We understand this nonlinearity in a quiescent black hole; there it is responsible for holding the hole together; it is the hole's "glue." But we do not understand what the nonlinearity does, how it behaves, what its effects are, when the strong curvature is violently dynamical. The merger and vibration of two holes is a promising "laboratory" in which to seek such understanding. The understanding can come through hand-in-hand cooperation between experimental physicists who monitor the symphonic ripples from coalescing holes in the distant Universe and theoretical physicists who simulate the coalescence on supercomputers.

Nonlinearity and Its Consequences

A quantity is called *linear* if its total size is the sum of its parts; otherwise it is *nonlinear.*

My family income is linear: It is the sum of my wife's salary and my own. The amount of money I have in my retirement fund is nonlinear: It is not the sum of all the contributions I have invested in the past; rather, it is far greater than that sum, because each contribution started earning interest when it was invested, and each bit of interest in turn earned interest of its own.

The volume of water flowing in a sewer pipe is linear: It is the sum of the contributions from all the homes that feed into the pipe. The volume of snow flowing in an avalanche is nonlinear: A tiny trickle of snow can trigger a whole mountainside of snow to start sliding.

Linear phenomena are simple, easy to analyze, easy to predict. Nonlinear phenomena are complex and hard to predict. Linear phenomena exhibit only a few types of behaviors; they are easy to categorize. Nonlinear phenomena exhibit great richness—a richness that scientists and engineers have only appreciated in recent years, as they have begun to confront a type of nonlinear behavior called *chaos.* (For a beautiful introduction to the concept of chaos see Gleick, 1987.)

When spacetime curvature is weak (as in the solar system), it is very nearly linear; for example, the tides on the Earth's oceans are the sum of the tides produced by the Moon's spacetime curvature (tidal gravity) and the tides produced by the Sun. By contrast, when spacetime curvature is

To achieve this understanding will require monitoring the holes' symphonic ripples of curvature. How can they be monitored? The key is the physical nature of the curvature: Spacetime curvature is the same thing as tidal gravity. The spacetime curvature produced by the Moon raises tides in the Earth's oceans (Figure 10.3a), and the ripples of spacetime curvature in a gravitational wave should similarly raise ocean tides (Figure 10.3b).

General relativity insists, however, that the ocean tides raised by the Moon and those raised by a gravitational wave differ in three major ways. The first difference is propagation. The gravitational wave's tidal forces (curvature ripples) are analogous to light waves or radio waves: They travel from their source to the Earth at the speed of light, oscillating as they travel. The Moon's tidal forces, by contrast, are like the electric field of a charged body. Just as the electric field is attached firmly to the charged body and the body carries it around, always

strong (as in the big bang and near a black hole), Einstein's general relativistic laws of gravity predict that the curvature should be extremely nonlinear—among the most nonlinear phenomena in the Universe. However, as yet we possess almost no experimental or observational data to show us the effects of gravitational nonlinearity, and we are so inept at solving Einstein's equation that our solutions have taught us about the nonlinearity only in simple situations—for example, around a quiescent, spinning black hole.

A quiescent black hole owes its existence to gravitational nonlinearity; without the gravitational nonlinearity, the hole could not hold itself together, just as without gaseous nonlinearities, the great red spot on the planet Jupiter could not hold itself together. When the imploding star that creates a black hole disappears through the hole's horizon, the star loses its ability to influence the hole in any way; most important, the star's gravity can no longer hold the hole together. The hole then continues to exist solely because of gravitational nonlinearity: The hole's spacetime curvature continuously regenerates itself nonlinearly, without the aid of the star; and the self-generated curvature acts as a nonlinear "glue" to bind itself together.

The quiescent black hole whets our appetites to learn more. What other phenomena can gravitational nonlinearity produce? Some answers may come from monitoring and decoding the ripples of spacetime curvature produced by coalescing black holes. We there might see chaotic, bizarre behaviors that we never anticipated.

sticking out of itself like quills out of a hedgehog, so also the tidal forces are attached firmly to the Moon, and the Moon carries them around, sticking out of itself in a never-changing way, ready always to grab hold of and squeeze and stretch anything that comes into the Moon's vicinity. The Moon's tidal forces squeeze and stretch the Earth's oceans in a way that seems to change every few hours only because the Earth rotates through them. If the Earth did not rotate, the squeeze and stretch would be constant, unchanging.

The second difference is the direction of the tides (Figures 10.3a,b): The Moon produces tidal forces in all spatial directions. It stretches the oceans in the *longitudinal* direction (toward and away from the Moon), and it squeezes the oceans in *transverse* directions (perpendicular to the Moon's direction). By contrast, a gravitational wave produces no tidal forces at all in the longitudinal direction (along the direction of the wave's propagation). However, in the transverse plane, the wave

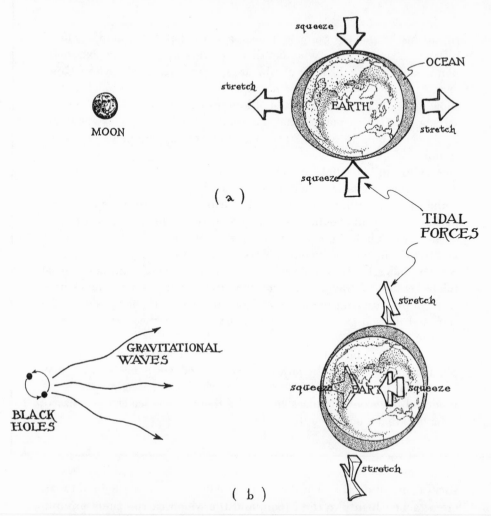

10.3 The tidal forces produced by the Moon and by a gravitational wave. (a) The Moon's tidal forces stretch and squeeze the Earth's oceans; the stretch is longitudinal, the squeeze is transverse. (b) A gravitational wave's tidal forces stretch and squeeze the Earth's oceans; the forces are entirely transverse, with a stretch along one transverse direction and a squeeze along the other.

stretches the oceans in one direction (the up–down direction in Figure 10.3b) and squeezes along the other direction (the front–back direction in Figure 10.3b). This stretch and squeeze is oscillatory. As a crest of the wave passes, the stretch is up–down, the squeeze is front–back; as a trough of the wave passes, there is a reversal to up–down squeeze and front–back stretch; as the next crest arrives, there is a reversal again to up–down stretch and front–back squeeze.

The third difference between the Moon's tides and those of a gravitational wave is their size. The Moon produces tides roughly 1 meter in size, so the difference between high tide and low tide is about 2 meters. By contrast, the gravitational waves from coalescing black holes should produce tides in the Earth's oceans no larger than about 10^{-14} meter, which is 10^{-21} of the size of the Earth (and 1/10,000 the size of a single atom, and just 10 times larger than an atom's nucleus). Since tidal forces are proportional to the size of the object on which they act (Chapter 2), the waves will tidally distort *any* object by about 10^{-21} of its size. In this sense, 10^{-21} is the *strength of the waves* when they arrive at Earth.

Why are the waves so weak? Because the coalescing holes are so far away. The strength of a gravitational wave, like the strength of a light wave, dies out inversely with the distance traveled. When the waves are still close to the holes, their strength is roughly 1; that is, they squeeze and stretch an object by about as much as the object's size; humans would be killed by so strong a stretch and squeeze. However, when the waves have reached Earth, their strength is reduced to roughly (1/30 of the holes' circumference) / (the distance the waves have traveled).[1] For holes that weigh about 10 times as much as the Sun and are a billion light-years away, this wave strength is ($\frac{1}{30}$) × (180 kilometers for the horizon circumference)/(a billion light-years for the distance to Earth) $\simeq 10^{-21}$. Therefore, the waves distort the Earth's oceans by 10^{-21} × (10^7 meters for the Earth's size) = 10^{-14} meter, or 10 times the diameter of an atomic nucleus.

It is utterly hopeless to think of measuring such a tiny tide on the Earth's turbulent ocean. Not quite so hopeless, however, are the prospects for measuring the gravitational wave's tidal forces on a carefully designed laboratory instrument—a *gravitational-wave detector*.

Bars

Joseph Weber was the first person with sufficient insight to realize that it is *not* utterly hopeless to try to detect gravitational waves. A graduate of the U.S. Naval Academy in 1940 with a bachelor's degree

---

1. The factor $\frac{1}{30}$ comes from detailed calculations with the Einstein field equation. It includes a factor $1/(2\pi)$, which is approximately $\frac{1}{6}$, to convert the hole's circumference into a radius, and an additional factor $\frac{1}{5}$ that arises from details of the Einstein field equation.

in engineering, Weber served in World War II on the aircraft carrier *Lexington*, until it was sunk in the Battle of the Coral Sea, and then became commanding officer of Submarine Chaser No. 690; and he led Brigadier General Theodore Roosevelt, Jr., and 1900 Rangers onto the beach in the 1943 invasion of Italy. After the war he became head of the electronic countermeasures section of the Bureau of Ships for the U.S. Navy. His reputation for mastery of radio and radar technology was so great that in 1948 he was offered and accepted the position of full professor of electrical engineering at the University of Maryland— full professor at age twenty-nine, and with no more college education than a bachelor's degree.

While teaching electrical engineering at Maryland, Weber prepared for a career change: He worked toward, and completed, a Ph.D. in physics at Catholic University, in part under the same person as had been John Wheeler's Ph.D. adviser, Karl Herzfeld. From Herzfeld, Weber learned enough about the physics of atoms, molecules, and radiation to invent, in 1951, one version of the mechanism by which lasers work, but he did not have the resources to demonstrate his concept experimentally. While Weber was publishing his concept, two other research groups, one at Columbia University led by Charles Townes and the other in Moscow led by Nikolai Gennadievich Basov and Aleksandr Michailovich Prokharov, independently invented alternative versions of the mechanism, and then went on to construct working lasers.[2] Though Weber's paper had been the first publication on the mechanism, he received hardly any credit; the Nobel Prize and patents went to the Columbia and Moscow scientists. Disappointed, but maintaining close friendships with Townes and Basov, Weber sought a new research direction.

As part of his search, Weber spent a year in John Wheeler's group, became an expert on general relativity, and with Wheeler did theoretical research on general relativity's predictions of the properties of gravitational waves. By 1957, he had found his new direction. He would embark on the world's first effort to build apparatus for detecting and monitoring gravitational waves.

Through late 1957, all of 1958, and early 1959, Weber struggled to invent every scheme he could for detecting gravitational waves. This

---

2. Their lasers actually produced microwaves (short-wavelength radio waves) rather than light, and thus were called *masers* rather than "lasers." "Real" lasers, the kind that produce light, were not successfully constructed until several years later.

was a pen, paper, and brainpower exercise, not experimental. He filled four 300-page notebooks with ideas, possible detector designs, and calculations of the expected performance of each design. One idea after another he cast aside as not promising. One design after another failed to give high sensitivity. But a few held promise; and of them, Weber ultimately chose a cylindrical aluminum bar about 2 meters long, a half meter in diameter, and a ton in weight, oriented broadside to the incoming waves (Figure 10.4 below).

As the waves' tidal force oscillates, it should first compress, then stretch, then compress such a bar's ends. The bar has a natural mode of vibration which can respond resonantly to this oscillating tidal force, a mode in which its ends vibrate in and out relative to its center. That natural mode, like the ringing of a bell or tuning fork or wine glass, has a well-defined frequency. Just as a bell or tuning fork or wine glass can be made to ring sympathetically by sound waves that match its natural frequency, so the bar can be made to vibrate sympathetically by oscillating tidal forces that match its natural frequency. To use such a bar as a gravitational-wave detector, then, one should adjust its size so its natural frequency will match that of the incoming gravitational waves.

What frequency will that be? In 1959, when Weber embarked on this project, few people believed in black holes (Chapter 6), and the believers understood only very little about a hole's properties. Nobody then imagined that holes could collide and coalesce and eject ripples of spacetime curvature with encoded histories of their collisions. Nor could anyone give much hopeful guidance about other sources of gravitational waves.

So Weber embarked on his effort nearly blind. His sole guide was a crude (but correct) argument that the gravitational waves probably would have frequencies below about 10,000 Hertz (10,000 cycles per second)—that being the orbital frequency of an object which moves at the speed of light (the fastest possible) around the most compact conceivable star: one with size near the critical circumference. So Weber designed the best detectors he could, letting their resonant frequencies fall wherever they might below 10,000 Hertz, and hoped that the Universe would provide waves at his chosen frequencies. He was lucky. The resonant frequencies of his bars were about 1000 Hertz (1000 cycles of oscillation per second), and it turns out that some of the waves from coalescing black holes should oscillate at just such frequencies, as should some of the waves from supernova explosions and from coalescing pairs of neutron stars.

The most challenging aspect of Weber's project was to invent a *sensor* for monitoring his bars' vibrations. Those wave-induced vibrations, he expected, would be tiny: smaller than the diameter of the nucleus of an atom [but he did not know, in the 1960s, how very tiny: just $10^{-21} \times$ (the 2-meter length of his bars) $\simeq 10^{-21}$ meter or one-millionth the diameter of the nucleus of an atom, according to more recent estimates]. To most physicists of the late 1950s and the 1960s, even one-tenth of the diameter of an atomic nucleus looked impossibly

10.4 Joseph Weber, demonstrating the piezoelectric crystals glued around the middle of his aluminum bar; ca. 1973. Gravitational waves should drive the bar's end-to-end vibrations, and those vibrations should squeeze the crystals in and out so they produce oscillating voltages that are detected electronically. [Photo by James P. Blair, courtesy the National Geographic Society.]

difficult to measure. Not so to Weber. He invented a sensor that was up to the task.

Weber's sensor was based on the *piezoelectric effect,* in which certain kinds of materials (certain crystals and ceramics), when squeezed slightly, develop electric voltages from one end to the other. Weber would have liked to make his bar from such a material, but these materials were far too expensive, so he did the next best thing: He made his bar from aluminum, and he then glued piezoelectric crystals around the bar's middle (Figure 10.4). As the bar vibrated, its surface squeezed and stretched the crystals, each crystal developed an oscillating voltage, and Weber strung the crystals together one after another in an electric circuit so their tiny oscillating voltages would add up to a large enough voltage for electronic detection, even when the bar's vibrations were only one-tenth the diameter of the nucleus of an atom.

In the early 1960s, Weber was a lonely figure, the only experimental physicist in the world seeking gravitational waves. With his bitter aftertaste of laser competition, he enjoyed the loneliness. However, in the early 1970s, his impressive sensitivities and evidence that he might actually be detecting waves (which, in retrospect, I am convinced he was not) attracted dozens of other experimenters, and by the 1980s more than a hundred talented experimenters were engaged in a competition with him to make gravitational-wave astronomy a reality.

I first met Weber on a hillside opposite Mont Blanc in the French Alps, in the summer of 1963, four years after he embarked on his project to detect gravitational waves. I was a graduate student, just beginning research in relativity, and along with thirty-five other students from around the world I had come to the Alps for an intensive two-month summer school focusing solely on Einstein's general relativistic laws of gravity. Our teachers were the world's greatest relativity experts—John Wheeler, Roger Penrose, Charles Misner, Bryce De-Witt, Joseph Weber, and others—and we learned from them in lectures and private conversations, with the glistening snows of the Agui de Midi and Mont Blanc towering high in the sky above us, belled cows grazing in brilliant green pastures around us, and the picturesque village of Les Houches several hundred meters below us, at the foot of our school's hillside.

In this glorious setting, Weber lectured about gravitational waves and his project to detect them, and I listened, fascinated. Between lectures Weber and I conversed about physics, life, and mountain climbing, and I came to regard him as a kindred soul. We were both loners; neither of us enjoyed intense competition or vigorous intellectual give-and-take. We both preferred to wrestle with a problem on our own, seeking advice and ideas occasionally from friends, but not being buffeted by others who were trying to beat us to a new insight or discovery.

Over the next decade, as research on black holes heated up and entered its golden age (Chapter 7), I began to find black-hole research distasteful—too much intensity, too much competition, too much rough-and-tumble. So I cast about for another area of research, one with more elbow room, into which I could put most of my effort while still working on black holes and other things part time. Inspired by Weber, I chose gravitational waves.

Like Weber, I saw gravitational waves as an infant research field with a bright future. By entering the field in its infancy, I could have the fun of helping mold it, I could lay foundations on which others later would build, and I could do so without others breathing down my neck, since most other relativity theorists were then focusing on black holes.

For Weber, the foundations to be laid were experimental: the invention, construction, and continual improvement of detectors. For me, they were theoretical: try to understand what Einstein's gravitational laws have to say about how gravitational waves are produced, how they push back on their sources as they depart, and how they propagate; try to figure out which kinds of astronomical objects will produce the Universe's strongest waves, how strong their waves will be, and with what frequencies they will oscillate; invent mathematical tools for computing the details of the encoded symphonies produced by these objects, so when Weber and others ultimately detect the waves, theory and experiment can be compared.

In 1969 I spent six weeks in Moscow, at Zel'dovich's invitation. One day Zel'dovich took time out from bombarding me and others with new ideas (Chapters 7 and 12), and drove me over to Moscow University to introduce me to a young experimental physicist, Vladimir Braginsky. Braginsky, stimulated by Weber, had been working for several years to develop techniques for gravitational-wave detection; he was the first experimenter after Weber to enter the field. He was also in the midst of other fascinating experiments: a search for *quarks* (a fundamental building block of protons and neutrons), and an experiment to test Einstein's assertion that all objects, no matter what their composition, fall with the same acceleration in a gravitational field (an assertion that underlies Einstein's description of gravity as spacetime curvature).

I was impressed. Braginsky was clever, deep, and had excellent taste in physics; and he was warm and forthright, as easy to talk to about politics as about science. We quickly became close friends and learned to respect each other's world views. For me, a liberal Democrat in the American spectrum, the freedom of the individual was paramount over all other considerations. No government should have the right to dictate how one lives one's life. For Braginsky, a nondoctrinaire Communist, the responsibility of the individual to society was paramount. We are our brothers' keepers, and well we should be in a world where evil people like Joseph Stalin can gain control if we are not vigilant.

Left: Joseph Weber, Kip Thorne, and Tony Tyson at a conference on gravitational radiation in Warsaw, Poland, September 1973. *Right:* Vladimir Braginsky and Kip Thorne, in Pasadena, California, October 1984. [Left: photo by Marek Holzman, courtesy Andrzej Trautman; right: courtesy Valentin N. Rudenko.]

Braginsky had foresight that nobody else possessed. During our 1969 meeting, and then again in 1971 and 1972, he warned me that the bars being used to search for gravitational waves have a fundamental, ultimate limitation. That limitation, he told me, comes from the laws of quantum mechanics. Although we normally think of quantum mechanics as governing tiny objects such as electrons, atoms, and molecules, if we make sufficiently precise measurements on the vibrations of a one-ton bar, we should see those vibrations also behave quantum mechanically, and their quantum mechanical behavior will ultimately cause problems for gravitational-wave detection. Braginsky had convinced himself of this by calculating the ultimate performance of Weber's piezoelectric crystals and of several other kinds of sensors that one might use to measure a bar's vibrations.

I didn't understand what Braginsky was talking about; I didn't understand his reasoning, I didn't understand his conclusion, I didn't understand its importance, and I didn't pay much attention. Other things he was teaching me seemed much more important: From him I was learning how to think about experiments, how to design experimental apparatus, how to predict the noise that will plague the apparatus, and how to suppress the noise so the apparatus will succeed in its task—and from me, Braginsky was learning how to think about Einstein's laws of gravity, how to identify their predictions. We were rapidly becoming a team, each bringing to our joint enterprise his own expertise; and over the next two decades, together we would have great fun and make a few discoveries.

Each year in the early and mid-1970s, when we saw each other in Moscow or Pasadena or Copenhagen or Rome or wherever, Braginsky repeated his warning about quantum mechanical trouble for gravitational-wave detectors, and each year I again did not understand. His warning was somewhat muddled because he himself did not understand fully what was going on. However, in 1976, after Braginsky, and independently Robin Giffard at Stanford University, managed to make the warning more clear, I suddenly understood. The warning was serious, I finally realized; the ultimate sensitivity of a bar detector is severely limited by the *uncertainty principle*.

The uncertainty principle is a fundamental feature of the laws of quantum mechanics. It says that, if you make a highly accurate measurement of the position of an object, then in the process of your measurement you will necessarily kick the object, thereby perturbing

Box 10.2
The Uncertainty Principle and
Wave/Particle Duality

The uncertainty principle is intimately related to wave/particle duality (Box 4.1)—that is, to the propensity of particles to act sometimes like waves and sometimes like particles.

If you measure the position of a particle (or any other object, for example, the end of a bar) and learn that it is somewhere inside some error box, then regardless of what the particle's wave might have looked like before the measurement, during the measurement the measuring apparatus will "kick" the wave and thereby confine it inside the error box's interior. The wave, thereby, will acquire a confined form something like the following:

(figure: wave confined inside error box, with max, min, min, error-box, WAVE axis, DISTANCE axis)

Such a confined wave contains many different wavelengths, ranging from the size of the box itself (marked *max* above) to the tiny size of the corners at which the wave begins and ends (marked *min*). More specifically, the confined wave can be constructed by adding together, that is, superimposing, the following oscillatory waves, which have wavelengths ranging from *max* down to *min:*

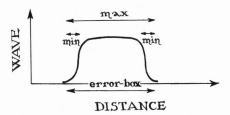

Now, recall that the shorter the wavelength of the wave's oscillations, the larger the energy of the particle, and thus also the larger the particle's velocity. Since the measurement has given the wave a range of wavelengths, the particle's energy and velocity might now be anywhere in the corresponding ranges; in other words, its energy and velocity are uncertain.

(continued next page)

(Box 10.2 continued)

To recapitulate, the measurement confined the particle's wave to the error box (first diagram above); this made the wave consist of a range of wavelengths (second diagram); that range of wavelengths corresponds to a range of energy and velocity; and the velocity is therefore uncertain. No matter how hard you try, you cannot avoid producing this velocity uncertainty when you measure the particle's position. Moreover, when this chain of reasoning is examined in greater depth, it predicts that the more accurate your measurement, that is, the smaller your error box, the larger the ranges of wavelengths and velocity, and thus the larger the uncertainty in the particle's velocity.

the object's velocity in a random, unpredictable way. The more accurate your position measurement is, the more strongly and unpredictably you must perturb the object's velocity. No matter how clever you are in designing your measurement, you cannot circumvent this innate uncertainty. (See Box 10.2.)

The uncertainty principle governs not only measurements of microscopic objects such as electrons, atoms, and molecules; it also governs measurements of large objects. However, because a large object has large inertia, a measurement's kick will perturb its velocity only slightly. (The velocity perturbation will be inversely proportional to the object's mass.)

The uncertainty principle, when applied to a gravitational-wave detector, says that the more accurately a sensor measures the position of the end or side of a vibrating bar, the more strongly and randomly the measurement must kick the bar.

For an inaccurate sensor, the kick can be tiny and unimportant, but because the sensor was inaccurate, you do not know very well the amplitude of the bar's vibrations and thus cannot monitor weak gravitational waves.

For an extremely accurate sensor, the kick is so enormous that it strongly changes the bar's vibrations. These large, unknowable changes thus mask the effects of any gravitational wave you might try to detect.

Somewhere between these two extremes there is an optimal accuracy for the sensor: an accuracy neither so poor that you learn little nor so great that the unknowable kick is strong. At that optimal accuracy, which is now called *Braginsky's standard quantum limit*, the effect of the kick is just barely as debilitating as the errors made by the sensor. No sensor can monitor the bar's vibrations more accurately than this standard quantum limit. How small is this limit? For a 2-meter-long, 1-ton bar, it is about 100,000 times smaller than the nucleus of an atom.

In the 1960s, nobody seriously contemplated the need for such accurate measurements, because nobody understood very clearly just how weak should be the gravitational waves from black holes and other astronomical bodies. But by the mid-1970s, spurred on by Weber's experimental project, I and other theorists had begun to figure out how strong the strongest waves were likely to be. Roughly 10^{-21} was the answer, and this meant the waves would make a 2-meter bar vibrate with an amplitude of only $10^{-21} \times$ (2 meters), or about a millionth the diameter of the nucleus of an atom. If these estimates were correct (and we knew they were highly uncertain), then the gravitational-wave signal would be *ten times smaller than Braginsky's standard quantum limit,* and therefore could not possibly be detected using a bar and any known kind of sensor.

Though this was extremely worrisome, all was not lost. Braginsky's deep intuition told him that, if experimenters were especially clever, they ought to be able to circumvent his standard quantum limit. There ought to be a new way to design a sensor, he argued, so that its unknowable and unavoidable kick does *not* hide the influence of the gravitational waves on the bar. To such a sensor Braginsky gave the name *quantum nondemolition*[3]; "quantum" because the sensor's kick is demanded by the laws of quantum mechanics, "nondemolition" because the sensor would be so configured that the kick would not demolish the thing you are trying to measure, the influence of the waves on the bar. Braginsky did not have a workable design for a quantum nondemolition sensor, but his intuition told him that such a sensor should be possible.

This time I listened, carefully; and over the next two years I and my group at Caltech and Braginsky and his group in Moscow both struggled, on and off, to devise a quantum nondemolition sensor.

We both found the answer simultaneously in the autumn of 1977— but by very different routes. I remember vividly my excitement when

3. Braginsky has a remarkable mastery of the nuances of the English language; he can construct an eloquent English phrase to describe a new idea far more readily than most Americans or Britons.

the idea occurred to Carlton Caves and me[4] in an intense discussion over lunch at the Greasy (Caltech's student cafeteria). And I recall the bittersweet taste of learning that Braginsky, Yuri Vorontsov, and Far-hid Khalili had had a significant piece of the same idea in Moscow at essentially the same time—bitter because I get great satisfaction from being the first to discover something new; sweet because I am so fond of Braginsky and thus get pleasure from sharing discoveries with him.

Our full quantum nondemolition idea is rather abstract and permits a wide variety of sensor designs for circumventing Braginsky's standard quantum limit. The idea's abstractness, however, makes it difficult to explain, so here I shall describe just one (not very practical) example of a quantum nondemolition sensor.[5] This example has been called, by Braginsky, a *stroboscopic sensor.*

A stroboscopic sensor relies on a special property of a bar's vibrations: If the bar is given a very sharp, unknown kick, its amplitude of vibration will change, but no matter what that amplitude change is, precisely one period of oscillation after the kick the bar's vibrating end will return to the same position as it had at the moment of the kick (black dots in Figure 10.5). At least this is true if a gravitational wave (or some other force) has not squeezed or stretched the bar in the meantime. If a wave (or other force) *has* squeezed the bar in the meantime, then the bar's position one period later will be changed.

To detect the wave, then, one should build a sensor that makes stroboscopic measurements of the bar's vibrating ends, a sensor that measures the position of the bar's ends quickly once each period of vibration. Such a sensor will kick the bar in each measurement, but the kicks will not change the position of the bar's ends at the times of subsequent measurements. If the position is found to have changed, then a gravitational wave (or some other force) must have squeezed the bar.

Although quantum nondemolition sensors solved the problem of Braginsky's standard quantum limit, by the mid-1980s I had become rather pessimistic about the prospects for bar detectors to bring gravitational-wave astronomy to fruition. My pessimism had two causes.

4. A key foundation for our idea came from a colleague, William Unruh, at the University of British Columbia. The development of the idea and its consequences was carried out jointly by Caves, me, and three others who were gathered around the lunch table with us when we discovered it: Ronald Drever, Vernon Sandberg, and Mark Zimmermann.

5. The full idea is described by Caves et al. (1980) and by Braginsky, Vorontsov, and Thorne (1980).

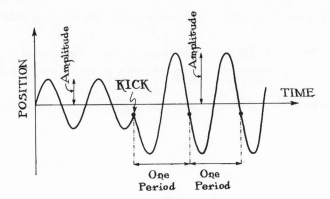

10.5 The principle underlying a stroboscopic quantum nondemolition measurement. Plotted vertically is the position of the end of a vibrating bar; plotted horizontally is time. If a quick, highly precise measurement of the position is made at the time marked KICK, the sensor that makes the measurement will give the bar a sudden, unknowable kick, thereby changing the bar's amplitude of vibration in an unknown way. However, there will be no change of the position of the bar's end precisely one period after the kick, or two periods, or three periods. Those positions will be the same as at the time of the kick and will be completely independent of the kick.

First, although the bars built by Weber, by Braginsky, and by others had achieved far better sensitivities than anyone had dreamed possible in the 1950s, they were still only able to detect with confidence waves of strength 10^{-17} or larger. This was 10,000 times too poor for success, if I and others had correctly estimated the strengths of the waves arriving at Earth. This by itself was not serious, since the march of technology has often produced 10,000-fold improvements in instruments over times of twenty years or less. [One example was the angular resolution of the best radio telescopes, which improved from tens of degrees in the mid-1940s to a few arc seconds in the mid-1960s (Chapter 9). Another was the sensitivity of astronomical X-ray detectors, which improved by a factor of 10^{10} between 1958 and 1978, that is, at an average rate of 10,000 every eight years (Chapter 8).] However, the rate of improvement of the bars was so slow, and projections of the future technology and techniques were so modest, that there seemed no reasonable way for a 10,000-fold improvement to be made in the foreseeable future. Success, thus, would likely hinge on the waves being stronger than the 10^{-21} estimates—a real possibility, but not one that anybody was happy to rely on.

Second, even if the bars did succeed in detecting gravitational waves, they would have enormous difficulty in decoding the waves' symphonic signals, and in fact would probably fail. The reason was simple: Just as a tuning fork or wine glass responds sympathetically only to a sound whose frequency is close to its natural frequency, so a bar would respond only to gravitational waves whose frequency is near the bar's natural frequency; in technical language, the bar detector has a *narrow bandwidth* (the bandwidth being the band of frequencies to which it responds). But the waves' symphonic information should typically be encoded in a very wide band of frequencies. To extract the waves' information, then, would require a "xylophone" of many bars, each covering a different, tiny portion of the signal's frequencies. How many bars in the xylophone? For the types of bars then being planned and constructed, several thousand—far too many to be practical. In principle it would be possible to widen the bars' bandwidths and thereby manage with, say, a dozen bars, but to do so would require major technical advances beyond those for reaching a sensitivity of 10^{-21}.

Although I did not say much in public in the 1980s about my pessimistic outlook, in private I regarded it as tragic because of the great effort that Weber, Braginsky, and my other friends and colleagues had put into bars, and also because I had become convinced that gravitational radiation has the potential to produce a revolution in our knowledge of the Universe.

LIGO

To understand the revolution that the detection and deciphering of gravitational waves might bring, let us recall the details of a previous revolution: the one created by the development of X-ray and radio telescopes (Chapters 8 and 9).

In the 1930s, before the advent of radio astronomy and X-ray astronomy, our knowledge of the Universe came almost entirely from light. Light showed it to be a serene and quiescent Universe, a Universe dominated by stars and planets that wheel smoothly in their orbits, shining steadily and requiring millions or billions of years to change in discernible ways.

This tranquil view of the Universe was shattered, in the 1950s, 1960s, and 1970s, when radio-wave and X-ray observations showed us our Universe's violent side: jets ejected from galactic nuclei, quasars with fluctuating luminosities far brighter than our galaxy, pulsars with intense beams shining off their surfaces and rotating at high speeds. The brightest objects seen by optical telescopes were the Sun, the planets, and a few nearby, quiescent stars. The brightest objects seen by radio telescopes were violent explosions in the cores of distant galaxies, powered (presumably) by gigantic black holes. The brightest objects seen by X-ray telescopes were small black holes and neutron stars accreting hot gas from binary companions.

What was it about radio waves and X-rays that enabled them to create such a spectacular revolution? The key was the fact that they brought us very different kinds of information than is brought by light: Light, with its wavelength of a half micron, was emitted primarily by hot atoms residing in the atmospheres of stars and planets, and it thus taught us about those atmospheres. The radio waves, with their 10-million-fold greater wavelengths, were emitted primarily by near-light-speed electrons spiraling in magnetic fields, and they thus taught us about the magnetized jets shooting out of galactic nuclei, about the gigantic, magnetized intergalactic lobes that the jets feed, and about the magnetized beams of pulsars. The X-rays, with their thousand-fold shorter wavelengths than light, were produced mostly by high-speed electrons in ultra-hot gas accreting onto black holes and neutron stars, and they thus taught us directly about the accreting gas and indirectly about the holes and neutron stars.

The differences between light, on the one hand, and radio waves and X-rays, on the other, are pale compared to the differences between the electromagnetic waves (light, radio, infrared, ultraviolet, X-ray, and gamma ray) of modern astronomy and gravitational waves. Correspondingly, gravitational waves might revolutionize our understanding of the Universe even more than did radio waves and X-rays. Among the differences between electromagnetic waves and gravitational waves, and their consequences, are these[6]:

- The gravitational waves should be emitted most strongly by large-scale, coherent vibrations of spacetime curvature (for example, the collision and coalescence of two black holes) and by large-scale,

6. These differences, their consequences, and the details of the waves to be expected from various astrophysical sources have been elucidated by a number of theorists including, among others, Thibault Damour in Paris, Leonid Grishchuk in Moscow, Takashi Nakamura in Kyoto, Bernard Schutz in Wales, Stuart Shapiro in Ithaca, New York, Clifford Will in St. Louis, and me.

coherent motions of huge amounts of matter (for example, the implosion of the core of a star that triggers a supernova, or the inspiral and merger of two neutron stars that are orbiting each other). Therefore, gravitational waves should show us the motions of huge curvatures and huge masses. By contrast, cosmic electromagnetic waves are usually emitted individually and separately by enormous numbers of individual and separate atoms or electrons; and these individual electromagnetic waves, each oscillating in a slightly different manner, then superimpose on each other to produce the total wave that an astronomer measures. As a result, from electromagnetic waves we learn primarily about the temperature, density, and magnetic fields experienced by the emitting atoms and electrons.

· Gravitational waves are emitted most strongly in regions of space where gravity is so intense that Newton's description fails and must be replaced by Einstein's, and where huge amounts of matter or spacetime curvature move or vibrate or swirl at near the speed of light. Examples are the big bang origin of the Universe, the collisions of black holes, and the pulsations of newborn neutron stars at the centers of supernova explosions. Since these strong-gravity regions are typically surrounded by thick layers of matter that absorb electromagnetic waves (but do not absorb gravitational waves), the strong-gravity regions cannot send us electromagnetic waves. The electromagnetic waves seen by astronomers come, by contrast, almost entirely from weak-gravity, low-velocity regions; for example, the surfaces of stars and supernovae.

These differences suggest that the objects whose symphonies we might study with gravitational-wave detectors will be largely invisible in light, radio waves, and X-rays; and the objects that astronomers now study in light, radio waves, and X-rays will be largely invisible in gravitational waves. The gravitational Universe should thus look extremely different from the electromagnetic Universe; from gravitational waves we should learn things that we will never learn electromagnetically. This is why gravitational waves are likely to revolutionize our understanding of the Universe.

It might be argued that our present electromagnetically based understanding of the Universe is so complete compared with the optically based understanding of the 1930s that a gravitational-wave revolution will be far less spectacular than was the radio-wave/X-ray revolution.

This seems to me unlikely. I am painfully aware of our lack of understanding when I contemplate the sorry state of present estimates of the gravitational waves bathing the Earth. For each type of gravitational-wave source that has been thought about, with the exception of binary stars and their coalescences, either the strength of the source's waves for a given distance from Earth is uncertain by several factors of 10, or the rate of occurrence of that type of source (and thus also the distance to the nearest one) is uncertain by several factors of 10, or the very existence of the source is uncertain.

These uncertainties cause great frustration in the planning and design of gravitational-wave detectors. That is the downside. The upside is the fact that, if and when gravitational waves are ultimately detected and studied, we may be rewarded with major surprises.

In 1976 I had not yet become pessimistic about bar detectors. On the contrary, I was highly optimistic. The first generation of bar detectors had recently reached fruition and had operated with a sensitivity that was remarkable compared to what one might have expected; Braginsky and others had invented a number of clever and promising ideas for huge future improvements; and I and others were just beginning to realize that gravitational waves might revolutionize our understanding of the Universe.

My enthusiasm and optimism drove me, one evening in November 1976, to wander the streets of Pasadena until late into the night, struggling with myself over whether to propose that Caltech create a project to detect gravitational waves. The arguments in favor were obvious: for science in general, the enormous intellectual payoff if the project succeeded; for Caltech, the opportunity to get in on the ground floor of an exciting new field; for me, the possibility to have a team of experimenters at my home institution with whom to interact, instead of relying primarily on Braginsky and his team on the other side of the world, and the possibility to play a more central role than I could commuting to Moscow (and thereby have more fun). The argument against was also obvious: The project would be risky; to succeed, it would require large resources from Caltech and the U.S. National Science Foundation and enormous time and energy from me and others; and after all that investment, it might fail. It was much more risky than Caltech's entry into radio astronomy twenty-three years earlier (Chapter 9).

After many hours of introspection, the lure of the payoffs won me over. And after several months studying the risks and payoffs, Caltech's physics and astronomy faculty and administration unanimously approved my proposal—subject to two conditions. We would have to find an outstanding experimental physicist to lead the project, and the project would have to be large enough and strong enough to have a good chance of success. This meant, we believed, much larger and stronger than Weber's effort at the University of Maryland or Braginsky's effort in Moscow or any of the other gravitational-wave efforts then under way.

The first step was finding a leader. I flew to Moscow to ask Braginsky's advice and feel him out about taking the post. My feeler tore him every which way. He was torn between the far better technology he would have in America and the greater craftsmanship of the technicians in Moscow (for example, intricate glassblowing was almost a lost art in America, but not in Moscow). He was torn between the need to build a project from scratch in America and the crazy impediments that the inefficient, bureaucracy-bound Soviet system kept putting in the way of his project in Moscow. He was torn between loyalty to his native land and disgust with his native land, and between his feelings that life in America is barbaric because of the way we treat our poor and our lack of medical care for everyone and his feelings that life in Moscow is miserable because of the power of incompetent officials. He was torn between the freedom and wealth of America and fear of KGB retribution against family and friends and perhaps even himself if he "defected." In the end he said no, and recommended instead Ronald Drever of Glasgow University.

Others I consulted were also enthusiastic about Drever. Like Braginsky, he was highly creative, inventive, and tenacious—traits that would be essential for success of the project. The Caltech faculty and administration gathered all the information they could about Drever and other possible leaders, selected Drever, and invited him to join the Caltech faculty and initiate the project. Drever, like Braginsky, was torn, but in the end he said yes. We were off and running.

I had presumed, when proposing the project, that like Weber and Braginsky, Caltech would focus on building bar detectors. Fortunately (in retrospect) Drever insisted on a radically different direction. In Glasgow he had worked with bar detectors for five years, and he could see their limitations. Much more promising, he thought, were *interferometric* gravitational-wave detectors (*interferometers* for short— though they are radically different from the radio interferometers of Chapter 9).

Interferometers for gravitational-wave detection had first been conceived of in primitive form in 1962 by two Russian friends of Braginsky's, Mikhail Gertsenshtein and V.I. Pustovoit, and independently
in 1964 by Joseph Weber. Unaware of these early ideas, Rainer Weiss
devised a more mature variant of an interferometric detector in 1969,
and then he and his MIT group went on to design and build one in the
early 1970s, as did Robert Forward and colleagues at Hughes Research
Laboratories in Malibu, California. Forward's detector was the first to
operate successfully. By the late 1970s, these interferometric detectors
had become a serious alternative to bars, and Drever had added his own
clever twists to their design.

Figure 10.6 shows the basic idea behind an interferometric gravitational-wave detector. Three masses hang by wires from overhead supports at the corner and ends of an "L" (Figure 10.6a). When the first
crest of a gravitational wave enters the laboratory from overhead or
underfoot, its tidal forces should stretch the masses apart along one arm
of the "L" while squeezing them together along the other arm. The
result will be an increase in the length L_1 of the first arm (that is, in the
distance between the arm's two masses) and a decrease in the length L_2
of the second arm. When the wave's first crest has passed and its first
trough arrives, the directions of stretch and squeeze will be changed: L_1
will decrease and L_2 will increase. By monitoring the arm-length difference, $L_1 - L_2$, one can seek gravitational waves.

10.6 A laser interferometric gravitational-wave detector. This instrument is
very similar to the one used by Michelson and Morley in 1887 to search for
motion of the Earth through the aether (Chapter 1). See the text for a detailed
explanation.

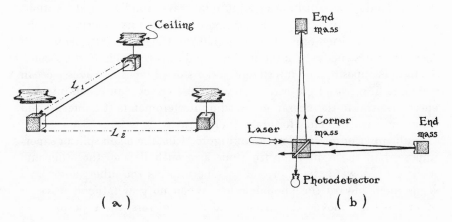

(a)   (b)

The difference $L_1 - L_2$ is monitored using *interferometry* (Figure 10.6b and Box 10.3). A laser beam shines onto a *beam splitter* that rides on the corner mass. The beam splitter reflects half of the beam and transmits half, and thereby splits the beam in two. The two beams go down the two arms of the interferometer and bounce off mirrors that ride on the arms' end masses, and then return to the beam splitter. The splitter half-transmits and half-reflects each of the beams, so part of each beam's light is combined with part from the other and goes back toward the laser, and the other parts of the two beams are combined and go toward the photodetector. When no gravitational wave is present, the contributions from the two arms interfere in such a way (Box 10.3) that all the net light goes back toward the laser and none toward

---

Box 10.3

Interference and Interferometry

Whenever two or more waves propagate through the same region of space, they superimpose on each other "linearly" (Box 10.1); that is, they add. For example, the following dotted wave and dashed wave superimpose to produce the heavy solid wave:

Notice that at locations such as *A* where a trough of one wave (dotted) superimposes on a crest of the other (dashed), the waves cancel, at least in part, to produce a vanishing or weak total wave (solid); and at locations such as *B* where two troughs superimpose or two crests superimpose, the waves reinforce each other. One says that the waves are *interfering* with each other, destructively in the first case and constructively in the second. Such superimposing and interference occurs in all types of waves—ocean waves, radio waves, light waves, gravitational waves—and such interference is central to the operation of radio interferometers (Chapter 9) and interferometric detectors for gravitational waves.

In the interferometric detector of Figure 10.6b, the beam splitter superimposes half the light wave from one arm on half from the other and sends them toward the laser, and it superimposes the other halves and sends them toward the photodetector. When no gravitational wave or other force has moved the masses and their mirrors, the superimposed

the photodetector. If a gravitational wave slightly changes $L_1 - L_2$, the two beams will then travel slightly different distances in their two arms and will interfere slightly differently—a tiny amount of their combined light will now go into the photodetector. By monitoring the amount of light reaching the photodetector, one can monitor the arm-length difference $L_1 - L_2$, and thereby monitor gravitational waves.

It is interesting to compare a bar detector with an interferometer. The bar detector uses the vibrations of a single, solid cylinder to monitor the tidal forces of a gravitational wave. The interferometric detector uses the relative motions of masses hung from wires to monitor the tidal forces.

light waves have the following forms, where the dashed curve shows the wave from arm 1, the dotted curve the wave from arm 2, and the solid curve the superimposed, total wave:

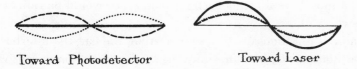

Toward Photodetector   Toward Laser

Toward the photodetector, the waves interfere perfectly destructively, so the total, superimposed wave vanishes, which means that the photodetector sees no light at all. When a gravitational wave or other force has lengthened one arm slightly and shortened the other, then the beam from the one arm arrives at the beam splitter with a slight delay relative to the other, and the superimposed waves therefore look like this:

Toward Photodetector   Toward Laser

The destructive interference in the photodetector's direction is no longer perfect; the photodetector receives some light. The amount it receives is proportional to the arm length difference, $L_1 - L_2$, which in turn is proportional to the gravitational-wave signal.

The bar detector uses an electrical sensor (for example, piezoelectric crystals squeezed by the bar) to monitor the bar's wave-induced vibrations. The interferometric detector uses interfering light beams to monitor its masses' wave-induced motions.

The bar responds sympathetically only to gravitational waves over a narrow frequency band, and therefore, decoding the waves' symphony would require a xylophone of many bars. The interferometer's masses wiggle back and forth in response to waves of *all* frequencies higher than about one cycle per second,[7] and therefore the interferometer has a wide bandwidth; three or four interferometers are sufficient to fully decode the symphony.

By making the interferometer's arms a thousand times longer than the bar (a few kilometers rather than a few meters), one can make the waves' tidal forces a thousand times bigger and thus improve the sensitivity of the instrument a thousand-fold.[8] The bar, by contrast, cannot be lengthened much. A kilometer-long bar would have a natural frequency less than one cycle per second and thus would not operate at the frequencies where we think the most interesting sources lie. Moreover, at such a low frequency, one must launch the bar into space to isolate it from vibrations of the ground and from the fluctuating gravity of the Earth's atmosphere. Putting such a bar in space would be ridiculously expensive.

Because it is a thousand times longer than the bar, the interferometer is a thousand times more immune to the "kick" produced by the measurement process. This immunity means that the interferometer does *not* need to circumvent the kick with the aid of a (difficult to construct) quantum nondemolition sensor. The bar, by contrast, can detect the expected waves only if it employs quantum nondemolition.

If the interferometer has such great advantages over the bar (far larger bandwidth and far larger potential sensitivity), then why didn't Braginsky, Weber, and others build interferometers instead of bars? When I asked Braginsky in the mid-1970s, he replied that bar detectors are simple, while interferometers are horrendously complex. A small, intimate team like his in Moscow had a reasonable chance of making bar detectors work well enough to discover gravitational waves. However, to construct, debug, and operate interferometric detectors success-

7. Below about one cycle per second, the wires that suspend the masses prevent them from wiggling in response to the waves.

8. Actually, the details of the improvement are far more complicated than this, and the resulting sensitivity enhancement is far more difficult to achieve than these words suggest; however, this description is roughly correct.

fully would require a huge team and large amounts of money—and Braginsky doubted whether, even with such a team and such money, so complex a detector could succeed.

Ten years later, as the painful evidence mounted that bars would have great difficulty reaching 10^{-21} sensitivity, Braginsky visited Caltech and was impressed with the progress that Drever's team had achieved with interferometers. Interferometers, he concluded, will probably succeed after all. But the huge team and large money required for success were not to his taste; so upon returning to Moscow, he redirected most of his own team's efforts away from gravitational-wave detection. (Elsewhere in the world bars have continued to be developed, which is fortunate; they are cheap compared to interferometers, for now they are more sensitive, and in the long run they might play special roles at high gravity-wave frequencies.)

Wherein lies the complexity of interferometric detectors? After all, the basic idea, as described in Figure 10.6, looks reasonably simple.

In fact, Figure 10.6 is a gross oversimplification because it ignores an enormous number of pitfalls. The tricks required to avoid these pitfalls make an interferometer into a very complex instrument. For example, the laser beam must point in precisely the right direction and have precisely the right shape and wavelength to fit into the interferometer perfectly; and its wavelength and intensity must not fluctuate. After the beam is split in half, the two beams must bounce back and forth in the two arms not just once as in Figure 10.6, but many times, so as to increase their sensitivity to the wiggling masses' motions, and after these many bounces, they must meet each other perfectly back at the beam splitter. Each mass must be continually controlled so its mirrors point in precisely the right directions and do not swing as a result of vibrations of the floor, and this must be done without masking the mass's gravitational-wave-induced wiggles. To achieve perfection in all these ways, and in many many more, requires continuously monitoring many different pieces of the interferometer and its light beams, and continuously applying feedback forces to keep them perfect.

One gets some impression of these complications from a photograph (Figure 10.7) of a 40-meter-long *prototype* interferometric detector that Drever's team has built at Caltech—a prototype which itself is far simpler than the full-scale, several-kilometer-long interferometers that are required for success.

10.7   The Caltech 40-meter interferometric prototype gravitational-wave detector, ca. 1989. The table in front and the front caged vacuum chamber hold lasers and devices to prepare the laser light for entry into the interferometer. The central mass resides in the second caged vacuum chamber—the chamber above which a dangling rope can be seen faintly. The end masses are 40 meters away, down the two corridors. The two arms' laser beams shine down the larger of the two vacuum pipes that extend the lengths of the corridors. [Courtesy LIGO Project, California Institute of Technology.]

During the early 1980s four teams of experimental physicists struggled to develop tools and techniques for interferometric detectors: Drever's Caltech team, the team he had founded at Glasgow (now led by James Hough), Rainer Weiss's team at MIT, and a team founded by Hans Billing at the Max Planck Institut in Munich, Germany. The teams were small and intimate, and they worked more or less independently,[9] pursuing their own approaches to the design of interferometric detectors. Within each team the individual scientists had free rein to invent new ideas and pursue them as they wished and for as long as they wished; coordination was very loose. This is just the kind of culture that inventive scientists love and thrive on, the culture that Bra-

9. Though with a close link, through Drever, between the Glasgow and Caltech teams.

ginsky craves, a culture in which loners like me are happiest. But it is not a culture capable of designing, constructing, debugging, and operating large, complex scientific instruments like the several-kilometer-long interferometers required for success.

To design in detail the many complex pieces of such interferometers, to make them all fit together and work together properly, and to keep costs under control and bring the interferometers to completion within a reasonable time require a different culture: a culture of tight coordination, with subgroups of each team focusing on well-defined tasks and a single director making decisions about what tasks will be done when and by whom.

The road from freewheeling independence to tight coordination is a painful one. The world's biology community is traveling that road, with cries of anguish along the way, as it moves toward sequencing the human genome. And we gravitational-wave physicists have been traveling that road since 1984, with no less pain and anguish. I am confident, however, that the excitement, pleasure, and scientific payoff of detecting the waves and deciphering their symphonies will one day make the pain and anguish fade in our memories.

The first sharp turn on our painful road was a 1984 shotgun marriage between the Caltech and MIT teams—each of which by then had about eight members. Richard Isaacson of the U.S. National Science Foundation (NSF) held the shotgun and demanded, as the price of the taxpayers' financial support, a tight marriage in which Caltech and MIT scientists jointly developed the interferometers. Drever (resisting like mad) and Weiss (willingly accepting the inevitable) said their vows, and I became the marriage counselor, the man with the task of forging consensus when Drever pulled in one direction and Weiss in another. It was a rocky marriage, emotionally draining for all; but gradually we began to work together.

The second sharp turn came in November 1986. A committee of eminent physicists—experts in all the technologies we need and experts in the organization and management of large scientific projects—spent an entire week with us, scrutinizing our progress and plans, and then reported to NSF. Our progress got high marks, our plans got high marks, and our prospects for success—for detecting waves and deciphering their symphonies—were rated as high. But our culture was rated as awful; we were still immersed in the loosely knit, freewheeling culture of our birth, and we could never succeed that way, NSF was told. Replace the Drever–Weiss–Thorne troika by a single director, the committee insisted—a director who can mold talented individualists into a tightly knit and effective team and can organize the project and make firm, wise decisions at every major juncture.

Out came the shotgun again. If you want your project to continue, NSF's Isaacson told us, you must find that director and learn to work with him like a football team works with a great coach or an orchestra with a great conductor.

We were lucky. In the midst of our search, Robbie Vogt got fired.

Vogt, a brilliant, strong-willed experimental physicist, had directed projects to construct and operate scientific instruments on spacecraft, had directed the construction of a huge millimeter-wavelength astronomical interferometer, and had reorganized the scientific research environment of NASA's Jet Propulsion Laboratory (which carries out

A portion of the Caltech/MIT team of LIGO scientists in late 1991. *Left:* Some Caltech members of the team, counterclockwise from upper left: Aaron Gillespie, Fred Raab, Maggie Taylor, Seiji Kawamura, Robbie Vogt, Ronald Drever, Lisa Sievers, Alex Abramovici, Bob Spero, Mike Zucker. *Right:* Some MIT members of the team, counterclockwise from upper left: Joe Kovalik, Yaron Hefetz, Nergis Mavalvala, Rainer Weiss, David Schumaker, Joe Giaime. [Left: courtesy Ken Rogers/ Black Star; right: courtesy Erik L. Simmons.]

most of the American planetary exploration program)—and he then had become Caltech's provost. As provost, though remarkably effective, Vogt battled vigorously with Caltech's president, Marvin Goldberger, over how to run Caltech—and after several years of battle, Goldberger fired him. Vogt was not temperamentally suited to working *under* others when he disagreed profoundly with their judgments; but on top, he was superb. He was just the director, the conductor, the coach that we needed. If anybody could mold us into a tightly knit team, he could.

"It will be painful working with Robbie," a former member of his millimeter team told us. "You will emerge bruised and scarred, but it will be worth it. Your project will succeed."

For several months Drever, Weiss, I, and others pleaded with Vogt to take the directorship. He finally accepted; and, as promised, six years later our Caltech/MIT team is bruised and scarred, but effective, powerful, tightly knit, and growing rapidly toward the critical size (about fifty scientists and engineers) required for success. Success, however, will not depend on us alone. Under Vogt's plan important inputs to our core effort will come from other scientists[10] who, by being only loosely associated with us, can maintain the individualistic, free-wheeling style that we have left behind.

A key to success in our endeavor will be the construction and operation of a national scientific facility called the *Laser Interferometer Gravitational-Wave Observatory,* or *LIGO.* The LIGO will consist of two L-shaped vacuum systems, one near Hanford, Washington, and the other near Livingston, Louisiana, in which physicists will develop and operate many successive generations of ever-improving interferometers; see Figure 10.8.

Why two facilities instead of one? Because Earth-bound gravitational-wave detectors always have ill-understood noise that simulates gravitational-wave bursts; for example, the wire that suspends a mass can creak slightly for no apparent reason, thereby shaking the mass and simulating the tidal force of a wave. However, such noise almost never happens simultaneously in two independent detectors, far apart. Thus, to be sure that an apparent signal is due to gravitational waves rather than noise, one must verify that it occurs in two such detectors. With

---

10. These, as of 1993, include Braginsky's group in Moscow, a group led by Bob Byers at Stanford University, a group led by Jim Faller at the University of Colorado, a group led by Peter Saulson at Syracuse University, and a group led by Sam Finn at Northwestern University.

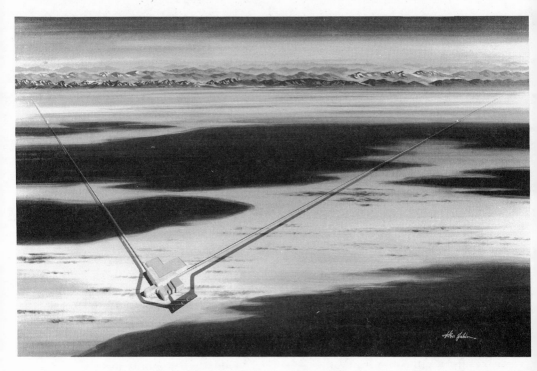

10.8 Artist's conception of LIGO's L-shaped vacuum system and the experimental facilities at the corner of the L, near Hanford, Washington. [Courtesy LIGO Project, California Institute of Technology.]

only one detector, gravitational waves cannot be detected and monitored.

Although two facilities are sufficient to detect a gravitational wave, at least three and preferably four are required, at widely separated sites, to fully decode the wave's symphony, that is, to extract all the information the wave carries. A joint French/Italian team will build the third facility, named VIRGO,[11] near Pisa, Italy. VIRGO and LIGO together will form an international network for extracting the full information. Teams in Britain, Germany, Japan, and Australia are seeking funds to build additional facilities for the network.

It might seem audacious to construct such an ambitious network for a type of wave that nobody has ever seen. Actually, it is not audacious at all. Gravitational waves have already been proved to exist by astronomical observations for which Joseph Taylor and Russell Hulse of

11. It is named for the Virgo cluster of galaxies, from which waves might be detected.

Princeton University won the 1993 Nobel Prize. Taylor and Hulse, using a radio telescope, found two neutron stars, one of them a pulsar, which orbit each other once each 8 hours; and by exquisitely accurate radio measurements, they verified that the stars are spiraling together at precisely the rate (2.7 parts in a billion per year) that Einstein's laws predict they should, due to being continually kicked by gravitational waves that they emit into the Universe. Nothing else, only tiny gravitational-wave kicks, can explain the stars' inspiral.

What will gravitational-wave astronomy be like in the early 2000s? The following scenario is plausible:

By 2007, eight interferometers, each several kilometers long, are in full-time operation, scanning the skies for incoming bursts of gravitational waves. Two are operating in the vacuum facility in Pisa, Italy, two in Livingston, Louisiana, in the southeastern United States, two in Hanford, Washington, in the northwestern United States, and two in Japan. Of the two interferometers at each site, one is a "workhorse" instrument that monitors a wave's oscillations between about 10 cycles per second and 1000; the other, only recently developed and installed, is an advanced, "specialty" interferometer that zeroes in on oscillations between 1000 and 3000 cycles per second.

A train of gravitational waves sweeps into the solar system from a distant, cosmic source. Each wave crest hits the Japanese detectors first, then sweeps through the Earth to the Washington detectors, then Louisiana, and finally Italy. For roughly a minute, crest is followed by trough is followed by crest. The masses in each detector wiggle ever so slightly, perturbing their laser beams and hence perturbing the light that enters the detector's photodiode. The eight photodiode outputs are transmitted by satellite links to a central computer, which alerts a team of scientists that another minute-long gravitational-wave burst has arrived at Earth, the third one this week. The computer combines the eight detectors' outputs to produce four things: a best-guess location for the burst's source on the sky; an error box for that best-guess location; and two *waveforms*—two oscillating curves, analogous to the oscillating curve that you obtain if you examine the sounds of a symphony on an oscilloscope. The history of the source is encoded in these waveforms (Figure 10.9).

There are two waveforms because a gravitational wave has two *polarizations.* If the wave travels vertically through an interferometer, one polarization describes tidal forces that oscillate along the east–west and north–south directions; the other describes tidal forces oscillating along the northeast–southwest and northwest–southeast directions. Each detector, with its own orientation, feels some combination of these two polarizations; and from the eight detector outputs, the computer reconstructs the two waveforms.

The computer then compares the waveforms with those in a large catalog, much as a bird watcher identifies a bird by comparing it with pictures in a book. The catalog has been produced by simulations of sources on computers, and by five years of previous experience monitoring gravitational waves from colliding and coalescing black holes, colliding and coalescing neutron stars, spinning neutron stars (pulsars), and supernova explosions. The identification of this burst is easy (some others, for example, from supernovae, are far harder). The waveforms show the unmistakable, unique signature of two black holes coalescing.

10.9   One of the two waveforms produced by the coalescence of two black holes. The wave is plotted upward in units of 10^{-21}; time is plotted horizontally in units of seconds. The first graph shows only the last 0.1 second of the inspiral part of the waveform; the preceding minute of the waveform is similar, with gradually increasing amplitude and frequency. The second graph shows the last 0.01 second, on a stretched-out scale. The *Inspiral* and *Ringdown* segments of the waveform are well understood, in 1993, from solutions of the Einstein field equation. The coalescence segment is not at all understood (the curve shown is my own fantasy); future supercomputer simulations will attempt to compute it. In the text these simulations are presumed to have been successful in the early twenty-first century.

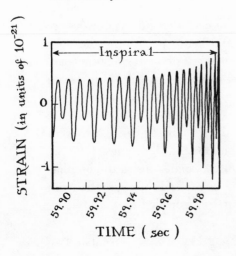

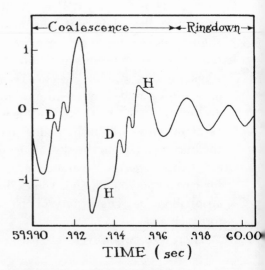

The waveforms have three segments:

- The minute-long first segment (of which only the last 0.1 second is shown in Figure 10.9) has oscillating strains that gradually grow in amplitude and frequency; these are precisely the waveforms expected from the *inspiral* of two objects in a binary orbit. The fact that alternate waves are smaller and larger indicates that the orbit is somewhat elliptical rather than circular.
- The 0.01-second-long middle segment matches almost perfectly the waveforms predicted by recent (early twenty-first century) supercomputer simulations of the *coalescence* of two black holes to form one; according to the simulations, the humps marked "H" signal the touching and merging of the holes' horizons. The double wiggles marked "D," however, are a new discovery, the first one made by the new specialty interferometers. The older, workhorse interferometers had never been able to detect these wiggles because of their high frequency, and they had never yet been seen in any supercomputer simulations. They are a new challenge for theorists to explain. They might be the first hints of some previously unsuspected quirk in the nonlinear vibrations of the colliding holes' spacetime curvature. Theorists, intrigued by this prospect, will go back to their simulations and search for signs of such doublet wiggles.
- The 0.03-second-long third segment (of which only the beginning is shown in Figure 10.9) consists of oscillations with fixed frequency and gradually dying amplitude. This is precisely the waveform expected when a deformed black hole pulsates to shake off its deformations, that is, as it *rings down* like a struck bell. The pulsations consist of two dumbbell-type protrusions that circulate around and around the hole's equator and gradually die out as ripples of curvature carry away their energy (Figure 10.2 above).

From the details of the waveforms, the computer extracts not only the history of the inspiral, coalescence, and ringdown; it also extracts the masses and spin rates of the initial holes and the final hole. The initial holes each weighed 25 times what the Sun weighs, and were slowly spinning. The final hole weighs 46 times what the Sun weighs and is spinning at 97 percent of the maximum allowed rate. Four solar masses' worth of energy ($2 \times 25 - 46 = 4$) were converted into ripples of curvature and carried away by the waves. The total surface area of the initial holes was 136,000 square kilometers. The total surface area of the final hole is larger, as demanded by the second law of black-hole mechanics (Chapter 12): 144,000 square kilometers. The waveforms also reveal the distance of the hole from Earth: 1 billion light-years, a result accurate to about 20 percent. The waveforms tell us that we on Earth were looking down nearly perpendicularly onto the plane of the orbit, and are now looking down the north pole of the spinning hole; and they show that the holes' orbit had an eccentricity (elongation) of 30 percent.

The computer determines the holes' location on the sky from the wave crests' times of arrival in Japan, Washington, Louisiana, and Italy. Since Japan was hit first, the holes were more or less overhead in Japan, and underfoot in America and Europe. A detailed analysis of the arrival times gives a best-guess location for the source, and an error box around that location of 1 degree in size. Had the holes been smaller, their waveforms would have oscillated more rapidly and the error box would have been tighter, but for these big holes 1 degree is the best the network can do. In another ten years, when an interferometric detector is operating on the Moon, the error boxes will be reduced in size along one side by a factor of 100.

Because the holes' orbit was elongated, the computer concludes that the two holes were captured into orbit around each other only a few hours before they coalesced and emitted the burst. (If they had been orbiting each other for longer than a few hours, the push of gravitational waves departing from the binary would have made their orbit circular.) Recent capture means the holes were probably in a dense cluster of black holes and massive stars at the center of some galaxy.

The computer therefore examines catalogs of optical galaxies, radio galaxies, and X-ray galaxies, searching for any that reside in the 1-degree error box, are between 0.8 and 1.2 billion light-years from Earth, and have peculiar cores. Forty candidates are found and turned over to astronomers. For the next few years these forty candidates will be studied in detail, with radio, millimeter, infrared, optical, ultraviolet, X-ray, and gamma-ray telescopes. Gradually it will become clear that one of the candidate galaxies has a core in which a massive agglomerate of gas and stars was beginning, when the light we now see left the galaxy, a million-year-long phase of violent evolution—an evolution that will trigger the birth of a gigantic black hole, and then a quasar. Thanks to the burst of gravitational waves which identified this specific galaxy as interesting, astronomers can now begin to unravel the details of how gigantic black holes are born.

11

What Is Reality?

*in which spacetime is viewed as
curved on Sundays and flat on Mondays,
and horizons are made from
vacuum on Sundays and charge on Mondays,
but Sunday's experiments and Monday's experiments
agree in all details*

Is spacetime *really* curved? Isn't it conceivable that spacetime is actually flat, but the clocks and rulers with which we measure it, and which we regard as *perfect* in the sense of Box 11.1, are actually rubbery? Might not even the most perfect of clocks slow down or speed up, and the most perfect of rulers shrink or expand, as we move them from point to point and change their orientations? Wouldn't such distortions of our clocks and rulers make a truly flat spacetime appear to be curved?

Yes.

Figure 11.1 gives a concrete example: the measurement of circumferences and radii around a nonspinning black hole. On the left is shown an embedding diagram for the hole's curved space. The space is curved in this diagram because we have chosen to define distances as though our rulers were *not* rubbery, as though they always hold their lengths fixed no matter where we place them and how we orient them. The rulers show the hole's horizon to have a circumference of 100 kilometers. A circle of twice this circumference, 200 kilometers, is drawn around the hole, and the radial distance from the horizon to that circle

---

Box 11.1
Perfection of Rulers and Clocks

By "perfect clocks" and "perfect rulers" I shall mean, in this book, clocks and rulers that are perfect in the sense that the world's best clock makers and ruler makers understand: Perfection is to be judged by comparison with the behaviors of atoms and molecules.

More specifically, perfect clocks must tick at a uniform rate when compared with the oscillations of atoms and molecules. The world's best atomic clocks are designed to do just that. Since the oscillations of atoms and molecules are controlled by what I called in earlier chapters the "rate of flow of time," this means that perfect clocks measure the "time" part of Einstein's curved spacetime.

The markings on perfect rulers must have uniform and standard spacings when compared to the wavelengths of the light emitted by atoms and molecules, for example, uniform spacings relative to the "21-centimeter-wavelength" light emitted by hydrogen molecules. This is equivalent to requiring that when one holds a ruler at some fixed, standard temperature (say, zero degrees Celsius), it contain always the same fixed number of atoms along its length between markings; and this, in turn, guarantees that perfect rulers measure the spatial lengths of Einstein's curved spacetime.

The body of this chapter introduces the concept of "true" times and "true" lengths. These are not necessarily the times and lengths measured by perfect clocks and perfect rulers, that is, not necessarily the times and lengths based on atomic and molecular standards, that is, not necessarily the times and lengths embodied in Einstein's curved spacetime.

---

is measured with a perfect ruler; the result is 37 kilometers. If space were flat, that radial distance would have to be the radius of the outside circle, $200/2\pi$ kilometers, minus the radius of the horizon, $100/2\pi$ kilometers; that is, it would have to be $200/2\pi - 100/2\pi = 16$ kilometers (approximately). To accommodate the radial distance's far larger, 37-kilometer size, the surface must have the curved, trumpet-horn shape shown in the diagram.

If space is actually flat around the black hole, but our perfect rulers are rubbery and thereby fool us into thinking space is curved, then the true geometry of space must be as shown on the right in Figure 11.1, and the true distance between the horizon and the circle must be 16 kilometers, as demanded by the flat-geometry laws of Euclid. However, general relativity insists that our perfect rulers not measure this true distance. Take a ruler and lay it down circumferentially around the hole just outside the horizon (curved thick black strip with ruler markings in right part of Figure 11.1). When oriented circumferentially like this, it does measure correctly the true distance. Cut the ruler off at 37 kilometers length, as shown. It now encompasses 37 percent of the distance around the hole. Then turn the ruler so it is oriented radially (straight thick black strip with ruler markings in Figure 11.1). As it is turned, general relativity requires that it shrink. When pointed

11.1 Length measurements in the vicinity of a black hole from two different viewpoints. *Left:* Spacetime is regarded as truly curved, and perfect rulers measure precisely the lengths of the true spacetime. *Right:* Spacetime is regarded as truly flat and perfect rulers are rubbery. A 37-km-long perfect ruler, when oriented in a circumferential direction, measures precisely the lengths of the true, flat spacetime. However, when oriented radially, it shrinks by an amount that is greater the nearer it is to the hole, and therefore it reports radial lengths that are larger than the true ones (it reports 37 km rather than the true 16 km in the case shown).

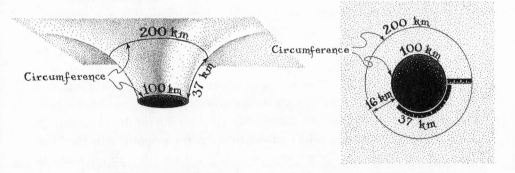

CURVED SPACETIME   FLAT SPACETIME

radially, its true length must have shrunk to 16 kilometers, so it will reach precisely from the horizon to the outer circle. However, the scale on its shrunken surface must claim that its length is still 37 kilometers, and therefore that the distance between horizon and circle is 37 kilometers. People like Einstein who are unaware of the ruler's rubbery nature, and thus believe its inaccurate measurement, conclude that space is curved. However, people like you and me, who understand the rubberiness, know that the ruler has shrunk and that space is really flat.

What could possibly make the ruler shrink, when its orientation changes? Gravity, of course. In the flat space of the right half of Figure 11.1 there resides a gravitational field that controls the sizes of fundamental particles, atomic nuclei, atoms, molecules, everything, and forces them all to shrink when laid out radially. The amount of shrinkage is great near a black hole, and smaller farther away, because the shrinkage-controlling gravitational field is generated by the hole, and its influence declines with distance.

The shrinkage-controlling gravitational field has other effects. When a photon or any other particle flies past the hole, this field pulls on it and deflects its trajectory. The trajectory is bent around the hole; it is curved, as measured in the hole's true, flat spacetime geometry. However, people like Einstein, who take seriously the measurements of their rubbery rulers and clocks, regard the photon as moving along a straight line through curved spacetime.

What is the real, genuine truth? Is spacetime really flat, as the above paragraphs suggest, or is it really curved? To a physicist like me this is an uninteresting question because it has no physical consequences. Both viewpoints, curved spacetime and flat, give precisely the same predictions for any measurements performed with perfect rulers and clocks, and also (it turns out) the same predictions for any measurements performed with any kind of physical apparatus whatsoever. For example, both viewpoints agree that the radial distance between the horizon and the circle in Figure 11.1, *as measured by a perfect ruler,* is 37 kilometers. They disagree as to whether that measured distance is the "real" distance, but such a disagreement is a matter of philosophy, not physics. Since the two viewpoints agree on the results of all experiments, they are physically equivalent. Which viewpoint tells the "real truth" is irrelevant for experiments; it is a matter for philosophers to debate, not physicists. Moreover, physicists can *and do* use the two viewpoints interchangeably when trying to deduce the predictions of general relativity.

The mental processes by which a theoretical physicist works are beautifully described by Thomas Kuhn's concept of a *paradigm.* Kuhn, who received his Ph.D. in physics from Harvard in 1949 and then became an eminent historian and philosopher of science, introduced the concept of a paradigm in his 1962 book *The Structure of Scientific Revolutions*—one of the most insightful books I have ever read.

A paradigm is a complete set of tools that a community of scientists uses in its research on some topic, and in communicating the results of its research to others. The curved spacetime viewpoint on general relativity is one paradigm; the flat spacetime viewpoint is another. Each of these paradigms includes three basic elements: a set of mathematically formulated *laws* of physics; a set of *pictures* (mental pictures, verbal pictures, drawings on paper) which give us insight into the laws and help us communicate with each other; and a set of *exemplars*—past calculations and solved problems, either in textbooks or in published scientific articles, which the community of relativity experts agrees were correctly done and were interesting, and which we use as patterns for our future calculations.

The *curved spacetime paradigm* is based on three sets of mathematically formulated laws: Einstein's field equation, which describes how matter generates the curvature of spacetime; the laws which tell us that perfect rulers and perfect clocks measure the lengths and the times of Einstein's curved spacetime; and the laws which tell us how matter and fields move through curved spacetime, for example, that freely moving bodies travel along straight lines (geodesics). The *flat spacetime paradigm* is also based on three sets of laws: a law describing how matter, in flat spacetime, generates the gravitational field; laws describing how that field controls the shrinkage of perfect rulers and the dilation of the ticking rates of perfect clocks; and laws describing how the gravitational field also controls the motions of particles and fields through flat spacetime.

The pictures in the curved spacetime paradigm include the embedding diagrams drawn in this book (for example, the left half of Figure 11.1) and the verbal descriptions of spacetime curvature around black holes (for example, the "tornado-like swirl of space around a spinning black hole"). The pictures in the flat spacetime paradigm include the right half of Figure 11.1, with the ruler that shrinks when it turns from circumferential orientation to radial, and the verbal description of "a gravitational field controlling the shrinkage of rulers."

The exemplars of the curved spacetime paradigm include the calculation, found in most relativity textbooks, by which one derives Schwarzschild's solution to the Einstein field equation, and the calculations by which Israel, Carter, Hawking, and others deduced that a black hole has no "hair." The flat spacetime exemplars include textbook calculations of how the mass of a black hole or other body changes when gravitational waves are captured by it, and calculations by Clifford Will, Thibault Damour, and others of how neutron stars orbiting each other generate gravitational waves (waves of shrinkage-producing field).

Each piece of a paradigm—its laws, its pictures, and its exemplars— is crucial to my own mental processes when I'm doing research. The pictures (mental and verbal as well as on paper) act as a general compass. They give me intuition as to how the Universe probably behaves; I manipulate them, along with mathematical doodlings, in search of interesting new insights. If I find, from the pictures and doodlings, an insight worth pursuing (for example, the hoop conjecture in Chapter 7), I then try to verify or refute it by careful mathematical calculations based on the paradigm's mathematically formulated laws of physics. I pattern my careful calculations after the paradigm's exemplars. They tell me what level of calculational precision is likely to be needed for reliable results. (If the precision is too poor, the results may be wrong; if the precision is too high, the calculations may eat up valuable time unnecessarily.) The exemplars also tell me what kinds of mathematical manipulations are likely to get me through the morass of mathematical symbols to my goal. Pictures also guide the calculations; they help me find shortcuts and avoid blind alleys. If the calculations verify or at least make plausible my new insight, I then communicate the insight to relativity experts by a mixture of pictures and calculations, and I communicate to others, such as readers of this book, solely with pictures—verbal pictures and drawings.

The flat spacetime paradigm's laws of physics can be derived, mathematically, from the curved spacetime paradigm's laws, and conversely. This means that the two sets of laws are different *mathematical representations* of the same physical phenomena, in somewhat the same sense as 0.001 and $\frac{1}{1000}$ are different mathematical representations of the same number. However, the mathematical formulas for the laws look very different in the two representations, and the pictures and exemplars that accompany the two sets of laws look very different.

As an example, in the curved spacetime paradigm, the verbal picture of Einstein's field equation is the statement that "mass generates the curvature of spacetime." When translated into the language of the flat spacetime paradigm, this field equation is described by the verbal picture "mass generates the gravitational field that governs the shrinkage of rulers and the dilation of the ticking of clocks." Although the two versions of the Einstein field equation are mathematically equivalent, their verbal pictures differ profoundly.

It is extremely useful, in relativity research, to have both paradigms at one's fingertips. Some problems are solved most easily and quickly using the curved spacetime paradigm; others, using flat spacetime. Black-hole problems (for example, the discovery that a black hole has no hair) are most amenable to curved spacetime techniques; gravitational-wave problems (for example, computing the waves produced when two neutron stars orbit each other) are most amenable to flat spacetime techniques. Theoretical physicists, as they mature, gradually build up insight into which paradigm will be best for which situation, and they learn to flip their minds back and forth from one paradigm to the other, as needed. They may regard spacetime as curved on Sunday, when thinking about black holes, and as flat on Monday, when thinking about gravitational waves. This mind-flip is similar to that which one experiences when looking at a drawing by M. C. Escher, for example, Figure 11.2.

Since the laws that underlie the two paradigms are mathematically equivalent, we can be sure that when the same physical situation is analyzed using both paradigms, the predictions for the results of experiments will be identically the same. We thus are free to use the paradigm that best suits us in any given situation.

This freedom carries power. That is why physicists were not content with Einstein's curved spacetime paradigm, and have developed the flat spacetime paradigm as a supplement to it.

Newton's description of gravity is yet another paradigm. It regards space and time as absolute, and gravity as a force that acts instantaneously between two bodies ("action at a distance," Chapters 1 and 2).

The Newtonian paradigm for gravity, of course, is *not* equivalent to Einstein's curved spacetime paradigm; the two give different predictions for the outcomes of experiments. Thomas Kuhn uses the phrase *scientific revolution* to describe the intellectual struggle by which Einstein invented his paradigm and convinced his colleagues that it gives a more nearly correct description of gravity than the Newtonian para-

11.2 A drawing by M. C. Escher. One can experience a *mind-flip* by looking at this drawing, first from one point of view (for example, with the flowing stream at the same height as the waterfall's top) and then from another (with the stream at the height of the waterfall's bottom). This mind-flip is somewhat like the one a theoretical physicist experiences when switching from the curved spacetime paradigm to the flat spacetime paradigm.

digm (Chapter 2). Physicists' invention of the flat spacetime paradigm was *not* a scientific revolution in this Kuhnian sense, because the flat spacetime paradigm and the curved spacetime paradigm give precisely the same predictions.

When gravity is weak, the predictions of the Newtonian paradigm and Einstein's curved spacetime paradigm are almost identical, and correspondingly the two paradigms are very nearly mathematically equivalent. Thus it is that, when studying gravity in the solar system, physicists often switch back and forth with impunity between the Newtonian paradigm, the curved spacetime paradigm, and also the flat spacetime paradigm, using at any time whichever one strikes their fancy or seems the more insightful.[1]

Sometimes people new to a field of research are more open-minded than the old hands. Such was the case in the 1970s, when new people had insights that led to a new paradigm for black holes, the *membrane paradigm.*

In 1971 Richard Hanni, an undergraduate at Princeton University, together with Remo Ruffini, a postdoc, noticed that a black hole's horizon can behave somewhat like an electrically conducting sphere. To understand this peculiar behavior, recall that a positively charged metal pellet carries an electric field which repels protons but attracts electrons. The pellet's electric field can be described by field lines, analogous to those of a magnetic field. The electric field lines point in the direction of the force that the field exerts on a proton (and oppositely to the force exerted on an electron), and the density of field lines is proportional to the strength of the force. If the pellet is alone in flat spacetime, its electric field lines point radially outward (Figure 11.3a). Correspondingly, the electric force on a proton points radially away from the pellet, and since the density of field lines decreases inversely with the square of the distance from the pellet, the electric force on a proton also decreases inversely with the square of the distance.

Now bring the pellet close to a metal sphere (Figure 11.3b). The sphere's metal surface is made of electrons that can move about on the sphere freely, and positively charged ions that cannot. The pellet's electric field pulls a number of the sphere's electrons into the pellet's

---

1. Compare with the last section of Chapter 1, "The Nature of Physical Law."

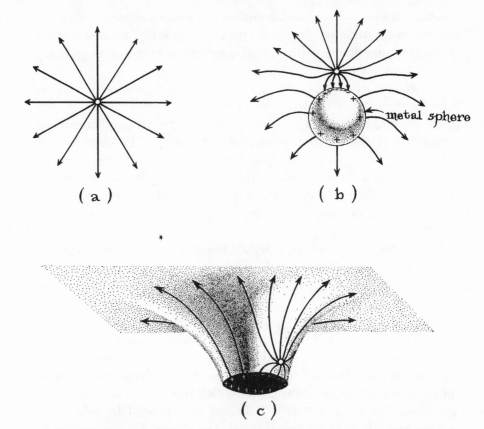

11.3 (a) The electric field lines produced by a positively charged metal pellet at rest, alone, in flat spacetime. (b) The electric field lines when the pellet is at rest just above an electrically conducting, metal sphere in flat spacetime. The pellet's electric field polarizes the sphere. (c) The electric field lines when the pellet is at rest just above the horizon of a black hole. The pellet's electric field appears to polarize the horizon.

vicinity, leaving excess ions everywhere else on the sphere; in other words, it *polarizes*[2] the sphere.

In 1971 Hanni and Ruffini, and independently Robert Wald of Princeton University and Jeff Cohen of the Princeton Institute for Advanced Study, computed the shapes of the electric field lines produced by a charged pellet near the horizon of a nonspinning black hole.

2. This is a different usage of the word "polarize" from that of "polarized gravitational waves" and "polarized light" (Chapter 10).

Their computations, based on the standard curved spacetime paradigm, revealed that the curvature of spacetime distorts the field lines in the manner shown in Figure 11.3c. Hanni and Ruffini, noticing the similarity to the field lines in Figure 11.3b [look at diagram (c) from below, and it will be nearly the same as diagram (b)], suggested that we can think of a black hole's horizon in the same manner as we think of a metal sphere; that is, we can regard the horizon as a thin membrane composed of positively and negatively charged particles, a membrane similar to the sphere's metal. Normally there are equal numbers of positive and negative particles everywhere on the membrane, that is, there is no net charge on any region of the membrane. However, when the pellet is brought near the horizon, excess negative particles move into the region below the pellet, leaving excess positive particles everywhere else on the membrane; the horizon's membrane thereby gets polarized; and the total set of field lines produced by the pellet's charges and the horizon's charges takes the form of diagram (c).

When I, as an old hand at relativity theory, heard this story, I thought it ludicrous. General relativity insists that, if one falls into a black hole, one will encounter nothing at the horizon except spacetime curvature. One will see no membrane and no charged particles. Thus, the Hanni–Ruffini description of why the pellet's electric field lines are bent can have no basis in reality. It is pure fiction. The cause of the field lines' bending, I was sure, is spacetime curvature and nothing else: The field lines bend down toward the horizon in diagram (c) solely because tidal gravity pulls on them, and not because they are being attracted to some polarized charge in the horizon. The horizon cannot possess any such polarized charge; I was sure of it. I was wrong.

Five years later Roger Blandford and a graduate student, Roman Znajek, at Cambridge University discovered that magnetic fields can extract the spin energy of a black hole and use it to power jets (the *Blandford–Znajek process*, Chapter 9 and Figure 11.4a). Blandford and Znajek also found by curved spacetime calculations that, as the energy is extracted, electric currents flow into the horizon near the hole's poles (in the form of positively charged particles falling inward), and currents flow out of the horizon near the equator (in the form of negatively charged particles falling inward). It was as though the black hole were part of an electric circuit.

The calculations showed, moreover, that the hole behaved as though it were a voltage generator in the circuit (Figure 11.4b). This black-hole voltage generator drove current out of the horizon's equator, then up magnetic field lines to a large distance from the hole, then through *plasma* (hot, electrically conducting gas) to other field lines near the hole's spin axis, then down those field lines and into the horizon. The magnetic field lines were the wires of the electric circuit, the plasma was the load that extracts power from the circuit, and the spinning hole was the power source.

From this viewpoint (Figure 11.4b), it is the power carried by the circuit that accelerates the plasma to form jets. From the viewpoint of

11.4 Two viewpoints on the *Blandford–Znajek process* by which a spinning, magnetized black hole can produce jets. (a) The hole's spin creates a swirl of space which forces magnetic fields threading the hole to spin. The spinning fields' centrifugal forces then accelerate plasma to high speeds (compare with Figure 9.7d). (b) The magnetic fields and the swirl of space together generate a large voltage difference between the hole's poles and equator; in effect, the hole becomes a voltage and power generator. This voltage drives current to flow in a circuit. The circuit carries electrical power from the black hole to the plasma, and that power accelerates the plasma to high speeds.

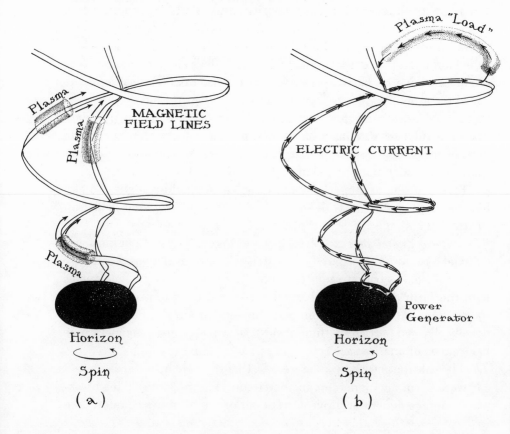

Chapter 9 (Figure 11.4a), it is the spinning magnetic field lines, whipping around and around, that accelerate the plasma. The two viewpoints are just different ways of looking at the same thing. The power comes ultimately from the hole's spin in both cases. Whether one thinks of the power as carried by the circuit or as carried by the spinning field lines is a matter of taste.

The electric circuit description, although based on the standard curved spacetime laws of physics, was totally unexpected, and the flow of current through the black hole—inward near the poles and outward near the equator—seemed very peculiar. During 1977 and 1978, Znajek and, independently, Thibault Damour (also a graduate student, but in Paris rather than Cambridge) puzzled over this peculiarity. While trying to understand it, they independently translated the curved spacetime equations, which describe the spinning hole and its plasma and magnetic field, into an unfamiliar form with an intriguing pictorial interpretation: The current, when it reaches the horizon, does not enter the hole. Instead, it attaches itself to the horizon, where it is carried by the kinds of horizon charges previously imagined by Hanni and Ruffini. This horizon current flows from the pole to the equator, where it exits up the magnetic field lines. Moreover, Znajek and Damour discovered, the laws that govern the horizon's charge and current are elegant versions of the flat spacetime laws of electricity and magnetism: They are Gauss's law, Ampère's law, Ohm's law, and the law of charge conservation (Figure 11.5).

Znajek and Damour did not assert that a being who falls into the black hole will encounter a membrane-like horizon with electric charges and currents. Rather, they asserted that if one wishes to figure out how electricity, magnetism, and plasmas behave outside a black hole, it is useful to regard the horizon as a membrane with charges and currents.

When I read the technical articles by Znajek and Damour, I suddenly understood: They, and Hanni and Ruffini before them, were discovering the foundations of a new paradigm for black holes. The paradigm was fascinating. It captivated me. Unable to resist its allure, I spent much of the 1980s, together with Richard Price, Douglas Macdonald, Ian Redmount, Wai-Mo Suen, Ronald Crowley, and others, bringing it into a polished form and writing a book on it, *Black Holes: The Membrane Paradigm*.

The laws of black-hole physics, written in this membrane paradigm, are completely equivalent to the corresponding laws of the curved

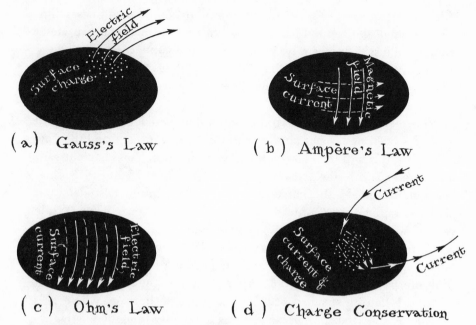

(a) Gauss's Law   **(b) Ampère's Law**

(c) Ohm's Law   **(d) Charge Conservation**

11.5   The laws governing electric charge and current on a black hole's membrane-like horizon: (a) Gauss's law—the horizon has precisely the right amount of surface charge to terminate all electric field lines which intersect the horizon, so they do not extend into the hole's interior; compare with Figure 11.3. (b) Ampère's law—the horizon has precisely the right amount of surface current to terminate that portion of the magnetic field which is parallel to the horizon, so there is no parallel field below the horizon. (c) Ohm's law—the surface current is proportional to the part of the electric field which is tangential to the surface; the proportionality constant is a resistivity of 377 ohms. (d) Charge conservation—no charge is ever lost or created; all positive charge that enters the horizon from the outside Universe becomes attached to the horizon, and moves around on it, until it exits back into the outside Universe (in the form of negative charge falling inward to neutralize the positive charge).

spacetime paradigm—so long as one restricts attention to the hole's exterior. Consequently, the two paradigms give precisely the same predictions for the outcomes of all experiments or observations that anyone might make outside a black hole—including all astronomical observations made from Earth. When thinking about astronomy and astrophysics, I find it useful to keep both paradigms at hand, membrane and curved spacetime, and to do Escher-type mind-flips back and forth between them. The curved spacetime paradigm, with its horizons made from curved empty spacetime, may be useful on Sunday, when I am puzzling over the pulsations of black holes. The membrane paradigm, with horizons made from electrically charged membranes, may be useful on Monday, when I am puzzling over a black hole's production of jets. And since the predictions of the two paradigms are guaranteed to be the same, I can use each day whichever one best suits my needs.

Not so inside a black hole. Any being who falls into a hole will discover that the horizon is *not* a charge-endowed membrane, and that inside the hole the membrane paradigm completely loses its power. However, infalling beings pay a price to discover this: They cannot publish their discovery in the scientific journals of the outside Universe.

12

Black Holes
Evaporate

*in which a black-hole horizon
is clothed in an atmosphere
of radiation and hot particles
that slowly evaporate,
and the hole shrinks
and then explodes*

Black Holes Grow

The Idea hit Stephen Hawking one evening in November 1970, as he was preparing for bed. It hit with such force that he was left almost gasping for air. Never before or since has an idea come to him so quickly.

Preparing for bed was not easy. Hawking's body is afflicted with amyotrophic lateral sclerosis (ALS), a disease that gradually destroys the nerves which control the body's muscles and leaves the muscles, one after another, to waste away in disuse. He moved slowly, with legs wobbling and at least one hand always firmly grasping a countertop or bedpost, as he brushed his teeth, disrobed, struggled into his pajamas, and climbed into bed. That evening he moved even more slowly than usual, since his mind was preoccupied with the Idea. The Idea excited him. He was ecstatic, but he didn't tell his wife, Jane; that would have made him most unpopular, since he was supposed to be concentrating on getting to bed.

He lay awake for many hours that night. He couldn't sleep. His mind kept roaming over the Idea's ramifications, its connections to other things.

The Idea had been triggered by a simple question. How much gravitational radiation (ripples of spacetime curvature) can two black holes produce, when they collide and coalesce to form a single hole? Hawking had been vaguely aware for some time that the single final hole would have to be larger, in some sense, than the "sum" of the two original holes, but in what sense, and what could that tell him about the amount of gravitational radiation produced?

Then, as he was preparing for bed, it had hit him. Suddenly, a series of mental pictures and diagrams had coalesced in his mind to produce the Idea: It was the area of the hole's horizon that would be larger. He was sure of it; the pictures and diagrams had coalesced into an unequivocal, mathematical proof. No matter what the masses of the two original holes might be (the same or very different), and no matter how the holes might spin (in the same direction or opposite or not at all), and no matter how the holes might collide (head-on or at a glancing angle), *the area of the final hole's horizon must always be larger than the sum of the areas of the original holes' horizons.* So what? So a lot, Hawking realized as his mind roamed over the ramifications of this *area-increase theorem.*

First of all, in order for the final hole's horizon to have a large area, the final hole must have a large mass (or equivalently a large energy), which means that not too much energy could have been ejected as gravitational radiation. But "not too much" was still quite a bit. By combining his new area-increase theorem with an equation that describes the mass of a black hole in terms of its surface area and spin, Hawking deduced that as much as 50 percent of the mass of the two original holes could be converted to gravitational-wave energy, leaving as little as 50 percent behind in the mass of the final hole.[1]

There were other ramifications Hawking realized in the months that followed his sleepless November night. Most important, perhaps,

---

1. It might seem counterintuitive that Hawking's area-increase theorem permits *any* of the holes' mass at all to be emitted as gravitational waves. Readers comfortable with algebra may find satisfaction in the example of two nonspinning holes that coalesce to produce a single, larger nonspinning hole. The surface area of a nonspinning hole is proportional to the square of its horizon circumference, which in turn is proportional to the square of the hole's mass. Thus, Hawking's theorem insists that the sum of the squares of the initial holes' masses must exceed the square of the final hole's mass. A little algebra shows that this constraint on the masses permits the final hole's mass to be less than the sum of the initial holes' masses, and thus permits some of the initial masses to be emitted as gravitational waves.

was a new answer to the question of how to *define* the concept of a hole's horizon when the hole is "dynamical," that is, when it is vibrating wildly (as it must during collisions), or when it is growing rapidly (as it will when it is first being created by an imploding star).

Precise and fruitful definitions are essential to physics research. Only after Hermann Minkowski had *defined* the absolute interval between two events (Box 2.1) could he deduce that, although space and time are "relative," they are unified into an "absolute" spacetime. Only after Einstein had *defined* the trajectories of freely falling particles to be straight lines (Figure 2.2) could he deduce that spacetime is curved (Figure 2.5), and thereby develop his laws of general relativity. And only after Hawking had *defined* the concept of a dynamical hole's horizon could he and others explore in detail how black holes change when pummeled by collisions or by infalling debris.

Before November 1970, most physicists, following Roger Penrose's lead, had thought of a hole's horizon as "the outermost location where photons trying to escape the hole get pulled inward by gravity." This old definition of the horizon was an intellectual blind alley, Hawking realized in the ensuing months, and to brand it as such he gave it a new, slightly contemptuous name, a name that would stick. He called it the *apparent horizon.*[2]

Hawking's contempt had several roots. First, the apparent horizon is a relative concept, not an absolute one. Its location depends on the observers' reference frame; observers falling into the hole might see it at a different location from observers at rest outside the hole. Second, when matter falls into the hole, the apparent horizon can jump suddenly, without warning, from one location to another—a rather bizarre behavior, one not conducive to easy insights. Third and most important, the apparent horizon had no connection at all to the flash of congealing mental pictures and diagrams that had produced Hawking's New Idea.

Hawking's new definition of the horizon, by contrast, was absolute (the same in all reference frames), not relative, so he called it the *absolute horizon.* This absolute horizon is beautiful, Hawking thought. It has a beautiful definition: It is "the boundary in spacetime between events (outside the horizon) that can send signals to the distant Universe and those (inside the horizon) that cannot." And it has a beautiful evolution: When a hole eats matter or collides with another hole or

2. A more precise definition of the apparent horizon is given in Box 12.1 below.

Box 12.1

Absolute and Apparent Horizons for a Newborn Black Hole

The spacetime diagrams shown below describe the implosion of a spherical star to form a spherical black hole; compare with Figure 6.7. The dotted curves are *outgoing light rays;* in other words, they are the world lines (trajectories through spacetime) of photons—the fastest signals that can be sent radially outward, toward the distant Universe. For optimal escape, the photons are idealized as not being absorbed or scattered at all by the star's matter.

The *apparent horizon* (left diagram) is the outermost location where outgoing light rays, trying to escape the hole, get pulled inward toward the singularity (for example, the outgoing rays QQ' and RR'). The apparent horizon is created suddenly, full-sized, at E, where the star's surface shrinks through the critical circumference. The *absolute horizon* (right diagram) is the boundary between events that *can* send signals to the distant Universe (for example, events P and S which send signals along the light rays PP' and SS') and events that *cannot* send signals to the distant Universe (for example, Q and R). The absolute horizon is created at the star's center, at the event labeled C, well before the star's surface shrinks through the critical circumference. The absolute horizon is just a point when created, but it then expands smoothly, like a balloon being blown up, and emerges through the star's surface precisely when the surface shrinks through the critical circumference (the circle labeled E). It then stops expanding, and thereafter coincides with the suddenly created apparent horizon.

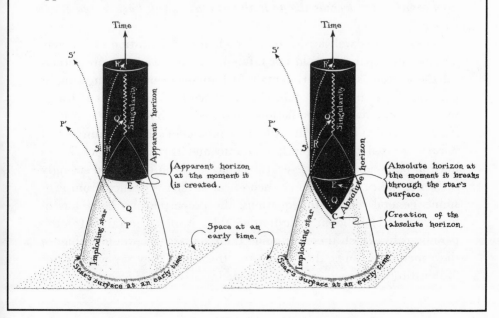

does anything at all, its absolute horizon changes shape and size in a smooth, continuous way, instead of a sudden, jumping way (Box 12.1). Most important, the absolute horizon meshed perfectly with Hawking's New Idea:

Hawking could see, in his congealed mental pictures and diagrams, that the areas of absolute horizons (but not necessarily apparent horizons) will increase not only when black holes collide and coalesce, but also when they are being born, when matter or gravitational waves fall into them, when the gravity of other objects in the Universe raises tides on them, and when rotational energy is being extracted from the swirl of space just outside their horizons. Indeed, the areas of absolute horizons will almost *always* increase, and can never decrease. The physical reason is simple: Everything that a hole encounters sends energy inward through its absolute horizon, and there is no way that any energy can come back out. Since all forms of energy produce gravity, this means that the hole's gravity is continually being strengthened, and correspondingly, its surface area is continually growing.

Hawking's conclusion, stated more precisely, was this: *In any region of space, and at any moment of time (as measured in anyone's reference frame), measure the areas of all the absolute horizons of all black holes and add the areas together to get a total area. Then wait however long you might wish, and again measure the areas of all the absolute horizons and add them. If no black holes have moved out through the "walls" of your region of space between the measurements, then the total horizon area cannot have decreased, and it almost always will have increased, at least a little bit.*

Hawking was well aware that the choice of definition of horizon, absolute or apparent, could not influence in any way any predictions for the outcomes of experiments that humans or other beings might perform; for example, it could not influence predictions of the waveforms of gravitational radiation produced in black-hole collisions (Chapter 10), nor could it influence predictions of the number of X-rays emitted by hot gas falling into and through a black hole's horizon (Chapter 8). However, the choice of definition could strongly influence the *ease* with which theoretical physicists deduce, from Einstein's general relativistic equations, the properties and behaviors of black holes. The chosen definition would become a central tool in the paradigm by which theorists guide their research; it would influence their mental pictures, their diagrams, the words they say when communicating with each other, and their intuitive leaps of insight. And for this purpose, Hawking believed, the new, absolute horizon, with its smoothly increasing area, would be superior to the old, apparent horizon, with its discontinuous jumps in size.

Stephen Hawking was not the first physicist to think about absolute horizons and discover their area increase. Roger Penrose at Oxford University, and Werner Israel at the University of Alberta, Canada, had already done so, before Hawking's sleepness November night. In fact, Hawking's insights were based largely on foundations laid by Penrose (Chapter 13). However, neither Penrose nor Israel had recognized the importance or the power of the area-increase theorem, so neither had published it. Why? Because they were mentally locked into regarding the apparent horizon as the hole's surface and the absolute horizon as just some rather unimportant auxiliary concept, and therefore they thought that the increase of the absolute horizon's area was not very interesting. Just how terribly wrong they were will become clear as this chapter progresses.

Why were Penrose and Israel so wedded to the apparent horizon? Because it had already played a central role in an amazing discovery: Penrose's 1964 discovery that the laws of general relativity force every black hole to have a singularity at its center. I shall describe Penrose's discovery and the nature of singularities in the next chapter. For now, the main point is that the apparent horizon had proved its power, and Penrose and Israel, blinded by that power, could not conceive of jettisoning the apparent horizon as the definition of a black hole's surface.

They especially could not conceive of jettisoning it in favor of the absolute horizon. Why? Because the absolute horizon—paradoxically, it might seem—violates our cherished notion that an effect should not precede its cause. When matter falls into a black hole, the absolute horizon starts to grow ("effect") before the matter reaches it ("cause"). The horizon grows in anticipation that the matter will soon be swallowed and will increase the hole's gravitational pull (Box 12.2).

Penrose and Israel knew the origin of this seeming paradox. The very definition of the absolute horizon depends on what will happen in the future: on whether or not signals will ultimately escape to the distant Universe. In the terminology of philosophers, it is a *teleological* definition (a definition that relies on "final causes"), and it forces the horizon's evolution to be teleological. Since teleological viewpoints have rarely if ever been useful in modern physics, Penrose and Israel were dubious about the merits of the absolute horizon.

Box 12.2
Evolution of an Accreting Hole's Apparent and Absolute Horizons

The spacetime diagram below illustrates the jerky evolution of the apparent horizon and the teleological evolution of the absolute horizon. At some initial moment of time (on a horizontal slice near the bottom of the diagram), an old, nonspinning black hole is surrounded by a thin, spherical shell of matter. The shell is like the rubber of a balloon, and the hole is like a pit at the balloon's center. The hole's gravity pulls on the shell (the balloon's rubber), forcing it to shrink and ultimately be swallowed by the hole (the pit). The *apparent horizon* (the outermost location at which outgoing light rays—shown dotted—are being pulled inward) jumps outward suddenly, and discontinuously, at the moment when the shrinking shell reaches the location of the final hole's critical circumference. The *absolute horizon* (the boundary between events that can and cannot send outgoing light rays to the distant Universe) starts to expand *before* the hole swallows the shell. It expands *in anticipation* of swallowing, and then, just as the hole swallows, it comes to rest at the same location as the jumping apparent horizon.

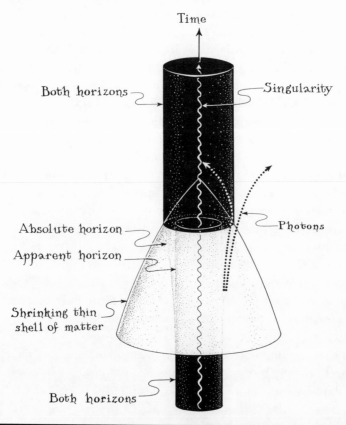

Hawking is a bold thinker. He is far more willing than most physicists to take off in radical new directions, if those directions "smell" right. The absolute horizon smelled right to him, so despite its radical nature, he embraced it, and his embrace paid off. Within a few months, Hawking and James Hartle were able to derive, from Einstein's general relativity laws, a set of elegant equations that describe how the absolute horizon continuously and smoothly expands and changes its shape, in anticipation of swallowing infalling debris or gravitational waves, or in anticipation of being pulled on by the gravity of other bodies.

In November 1970, Stephen Hawking was just beginning to reach full stride as a physicist. He had made several important discoveries already, but he was not yet a dominant figure. As we move on through this chapter, we shall watch him become dominant.

How, with his severe disability, has Hawking been able to out-think and out-intuit his leading colleague-competitors, people like Roger Penrose, Werner Israel, and (as we shall see) Yakov Borisovich Zel'dovich? They had the use of their hands; they could draw pictures and perform many-page-long calculations on paper—calculations in which one records many complex intermediate results along the way, and then goes back, picks them up one by one, and combines them to get a final result; calculations that I cannot conceive of anyone doing in his head. By the early 1970s, Hawking's hands were largely paralyzed; he could neither draw pictures nor write down equations. His research had to be done entirely in his head.

Because the loss of control over his hands was so gradual, Hawking has had plenty of time to adapt. He has gradually trained his mind to think in a manner different from the minds of other physicists: He thinks in new types of intuitive mental pictures and mental equations that, for him, have replaced paper-and-pen drawings and written equations. Hawking's mental pictures and mental equations have turned out to be more powerful, for some kinds of problems, than the old paper-and-pen ones, and less powerful for others, and he has gradually learned to concentrate on problems for which his new methods give greater power, a power that nobody else can begin to match.

Hawking's disability has helped him in other ways. As he himself has often commented, it has freed him from the responsibility of lecturing to university students, and he thus has had far more free time

Stephen Hawking with his wife Jane and their son Timothy in Cambridge, England, in 1980. [Photo by Kip Thorne.]

for research than his more healthy colleagues. More important, perhaps, his disease in some ways has improved his attitude toward life.

Hawking contracted ALS in 1963, soon after he began graduate school at Cambridge University. ALS is a catch-all name for a variety of motor neuron diseases, most of which kill fairly quickly. Thinking he had only a few years to live, Hawking at first lost his enthusiasm for life and physics. However, by the winter of 1964–65, it became apparent that his was a rare variant of ALS, a variant that saps the central nervous system's control of muscles over many years' time, not just a few. Suddenly life seemed wonderful. He returned to physics with greater vigor and enthusiasm than he had ever had as a healthy, devil-may-care undergraduate student; and with his new lease on life, he married Jane Wilde, whom he had met shortly after contracting ALS and with whom he had fallen in love during the early phases of his disease.

Stephen's marriage to Jane was essential to his success and happiness in the 1960s and 1970s and into the 1980s. She made for them a normal home and a normal life in the midst of physical adversity.

The happiest smile I ever saw in my life was Stephen's the evening in August 1972 in the French Alps when Jane, I, and the Hawkings' two oldest children, Robert and Lucy, returned from a day's excursion into the mountains. Through foolishness we had missed the last ski lift down the mountain, and had been forced to descend about 1000 meters on foot. Stephen, who had fretted about our tardiness, broke out into an enormous smile, and tears came to his eyes, as he saw Jane, Robert, and Lucy enter the dining room where he was poking at his evening meal, unable to eat.

Hawking lost the use of his limbs and then his voice very gradually. In June 1965, when we first met, he walked with a cane and his voice was only slightly shaky. By 1970 he required a four-legged walker. By 1972 he was confined to a motorized wheelchair and had largely lost the ability to write, but he could still feed himself with some ease, and most native English speakers could still understand his speech, though with difficulty. By 1975 he could no longer feed himself, and only people accustomed to his speech could understand it. By 1981 even I was having severe difficulty understanding him unless we were in an absolutely quiet room; only people who were with him a lot could understand with ease. By 1985 his lungs would not remain clear of fluid of their own accord, and he had to have a tracheostomy so they could be cleared regularly by suctioning. The price was high: He completely lost his voice. To compensate, he acquired a computer-driven voice synthesizer with an American accent for which he would apologize sheepishly. He controls the computer by a simple switch clutched in one hand, which he squeezes as a menu of words scrolls by on the computer screen. Grabbing one word after another from the scrolling menu with his switch, he builds up his sentences. It is painfully slow, but effective; he can produce no more than one short sentence per minute, but his sentences are enunciated clearly by the synthesizer, and are often pearls.

As his speech deteriorated, Hawking learned to make every sentence count. He found ways to express his ideas that were clearer and more succinct than the ways he had used in the early years of his disease. With clarity and succinctness of expression came improved clarity of thought, and greater impact on his colleagues—but also a tendency to seem oracular: When he issues a pronouncement on some deep question, we, his colleagues, sometimes cannot be sure, until after much thought and calculation of our own, whether he is just speculating or has strong evidence. He sometimes doesn't tell us, and we occasionally wonder whether he, with his absolutely unique insights, is playing games with us. He does, after all, still retain a streak of the impishness that made him popular in his undergraduate days at Oxford, and a sense of humor that rarely deserts him, even in times of trial. (Before his tracheostomy, when I began to have trouble understanding his speech, I sometimes found myself saying over and over again, as many as ten times, "Stephen, I still don't understand; please say it again." Showing a bit of frustration, he would continue to repeat himself until I suddenly understood: He was telling me a wonderfully funny, off-the-wall, one-line joke. When I finally caught it, he would grin with pleasure.)

Entropy

Having extolled Hawking's ability to out-think and out-intuit all his colleague-competitors, I must now confess that he has not managed to do so *all* the time, just most. Among his defeats, perhaps the most spectacular was at the hands of one of John Wheeler's graduate students, Jacob Bekenstein. But in the midst of that defeat, as we shall see, Hawking produced a far greater triumph: his discovery that black holes can evaporate. The tortuous route to that discovery will occupy much of the rest of this chapter.

The playing field on which Hawking was defeated was that of *black-hole thermodynamics*. Thermodynamics is the set of physical laws that govern the random, statistical behavior of large numbers of atoms, for example, the atoms that make up the air in a room or those that make up the entire Sun. The atoms' statistical behavior includes, among other things, their random jiggling caused by heat; and correspondingly, the laws of thermodynamics include, among other things, the laws that govern heat. Hence the name *thermo*dynamics.

A year before Hawking discovered his area theorem, Demetrios Christodoulou, a nineteen-year-old graduate student in Wheeler's Princeton group, noticed that the equations that describe slow changes in the properties of black holes (for example, when they slowly accrete gas) resemble some of the equations of thermodynamics. The resemblance was remarkable, but there was no reason to think it anything more than a coincidence.

This resemblance was strengthened by Hawking's area theorem: The area theorem closely resembled the *second law of thermodynamics*. In fact, the area theorem, as expressed earlier in this chapter, becomes the second law of thermodynamics if we merely replace the phrase "horizon areas" by the word "entropy": *In any region of space, and at any moment of time (as measured in anyone's reference frame), measure the total entropy of everything there. Then wait however long you might wish, and again measure the total entropy. If nothing has moved out through the "walls" of your region of space between the measurements, then the total entropy cannot have decreased, and it almost always will have increased, at least a little bit.*

What is this thing called "entropy" that increases? It is the amount of "randomness" in the chosen region of space, and the increase of entropy means that things are continually becoming more and more random.

Stated more precisely (see Box 12.3), *entropy is the logarithm of the number of ways that all the atoms and molecules in our chosen region can be distributed, without changing that region's macroscopic appearance.*[3] When there are many possible ways for the atoms and molecules to be distributed, there is a huge amount of microscopic randomness and the entropy is huge.

The law of entropy increase (the second law of thermodynamics) has great power. As an example, suppose that we have a room containing air and a few crumpled-up newspapers. The air and paper together contain less entropy than they would have if the paper were burned in the air to form carbon dioxide, water vapor, and a bit of ash. In other words, when the room contains the original air and paper, there are fewer ways that its molecules can be randomly distributed than when it contains the final air, carbon dioxide, water vapor, and ash. That is why the paper burns naturally and easily if a spark ignites it, and why the burning cannot easily and naturally be reversed to create paper from carbon dioxide, water, ash, and air. Entropy increases during burning; entropy would decrease during unburning; thus, burning occurs and unburning does not.

Stephen Hawking noticed immediately, in November 1970, the remarkable similarity between the second law of thermodynamics and his law of area increase, but it was obvious to him that the similarity

---

3. The laws of quantum mechanics guarantee that the number of ways to distribute the atoms and molecules is always finite, and never infinite. In defining the entropy, physicists often multiply the logarithm of this number of ways by a constant that will be irrelevant to us, $\log_e 10 \times k$, where $\log_e 10$ is the "natural logarithm" of 10, that is, $2.30258 \ldots$, and k is "Boltzmann's constant," 1.38062×10^{-16} erg per degree Celsius. Throughout this book I shall ignore this constant.

Box 12.3

Entropy in a Child's Playroom

Imagine a square playroom containing 20 toys. The floor of the room is made of 100 large tiles (with 10 tiles running along each side), and a father has cleaned the room, throwing all the toys onto the northernmost row of tiles. The father cared not one whit which toys landed on which tiles, so they are all randomly scrambled. One measure of their randomness is the number of ways that they could have landed (each of which the father considers as equally satisfactory), that is, the number of ways that the 20 toys can be distributed over the 10 tiles of the northern row. This number turns out to be $10 \times 10 \times 10 \times \ldots \times 10$, with one factor of 10 for each toy; that is, 10^{20}.

This number, 10^{20}, is one description of the amount of randomness in the toys. However, it is a rather unwieldly description, since 10^{20} is such a big number. More easy to manipulate is the *logarithm* of 10^{20}, that is, the number of factors of 10 that must be multiplied together to get 10^{20}. The logarithm is 20; and *this logarithm of the number of ways the toys could be scattered over the tiles is the toys' entropy.*

Now suppose that a child comes into the room and plays with the toys, throwing them around with abandon, and then leaves. The father returns and sees a mess. The toys are now far more randomly distributed than before. Their entropy has increased. The father doesn't care just where each toy is; all he cares is that they have been scattered randomly throughout the room. How many different ways might they have been scattered? How many ways could the 20 toys be distributed over the 100 tiles? $100 \times 100 \times 100 \times \ldots \times 100$, with one factor of 100 for each toy; that is, $100^{20} = 10^{40}$ ways. The logarithm of this number is 40, so the child increased the toys' entropy from 20 to 40.

"Aha, but then the father cleans up the room and thereby reduces the toys' entropy back to 20," you might say. "Doesn't this violate the second law of thermodynamics?" No, not at all. The toys' entropy may be reduced by the father's cleaning, but the entropy in the father's body and in the room's air has increased: It took a lot of energy to throw the toys back onto the northernmost tiles, energy that the father got by "burning up" some of his body's fat. The burning converted neatly organized fat molecules into disorganized waste products, for example, the carbon dioxide that he exhaled randomly into the room; and the resulting increase in the father's and the room's entropy (the increase in the number of ways their atoms and molecules can be distributed) far more than made up for the decrease in the toys' entropy.

was a mere coincidence. One would have to be crazy, or at least a little dim-witted, to claim that the area of a hole's horizon in some sense *is* the hole's entropy, Hawking thought. After all, there is nothing at all random about a black hole. A black hole is just the opposite of random; it is simplicity incarnate. Once a black hole has settled down into a quiescent state (by emitting gravitational waves; Figure 7.4), it is left totally "hairless": *All* of its properties are precisely determined by just three numbers, its mass, its angular momentum, and its electric charge. The hole has no randomness whatsoever.

Jacob Bekenstein was not persuaded. It seemed likely to him that a black hole's area in some deep sense *is* its entropy—or, more precisely, its entropy multiplied by some constant. If not, Bekenstein reasoned, if black holes have vanishing entropy (no randomness at all) as Hawking claimed, then black holes could be used to decrease the entropy of the Universe and thereby violate the second law of thermodynamics. All one need do is bundle all the air molecules from some room into a small package and drop them into a black hole. The air molecules and all the entropy they carry will disappear from our Universe when the package enters the hole, and if the hole's entropy does not increase to compensate for this loss, then the total entropy of the Universe will have been reduced. This violation of the second law of thermodynamics would be highly unsatisfactory, Bekenstein argued. To preserve the second law, a black hole *must* possess an entropy that goes up when the package falls through its horizon, and the most promising candidate for that entropy, it seemed to Bekenstein, was the hole's surface area.

Not at all, Hawking responded. You can lose air molecules by throwing them down a black hole, and you can also lose entropy. That is just the nature of black holes. We will just have to accept this violation of the second law of thermodynamics, Hawking argued; the properties of black holes require it—and besides, it has no serious consequences at all. For example, although under ordinary circumstances a violation of the second law of thermodynamics might permit one to make a perpetual motion machine, when it is a black hole that causes the violation, no perpetual motion machine is possible. The violation is just a tiny peculiarity in the laws of physics, one that the laws presumably live with quite happily.

Bekenstein was not convinced.

All the world's black-hole experts lined up on Hawking's side—all, that is, except Bekenstein's mentor, John Wheeler. "Your idea is just crazy enough that it might be right," Wheeler told Bekenstein. With this encouragement, Bekenstein plowed forward and tightened up his conjecture. He estimated just how much a hole's entropy would have to grow, when a package of air is dropped into it, to preserve the second law of thermodynamics, and he estimated how much the plunging package would increase the horizon's area; and from these rough estimates, he deduced a relationship between entropy and area which, he thought, *might* always preserve the second law of thermodynamics: The entropy, he concluded, is approximately the horizon's area divided by a famous area associated with the (as yet ill-understood) laws of quantum gravity, the *Planck–Wheeler area,* 2.61×10^{-66} square centimeter.[4] (We shall learn the significance of the Planck–Wheeler area in the next two chapters.) For a 10-solar-mass hole, this entropy would be the hole's area, 11,000 square kilometers, divided by the Planck–Wheeler area, 2.61×10^{-66} square centimeter, which is roughly 10^{79}.

This is an enormous amount of entropy. It represents a huge amount of randomness. Where does this randomness reside? Inside the hole, Bekenstein conjectured. The hole's interior must contain a huge number of atoms or molecules or something, all randomly distributed, and the total number of ways they could be distributed must be[5] $10^{10^{79}}$.

Nonsense, responded most of the leading black-hole physicists, including Hawking and me. The hole's interior contains a singularity, not atoms or molecules.

Nevertheless, the similarity between the laws of thermodynamics and the properties of black holes was impressive.

In August 1972, with the golden age of black-hole research in full swing, the world's leading black-hole experts and about fifty students congregated in the French Alps for an intense month of lectures and joint research. The site was the same Les Houches summer school, on the same green hillside opposite Mont Blanc, at which nine years earlier (1963) I had been taught the intricacies of general relativity (Chapter 10). In 1963 I had been a student. Now, in 1972, I was supposed to be an expert. In the mornings we "experts" lectured to

---

4. This Planck–Wheeler area is given by the formula $G\hbar/c^3$, where $G = 6.670 \times 10^{-8}$ dyne-centimeter^2/gram^2 is Newton's gravitation constant, $\hbar = 1.055 \times 10^{-27}$ erg-second is Planck's quantum mechanical constant, and $c = 2.998 \times 10^{10}$ centimeter/second is the speed of light. For related issues, see Footnote 2 in Chapter 13, Footnote 6 in Chapter 14, and the associated discussions in the text of those chapters.

5. The logarithm of $10^{10^{79}}$ is 10^{79} (Bekenstein's conjectured entropy). Note that $10^{10^{79}}$ is a 1 with 10^{79} zeroes after it, that is, with nearly as many zeroes as there are atoms in the Universe.

each other and the students about the discoveries we had made during the past five years and about our current struggles toward new insights. During most afternoons we continued our current struggles: Igor Novikov and I closeted ourselves in a small log cabin and struggled to discover the laws that govern gas as it accretes into black holes and emits X-rays (Chapter 8), while on couches in the school's lounge my students Bill Press and Saul Teukolsky sought ways to discover whether a spinning black hole is stable against small perturbations (Chapter 7), and fifty meters above me on the hillside, James Bardeen, Brandon Carter, and Stephen Hawking joined forces to try to deduce from Einstein's general relativity equations the full set of laws that govern the evolution of black holes. The setting was idyllic, the physics delicious.

By the end of the month, Bardeen, Carter, and Hawking had consolidated their insights into a set of *laws of black-hole mechanics* that bore an amazing resemblance to the laws of thermodynamics. Each black-hole law, in fact, turned out to be identical to a thermodynamical law, if one only replaced the phrase "horizon area" by "entropy," and the phrase "horizon surface gravity" by "temperature." (The surface gravity, roughly speaking, is the strength of gravity's pull as felt by somebody at rest just above the horizon.)

When Bekenstein (who was one of the fifty students at the school) saw this perfect fit between the two sets of laws, he became more convinced than ever that the horizon area *is* the hole's entropy. Bardeen, Carter, Hawking, I, and the other experts, by contrast, saw in this fit a firm proof that the horizon area *cannot be* the hole's entropy in disguise. If it were, then similarly the surface gravity would have to be the hole's temperature in disguise, and that temperature would not be zero. However, the laws of thermodynamics insist that any and every object with a nonzero temperature must emit radiation, at least a little bit (that is how the radiators that warm some homes work), and everybody knew that black holes cannot emit anything. Radiation can fall into a black hole, but none can ever come out.

If Bekenstein had followed his intuition to its logical conclusion, he would have asserted that somehow a black hole *must* have a finite temperature and *must* emit radiation, and we today would look back on him as an astounding prophet. But Bekenstein waffled. He conceded that it was obvious a black hole cannot radiate, but he clung tenaciously to his faith in black-hole entropy.

Black Holes Radiate

The first hint that black holes, in fact, *can* radiate came from Yakov Borisovich Zel'dovich, in June 1971, fourteen months before the Les Houches summer school. However, nobody was paying any attention, and for this I bare the brunt of the shame since I was Zel'dovich's confidant and foil as he groped toward a radical new insight.

Zel'dovich had brought me to Moscow for my second several-week stint as a member of his research group. On my first stint, two years earlier, he had commandeered for me, in the midst of Moscow's housing crunch, a spacious private apartment on Shabolovka Street, near October Square. While some of my friends shared one-room apartments with their spouses, children, and a set of parents—one *room*, not one bedroom—I had had all to myself an apartment with bedroom, living room, kitchen, television, and elegant china. On this second stint I lived more modestly, in a single room at a hotel owned by the Soviet Academy of Sciences, down the street from my old apartment.

At 6:30 one morning, I was roused from my sleep by a phone call from Zel'dovich. "Come to my flat, Kip! I have a new idea about spinning black holes!" Knowing that coffee, tea, and pirozhki (pastries containing ground beef, fish, cabbage, jam, or eggs) would be waiting, I sloshed cold water on my face, threw on my clothes, grabbed my briefcase, dashed down five flights of stairs into the street, grabbed a crowded trolley, transferred to a trolley bus, and alighted at Number 2B Vorobyevskoye Shosse in the Lenin Hills, 10 kilometers south of the Kremlin. Number 4, next door, was the residence of Alexei Kosygin, the Premier of the U.S.S.R.[6]

I walked through an open gate in the eight-foot-high iron fence and entered a four-acre, forested yard surrounding the massive, squat apartment house Number 2B and its twin Number 2A, with their peeling yellow paint. As one reward for his contributions to Soviet nuclear might (Chapter 6), Zel'dovich had been given one of 2B's eight apartments: the southwest quarter of the second floor. The apartment was

6. Vorobyevskoye Shosse has since been renamed Kosygin Street, and its buildings have been renumbered. In the late 1980s Mikhail Gorbachev had a home at Number 10, several doors west of Zel'dovich.

enormous by Moscow standards, 1500 square feet; he shared it with his wife, Varvara Pavlova, one daughter, and a son-in-law.

Zel'dovich met me at the apartment door, with a warm grin on his face and the sounds of his bustling family emerging from back rooms. I removed my shoes, put on slippers from the pile beside the door, and followed him into the shabby but comfortable living/dining room, with its overstuffed couch and chairs. On one wall was a map of the world, with colored pins identifying all the places to which Zel'dovich had been invited (London, Princeton, Beijing, Bombay, Tokyo, and many more), and which the Soviet state, in its paranoid fear of losing nuclear secrets, had forbade him to visit.

Zel'dovich, his eyes dancing, sat me down at the long dining table dominating the room's center, and announced, "A spinning black hole must radiate. The departing radiation will kick back at the hole and gradually slow its spin, and then halt it. With the spin gone, the radiation will stop, and the hole will live forever thereafter in a perfectly spherical, nonspinning state."

"That's one of the craziest things I've ever heard," I asserted. (Open confrontation is not my style, but Zel'dovich thrived on it. He wanted it, he expected it, and he had brought me to Moscow in part to serve as a sparring partner, an opponent against whom to test ideas.) "How can you make such a crazy claim?" I asked. "Everyone knows that radiation can flow into a hole, but nothing, not even radiation, can come out."

Zel'dovich explained his reasoning: "A spinning metal sphere emits electromagnetic radiation, and so, similarly, a spinning black hole should emit gravitational waves."

A typical Zel'dovich proof, I thought to myself. Pure physical intuition, based on nothing more than analogy. Zel'dovich doesn't understand general relativity well enough to compute what a black hole should do, so instead he computes the behavior of a spinning metal sphere, he then asserts that a black hole will behave analogously, and he wakes me up at 6:30 A.M. to test his assertion.

However, I had already seen Zel'dovich make discoveries with little more basis than this; for example, his 1965 claim that when a mountainous star implodes, it produces a perfectly spherical black hole (Chapter 7), a claim that turned out to be right and that foretold the hairlessness of holes. I thus proceeded cautiously. "I had no idea that a spinning metal sphere emits electromagnetic radiation. How?"

"The radiation is so weak," Zel'dovich explained, "that nobody has ever observed it, nor predicted it before. However, it must occur. The metal sphere will radiate when electromagnetic *vacuum fluctuations* tickle it. Similarly, a black hole will radiate when gravitational vacuum fluctuations graze its horizon."

I was too dumb in 1971 to realize the deep significance of this remark, but several years later it would become clear. *All* previous theoretical studies of black holes had been based on Einstein's general relativistic laws, and those studies were unequivocal: A black hole cannot radiate. However, we theorists knew that general relativity is only an approximation to the true laws of gravity—an approximation that should be excellent when dealing with black holes, we thought, but an approximation nonetheless.[7] The true laws, we were sure, must

7. See the last section of Chapter 1: "The Nature of Physical Law."

Box 12.4
Vacuum Fluctuations

Vacuum fluctuations are, for electromagnetic and gravitational waves, what "claustrophobic degeneracy motions" are for electrons.

Recall (Chapter 4) that if one confines an electron to a small region of space, then no matter how hard one tries to slow it to a stop, the laws of quantum mechanics force the electron to continue moving randomly, unpredictably. This is the claustrophobic degeneracy motion that produces the pressure by which white-dwarf stars support themselves against their own gravitational squeeze.

Similarly, if one tries to remove all electromagnetic or gravitational oscillations from some region of space, one will never succeed. The laws of quantum mechanics insist that there always remain some random, unpredictable oscillations, that is, some random, unpredictable electromagnetic and gravitational waves. These are the vacuum fluctuations that (according to Zel'dovich) will "tickle" a spinning metal sphere or black hole and cause it to radiate.

These vacuum fluctuations cannot be stopped by removing their energy, because they contain, on average, no energy at all. At some locations and some moments of time they have positive energy that has been "borrowed" from other locations, and those other locations, as a result, have negative energy. Just as banks will not let customers maintain negative bank balances for long, so the laws of physics force the regions of negative energy to quickly suck energy out of their positive-energy neighbors, thereby restoring themselves to a zero or positive balance. This continual, random, borrowing and returning of energy is what drives the vacuum fluctuations.

Just as an electron's degeneracy motions become more vigorous when

be quantum mechanical, so we called them the laws of *quantum gravity*. Although those quantum gravity laws were only vaguely understood at best, John Wheeler had deduced in the 1950s that they must entail *gravitational vacuum fluctuations,* tiny, unpredictable fluctuations in the curvature of spacetime, fluctuations that remain even when spacetime is completely empty of all matter and one tries to remove all gravitational waves from it, that is, when it is a perfect vacuum (Box 12.4). Zel'dovich was claiming to foresee, from his electromagnetic analogy, that these gravitational vacuum fluctuations would cause spinning black holes to radiate. "But how?" I asked, puzzled.

Zel'dovich bounded to his feet, strode to a one-meter-square blackboard on the wall opposite his map, and began drawing a sketch and

one confines the electron to a smaller and smaller region (Chapter 4), so also the vacuum fluctuations of electromagnetic and gravitational waves are more vigorous in small regions than in large, that is, more vigorous for small wavelengths than for large. This, as we shall see in Chapter 13, has profound consequences for the nature of the singularities at the centers of black holes.

Electromagnetic vacuum fluctuations are well understood and are a common feature of everyday physics. For example, they play a key role in the operation of a fluorescent light tube. An electrical discharge excites mercury vapor atoms in the tube, and then random electromagnetic vacuum fluctuations tickle each excited atom, causing it, at some random time, to emit some of its excitation energy as an electromagnetic wave (a photon).* This emission is called *spontaneous* because, when it was first identified as a physical effect, physicists did not realize it was being triggered by vacuum fluctuations. As another example, inside a laser, random electromagnetic vacuum fluctuations interfere with the coherent laser light (interference in the sense of Box 10.3), thereby modulating the laser light in unpredictable ways. This causes the photons emerging from the laser to come out at random, unpredictable times, instead of uniformly one after another—a phenomenon called *photon shot noise.*

Gravitational vacuum fluctuations, by contrast with electromagnetic, have never yet been seen experimentally. Technology of the 1990s, with great effort, should be able to detect highly energetic gravitational waves from black-hole collisions (Chapter 10), but not the waves' far weaker vacuum fluctuations.

*This "primary" photon gets absorbed by a phosphor coating on the tube's walls, which in turn emits "secondary" photons that we see as light.

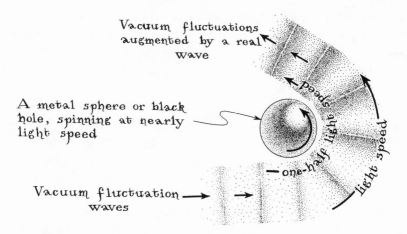

Vacuum fluctuations
augmented by a real
wave

A metal sphere or black
hole, spinning at nearly
light speed

—one-half light speed

Vacuum fluctuation
waves

12.1 Zel'dovich's mechanism by which vacuum fluctuations cause a spinning body to radiate.

talking at the same time. His sketch (Figure 12.1) showed a wave flowing toward a spinning object, skimming around its surface for a while, and then flowing away. The wave might be electromagnetic and the spinning body a metal sphere, Zel'dovich explained, or the wave might be gravitational and the body a black hole.

The incoming wave is not a "real" wave, Zel'dovich explained, but rather a vacuum fluctuation. As this fluctuational wave sweeps around the spinning body, it behaves like a line of ice skaters making a turn: The outer skaters must whip around at high speed while the inner ones move much more slowly; similarly, the wave's outer parts move at a very high speed, the speed of light, while its inner parts move much more slowly than light and, in fact, more slowly than the body's surface is spinning.[8] In such a situation, Zel'dovich asserted, the rapidly spinning body will grab hold of the fluctuational wave and accelerate it, much like a small boy accelerating a slingshot as he swings it faster and faster. The acceleration feeds some of the body's spin energy into the wave, amplifying it. The new, amplified portion of the wave is a "real wave" with positive total energy, while the original, unamplified portion remains a vacuum fluctuation with zero total energy (Box 12.4). The spinning body has thus used the vacuum fluctuation as a sort of catalyst for creating a real wave, and as a template for the shape of the

---

8. In technical language, the outer parts are in the "radiation zone" while the inner parts are in the "near zone."

real wave. This is similar, Zel'dovich pointed out, to the manner in which vacuum fluctuations cause a vibrating molecule to "spontaneously" emit light (Box 12.4).

Zel'dovich told me he had proved that a spinning metal sphere radiates in this way; his proof was based on the laws of *quantum electrodynamics*—that is, the well-known laws that arise from a marriage of quantum mechanics with Maxwell's laws of electromagnetism. Though he did not have a similar proof that a spinning black hole will radiate, he was quite sure by analogy that it must. In fact, he asserted, a spinning hole will radiate not only gravitational waves, but also electromagnetic waves (photons[9]), neutrinos, and all other forms of radiation that can exist in nature.

I was quite sure that Zel'dovich was wrong. Several hours later, with no agreement in sight, Zel'dovich offered me a wager. In the novels of Ernest Hemingway, Zel'dovich had read of White Horse scotch, an elegant and esoteric brand of whisky. If detailed calculations with the laws of physics showed that a spinning black hole radiates, then I was to bring Zel'dovich a bottle of White Horse scotch from America. If the calculations showed that there is no such radiation, Zel'dovich would give me a bottle of fine Georgian cognac.

I accepted the wager, but I knew it would not be settled quickly. To settle it would require understanding the marriage of general relativity and quantum mechanics far more deeply than anyone did in 1971.

Having made the wager, I soon forgot it. I have a lousy memory, and my own research was concentrated elsewhere. Zel'dovich, however, did not forget; several weeks after arguing with me, he wrote down his argument and submitted it for publication. The referee probably would have rejected his manuscript had it come from somebody else; his argument was too heuristic for acceptance. But Zel'dovich's reputation carried the day; his paper was published—and hardly anyone paid any attention. Black-hole radiation just seemed horribly implausible.

A year later, at the Les Houches summer school, we "experts" were still ignoring Zel'dovich's idea. I don't recall it being mentioned even once.[10]

9. Recall that photons and electromagnetic waves are different aspects of the same thing; see the discussion of wave/particle duality in Box 4.1.

10. This lack of interest was all the more remarkable because in the meantime, Charles Misner in America had shown that *real* waves (as opposed to Zel'dovich's vacuum fluctuations) can be amplified by a spinning hole in a manner analogous to Figure 12.2, and this amplification—to which Misner gave the name "superradiance"—was generating great interest.

Ⅰn September 1973, I was back in Moscow once again, this time accompanying Stephen Hawking and his wife Jane. This was Stephen's first trip to Moscow since his student days. He, Jane, and Zel'dovich (our Soviet host), uneasy about how to cope in Moscow with Stephen's special needs, thought it best that I, being familiar with Moscow and a close friend of Stephen's and Jane's, act as their companion, translator for physics conversations, and guide.

We stayed at the Hotel Rossiya, just off Red Square near the Kremlin. Although we ventured out nearly every day to give lectures at one institute or another, or to visit a museum or the opera or ballet, our interactions with Soviet physicists occurred for the most part in the Hawkings' two-room hotel suite, with its view of St. Basil's Cathedral. One after another, the Soviet Union's leading theoretical physicists came to the hotel to pay homage to Hawking and to converse.

Among the physicists who made repeated trips to Hawking's hotel room were Zel'dovich and his graduate student Alexi Starobinsky. Hawking found Zel'dovich and Starobinsky as fascinating as they did him. On one visit, Starobinsky described Zel'dovich's conjecture that a

Left: Stephen Hawking listening to a lecture at the Les Houches summer school in summer 1972. *Right:* Yakov Borisovich Zel'dovich at the blackboard in his apartment in Moscow in summer 1971. [Photos by Kip Thorne.]

spinning black hole should radiate, described a partial marriage of quantum mechanics with general relativity that he and Zel'dovich had developed (based on earlier, pioneering work by Bryce DeWitt, Leonard Parker, and others), and then described a proof, using this partial marriage, that a spinning hole does, indeed, radiate. Zel'dovich was well on his way toward winning his bet with me.

Of all the things Hawking learned from his conversations in Moscow, this one intrigued him most. However, he was skeptical of the manner in which Zel'dovich and Starobinsky had combined the laws of general relativity with the laws of quantum mechanics, so, after returning to Cambridge, he began to develop his own partial marriage of quantum mechanics and general relativity and use it to test Zel'dovich's claim that spinning holes should radiate.

In the meantime, several other physicists in America were doing the same thing, among them William Unruh (a recent student of Wheeler's) and Don Page (a student of mine). By early 1974 Unruh and Page, each in his own way, had tentatively confirmed Zel'dovich's prediction: A spinning hole should emit radiation until all of its spin energy has been used up and its emission stops. I would have to concede my bet.

Black Holes Shrink and Explode

Then came a bombshell. Stephen Hawking, first at a conference in England and then in a brief technical article in the journal *Nature*, announced an outrageous prediction, a prediction that conflicted with Zel'dovich, Starobinsky, Page, and Unruh. Hawking's calculations confirmed that a spinning black hole must radiate and slow its spin. However, they also predicted that, when the hole stops spinning, its radiation does *not* stop. With no spin left, and no spin energy left, the hole keeps on emitting radiation of all sorts (gravitational, electromagnetic, neutrino), and as it emits, it keeps on losing energy. Whereas the spin energy was stored in the swirl of space outside the horizon, the energy now being lost could come from only one place: from the hole's interior!

Equally amazing, Hawking's calculations predicted that the spectrum of the radiation (that is, the amount of energy radiated at each wavelength) is precisely like the spectrum of thermal radiation from a hot body. In other words, a black hole behaves precisely as though its horizon has a finite temperature, and that temperature, Hawking concluded, is proportional to the hole's surface gravity. This (if Hawking was right) was incontrovertible proof that the Bardeen–Carter–Hawking laws of black-hole mechanics *are* the laws of thermodynamics in disguise, and that, as Bekenstein had claimed two years earlier, a black hole has an entropy proportional to its surface area.

Hawking's calculations said more. Once the hole's spin has slowed, its entropy and the area of its horizon are proportional to its mass squared, while its temperature and surface gravity are proportional to its mass divided by its area, which means inversely proportional to its mass. Therefore, as the hole continues to emit radiation, converting mass into outflowing energy, its mass goes down, its entropy and area go down, and its temperature and surface gravity go up. The hole shrinks and becomes hotter. In effect, the hole is evaporating.

A hole that has recently formed by stellar implosion (and that thus has a mass larger than about 2 Suns) has a very low temperature: less than 3×10^{-8} degree above absolute zero (0.03 microkelvin). Therefore, the evaporation at first is very slow; so slow that the hole will require longer than 10^{67} years (10^{57} times the present age of the Universe) to shrink appreciably. However, as the hole shrinks and heats up, it will radiate more strongly and its evaporation will quicken. Finally, when the hole's mass has been reduced to somewhere between a thousand tons and 100 million tons (we are not sure where), and its horizon has shrunk to a fraction the size of an atomic nucleus, the hole will be so extremely hot (between a trillion and 100,000 trillion degrees) that it will explode violently, in a fraction of a second.

The world's dozen experts on the partial marriage of general relativity with quantum theory were quite sure that Hawking had made a mistake. His conclusion violated everything then known about black holes. Perhaps his partial marriage, which differed from other people's, was wrong; or perhaps he had the right marriage, but had made a mistake in his calculations.

For the next several years the experts minutely examined Hawking's version of the partial marriage and their own versions, Hawking's calculations of the waves from black holes and their own calculations. Gradually one expert after another came to agree with Hawking, and in the process they firmed up the partial marriage, producing a new set of physical laws. The new laws are called the *laws of quantum fields in curved spacetime* because they come from a partial marriage in which the black hole is regarded as a non–quantum mechanical, general relativistic, curved spacetime object, while the gravitational waves, electromagnetic waves, and other types of radiation are regarded as *quantum fields*—in other words, as waves that are subject to the laws of quantum mechanics and that therefore behave sometimes like waves and sometimes like particles (see Box 4.1). [A full marriage of general relativity and quantum theory, that is, the fully correct laws of quantum gravity, would treat everything, including the hole's curved spacetime, as quantum mechanical, that is, as subject to the uncertainty principle (Box 10.2), to wave/particle duality (Box 4.1), and to vacuum fluctuations (Box 12.4). We shall meet this full marriage and some of its implications in the next chapter.]

How was it possible to reach agreement on the fundamental laws of quantum fields in curved spacetime without any experiments to guide the choice of the laws? How could the experts claim near certainty that Hawking was right without experiments to check their claims? Their near certainty came from the requirement that the laws of quantum fields and the laws of curved spacetime be meshed in a totally consistent way. (If the meshing were not totally consistent, then the laws of physics, when manipulated in one manner, might make one prediction, for example, that black holes never radiate, and when manipulated in another manner, might make a different prediction, for example, that black holes must always radiate. The poor physicists, not knowing what to believe, might be put out of business.)

The new, meshed laws had to be consistent with general relativity's laws of curved spacetime in the absence of quantum fields and with the laws of quantum fields in the absence of spacetime curvature. This and the demand for a perfect mesh, analogous to the demand that the rows and columns of a crossword puzzle mesh perfectly, turned out to determine the form of the new laws almost[11] completely. If the laws could be meshed consistently at all (and they must be, if the physicists' approach to understanding the Universe makes any sense), then they could be meshed only in the manner described by the new, agreed-upon laws of quantum fields in curved spacetime.

11. The "almost" takes care of certain ambiguities in a procedure called "renormalization," by which one computes the net energy carried by vacuum fluctuations. These ambiguities, which were identified and codified by Robert Wald (a former student of Wheeler's), do not influence a black hole's evaporation, and they probably will not be resolved until the full quantum theory of gravity is in hand.

The requirement that the laws of physics mesh consistently is often used as a tool in the search for new laws. However, this consistency requirement has rarely exhibited such great power as here, in the arena of quantum fields in curved spacetime. For example, when Einstein was developing his laws of general relativity (Chapter 2), considerations of consistency could not and did not tell him his starting premise, that gravity is due to a curvature of spacetime; this starting premise came largely from Einstein's intuition. However, with this premise in hand, the requirement that the new general relativistic laws mesh consistently with Newton's laws of gravity when gravity is weak, and with the laws of special relativity when there is no gravity at all, determined the forms of the new laws almost uniquely; for example, it was the key to Einstein's discovery of his field equation.

In September 1975, I returned to Moscow for my fifth visit, bearing a bottle of White Horse scotch for Zel'dovich. To my surprise, I discovered that, although all the Western experts by now had agreed that Hawking was right and black holes can evaporate, nobody in Moscow believed Hawking's calculations or conclusions. Although several confirmations of Hawking's claims, derived by new, completely different methods, had been published during 1974 and 1975, those confirmations had had little impact in the U.S.S.R. Why? Because Zel'dovich and Starobinsky, the greatest Soviet experts, were disbelievers: They continued to maintain that, after a radiating black hole has lost all its spin, it must stop radiating, and it therefore cannot evaporate completely. I argued with Zel'dovich and Starobinsky, to no avail; they knew so much more about quantum fields in curved spacetime than I that although (as usual) I was quite sure I had truth on my side, I could not counter their arguments.

My return flight to America was scheduled for Tuesday, 23 September. On Monday evening, as I was packing my bags in my tiny room at the University Hotel, the telephone rang. It was Zel'dovich: "Come to my flat, Kip! I want to talk about black-hole evaporation!" Tight for time, I sought a taxi in front of the hotel. None was in sight, so in standard Muscovite fashion I flagged down a passing motorist and offered him five rubles to take me to Number 2B Vorobyevskoye Shosse. He nodded agreement and we were off, down back streets I had never traveled. My fear of being lost abated when we swung onto Vorobyevskoye Shosse. With a grateful "Spasibo!" I alighted in front of

2B, jogged through the gate and forested grounds, into the building, and up the stairs to the second floor, southwest corner.

Zel'dovich and Starobinsky greeted me at the door, grins on their faces and their hands above their heads. "We give up; Hawking is right; we were wrong!" For the next hour they described to me how their version of the laws of quantum fields in a black hole's curved spacetime, while seemingly different from Hawking's, was really completely equivalent. They had concluded black holes cannot evaporate because of an error in their calculations, not because of wrong laws. With the error corrected, they now agreed. There is no escape. The laws require that black holes evaporate.

There are several different ways to picture black-hole evaporation, corresponding to the several different ways to formulate the laws of quantum fields in a black hole's curved spacetime. However, all the ways acknowledge vacuum fluctuations as the ultimate source of the outflowing radiation. Perhaps the simplest pictorial description is one based on particles rather than waves:

Vacuum fluctuations, like "real," positive-energy waves, are subject to the laws of wave/particle duality (Box 4.1); that is, they have both wave aspects and particle aspects. The wave aspects we have met already (Box 12.4): The waves fluctuate randomly and unpredictably, with positive energy momentarily here, negative energy momentarily there, and zero energy on average. The particle aspect is embodied in the concept of *virtual particles*, that is, particles that flash into existence in pairs (two particles at a time), living momentarily on fluctuational energy borrowed from neighboring regions of space, and that then annihilate and disappear, giving their energy back to the neighboring regions. For electromagnetic vacuum fluctuations, the virtual particles are *virtual photons;* for gravitational vacuum fluctuations, they are *virtual gravitons.*[12]

---

12. Some readers may already be familiar with these concepts in the context of matter and antimatter, for example, an electron (which is a particle of matter) and a positron (its antiparticle). Just as the electromagnetic field is the field aspect of a photon, so also there exists an electron field which is the field aspect of the electron and the positron. At locations where the electron field's vacuum fluctuations are momentarily large, a virtual electron and a virtual positron are likely to flash into existence, as a pair; when the field fluctuates down, the electron and positron are likely to annihilate each other and disappear. The photon is its own antiparticle, so virtual photons flash in and out of existence in pairs, and similarly for gravitons.

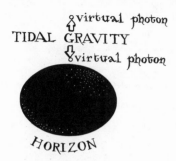

12.2   The mechanism of black-hole evaporation, as viewed by someone who is falling into the hole. *Left:* A black hole's tidal gravity pulls a pair of virtual photons apart, thereby feeding energy into them. *Right:* The virtual photons have acquired enough energy from tidal gravity to materialize, permanently, into real photons, one of which escapes from the hole while the other falls toward the hole's center.

The manner in which vacuum fluctuations cause black holes to evaporate is depicted in Figure 12.2. On the left is shown a pair of virtual photons near a black hole's horizon, as viewed in the reference frame of someone who is falling into the hole. The virtual photons can separate from each other easily, so long as they both remain in a region where the electromagnetic field has momentarily acquired positive energy. That region can have any size from tiny to huge, since vacuum fluctuations occur on all length scales; however, the region's size will always be about the same as the wavelength of its fluctuating electro- magnetic wave, so the virtual photons can move apart by only about one wavelength. If the wavelength happens to be about the same as the hole's circumference, then the virtual photons can easily separate from each other by a quarter of the circumference, as shown in the figure.

Tidal gravity near the horizon is very strong; it pulls the virtual photons apart with a huge force, thereby feeding great energy into them, as seen by the infalling observer who is halfway between the photons. The increase in photon energy is sufficient, by the time the photons are a quarter of a horizon circumference apart, to convert them into real long-lived photons (right half of Figure 12.2), and have enough energy left over to give back to the neighboring, negative-energy regions of space. The photons, now real, are liberated from each other. One is inside the horizon and lost forever from the external Universe. The other escapes from the hole, carrying away the energy (that is, the mass[13]) that the hole's tidal gravity gave to it. The hole, with its mass reduced, shrinks a bit.

This mechanism of emitting particles does not depend at all on the fact that the particles were photons, and their associated waves were electromagnetic. The mechanism will work equally well for all other forms of particle/wave (that is, for all other types of radiation—gravitational, neutrino, and so forth), and therefore a black hole radiates *all* types of radiation.

Before the virtual particles have materialized into real particles, they must stay closer together than roughly the wavelength of their waves. To acquire enough energy from the hole's tidal gravity to materialize, however, they must get as far apart as about a quarter of the circumference of the hole. This means that the wavelengths of the particle/waves that the hole emits will be about one-fourth the hole's circumference in size, and larger.

A black hole with mass twice as large as the Sun has a circumference of about 35 kilometers, and thus the particle/waves that it emits have wavelengths of about 9 kilometers and larger. These are enormous wavelengths compared to light or ordinary radio waves, but not much different from the lengths of the gravitational waves that the hole would emit if it were to collide with another hole.

During the early years of his career, Hawking tried to be very careful and rigorous in his research. He never asserted things to be true unless he could give a nearly airtight proof of them. However, by 1974 he had changed his attitude: "I would rather be right than rigorous," he told me firmly. Achieving high rigor requires much time. By 1974 Hawking had set for himself goals of understanding the full marriage of general relativity with quantum mechanics, and understanding the origin of the Universe—goals that to achieve would require enormous amounts of time and concentration. Perhaps feeling more finite than

---

13. Recall that, since mass and energy are totally convertible into each other, they are really just different names for the same concept.

other people feel because of his life-shortening disease, Hawking felt he could not afford to dally with his discoveries long enough to achieve high rigor, nor could he afford to explore all the important features of his discoveries. He must push on at high speed.

Thus it was that Hawking, in 1974, having proved firmly that a black hole radiates as though it had a temperature proportional to its surface gravity, went on to assert, without real proof, that *all* of the other similarities between the laws of black-hole mechanics and the laws of thermodynamics were more than a coincidence: The black-hole laws *are the same thing as* the thermodynamic laws, but in disguise. From this assertion and his firmly proved relationship between temperature and surface gravity, Hawking inferred a precise relationship between the hole's entropy and its surface area: The entropy is 0.10857 . . . times[14] the surface area, divided by the Planck–Wheeler area. In other words, a 10-solar-mass, nonspinning hole has an entropy of 4.6×10^{78}, which is approximately the same as Bekenstein's conjecture.

Bekenstein, of course, was sure Hawking was right, and he glowed with pleasure. By the end of 1975, Zel'dovich, Starobinsky, I, and Hawking's other colleagues were also strongly inclined to agree. However, we would not feel fully satisfied until we understood the precise nature of a black hole's enormous randomness. There must be $10^{4.6 \times 10^{78}}$ ways to distribute *something* inside the black hole, without changing its external appearance (its mass, angular momentum, and charge), but what was that something? And how, in simple physical terms, could one understand the thermal behavior of a black hole—the fact that the hole behaves just like an ordinary body with temperature? As Hawking moved on to research on quantum gravity and the origin of the Universe, Paul Davies, Bill Unruh, Robert Wald, James York, I, and many others of his colleagues zeroed in on these issues. Gradually over the next ten years we arrived at the new understanding embodied in Figure 12.3.

Figure 12.3a depicts a black hole's vacuum fluctuations, as viewed by observers falling inward through the horizon. The vacuum fluctuations consist of pairs of virtual particles. Occasionally tidal gravity manages to give one of the plethora of pairs sufficient energy for its two virtual particles to become real, and for one of them to escape from the hole.

14. The peculiar factor 0.10857 . . . is actually $1/(4\log_e 10)$, where $\log_e 10 = 2.30258$ . . . results from my choice of "normalization" of the entropy; see Footnote 3 on page 423.

This was the viewpoint on vacuum fluctuations and black-hole evaporation discussed in Figure 12.2.

Figure 12.3b depicts a different viewpoint on the hole's vacuum fluctuations, the viewpoint of observers who reside just above the hole's horizon and are forever at rest relative to the horizon. To prevent themselves from being swallowed by the hole, such observers must accelerate hard, relative to falling observers—using a rocket engine or hanging by a rope. For this reason, these observers' viewpoint is called the "accelerated viewpoint." It is also the viewpoint of the "membrane paradigm" (Chapter 11).

Surprisingly, from the accelerated viewpoint, the vacuum fluctuations consist not of virtual particles flashing in and out of existence, but rather of real particles with positive energies and long lives; see Box 12.5. The real particles form a hot atmosphere around the hole, much like the atmosphere of the Sun. Associated with these real particles are

12.3 (a) Observers falling into a black hole (the two little men in space suits) see vacuum fluctuations near the hole's horizon to consist of pairs of virtual particles. (b) As viewed by observers just above the horizon and at rest relative to the horizon (the little man hanging by a rope and the little man blasting his rocket engine), the vacuum fluctuations consist of a hot atmosphere of real particles; this is the "accelerated viewpoint." (c) The atmosphere's particles, in the accelerated viewpoint, appear to be emitted by a hot, membrane-like horizon. They fly upward short distances, and most are then pulled back into the horizon. However, a few of the particles manage to escape the hole's grip and evaporate into outer space.

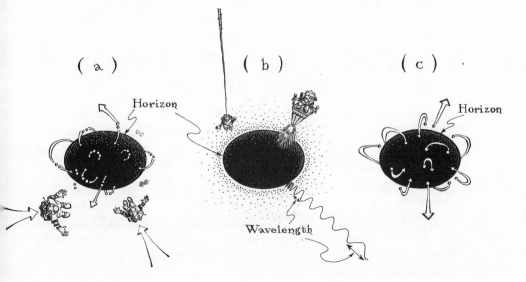

Box 12.5
Acceleration Radiation

In 1975, Wheeler's recent student William Unruh, and independently Paul Davies at King's College, London, discovered (using the laws of quantum fields in curved spacetime) that accelerated observers just above a black hole's horizon must see the vacuum fluctuations there not as virtual pairs of particles but rather as an atmosphere of real particles, an atmosphere that Unruh called "acceleration radiation."

This startling discovery revealed that *the concept of a real particle is relative*, not absolute; that is, it depends on one's reference frame. Observers in freely falling frames who plunge through the hole's horizon see no real particles outside the horizon, only virtual ones. Observers in accelerated frames who, by their acceleration, remain always above the horizon see a plethora of real particles.

How is this possible? How can one observer claim that the horizon is surrounded by an atmosphere of real particles and the other that it is not? The answer lies in the fact that the virtual particles' vacuum fluctuational waves are not confined solely to the region above the horizon; part of each fluctuational wave is inside the horizon and part is outside.

- The freely falling observers, who plunge through the horizon, can see both parts of the vacuum fluctuational wave, the part inside the horizon and the part outside; so such observers are well aware (by their measurements) that the wave is a mere vacuum fluctuation and correspondingly that its particles are virtual, not real.
- The accelerated observers, who remain always outside the horizon, can see only the outside part of the vacuum fluctuational wave, not the inside part; and correspondingly, by their measurements they are unable to discern that the wave is a mere vacuum fluctuation accompanied by virtual particles. Seeing only a part of the fluctuational wave, they mistake it for "the real thing"—a real wave accompanied by real particles, and as a result their measurements reveal all around the horizon an atmosphere of real particles.

That this atmosphere's real particles can gradually evaporate and fly off into the external Universe (Figure 12.3c) is an indication that the viewpoint of the accelerated observers is just as correct, that is, just as valid, as that of the freely falling observers: What the freely falling observers see as virtual pairs converted into real particles by tidal gravity, followed by evaporation of one of the real particles, the accelerated observers see simply as the evaporation of one of the particles that was always real and always populated the black hole's atmosphere. Both viewpoints are correct; they are the same physical situation, seen from two different reference frames.

real waves. As a particle moves upward through the atmosphere, gravity pulls on it, reducing its energy of motion; correspondingly, as a wave moves upward, it becomes gravitationally redshifted to longer and longer wavelengths (Figure 12.3b).

Figure 12.3c shows the motion of a few of the particles in a black-hole atmosphere, from the accelerated viewpoint. The particles appear to be emitted by the horizon; most fly upward a short distance and are then pulled back down to the horizon by the hole's strong gravity, but a few manage to escape the hole's grip. The escaping particles are the same ones as the infalling observers see materialize from virtual pairs (Figure 12.3a). They are Hawking's evaporating particles.

From the accelerated viewpoint, the horizon behaves like a high-temperature, membrane-like surface; it is the membrane of the "membrane paradigm" described in Chapter 11. Just as the Sun's hot surface emits particles (for example, the photons that make daylight on Earth), so the horizon's hot membrane emits particles: the particles that make up the hole's atmosphere, and the few that evaporate. The gravitational redshift reduces the particles' energy as they fly upward from the membrane, so although the membrane itself is extremely hot, the evaporating radiation is much cooler.

The accelerated viewpoint not only explains the sense in which a black hole is hot, it also accounts for the hole's enormous randomness. The following thought experiment (invented by me and my postdoc, Wojciech Zurek) explains how.

Throw into a black hole's atmosphere a small amount of material containing some small amount of energy (or, equivalently, mass), angular momentum (spin), and electric charge. From the atmosphere this material will continue on down through the horizon and into the hole. Once the material has entered the hole, it is impossible by examining the hole from outside to learn the nature of the injected material (whether it consisted of matter or of antimatter, of photons and heavy atoms, or of electrons and positrons), and it is impossible to learn just where the material was injected. Because a black hole has no "hair," all one can discover, by examining the hole from outside, are the total amounts of mass, angular momentum, and charge that entered the atmosphere.

Ask how many ways those amounts of mass, angular momentum, and charge *could* have been injected into the hole's hot atmosphere. This question is analogous to asking how many ways the child's toys *could* have been distributed over the tiles in the playroom of Box 12.3, and correspondingly, the logarithm of the number of ways to inject must be the increase in the atmosphere's entropy, as described by the standard laws of thermodynamics. By a fairly simple calculation, Zurek and I were able to show that this increase in thermodynamic entropy is precisely equal to $\frac{1}{4}$ times the increase in the horizon's area, divided by the Planck–Wheeler area; that is, it is precisely the increase in the horizon's area in disguise, the same disguise that Hawking inferred, in 1974, from the mathematical similarity of the laws of black-hole mechanics and the laws of thermodynamics.

The outcome of this thought experiment can be expressed succinctly as follows: *A black hole's entropy is the logarithm of the number of ways that the hole could have been made.* This means that there are $10^{4.6 \times 10^{78}}$ different ways to make a 10-solar-mass black hole whose entropy is 4.6×10^{78}. This explanation of the entropy was originally conjectured by Bekenstein in 1972, and a highly abstract proof was given by Hawking and his former student, Gary Gibbons, in 1977.

The thought experiment also shows the second law of thermodynamics in action. The energy, angular momentum, and charge that one throws into the hole's atmosphere can have any form at all; for example, they might be the roomful of air wrapped up in a bag, which we met earlier in this chapter while puzzling over the second law. When the bag is thrown into the hole's atmosphere, the entropy of the external Universe is reduced by the amount of entropy (randomness) in the bag. However, the entropy of the hole's atmosphere, and thence of the hole, goes up by more than the bag's entropy, so the total entropy of hole plus external Universe goes up. The second law of thermodynamics is obeyed.

Similarly, it turns out, when the black hole evaporates some particles, its own surface area and entropy typically go down; but the particles get distributed randomly in the external Universe, increasing its entropy by more than the hole's entropy loss. Again the second law is obeyed.

How long does it take for a black hole to evaporate and disappear? The answer depends on the hole's mass. The larger the hole, the lower its temperature, and thus the more weakly it emits particles and the more slowly it evaporates. The total lifetime, as worked out by Don Page in 1975 when he was jointly my student and Hawking's, is 1.2×10^{67} years if the hole's mass is twice that of the Sun. The lifetime is proportional to the cube of the hole's mass, so a 20-solar-mass hole has a life of 1.2×10^{70} years. These lifetimes are so enormous compared to the present age of the Universe, about 1×10^{10} years, that the evaporation is totally irrelevant for astrophysics. Nevertheless, the evaporation has been very important for our understanding of the marriage between general relativity and quantum mechanics; the struggle to understand the evaporation taught us the laws of quantum fields in curved spacetime.

Holes far less massive than 2 Suns, if they could exist, would evaporate far more rapidly than 10^{67} years. Such small holes cannot be formed in the Universe today because degeneracy pressures and nuclear pressures prevent small masses from imploding, even if one squeezes them with all the force the present-day Universe can muster (Chapters 4 and 5). However, such holes might have formed in the big bang, where matter experienced densities and pressures and gravitational squeezes that were enormously higher than in any modern-day star.

Detailed calculations by Hawking, Zel'dovich, Novikov, and others have shown that tiny lumps in the matter emerging from the big bang could have produced tiny black holes, if the lumps' matter had a rather soft equation of state (that is, had only small increases of pressure when squeezed). Powerful squeezing by other, adjacent matter in the very early Universe, like the squeezing of carbon in the jaws of a powerful anvil to form diamond, could have made the tiny lumps implode to produce tiny holes.

A promising way to search for such tiny *primordial black holes* is by searching for the particles they produce when they evaporate. Black holes weighing less than about 500 billion kilograms (5×10^{14} grams, the weight of a modest mountain) should have evaporated completely away by now, and black holes a few times heavier than this should still be evaporating strongly. Such black holes have horizons about the size of an atomic nucleus.

A large portion of the energy emitted in the evaporation of such holes should now be in the form of gamma rays (high-energy photons) traveling randomly through the Universe. Such gamma rays do exist, but in amounts and with properties that are readily explained in other ways. The absence of excess gamma rays tells us (according to calculations by Hawking and Page) that there now are no more than about 300 tiny, strongly evaporating black holes in each cubic light-year of space; and this, in turn, tells us that matter in the big bang cannot have had an extremely soft equation of state.

Skeptics will argue that the absence of excess gamma rays might have another interpretation: Perhaps many small black holes were formed in the big bang, but we physicists understand quantum fields in curved spacetime far less well than we think we do, and thus we are misleading ourselves when we believe that black holes evaporate. I and my colleagues resist such skepticism because of the seeming perfection with which the standard laws of curved spacetime and the standard laws of quantum fields mesh to give us a nearly *unique* set of laws for quantum fields in curved spacetime. Nevertheless, we would feel rather more comfortable if astronomers could find observational evidence of black-hole evaporation.

13

Inside
Black Holes

in which physicists, wrestling with Einstein's equation,
seek the secret of what is inside a black hole:
a route into another universe?
a singularity with infinite tidal gravity?
the end of space and time, and birth of quantum foam?

Singularities and Other Universes

What is inside a black hole?

How can we know, and why should we care? No signal can ever emerge from the hole to tell us the answer. No intrepid explorer who might enter the hole to find out can ever come back and tell us, or ever transmit the answer to us. Whatever may be in the hole's core can never reach out and influence our Universe in any way.

Human curiosity is hardly satisfied by such replies. Especially not when we have tools that can tell us the answer: the laws of physics.

John Archibald Wheeler taught us the importance of the quest to understand a black hole's core. In the 1950s he posed "the issue of the final state" of gravitational implosion as a holy grail for theoretical physics, one that might teach us details of the "fiery marriage" of general relativity with quantum mechanics. When J. Robert Oppenheimer insisted that the final state is hidden from view by a horizon, Wheeler resisted (Chapter 6)—not least, I suspect, because of his anguish at losing the possibility to see the fiery marriage in action from outside the horizon.

After accepting the horizon, Wheeler retained his conviction that understanding the hole's core was a holy grail worth pursuing. Just as struggling to understand the evaporation of black holes has helped us to discover a partial marriage of quantum mechanics with general relativity (Chapter 12), so struggling to understand a black hole's core might help us to discover the full marriage; it might lead us to the full laws of quantum gravity. And perhaps the nature of the core will hold the keys to other mysteries of the Universe: There is a similarity between the "big crunch" implosion in which, eons hence, our Universe might die, and the implosion of the star that creates a black hole's core. By coming to grips with the one, we might learn about the other.

For thirty-five years physicists have pursued Wheeler's holy grail, but with only modest success. We do not yet know for certain what inhabits a hole's core, and the struggle to understand has not yet taught us with clarity the laws of quantum gravity. But we have learned much—not least that whatever is inside a black hole's core *is indeed* intimately connected with the laws of quantum gravity.

This chapter describes a few of the more interesting twists and turns in the quest for Wheeler's holy grail, and where the quest has led thus far.

The first, tentative answer to "What is inside a black hole?" came from J. Robert Oppenheimer and Hartland Snyder, in their classic 1939 calculation of the implosion of a spherical star (Chapter 6). Although the answer was contained in the equations they published, Oppenheimer and Snyder chose not to discuss it. Perhaps they feared it would only add fuel to the controversy over their prediction that the imploding star "cuts itself off from the rest of the Universe" (that is, forms a black hole). Perhaps Oppenheimer's innate scientific conservatism, his unwillingness to speculate, kept them quiet. Whatever the reason, they said nothing. But their equations spoke.

After creating a black-hole horizon around itself, their equations said, the spherical star continues imploding, inexorably, to infinite density and zero volume, whereupon it creates and merges into a *spacetime singularity*.

A singularity is a region where—according to the laws of general relativity—the curvature of spacetime becomes infinitely large, and spacetime ceases to exist. Since tidal gravity is a manifestation of spacetime curvature (Chapter 2), a singularity is also a region of infinite tidal gravity, that is, a region where gravity stretches all objects infinitely along some directions and squeezes them infinitely along others.

One can conceive of a variety of different kinds of spacetime singularities, each with its own peculiar form of tidal stretch and squeeze, and we shall meet several different kinds in this chapter.

The singularity predicted by the Oppenheimer–Snyder calculations is a very simple one. Its tidal gravity has essentially the same form as the Earth's or Moon's or Sun's; that is, the same form as the tidal gravity that creates the tides on the Earth's oceans (Box 2.5): The singularity stretches all objects radially (in the direction toward and away from itself), and squeezes all objects transversely.

Imagine an astronaut falling feet first into the kind of black hole described by Oppenheimer and Snyder's equations. The larger the hole, the longer he can survive, so for maximum longevity, let the hole be among the largest that inhabit the cores of quasars (Chapter 9): 10 billion solar masses. Then the falling astronaut crosses the horizon and enters the hole about 20 hours before his final death, but as he enters, he is still too far from the singularity to feel its tidal gravity. As he continues to fall faster and faster, coming closer and closer to the singularity, the tidal gravity grows stronger and stronger until, just 1 second before the singularity, he begins to feel it stretching his feet and head apart and squeezing him from the sides (bottom picture in Figure 13.1). At first, the stretch and squeeze are only mildly annoying, but they continue to grow until, a few hundredths of a second before the singularity (middle picture), they get so strong that his bones and flesh can no longer resist. His body comes apart and he dies. In the last hundredth second, the stretch and squeeze continue mounting, and as he reaches the singularity, they become infinitely strong, first at his feet, then at his trunk, then at his head; his body gets infinitely distended; and then, according to general relativity, he merges with and becomes part of the singularity.

It is utterly impossible for the astronaut to move on through the singularity and come out the other side because, according to general relativity, there is no "other side." Space, time, and spacetime cease to exist at the singularity. The singularity is a sharp edge, much like the edge of a sheet of paper. There is no paper beyond its edge; there is no spacetime beyond the singularity. But there the similarity ends. An ant on the paper can go right up to the edge and then back away, but nothing can back away from the singularity; all astronauts, particles, waves, whatever, that hit it are instantaneously destroyed, according to Einstein's general relativistic laws.

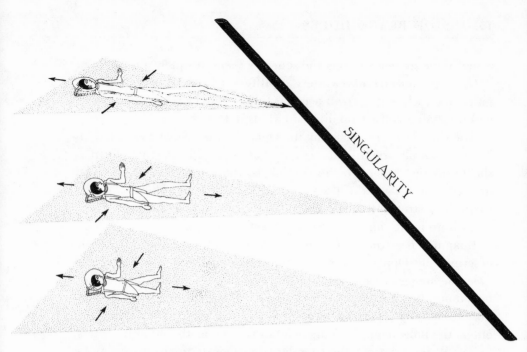

13.1 Spacetime diagram depicting the feet-first fall of an astronaut into the singularity at a black hole's center, according to the Oppenheimer–Snyder calculations. As in all previous spacetime diagrams (for example, Figure 6.7), one spatial dimension is missing; that is why the astronaut looks two-dimensional rather than three-dimensional. The singularity is tilted in this diagram, in contrast to its vertical position in Figure 6.7 and Box 12.1, because the time plotted upward and the space plotted horizontally here are different from there. Here they are the astronaut's own time and space; there they were Finkelstein's.

The mechanism of destruction is not fully clear in Figure 13.1, because the figure ignores the curvature of space. In fact, as the astronaut's body reaches the singularity, it gets stretched out to truly infinite length and squashed transversely to truly zero size. The extreme curvature of space near the singularity permits him to become infinitely long without shoving his head out through the hole's horizon. His head and feet are both pulled into the singularity, but they are pulled in infinitely far apart.

Not only is an astronaut stretched and squeezed infinitely at the singularity, according to the Oppenheimer–Snyder equations; all forms of matter are infinitely stretched and squeezed—even an individual atom; even the electrons, protons, and neutrons that make up atoms; even the quarks that make up protons and neutrons.

Is there any way for the astronaut to escape this infinite stretch and squeeze? No, not after he has crossed the horizon. Everywhere inside the horizon, according to the Oppenheimer–Snyder equations, gravity is so strong (spacetime is so strongly warped) that time itself (everyone's time) flows into the singularity.[1] Since the astronaut, like anyone else, must move inexorably forward in time, he is driven with the flow of time into the singularity. No matter what he does, no matter how he blasts his rocket engines, the astronaut cannot avoid the singularity's infinite stretch and squeeze.

Whenever we physicists see our equations predict something infinite, we become suspicious of the equations. Almost nothing in the real Universe ever gets truly infinite (we think). Therefore, an infinity is almost always a sign of a mistake.

The singularity's infinite stretch and squeeze was no exception. Those few physicists who studied Oppenheimer and Snyder's publication during the 1950s and early 1960s agreed unanimously that something was wrong. But there the unanimity stopped.

One group, led vigorously by John Wheeler, identified the infinite stretch and squeeze as an unequivocal message that general relativity fails inside a black hole, at the endpoint of stellar implosion. Quantum mechanics should prevent tidal gravity from becoming truly infinite there, Wheeler asserted; but how? To learn the answer, Wheeler argued, would require marrying the laws of quantum mechanics with the laws of tidal gravity, that is, with Einstein's general relativistic laws of curved spacetime. The progeny of that marriage, the laws of quantum gravity, must govern the singularity, Wheeler claimed; and these new laws might create new physical phenomena inside the black hole, phenomena unlike any we have ever met.

A second group, led by Isaac Markovich Khalatnikov and Evgeny Michailovich Lifshitz (members of Lev Landau's Moscow research group), saw the infinite stretch and squeeze as a warning that Oppenheimer and Snyder's idealized model of an imploding star could not be trusted. Recall that Oppenheimer and Snyder required, as a foundation for their calculations, that the star be precisely spherical and nonspinning and have uniform density, zero pressure, no shock waves, no ejected matter, and no outpouring radiation (Figure 13.2). These extreme idealizations were responsible for the singularity, Khalatnikov and Lifshitz argued. Every real star has tiny, random deformations (tiny, random nonuniformities in its shape, velocity, density, and pressure), and as the star implodes, they claimed, *these deformations will*

---

1. In technical jargon, we say that the singularity is "spacelike."

grow large and halt the implosion before a singularity can form. Similarly, Khalatnikov and Lifshitz asserted, random deformations will halt the big crunch implosion of our entire Universe eons hence, and thereby save the Universe from destruction in a singularity.

Khalatnikov and Lifshitz came to these views in 1961 by asking themselves whether, according to Einstein's general relativistic laws, singularities are *stable against small perturbations.* In other words, they posed the same question for singularities as we met in Chapter 7 for black holes: If, in solving Einstein's field equation, we alter, in small but random ways, the shape of the imploding star or Universe and the velocity and density and pressure of its material, and if we insert into the material tiny but random amounts of gravitational radiation, how will these changes (these *perturbations*) affect the implosion's predicted endpoint?

For the black hole's horizon, as we saw in Chapter 7, the perturbations make no difference. The perturbed, imploding star still forms a horizon, and although the horizon is deformed at first, all its deformations quickly get radiated away, leaving behind a completely "hairless" black hole. In other words, the horizon is *stable* against small perturbations.

13.2   (Same as Figure 6.3.) *Left:* Physical phenomena in a realistic, imploding star. *Right:* The idealizations which Oppenheimer and Snyder made in order to compute stellar implosion. For a detailed discussion see Chapter 6.

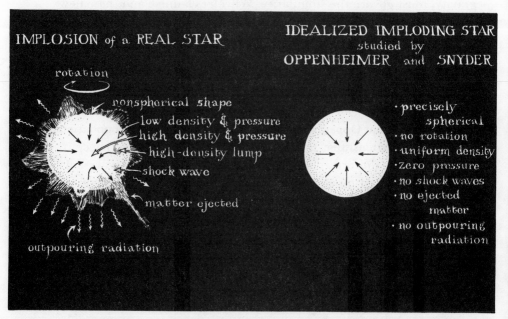

Not so for the singularity at the hole's center or in the Universe's final crunch, Khalatnikov and Lifshitz concluded. Their calculations seemed to show that tiny, random perturbations will grow large when the imploding matter attempts to create a singularity; they will grow so large, in fact, that they will prevent the singularity from forming. Presumably (though the calculations could not say for sure), the perturbations will halt the implosion and transform it into an explosion.

How could perturbations possibly reverse the implosion? The physical mechanism was not at all clear in the Khalatnikov–Lifshitz calculations. However, other calculations using Newton's laws of gravity, which are far easier than calculations using Einstein's laws, give hints. For example (see Figure 13.3), if gravity were weak enough inside an imploding star for Newton's laws to be accurate, and if the star's pressure were too small to be important, then small perturbations would cause different atoms to implode toward slightly different points near the star's center. Most of the imploding atoms would miss the center by some small amount and would swing around the center and fly back out, thereby converting the implosion into an explosion. It seemed conceivable that, even though Newton's laws of gravity fail inside a black hole, some mechanism analogous to this might convert the implosion into an explosion.

13.3 One mechanism for converting a star's implosion into an explosion, when gravity is weak enough that Newton's laws are accurate, and when internal pressure is weak enough to be unimportant. If the imploding star is slightly deformed ("perturbed"), its atoms implode toward slightly different points, swing around each other, and then fly back out.

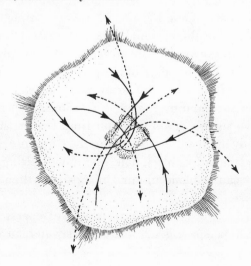

I joined John Wheeler's research group as a graduate student in 1962, shortly after Khalatnikov and Lifshitz had published their calculation, and shortly after Lifshitz together with Landau had enshrined the calculation and its "no singularity" conclusion in a famous textbook, *The Classical Theory of Fields*. I recall vividly Wheeler encouraging his research group to study the calculation. If it is right, its consequences are profound, he told us. Unfortunately, the calculation was extremely long and complicated, and the published details were too sketchy to permit us to check them—and Khalatnikov and Lifshitz were confined within the Soviet Union's iron curtain, so we could not sit down with them and discuss the details.

Nevertheless, we began to contemplate the possibility that the imploding Universe, upon reaching some very small size, *might* "bounce" and reexplode in a new "big bang," and similarly that an imploding star, after sinking inside its horizon, *might* bounce and reexplode.

But where could the star go if it reexplodes? It surely could not explode back out through the hole's horizon. Einstein's laws of gravity forbid anything (except virtual particles) to fly out of the horizon. There was another possibility, however: *The star might manage to explode into some other region of our Universe, or even into another universe.*

Figure 13.4 depicts such an implosion and reexplosion using a sequence of embedding diagrams. (Embedding diagrams, which are quite different from spacetime diagrams, were introduced in Figures 3.2 and 3.3.)

Each diagram in Figure 13.4 depicts our Universe's curved space, and the curved space of another universe, as two-dimensional surfaces embedded in a higher-dimensional *hyperspace*. [Recall that hyperspace is a figment of the physicists' imagination: We, as humans, are confined always to live in the space of our own Universe (or, if we can get there, the space of the other universe); we can never get out of those spaces into the surrounding higher-dimensional hyperspace, nor can we ever receive signals or information from hyperspace. The hyperspace serves only as an aid in visualizing the curvature of space around the imploding star and its black hole, and in visualizing the manner in which the star can implode in our Universe and then reexplode into another universe.]

In Figure 13.4, the two universes are like separate islands in an ocean and the hyperspace is like the ocean's water. Just as there is no land connection between the islands, so there is no space connection between the universes.

The sequence of diagrams in Figure 13.4 depicts the star's evolution. The star, in our Universe, is beginning to implode in diagram (a). In (b) the star has formed a black-hole horizon around itself and is continuing to implode. In (c) and (d) the star's highly compressed matter curves space up tightly around the star, forming a little, closed universe that resembles the surface of a balloon; and this new, little universe

13.4 Embedding diagrams depicting a conceivable (though, as it turns out later in this chapter, a very *unlikely*) fate of the star that implodes to form a black hole. The eight diagrams, (a) through (h), are a sequence of snapshots showing the evolution of the star and the geometry of space. The star implodes in our Universe (a), and forms a black-hole horizon around itself (b). Then deep inside the hole the region of space containing the star pinches off from our Universe and forms a small, closed universe with no connection to anything else (c). That closed universe then moves through hyperspace (d, e) and attaches itself to another large universe (f); and the star then explodes outward into that other universe (g, h).

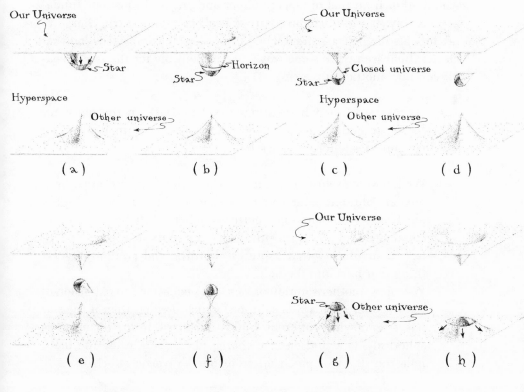

pinches off from our Universe and moves, alone, out into hyperspace. (This is somewhat analogous to natives on one of the islands building a little boat and setting sail across the ocean.) In (d) and (e) the little universe, with the star inside, moves through hyperspace from our big Universe to the other big universe (like the boat sailing from one island to another). In (f) the little universe attaches itself to the other large universe (like the boat landing at the other island), and continues to expand, disgorging the star. In (g) and (h) the star explodes into the other universe.

I am uncomfortably aware that this scenario sounds like pure science fiction. However, just as black holes were a natural outgrowth of Schwarzschild's solution to the Einstein field equation (Chapter 3), so also this scenario is a natural outgrowth of another solution to the Einstein equation, a solution found in 1916–18 by Hans Reissner and Gunnar Nordström but not fully understood by them. In 1960 two of Wheeler's students, Dieter Brill and John Graves, deciphered the physical meaning of the Reissner–Nordström solution, and it soon became obvious that, with modest changes, the Reissner–Nordström solution would describe the imploding/exploding star of Figure 13.4. This star would differ from that of Oppenheimer and Snyder in just one fundamental way: It would contain within itself enough electric charge to produce a strong electric field when it gets highly compacted, and that electric field seemed in some way to be responsible for the star's reexplosion into another universe.

Let us take stock of where things stood in 1964, in the quest for Wheeler's holy grail—the quest to understand the ultimate fate of a star that implodes to form a black hole:

1. We knew one solution of Einstein's equation (the Oppenheimer–Snyder solution) which predicts that, if the star has a highly idealized form, including a perfectly spherical shape, then it will create a singularity with infinite tidal gravity at the hole's center—a singularity that captures, destroys, and swallows everything that falls into the hole.

2. We knew another solution of Einstein's equation (an extension of the Reissner–Nordström solution) which predicts that, if the star has a somewhat different highly idealized form, including a spherical shape and electric charge, then deep inside the black hole the star will pinch off from our Universe, attach itself to

  another universe (or to a distant region of our own Universe), and
  there reexplode.
3. It was far from clear which, if either, of these solutions was
  "stable against small, random perturbations" and thus was a candidate for occurring in the real Universe.
4. Khalatnikov and Lifshitz had claimed to prove, however, that
  singularities are *always* unstable against small perturbations and
  thus never occur, and therefore the Oppenheimer–Snyder singularity could never occur in our real Universe.
5. In Princeton, at least, there was some skepticism about the Khalatnikov–Lifshitz claim. This skepticism may have been driven
  in part by Wheeler's desire for singularities, since they would be
  a "marrying" place for general relativity and quantum mechanics.

Nineteen sixty-four was a watershed year. It was the year that Roger
Penrose revolutionized the mathematical tools that we use to analyze
the properties of spacetime. His revolution was so important, and had
such great impact on the quest for Wheeler's holy grail, that I shall
digress for a few pages to describe his revolution and describe Penrose
himself.

Penrose's Revolution

Roger Penrose grew up in a British medical family; his mother was a
physician, his father was an eminent professor of human genetics at
University College in London, and his parents wanted at least one of
their four children to follow in their footsteps with a medical career.
Roger's older brother Oliver was a dead loss; from an early age he was
intent on a career in physics (and in fact would go on to become one of
the world's leading researchers in statistical physics—the study of the
behaviors of huge numbers of interacting atoms). Roger's younger
brother Jonathon was also a dead loss; all he wanted to do was play
chess (and in fact he would go on to become the British chess champion
for seven years running). Roger's little sister Shirley was much too
young, when Roger was choosing a career, to show inclinations in any
direction (though she ultimately would delight her parents by becoming a physician). That left Roger as his parents' greatest hope.

At age sixteen Roger, like all the others in his class, was interviewed by the school's headmaster. It was time to decide the topics for his last two years of pre-college study. "I'd like to do mathematics, chemistry, and biology," he told the headmaster. "No. Impossible. You cannot combine biology with mathematics. It must be one or the other," the headmaster proclaimed. Mathematics was more precious to Roger than biology. "All right, I'll do mathematics, chemistry, and physics," he said. When Roger got home that evening his parents were furious. They accused Roger of keeping bad company. Biology was essential to a medical career; how could he give it up?

Two years later came the decision of what to study in college. "I proposed to go to University College, London, and study for a mathematics degree," Roger recalls. "My father didn't approve at all. Mathematics might be all right for people who couldn't do anything else, but it wasn't the right thing to make a real career of." Roger was insistent, so his father arranged for one of the College's mathematicians to give him a special test. The mathematician invited Roger to take all day on the test, and warned him that he probably would be able to solve only one or two of the problems. When Roger solved all twelve problems correctly in a few hours, his father capitulated. Roger could study mathematics.

Roger initially had no intention of applying his mathematics to physics. It was pure math that interested him. But he got seduced.

The seduction began in 1952, when Roger as a fourth-year university student in London listened to a series of radio talks on cosmology by Fred Hoyle. The talks were fascinating, stimulating—and a bit confusing. A few of the things Hoyle said didn't quite make sense. One day Roger took the train up to Cambridge to visit his brother Oliver, who was studying physics there. At the end of the day, over dinner at the Kingswood restaurant, Roger discovered that Dennis Sciama, Oliver's officemate, was studying the Bondi–Gold–Hoyle steady-state theory of the Universe. How wonderful! Maybe Sciama could resolve Roger's confusion. "Hoyle says that according to the steady-state theory the expansion of the Universe will drive a distant galaxy out of sight; the galaxy will move out of the observable part of our Universe. But I don't see how this can be so." Roger pulled out a pen and began drawing a spacetime diagram on a napkin. "This diagram makes me think that the galaxy will become dimmer and dimmer, redder and redder, but will never quite disappear. What am I doing wrong?"

Sciama was taken aback. Never had he seen such power in a spacetime diagram. Penrose was right; Hoyle had to be wrong. More important, Oliver's little brother was phenomenal.

Thereupon Dennis Sciama began with Roger Penrose the pattern he would continue with his own students in the 1960s (Stephen Hawking, George Ellis, Brandon Carter, Martin Rees, and others; see Chapter 7). He pulled Penrose into long discussions, sessions of many hours' length, about the exciting things happening in physics. Sciama knew everything that was going on; he infused Penrose with his enthusiasm, with the excitement of it all. Soon Penrose was hooked. He would complete his Ph.D. in mathematics, but the quest to understand the Universe henceforth would drive him forward. He would spend the coming decades with one foot firmly planted in mathematics, the other in physics.

Roger Penrose, ca. 1964. [Photo by Godfrey Argent for the National Portrait Gallery of Britain and the Royal Society of London; courtesy Godfrey Argent.]

New ideas often arrive at the oddest moments, at moments when one is least expecting them. I suppose this is because they come from one's subconscious mind, and the subconscious performs most effectively when the conscious part of the mind is not in high gear. A good example was Stephen Hawking's 1970 discovery, as he was getting ready for bed, that the areas of black-hole horizons must always increase (Chapter 12). Another example is a discovery by Roger Penrose that changed our understanding of what is inside a black hole.

One day in the late autumn of 1964, Penrose, by then a professor at Birkbeck College in London, was walking toward his office with a friend, Ivor Robinson. For the past year, ever since quasars were discovered and astronomers began speculating that they are powered by stellar implosion (Chapter 9), Penrose had been trying to figure out whether singularities are created by realistic, randomly deformed, imploding stars. As he walked and talked with Robinson, his subconscious was mulling over the pieces of this puzzle—pieces with which his conscious mind had struggled for many many hours.

As Penrose recalls it, "My conversation with Robinson stopped momentarily as we crossed a side road, and resumed again at the other side. Evidently, during those few moments an idea occurred to me, but then the ensuing conversation blotted it from my mind! Later in the day, after Robinson had left, I returned to my office. I remember having an odd feeling of elation that I could not account for. I began going through in my mind all the various things that had happened to me during the day, in an attempt to find what it was that had caused this elation. After eliminating numerous inadequate possibilities, I finally brought to mind the thought that I had had while crossing the street."

The thought was beautiful, unlike anything ever seen before in relativity physics. Carefully over the next few weeks Penrose manipulated it, looking at it from this direction and then from that, working through the details, making it as concrete and mathematically precise as he could. With all details in hand, he wrote a short article for publication in the journal *Physical Review Letters*, describing the issue of singularities in stellar implosion, and then proving a mathematical theorem.

Penrose's theorem said *roughly* this: Suppose that a star—any kind of star whatsoever—implodes so far that its gravity becomes strong enough to form an *apparent horizon*, that is, strong enough to pull outgoing light rays back inward (Box 12.1). After this happens, nothing can prevent the gravity from growing so strong that it creates a singularity. Consequently (since black holes always have apparent horizons), *every black hole must have a singularity inside itself.*

The most amazing thing about this *singularity theorem* was its sweeping power. It dealt not solely with idealized imploding stars that have special, idealized properties (such as being precisely spherical or having no pressure); and it dealt not solely with stars whose initial random deformations are tiny. Instead, it dealt with every imploding star imaginable, and thus, undoubtedly, with the real imploding stars that inhabit our real Universe.

Penrose's singularity theorem acquired its amazing power from a new mathematical tool that he used in its proof, a tool that no physicist had ever before used in calculations about curved spacetime, that is, in general relativistic calculations: *topology.*

Topology is a branch of mathematics that deals with the qualitative ways in which things are connected to each other or to themselves. For example, a coffee cup and a doughnut "have the same topology" because (if they are both made from putty) we can smoothly and continuously deform one into the other without tearing it, that is, without changing any connections (Figure 13.5a). By contrast, a sphere has a different topology from a doughnut; to deform a sphere into a doughnut, we must tear a hole in it, thereby changing how it is connected to itself (Figure 13.5b).

Topology cares *only* about connections, and *not* about shapes or sizes or curvatures. For example, the doughnut and the coffee cup have very different shapes and curvatures, but they have the same topology.

We physicists, before Penrose's singularity theorem, ignored topology because we were fixated on the fact that spacetime *curvature* is the central concept of general relativity, and topology cannot tell us anything about curvature. (Indeed, because Penrose's theorem was based so strongly on topology, it told us nothing about the singularity's curvature, that is, nothing about the details of its tidal gravity. The theorem simply told us that somewhere inside the black hole, spacetime comes to an end, and anything that reaches that end gets destroyed. *How* it gets destroyed was the province of curvature; *that* it gets destroyed—that there is an end to spacetime—was the province of topology.)

If we physicists, before Penrose, had only looked beyond the issue of curvature, we would have realized that relativity *does* deal with ques-

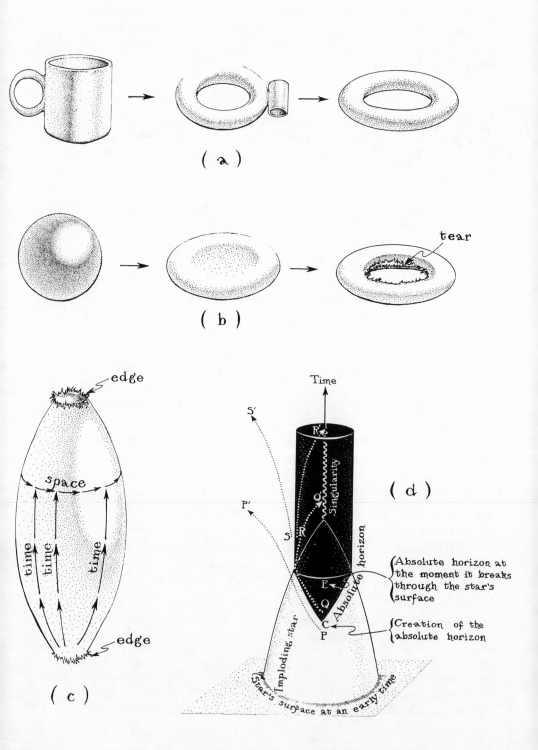

(a)

(b)

tear

edge

space

time   time   time

edge

(c)

Time

S'

P'

R'

Singularity

Q

S R

Absolute horizon

E

Q

C

Absolute horizon

P

Imploding star

Absolute horizon at
the moment it breaks
through the star's
surface

Creation of the
absolute horizon

Star's surface at an early time

(d)

tions of topology, questions such as "Does spacetime come to an end (does it have an edge beyond which spacetime ceases to exist)?" (Figure 13.5c) and "Which regions of spacetime can send signals to each other, and which cannot?" (Figure 13.5d). The first of these topological questions is central to singularities; the second is central to the formation and existence of black holes and also to *cosmology* (to the large-scale structure and evolution of the Universe).

These topological issues are so important, and the mathematical tools of topology are so powerful in dealing with them, that by introducing us to topology, Penrose triggered a revolution in our research.

Taking off from Penrose's seminal ideas, during the middle and late 1960s Penrose, Hawking, Robert Geroch, George Ellis, and other physicists created a powerful set of combined topological and geometrical tools for general relativity calculations, tools that are now called *global methods.* Using these methods, Hawking and Penrose in 1970 proved—without any idealizing assumptions—that our Universe must have had a spacetime singularity at the beginning of its big bang expansion, and if it one day recollapses, it must produce a singularity in its big crunch. And using these global methods, Hawking in 1970 invented the concept of a black hole's absolute horizon and proved that the surface areas of absolute horizons always increase (Chapter 12).

Let us return, now, to 1965. The stage was set for a momentous confrontation. Isaac Khalatnikov and Evgeny Lifshitz in Moscow had proved (or so they thought) that when a real star, with random internal deformations, implodes to form a black hole it *cannot* create a singularity at the hole's center, while Roger Penrose in England had proved that every black hole *must* have a singularity at its center.

13.5 All of the following issues deal with the nature of the connections between points; that is, they are topological issues. (a) A coffee cup (left) and a doughnut (right) can be deformed into each other smoothly and continuously without tearing, in other words, without changing the qualitative nature of any of the connections between points. They thus have the same topology. (b) To deform a sphere (left) into a doughnut (right), one must tear a hole in it. (c) The spacetime shown here has two sharp edges [analogous to the tear in (b)]: one edge at which time begins (analogous to the big bang beginning of our Universe), and one at which time ends (analogous to the big crunch). One can also conceive of a universe that has existed for all time and will always continue to exist; such a universe's spacetime would have no edges. (d) The blackened region of spacetime is the interior of a black hole; the white region is the exterior (see Box 12.1). Points in the interior cannot send any signals to points in the exterior.

The lecture hall seated 250 and was filled to overflowing as Isaac Khalatnikov rose to speak. It was a warm summer day in 1965, and the world's leading relativity researchers had gathered in London for the Third International Conference on General Relativity and Gravitation. This was the first opportunity, at such a worldwide gathering, for Isaac Khalatnikov and Evgeny Lifshitz to present the details of their proof that black holes do not contain singularities.

Permission to travel beyond the iron curtain was granted and withdrawn with relative capriciousness in the Soviet Union during the decades between Stalin's death and the Gorbachev era. Lifshitz, though Jewish, had traveled rather freely in the late 1950s, but he was now on a travel blacklist and would remain so until 1976. Khalatnikov had two strikes against him; he was Jewish, and he had never yet traveled abroad. (Permission for one's first trip was exceedingly difficult to win.) However, after a vigorous struggle, including a telephone call in his behalf from the vice-president of the Academy of Sciences, Nikolai Nikolaievich Semenov, to the Central Committee of the Communist party, Khalatnikov had finally won permission to come to London.

As he spoke in the packed London lecture hall, dragging a microphone with him, Khalatnikov wrote equations all over the blackboard, which extended the entire 50-foot width of the room. His were not topological methods; they were the standard, equation-intensive methods that physicists had used for decades when analyzing spacetime curvature. Khalatnikov demonstrated mathematically that random perturbations must grow as a star implodes. This meant, he asserted, that if the implosion is to form a singularity, it must be one with completely random deformations in its spacetime curvature. He then described how he and Lifshitz had searched, among all types of singularities permitted by the laws of general relativity, for one with completely random curvature deformations. He exhibited, mathematically, one type of singularity after another; he cataloged the types of singularities almost *ad nauseum.* Among them, none had completely random deformations. Therefore, he concluded—bringing his forty-minute lecture to a close—an imploding star with random perturbations cannot produce a singularity. The perturbations must save the star from destruction.

As the applause ended, Charles Misner, one of Wheeler's most brilliant former students, leaped up and objected strenuously. Excitedly, vigorously, and in rapid-fire English, Misner described the theorem that Penrose had proved a few months earlier. If Penrose's theorem was right, then Khalatnikov and Lifshitz must be wrong.

The Soviet delegation was confused and incensed. Misner's English was too fast to follow, and since Penrose's theorem relied on topological arguments that were alien to relativity experts, the Soviets regarded it as suspect. By contrast, the Khalatnikov–Lifshitz analysis was based on tried-and-true methods. Penrose, they asserted, was probably wrong.

During the next few years, relativity experts in East and West plumbed the depths of Penrose's analysis, and of the Khalatnikov–Lifshitz analysis. At first both analyses looked suspect; both had dangerous, potential flaws. Gradually, however, as the experts began to master and extend Penrose's topological techniques, they became convinced that Penrose was right.

In September 1969, while I was a visiting member of Zel'dovich's research team in Moscow, Evgeny Lifshitz came to me with a manu-

A dinner party in the apartment of Isaac Khalatnikov in Moscow, June 1971. Clockwise from left: Kip Thorne, John Wheeler, Isaac Khalatnikov, Evgeny Lifshitz, Khalatnikov's wife Valentina Nikolaievna, Vladimir Belinsky, and Khalatnikov's daughter Eleanora. [Courtesy Charles W. Misner.]

script that he and Khalatnikov had just written. "Please, Kip, take this manuscript back to America for me and submit it to *Physical Review Letters,*" he requested. He explained that any manuscript written in the U.S.S.R., regardless of its content, was automatically classified secret until declassified, and declassification would take three months. The ludicrous Soviet system permitted me or any other foreign visitor to read the manuscript while in Moscow, but the manuscript should not itself leave the country until passed by the censors. This manuscript was too precious, too urgent for such a ridiculous delay. It contained, Lifshitz explained to me, their capitulation, their confession of error: Penrose was right; they were wrong. In 1961 they had been unable to find, among the solutions to Einstein's field equation, any singularity with completely random deformations; but now, spurred by Penrose's theorem, they and a graduate student, Vladimir Belinsky, had managed to find one. This new singularity, they thought, must be the one that terminates the implosion of randomly deformed stars and that might someday destroy our Universe at the end of the big crunch. [And, indeed, in 1993 I think they probably were right. To this 1993 viewpoint, and to the nature of their new *BKL* ("Belinsky–Khalatnikov–Lifshitz") singularity, I shall return near the end of this chapter.]

For a theoretical physicist it is more than embarrassing to admit a major error in a published result. It is ego shattering. I should know. In 1966 I miscalculated the pulsations of white-dwarf stars, and two years later my wrong calculations briefly misled astronomers into thinking that the newly discovered pulsars might be pulsating white dwarfs. My error, when found, was significant enough to figure in an editorial in the British journal *Nature.* It was a bitter pill to swallow.

Though errors like this can be shattering for an American or European physicist, in the Soviet Union they were far worse. One's position in the pecking order of scientists was especially important in the Soviet Union; it determined such things as possibilities for travel abroad and election to the Academy of Sciences, which in turn brought privileges such as a near doubling of one's salary and a chauffeured limousine at one's beck and call. Thus it was that the temptation to try to hide or downplay mistakes, when mistakes occur, was greater for Soviet scientists than for Westerners. And thus it was that Lifshitz's plea for help was impressive. He wanted no delay in disseminating the truth, and his manuscript was forthright: It confessed the error and announced that future editions of *The Classical Theory of Fields* (the Landau–Lifshitz textbook on general relativity) would be modified to remove the claim that implosion does not produce singularities.

I carried the manuscript to America, hidden among my personal papers, and it was published. The Soviet authorities never noticed.

Why was it a British physicist (Penrose) and not an American or French or Soviet physicist who introduced topological methods into relativity research? And why was it that throughout the 1960s, topological methods were pursued with vigor and success by other British relativity physicists, but took hold much more slowly in America, France, the U.S.S.R., and elsewhere?

The reason, I suspect, was the undergraduate training of British theoretical physicists. They typically major in mathematics as undergraduates, then do Ph.D. research in departments of applied mathematics or departments of applied mathematics and theoretical physics. In America, by contrast, aspiring theoretical physicists typically major in physics as undergraduates, and then do Ph.D. research in physics departments. Thus, young British theoretical physicists are well versed in esoteric branches of mathematics which have not yet seen much physics application, but they may have a weak background in "gutsy" physics topics such as the behaviors of molecules, atoms, and atomic nuclei. By contrast, young American theoretical physicists know little mathematics beyond what their physics professors have taught them, but are deeply versed in the lore of molecules, atoms, and nuclei.

To a great extent, we Americans have dominated theoretical physics since World War II, and we have foisted on the world's physics community our scandalously low mathematical standards. Most of us use the mathematics of fifty years ago and are incapable of communicating with modern mathematicians. With our poor mathematical training, it was difficult for us Americans to absorb and start using the topological methods when Penrose introduced them.

French theoretical physicists, even more than the British, are well trained in mathematics. However, during the 1960s and 1970s French relativity theorists were so wrapped up in mathematical rigor (that is, perfection), and so deemphasized physical intuition, that they contributed little to our understanding of imploding stars and black holes. Their quest for rigor slowed them down to the point that, although they knew well the mathematics of topology, they could not compete with the British. They didn't even try; their attention was riveted elsewhere.

Lev Davidovich Landau, who was largely responsible for the strength of Soviet theoretical physics in the 1930s through 1960s, was also a source of Soviet resistance to topology: Landau had transfused theoretical physics from Western Europe to the U.S.S.R. in the 1930s (Chapter 5). As one tool in that transfusion, he had created a set of examinations on theoretical physics, called the "Theoretical Minimum," which he required be passed as an entree into his own research group. Anyone, regardless of educational background, could walk in off the street and take these examinations, but few could pass them. In the twenty-nine years of the Theoretical Minimum (1933–62) only forty-three passed, but a remarkable portion of those forty-three went on to make great physics discoveries.

Evgeny Michailovich Lifshitz *(left)* and Lev Davidovich Landau *(right)* in Landau's room in his flat at the Institute for Physical Problems, No. 2 Vorobyevskoye Shosse, Moscow, in 1954. [Courtesy Lifshitz's wife, Zinaida Ivanovna Lifshitz.]

Landau's Theoretical Minimum had included problems from *all* the branches of mathematics that Landau deemed important for theoretical physics. Topology was not among them. Calculus, complex variables, the qualitative theory of differential equations, group theory, and differential geometry were all covered; they would all be needed in a physicist's career. But topology would not be needed. Landau had nothing against topology; he just ignored it; it was irrelevant—and his view of its irrelevance became near gospel among most Soviet theoretical physicists in the 1940s through the 1960s.

This view was transmitted to theoretical physicists around the world by the set of textbooks, called *Course of Theoretical Physics,* that Landau and Lifshitz wrote. These became, worldwide, the most influential set of physics texts of the twentieth century, and like Landau's Theoretical Minimum examinations, they ignored topology.

Curiously, topological techniques were introduced into relativity research in an abortive way, long before Penrose's theorem, by two Soviet mathematicians in Leningrad: Aleksander Danilovich Aleksandrov and Revol't Ivanovich Pimenov. In 1950–59, Aleksandrov used topology to probe the "causal structure" of spacetime, that is, to study the relationships between regions of spacetime that can communicate with each other and those that cannot. This was just the type of topological analysis that would ultimately pay rich dividends in the theory of black holes. Aleksandrov built up a rather powerful and beautiful topological formalism, and in the mid-1950s that formalism was picked up and pushed further by Pimenov, a young colleague of Aleksandrov's.

But in the end this research led nowhere. Aleksandrov and Pimenov had little contact with physicists who specialize in gravitation. Such physicists would have known what kinds of calculations were useful and what were not. They might have told Aleksandrov and Pimenov that the big bang singularity or gravitational implosion of stars deserved probing with their formalism. But no such advice was to be had in Leningrad; the key physicists worked 600 kilometers southeast of Leningrad, in Moscow, and were ignorant of topology and topologists. The Aleksandrov–Pimenov formalism flowered, and then went dormant.

Its dormancy was forced by the fates of Aleksandrov and Pimenov: Aleksandrov became the rector (president) of Leningrad University, and had inadequate time for further research. Pimenov was arrested in 1957 for founding "an anti-Soviet group," was imprisoned for six years, and then after seven years of freedom was rearrested and sent into five years' exile in the Komi Republic, 1200 kilometers east of Leningrad.

I have never met Aleksandrov or Pimenov, but tales of Pimenov were still rippling through Leningrad's community of scientists when I visited there in 1971, a year after Pimenov's second arrest. Rumor had it that Pimenov viewed the Soviet government as morally corrupt, and, like many young people in America during the Vietnam War, he felt that, if he cooperated with the government, the government's corruption would rub off on him. The only way to feel morally clean was through civil disobedience. In America, civil disobedience meant refusing to register for the draft. For Pimenov, civil disobedience meant *samizdat.* Samizdat was the "self-publication" of forbidden manuscripts. Pimenov, it was rumored, would receive from friends a manuscript which had been forbidden for publication in the Soviet Union, he would type out a half-dozen copies using carbon paper, and he would then pass those copies on to other friends, who would repeat the process. Pimenov got caught, was convicted, and was sentenced to five years' exile in the Komi Republic, where he worked as a tree-feller and an electrician in a sawmill until the Komi Academy of Sciences took advantage of his exile and made him the chair of their mathematics department.

Finally able to do mathematics again, Pimenov continued his topological studies of spacetime. By then topology had taken firm root as a key tool for physicists' gravitation research, but Pimenov remained isolated from the leading physicists of his country. He never had the impact that, under other circumstances, he might have.

Roger Penrose, by contrast with Aleksandrov and Pimenov, lives with one foot firmly planted in the mathematics community and the other firmly planted in physics, and this has been a major source of his success.

Best Guesses

One might have thought that Penrose's singularity theorem would settle once and for all the question of what is inside a black hole. Not so. Instead it opened up a new set of questions—questions with which physicists have struggled, with only modest success, since the mid-1960s. Those questions, and our best 1993 answers (our "best guesses" is a better way to say it), are:

1. Does everything that enters the hole necessarily get swallowed by the singularity? We think so, but we're not sure.
2. Is there any route from inside the hole to another universe, or to another part of our own Universe? Very probably not, but we're not absolutely sure.
3. What is the fate of things that fall into the singularity? We think that things that fall in when the hole is quite young get torn apart by tidal gravity in a violent, chaotic way, before quantum gravity becomes important. However, things that fall into an old hole might survive unscathed until they come face-to-face with the laws of quantum gravity.

In the remainder of this chapter I shall explain these answers in more detail.

Recall that Oppenheimer and Snyder gave us a clear and unequivocal answer to our three questions: When the black hole is created by a highly idealized, spherical, imploding star, then (1) everything that enters the hole gets swallowed by the singularity; (2) nothing travels to another universe or another part of our Universe; (3) when nearing the singularity, everything experiences an infinitely growing radial stretch and transverse squeeze (Figure 13.1 above), and thereby gets destroyed.

This answer was pedagogically useful; it helped motivate calculations that brought deeper understanding. However, the deeper understanding (due to Khalatnikov and Lifshitz) showed that the Oppenheimer–Snyder answer is irrelevant to the real Universe in which we live, because the random deformations that occur in all real stars will completely change the hole's interior. The Oppenheimer–Snyder interior is "unstable against small perturbations."

The Reissner–Nordström type of solution to the Einstein field equation also gave a clear and unequivocal answer: When the black hole is created by a particular, highly idealized, spherical, electrically charged star, then the imploding star and other things that fall into the hole can travel, via a "little closed universe," from the hole's interior to another large universe (Figure 13.4).

This answer was also pedagogically useful (and has provided grist for the mills of many a science fiction writer). However, like the Oppenheimer–Snyder prediction, it has nothing to do with the real Universe in which we live because it is unstable against small perturbations. More specifically, in our real Universe, the black hole is continually bombarded by tiny electromagnetic vacuum fluctuations and by tiny amounts of radiation. As these fluctuations and radiation fall into the hole, the hole's gravity accelerates them to enormous energy, and they then explosively hit and destroy the little closed universe, just before the little universe begins its trip. This was conjectured by Penrose in 1968, and has since been verified in many different calculations, carried out by many different physicists.

Belinsky, Khalatnikov, and Lifshitz have given us yet another answer to our questions, and this one, being totally stable against small perturbations, is probably the "right" answer, the answer that applies to the real black holes that inhabit our Universe: *The star that forms the hole and everything that falls into the hole when the hole is young get torn apart by the tidal gravity of a BKL singularity.* (This is the kind of singularity that Belinsky, Khalatnikov, and Lifshitz discovered, as a solution of Einstein's equation, after Penrose convinced them that singularities must inhabit black holes.)

The tidal gravity of a BKL singularity is radically different from that of the Oppenheimer–Snyder singularity. The Oppenheimer–Snyder singularity stretches and squeezes an infalling astronaut (or anything else) in a steady but mounting way; the stretch is always radial, the squeeze is always transverse, and the strengths of stretch and squeeze grow steadily and smoothly (Figure 13.1). The BKL singularity, by contrast, is somewhat like the taffy-pulling machines that one sometimes sees in candy stores or at carnivals. It stretches and squeezes first in this direction, then that, then another, then another, and yet another. The stretch and squeeze oscillate with time in a random and chaotic way (as measured by the infalling astronaut), but on average they get stronger and stronger, and their oscillations get faster and faster as the astronaut gets closer and closer to the singularity. Charles Misner (who discovered this type of chaotically oscillating singularity independently of Belinsky, Khalatnikov, and Lifshitz) has called this a *mixmaster oscillation* because one can imagine it mixing up the astronaut's body parts in the way that a mixmaster or eggbeater mixes up the yolk and white of an egg. Figure 13.6 depicts a specific example of how the tidal forces might oscillate, but the precise sequence of oscillations is chaotically unpredictable.

In Misner's version of the mixmaster singularity, the oscillations were the same everywhere in space, at a particular moment of time (as measured, say, by the astronaut). Not so for the BKL singularity. Its

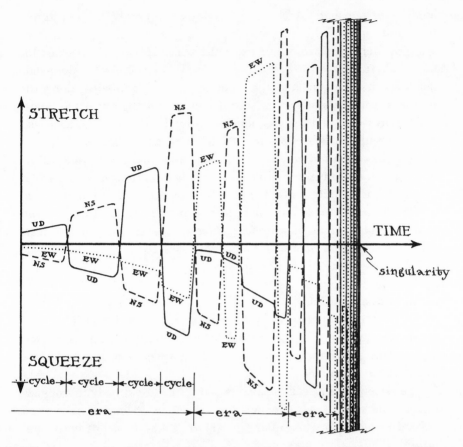

13.6 An example of how the tidal forces might oscillate with time in a BKL singularity. The tidal forces act in different manners along three different, perpendicular directions. These directions, for definiteness, are here called UD (for "up/down"), NS (for "north/south"), and EW (for "east/west"), and each of the three curves describes the behavior of the tidal force along one of these directions. Time is plotted horizontally. At any time when the UD curve is *above* the horizontal time axis, the tidal force is *stretching* along the UD direction, while at a time when the UD curve is *below* the axis, the UD tidal force is *squeezing*. The higher the curve above the axis, the stronger the stretch; the lower the curve below the axis, the stronger the squeeze. Notice the following: (i) At any moment of time there is a squeeze along two directions and a stretch along one. (ii) The tidal forces oscillate between stretch and squeeze; each oscillation is called a "cycle." (iii) The cycles are collected into "eras." During each era, one of the three directions is subjected to a fairly steady squeeze, while the other two oscillate between stretch and squeeze. (iv) When the era changes, there is a change of the steady direction. (v) As the singularity is approached, the oscillations become infinitely rapid and the tidal forces become infinitely strong. The details of the division of cycles into eras and the change of oscillation patterns at the beginning of each era are governed by what is sometimes called a "chaotic map."

oscillations are spatially chaotic as well as temporally chaotic, just as turbulent motions of the froth in a breaking ocean wave are chaotic in space as well as in time. For example, while the astronaut's head is being alternately stretched and squeezed ("pummeled") along the north/south direction, his right foot might be pummeled along the northeast/southwest direction, and his left foot along south-southeast/north-northwest; and the frequencies of oscillation of the pummeling might be quite different on his head, his left foot, and his right foot.

Einstein's equation predicts that, as the astronaut reaches the singularity, the tidal forces grow infinitely strong, and their chaotic oscillations become infinitely rapid. The astronaut dies and the atoms from which his body is made become infinitely and chaotically distorted and mixed—and then, at the moment when everything becomes infinite (the tidal strengths, the oscillation frequencies, the distortions, and the mixing), spacetime ceases to exist.

The laws of quantum mechanics object. They forbid the infinities. Very near the singularity, as best we understand it in 1993, the laws of quantum mechanics merge with Einstein's general relativistic laws and completely change the "rules of the game." The new rules are called *quantum gravity*.

The astronaut is already dead, his body parts are already thoroughly mixed, and the atoms of which he was made are already distorted beyond recognition when quantum gravity takes over. But nothing is infinite. The "game" goes on.

Just when does quantum gravity take over, and what does it do? As best we understand it in 1993 (and our understanding is rather poor), quantum gravity takes over when the oscillating tidal gravity (spacetime curvature) becomes so large that it completely deforms all objects in about 10^{-43} second or less.[2] Quantum gravity then radically changes the character of spacetime: It ruptures the unification of space and time into spacetime. It unglues space and time from each other, and then destroys time as a concept and destroys the definiteness of space. Time ceases to exist; no longer can we say that "this thing happens before that one," because without time, there is no concept of "before" or

---

2. 10^{-43} second is the *Planck–Wheeler time*. It is given (approximately) by the formula $\sqrt{G\hbar/c^5}$, where $G = 6.670 \times 10^{-8}$ dyne-centimeter^2/gram^2 is Newton's gravitation constant, $\hbar = 1.055 \times 10^{-27}$ erg-second is Planck's quantum mechanical constant, and $c = 2.998 \times 10^{10}$ centimeter/second is the speed of light. Note that the Planck–Wheeler time is equal to the square root of the Planck–Wheeler area (Chapter 12) divided by the speed of light.

"after." Space, the sole remaining remnant of what was once a unified spacetime, becomes a random, probabilistic froth, like soapsuds.

Before its rupture (that is, outside the singularity), spacetime is like a piece of wood impregnated with water. In this analogy, the wood represents space, the water represents time, and the two (wood and water; space and time) are tightly interwoven, unified. The singularity and the laws of quantum gravity that rule it are like a fire into which the water-impregnated wood is thrown. The fire boils the water out of the wood, leaving the wood alone and vulnerable; in the singularity, the laws of quantum gravity destroy time, leaving space alone and vulnerable. The fire then converts the wood into a froth of flakes and ashes; the laws of quantum gravity then convert space into a random, probabilistic froth.

This random, probabilistic froth is the thing of which the singularity is made, and the froth is governed by the laws of quantum gravity. In the froth, space does not have any definite shape (that is, any definite curvature, or even any definite topology). Instead, space has various probabilities for this, that, or another curvature and topology. For example, inside the singularity there might be a 0.1 percent probability for the curvature and topology of space to have the form shown in Figure 13.7a, and a 0.4 percent probability for the form in Figure 13.7b, and a 0.02 percent probability for the form in Figure 13.7c, and so on. This does *not* mean that space spends 0.1 percent of its *time* in the form (a), 0.4 percent of its *time* in the form (b), and 0.02 percent of its *time* in the form (c), because *there is no such thing as time inside the singularity*. And similarly, because there is no time, it is totally meaningless to ask whether space assumes the form (b) "before" or "after" it assumes the form (c). The only meaningful question one can ask of the singularity is, "What are the probabilities that the space of which you are made has the forms (a), (b), and (c)?" And the answers will be simply 0.1, 0.4, and 0.02 percent.

Because all conceivable curvatures and topologies are permitted inside the singularity, no matter how wild, one says that the singularity is made from a probabilistic foam. John Wheeler, who first argued that this must be the nature of space when the laws of quantum gravity hold sway, has called it *quantum foam*.

To recapitulate, at the center of a black hole, in the spacetime region where the oscillating BKL tidal forces reach their peak, there resides a singularity: a region in which time no longer exists, and space has given way to quantum foam.

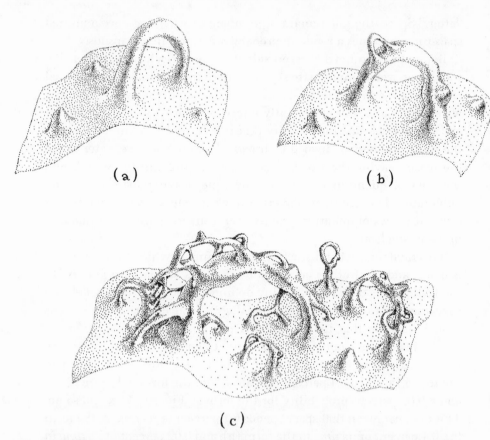

(a)   (b)

(c)

13.7 Embedding diagrams illustrating the quantum foam that is thought to
reside in the singularity inside a black hole. The geometry and topology of space
are not definite; instead, they are probabilistic. They might have, for example, a
0.1 percent probability for the form shown in (a), a 0.4 percent probability for (b),
a 0.02 percent probability for (c), and so on.

One task of the laws of quantum gravity is to govern the probabilities for the various curvatures and topologies within a black hole's
singularity. Another, presumably, is to determine the probabilities for
the singularity to give birth to "new universes," that is, to give birth to
new, classical (non-quantum) regions of spacetime, in the same sense
as the big bang singularity gave birth to our Universe some 15 billion
years ago.

How probable is it that a black hole's singularity will give birth to
"new universes"? We don't know. It might well never happen, or it
might be quite common—or we might be on completely the wrong
track in believing that singularities are made of quantum foam.

Clear answers might come in the next decade or two from research now being carried out by Stephen Hawking, James Hartle, and others, building on foundations laid by John Wheeler and Bryce DeWitt.[3]

Most everything in the Universe changes with age: Stars consume their fuel and die; the Earth gradually loses its atmosphere by evaporation into space and ultimately will become an airless, dead planet; and we humans grow wrinkled and wise.

The tidal forces deep inside a black hole, near its singularity, are no exception. They, too, must change with age, according to calculations done in 1991 by Werner Israel and Eric Poisson of the University of Alberta, and Amos Ori, a postdoc in my Caltech group (building on earlier work of Andrei Doroshkevich and Igor Novikov). When the hole is newborn, its interior tidal forces exhibit violent, chaotic, BKL-type oscillations (Figure 13.6 above). However, as the hole ages, the chaotic oscillations become tamer and gentler, and gradually disappear.

For example, an astronaut who falls into a 10-billion-solar-mass hole in the core of a quasar within the first few hours after the hole is born will be torn apart by wildly oscillating BKL tidal forces. However, a second astronaut, who waits until a day or two after the hole is born before plunging inside, will encounter much more gently oscillating tidal forces. The tidal stretch and squeeze are still large enough to kill the second astronaut, but being more gentle than the day before, the oscillating stretch and squeeze will allow the second astronaut to survive longer, and approach closer to the singularity before he dies, than did the first astronaut. A third astronaut, who waits until the hole is many years old before taking the plunge, will face an even gentler fate. The tidal forces surrounding the singularity have now become so tame and meek, according to Israel's, Poisson's, and Ori's calculations, that the astronaut will hardly feel them at all. He will survive, almost unscathed, right up to the edge of the probabilistic quantum gravity singularity. Only at the singularity's edge, just as he comes face-to-face with the laws of quantum gravity, will the astronaut be killed—and we cannot even be absolutely sure he gets killed then, since we do not really understand at all well the laws of quantum gravity and their consequences.

3. The above description is based on the Wheeler–DeWitt, Hawking–Hartle approach to formulating the laws of quantum gravity. Although theirs is but one of many approaches now being pursued, it is one to which I would give good odds of success.

This aging of a black hole's internal tidal forces is not inexorable. Whenever matter and radiation (or astronauts) fall into the hole, they will feed and energize the tidal forces, much like a hunk of meat thrown to a lion energizes him. The oscillatory stretch and squeeze near the singularity, having been fed, will grow stronger for a short while, and then will die out and become quiescent once again.

In the late 1950s and early 1960s John Wheeler had a dream, a hope, that we humans might one day be able to probe into a singularity and there see quantum gravity at work—that we might probe not only with mathematics and computer simulations, but also with real, physical observations and experiments. Oppenheimer and Snyder dashed that hope (Chapter 6). The horizon that they discovered forming around an imploding star hides the singularity from external view. If we remain forever outside the horizon, there is no way that we can probe the singularity. And if we plunge through the horizon of a huge old hole, and survive to meet the quantum gravity singularity face-to-face, there is no way we can transmit a description of our meeting back to Earth. Our transmission cannot escape from the hole; the horizon hides it.

Though Wheeler has long since renounced his dream and now vigorously champions the view that it is impossible to probe singularities, it is not at all certain that he is correct. It is conceivable that some extremely nonspherical stellar implosions produce *naked singularities*, that is, singularities that are not surrounded by horizons and that therefore can be observed and probed from the external Universe, even from Earth.

In the late 1960s, Roger Penrose searched hard, mathematically, for an example of an implosion that creates a naked singularity. His search came up empty. Whenever, in his equations, an implosion created a singularity, it also created a horizon around the singularity. Penrose was not surprised. After all, if a naked singularity were to form, then it seems reasonable to expect that, just before the singularity forms, light can escape from its vicinity; and if light can escape, then (it would seem) so can the material that is imploding to create the singularity; and if the imploding material can escape, then presumably the material's huge internal pressure will make it escape, thereby reversing the implosion and preventing the singularity from forming in the first place. So it seemed. However, neither Penrose's mathematical manipulations nor anybody else's were powerful enough to say for sure.

In 1969 Penrose, strongly convinced that naked singularities cannot form, but unable to prove it, proposed a conjecture, *the conjecture of cosmic censorship: No imploding object can ever form a naked singularity; if a singularity is formed, it must be clothed in a horizon so that we in the external Universe cannot see it.*

Members of the physics "establishment"—physicists like John Wheeler, whose viewpoints are the most influential—have embraced cosmic censorship and espouse it as almost surely correct. Nevertheless, nearly a quarter century after Penrose proposed it, cosmic censorship remains unproved; and recent computer simulations of the implosion of highly nonspherical stars suggest that it *might* even be wrong. Some implosions, according to these simulations by Stuart Shapiro and Saul Teukolsky of Cornell University, might actually create naked singularities. Might. Not will; just might.

Stephen Hawking is the epitome of the establishment these days, and John Preskill (a colleague of mine at Caltech) and I enjoy tweaking the establishment a bit. Therefore, in 1991 Preskill and I made a bet with Hawking (Figure 13.8). We bet that cosmic censorship is wrong; naked singularities *can* form in our Universe. Hawking bet that cosmic censorship is right; naked singularities can never form.

13.8 Bet between Stephen Hawking, John Preskill, and me on the correctness of Penrose's cosmic censorship conjecture.

Whereas Stephen W. Hawking firmly believes that naked singularities are an anathema and should be prohibited by the laws of classical physics,

And whereas John Preskill and Kip Thorne regard naked singularities as quantum gravitational objects that might exist unclothed by horizons, for all the Universe to see,

Therefore Hawking offers, and Preskill/Thorne accept, a wager with odds of 100 pounds stirling to 50 pounds stirling, that when any form of classical matter or field that is incapable of becoming singular in flat spacetime is coupled to general relativity via the classical Einstein equations, the result can never be a naked singularity.

The loser will reward the winner with clothing to cover the winner's nakedness. The clothing is to be embroidered with a suitable concessionary message.

Stephen W. Hawking   John P. Preskill & Kip S. Thorne
Pasadena, California, 24 September 1991

Just four months after agreeing to the bet, Hawking himself discovered mathematical evidence (but *not a firm proof*) that, when a black hole completes its evaporation (Chapter 12), it might not disappear entirely as he had previously expected, but instead it might leave behind a tiny naked singularity. Hawking announced this result to Preskill and me privately, a few days after he discovered it, at a dinner party at Preskill's home. However, when Preskill and I then pressed him to concede our bet, he refused on grounds of a technicality. The wording of our bet was very clear, he insisted: The bet was restricted to naked singularities whose formation is governed by the laws of classical (that is, not quantum) physics, including the laws of general relativity. However, the evaporation of black holes is a quantum mechanical phenomenon and is governed not by the laws of classical general relativity, but rather by the laws of quantum fields in curved spacetime, so any naked singularity that might result from black-hole evaporation is outside the realm of our bet, Hawking insisted (correctly). Nevertheless, a naked singularity, however it forms, would surely be a blow to the establishment!

Though we enjoy our bets, the issues we argue are deeply serious. If naked singularities can exist, then only the ill-understood laws of quantum gravity can tell us how they behave, what they might do to spacetime in their vicinities, and whether their actions can have a large effect on the Universe in which we live, or only a small one. Because naked singularities, if they can exist, might strongly influence our Universe, we want very much to understand whether cosmic censorship is correct, and what the laws of quantum gravity predict for the behaviors of singularities. The struggle to find out will not be quick or easy.

14

Wormholes
and Time Machines[1]

*in which the author seeks insight
into physical laws by asking:
can highly advanced civilizations
build wormholes through hyperspace
for rapid interstellar travel
and machines for traveling backward in time?*

Wormholes and Exotic Material

I had just taught my last class of the 1984–85 academic year and was sinking into my office chair to let the adrenaline subside, when the telephone rang. It was Carl Sagan, the Cornell University astrophysicist and a personal friend from way back. "Sorry to bother you, Kip," he said. "But I'm just finishing a novel about the human race's first contact with an extraterrestrial civilization, and I'm worried. I want the science to be as accurate as possible, and I'm afraid I may have got some of the gravitational physics wrong. Would you look at it and give me advice?" Of course I would. It would be interesting, since Carl is a clever guy. It might even be fun. Besides, how could I turn down this kind of request from a friend?

The novel arrived a couple of weeks later, a three-and-a-half-inch-thick stack of double-spaced typescript.

1. I have chosen to write this chapter solely from my own personal viewpoint. It therefore is much less objective than the rest of the book, and represents other people's research much less fairly and less completely than it does my own.

I slipped the stack into an overnight bag and threw the bag into the back seat of Linda's Bronco, when she picked me up for the long drive from Pasadena to Santa Cruz. Linda is my ex-wife; she, I, and our son Bret were on our way to see our daughter Kares graduate from college.

As Linda and Bret took turns driving, I read and thought. (Linda and Bret were accustomed to such introversion; they had lived with me for many years.) The novel was fun, but Carl, indeed, was in trouble. He had his heroine, Eleanor Arroway, plunge into a black hole near Earth, travel through hyperspace in the manner of Figure 13.4, and emerge an hour later near the star Vega, 26 light-years away. Carl, not being a relativity expert, was unfamiliar with the message of perturbation calculations[2]: *It is impossible to travel through hyperspace from a black hole's core to another part of our Universe.* Any black hole is continually being bombarded by tiny electromagnetic vacuum fluctuations and by tiny amounts of radiation. As these fluctuations and radiation fall into the hole, they get accelerated by the hole's gravity to enormous energy, and they then rain down explosively on any "little closed universe" or "tunnel" or other vehicle by which one might try to launch the trip through hyperspace. The calculations were unequivocal; any vehicle for hyperspace travel gets destroyed by the explosive "rain" before the trip can be launched. Carl's novel had to be changed.

During the return drive from Santa Cruz, somewhere west of Fresno on Interstate 5, a glimmer of an idea came to me. Maybe Carl could replace his black hole by a *wormhole* through hyperspace.

A wormhole is a hypothetical shortcut for travel between distant points in the Universe. The wormhole has two entrances called "mouths," one (for example) near Earth, and the other (for example) in orbit around Vega, 26 light-years away. The mouths are connected to each other by a tunnel through hyperspace (the wormhole) that might be only a kilometer long. If we enter the near-Earth mouth, we find ourselves in the tunnel. By traveling just one kilometer down the tunnel we reach the other mouth and emerge near Vega, 26 light-years away as measured in the external Universe.

Figure 14.1 depicts such a wormhole in an embedding diagram. This diagram, as is usual for embedding diagrams, idealizes our Universe as having only two spatial dimensions rather than three (see Figures 3.2 and 3.3). In the diagram the space of our Universe is depicted as a

2. See the "Best Guesses" section of Chapter 13.

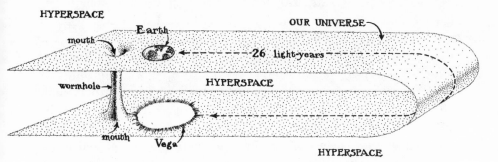

14.1 A 1-kilometer-long wormhole through hyperspace linking the Earth to the neighborhood of Vega, 26 light-years away. (Not drawn to scale.)

two-dimensional sheet. Just as an ant crawling over a sheet of paper is oblivious to whether the paper is lying flat or is gently folded, so we in our Universe are oblivious to whether our Universe is lying flat in hyperspace or is gently folded, as in the diagram. However, the gentle fold is important; it permits the Earth and Vega to be near each other in hyperspace so they can be connected by the short wormhole. With the wormhole in place, we, like an ant or worm crawling over the embedding diagram's surface, have two possible routes from Earth to Vega: the long, 26-light-year route through the external Universe, and the short, 1-kilometer route through the wormhole.

What would the wormhole's mouth look like, if it were on Earth, in front of us? In the diagram's two-dimensional universe the wormhole's mouth is drawn as a circle; therefore, in our three-dimensional Universe it would be the three-dimensional analogue of a circle; it would be a sphere. In fact, the mouth would look something like the spherical horizon of a nonrotating black hole, with one key exception: The horizon is a "one-way" surface; anything can go in, but nothing can come out. By contrast, the wormhole mouth is a "two-way" surface; we can cross it in both directions, inward into the wormhole, and back outward to the external Universe. Looking into the spherical mouth, we can see light from Vega; the light has entered the other mouth near Vega and has traveled through the wormhole, as though the wormhole were a light pipe or optical fiber, to the near-Earth mouth, where it now emerges and strikes us in the eyes.

Wormholes are not mere figments of a science fiction writer's imagination. They were discovered mathematically, as a solution to Einstein's field equation, in 1916, just a few months after Einstein formulated his field equation; and John Wheeler and his research group studied them extensively, by a variety of mathematical calculations, in the 1950s. However, none of the wormholes that had been found as solutions of Einstein's equation, prior to my trip down Interstate 5 in 1985, was suitable for Carl Sagan's novel, because none of them could be traversed safely. Each and every one of them was predicted to evolve with time in a very peculiar way: The wormhole is created at some moment of time, opens up briefly, and then pinches off and disappears—and its total life span from creation to pinch-off is so short that nothing whatsoever (no person, no radiation, no signal of any sort) can travel through it, from one mouth to the other. Anything that tries will get caught and destroyed in the pinch-off. Figure 14.2 shows a simple example.

Like most of my physicist colleagues, I have been skeptical of wormholes for decades. Not only does Einstein's field equation predict that wormholes live short lives if left to their own devices; their lives are made even shorter by random infalling bits of radiation: The radiation (according to calculations by Doug Eardley and Ian Redmount) gets accelerated to ultra-high energy by the wormhole's gravity, and as the energized radiation bombards the wormhole's throat, it triggers the throat to recontract and pinch off far faster than it would otherwise—so fast, in fact, that the wormhole has hardly any life at all.

There is another reason for skepticism. Whereas *black holes* are an inevitable consequence of stellar evolution (massive, slowly spinning stars, of just the sort that astronomers see in profusion in our galaxy, will implode to form black holes when they die), there is no analogous, natural way for a *wormhole* to be created. In fact, there is no reason at all to think that our Universe contains today *any* singularities of the sort that give birth to wormholes (Figure 14.2); and even if such singularities did exist, it is hard to understand how two of them could find each other in the vast reaches of hyperspace, so as to create a wormhole in the manner of Figure 14.2.

When one's friend needs help, one is willing to turn most anywhere that help might be found. Wormholes—despite my skepticism about them—seemed to be the only help in sight. Perhaps, it occurred to me on Interstate 5 somewhere west of Fresno, there is some way that an infinitely advanced civilization could hold a wormhole open, that is, prevent it from pinching off, so that Eleanor Arroway could travel

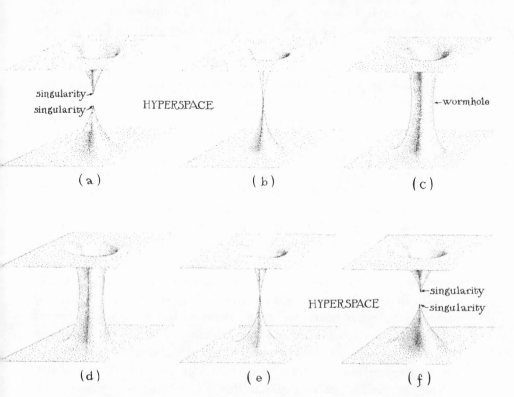

singularity→
singularity→

HYPERSPACE

←wormhole

(a)   (b)   (c)

HYPERSPACE

←singularity
←singularity

(d)   (e)   (f)

14.2 The evolution of a precisely spherical wormhole that has no material in its interior. (This evolution was discovered as a solution of Einstein's field equation in the mid-1950s by Martin Kruskal, a young associate of Wheeler's at Princeton University.) Initially (a) there is no wormhole; instead there is a singularity near Earth and one near Vega. Then, at some moment of time (b), the two singularities reach out through hyperspace, find each other, annihilate each other, and in the annihilation they create the wormhole. The wormhole grows in circumference (c), then begins to recontract (d), and pinches off (e), creating two singularities (f) similar to those in which the wormhole was born—but with one crucial exception. Each initial singularity (a) is like that of the big bang; time flows out of it, so it can give birth to something: the Universe in the case of the big bang, and the wormhole in this case. Each final singularity (f), by contrast, is like that of the big crunch (Chapter 13); time flows into it, so things get destroyed in it: the Universe in the case of the big crunch, and the wormhole in this case. Anything that tries to cross through the wormhole during its brief life gets caught in the pinch-off and, along with the wormhole itself, gets destroyed in the final singularities (f).

through it from Earth to Vega and back. I pulled out pen and paper and began to calculate. (Fortunately, Interstate 5 is very straight; I could calculate without getting carsick.)

To make the calculations easy, I idealized the wormhole as precisely spherical (so in Figure 14.1, where one of our Universe's three dimensions is suppressed, it is precisely circular in cross section). Then, by two pages of calculations based on the Einstein field equation, I discovered three things:

First, *the only way to hold the wormhole open is to thread the wormhole with some sort of material that pushes the wormhole's walls apart, gravitationally.* I shall call such material *exotic* because, as we shall see, it is quite different from any material that any human has ever yet met.

Second, I discovered that, just as the required exotic material must push the wormhole's walls outward, so also, whenever a beam of light passes through the material, the material will gravitationally push outward on the beam's light rays, prying them apart from each other. In other words, the exotic material will behave like a "defocusing lens"; it will gravitationally defocus the light beam. See Box 14.1.

Third, I learned from the Einstein field equation that, in order to gravitationally defocus light beams and gravitationally push the wormhole's walls apart, *the exotic material threading the wormhole must have a negative average energy density, as seen by a light beam traveling through it.* This requires a bit of explanation. Recall that gravity (spacetime curvature) is produced by mass (Box 2.6) and that mass and energy are equivalent (Box 5.2, where the equivalence is embodied in Einstein's famous equation $E = Mc^2$). This means that gravity can be thought of as produced by energy. Now, take the energy density of the material inside the wormhole (its energy per cubic centimeter), as measured by a light beam—that is, as measured by someone who travels through the wormhole at (nearly) the speed of light—and average that energy density along the light beam's trajectory. The resulting averaged energy density must be negative in order for the material to be able to defocus the light beam and hold the wormhole open—that is, in order for the wormhole's material to be "exotic."[3]

This does not necessarily mean that the exotic material has a negative energy as measured by someone at rest inside the wormhole. En-

---

3. In technical language, we say that the exotic material "violates the averaged weak energy condition."

Box 14.1

Holding a Wormhole Open: Exotic Material

Any spherical wormhole through which a beam of light can travel will gravitationally defocus the light beam. To see that this is so, imagine (as drawn below) that the beam is sent through a converging lens before it enters the wormhole, thereby making all its rays converge radially toward the wormhole's center. Then the rays will always continue to travel radially (how else could they possibly move?), which means that when they emerge from the other mouth, they are diverging radially outward, away from the wormhole's center, as shown. The beam has been defocused.

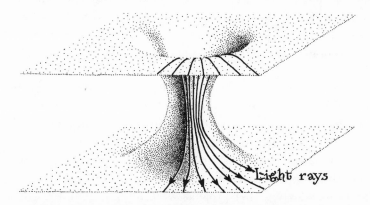

The wormhole's spacetime curvature, which causes the defocusing, is produced by the "exotic" material that threads through the wormhole and holds the wormhole open. Since spacetime curvature is equivalent to gravity, it in fact is the exotic material's gravity that defocuses the light beam. In other words, the exotic material gravitationally repels the beam's light rays, pushing them away from itself and hence away from each other, and thereby defocuses them.

This is precisely the opposite to what happens in a gravitational lens (Figure 8.2). There light from a distant star is focused by the gravitational pull of an intervening star or galaxy or black hole; here the light is defocused.

ergy density is a relative concept, not absolute; in one reference frame it may be negative, in another positive. The exotic material can have a negative energy density as measured in the reference frame of a light beam that travels through it, but a positive energy density as measured in the wormhole's reference frame. Nevertheless, because almost all forms of matter that we humans have ever encountered have positive average energy densities in *everyone's* reference frame, physicists have long suspected that exotic material cannot exist. Presumably the laws of physics forbid exotic material, we physicists have conjectured, but just *how* the laws of physics might do so was not at all clear.

Perhaps our prejudice against the existence of exotic material is wrong, I thought to myself as I rode down Interstate 5. Perhaps exotic material *can* exist. This was the only way I could see to help Carl. So upon reaching Pasadena, I wrote Carl a long letter, explaining why his heroine could not use black holes for rapid interstellar travel, and suggesting that she use wormholes instead, and that somebody in the novel discover that exotic material can really exist and can be used to hold the wormholes open. Carl accepted my suggestion with pleasure and incorporated it into the final version of his novel, *Contact.*[4]

It occurred to me, after offering Carl Sagan my comments, that his novel could serve as a pedagogical tool for students studying general relativity. As an aid for such students, during the autumn of 1985 Mike Morris (one of my own students) and I began to write a paper on the general relativistic equations for wormholes supported by exotic material, and those equations' connection to Sagan's novel.

We wrote slowly. Other projects were more urgent and got higher priority. By the winter of 1987–88, we had submitted our paper to the *American Journal of Physics,* but it was not yet published; and Morris, nearing the end of his Ph.D. training, was applying for postdoctoral positions. With his applications, Morris enclosed the manuscript of our paper. Don Page (a professor at Pennsylvania State University and a former student of mine and Hawking's) received the application, read our manuscript, and fired off a letter to Morris.

"Dear Mike, . . . it follows immediately from Proposition 9.2.8 of the book by Hawking & Ellis, plus the Einstein field equations, that *any* wormhole [requires exotic material to hold it open] . . . Sincerely, Don N. Page."

How stupid I felt. I had never studied global methods[5] (the topic of the Hawking and Ellis book) in any depth, and I was now paying the

---

4. See especially pages 347, 348, and 406 of *Contact* by Carl Sagan. There the exotic condition (negative average energy density as seen by light beams traveling through the wormhole) is expressed in a different, but equivalent way: As seen by someone at rest inside the wormhole, the material must have a large tension, along the radial direction, a tension that is bigger than the material's energy density.

5. Chapter 13.

price. I had deduced on Interstate 5, with modest labor, that to hold a precisely spherical wormhole open one must thread it with exotic material. However, now, using global methods and with even less labor, Page had deduced that to hold *any* wormhole open (a spherical wormhole, a cubical wormhole, a wormhole with random deformations), one must thread it with exotic material. I later learned that Dennis Gannon and C. W. Lee reached almost the same conclusion in 1975.

This discovery, that all wormholes require exotic material to hold them open, triggered much theoretical research during 1988–92. "Do the laws of physics permit exotic material to exist, and if so, under what circumstances?" This was the central issue.

A key to the answer had already been provided in the 1970s by Stephen Hawking. In 1970, when proving that the surface areas of black holes always increase (Chapter 12), Hawking had to assume that there is *no* exotic material near any black hole's horizon. If exotic material were in the horizon's vicinity, then Hawking's proof would fail, his theorem would fail, and the horizon's surface area could shrink. Hawking didn't worry much about this possibility, however; it seemed in 1970 a rather safe bet that exotic material cannot exist.

Then, in 1974, came a great surprise: Hawking inferred as a by-product of his discovery of black-hole evaporation (Chapter 12) that *vacuum fluctuations near a hole's horizon are exotic:* They have negative average energy density as seen by outgoing light beams near the hole's horizon. In fact, it is this exotic property of the vacuum fluctuations that permits the hole's horizon to shrink as the hole evaporates, in violation of Hawking's area-increase theorem. Because exotic material is so important for physics, I shall explain this in greater detail:

Recall the origin and nature of vacuum fluctuations, as discussed in Box 12.4: When one tries to remove all electric and magnetic fields from some region of space, that is, when one tries to create a perfect vacuum, there always remain a plethora of random, unpredictable electromagnetic oscillations—oscillations caused by a tug-of-war between the fields in adjacent regions of space. The fields "here" borrow energy from the fields "there," leaving the fields there with a deficit of energy, that is, leaving them momentarily with negative energy. The fields there then quickly grab the energy back and with it a little excess, driving their energy momentarily positive, and so it goes, onward and onward.

Under normal circumstances on Earth, the average energy of these vacuum fluctuations is zero. They spend equal amounts of time with energy deficits and energy excesses, and the average of deficit and excess vanishes. Not so near the horizon of an evaporating black hole, Hawking's 1974 calculations suggested. Near a horizon the average energy must be negative, at least as measured by light beams, which means that the vacuum fluctuations are exotic.

How this comes about was not deduced in detail until the early 1980s, when Don Page at Pennsylvania State University, Philip Candelas at Oxford, and many other physicists used the laws of quantum fields in curved spacetime to explore in great detail the influence of a hole's horizon on the vacuum fluctuations. They found that the horizon's influence is key. The horizon distorts the vacuum fluctuations away from the shapes they would have on Earth, and by this distortion it makes their average energy density negative, that is, it makes the fluctuations exotic.

Under what other circumstances will vacuum fluctuations be exotic? Can they ever be exotic inside a wormhole, and thereby hold the wormhole open? This was the central thrust of the research effort triggered by Page's noticing that the only way to hold *any* wormhole open is with exotic material.

The answer has not come easily, and is not entirely in hand. Gunnar Klinkhammer (a student of mine) has proved that in flat spacetime, that is, far from all gravitating objects, vacuum fluctuations can *never* be exotic—they can never have a negative average energy density as measured by light beams. On the other hand, Robert Wald (a former student of Wheeler's) and Ulvi Yurtsever (a former student of mine) have proved that in curved spacetime, under a very wide variety of circumstances, the curvature distorts the vacuum fluctuations and thereby makes them exotic.

Is a wormhole that is trying to pinch off such a circumstance? Can the curvature of the wormhole, by distorting the vacuum fluctuations, make them exotic and enable them to hold the wormhole open? We still do not know, as this book goes to press.

In early 1988, as theoretical studies of exotic material were getting under way, I began to recognize the power of the kind of research that Carl Sagan's phone call had triggered. Just as among all *real* physics experiments that an *experimenter* might do the ones most likely to yield deep new insights into the laws of physics are those that push on the laws the hardest, then similarly, among all *thought* experiments that a *theorist* might study, when probing laws that are beyond the reaches of modern technology, the ones most likely to yield deep new insights are those that push the hardest. And no type of thought experiment pushes the laws of physics harder than the type triggered by Carl Sagan's phone call to me—thought experiments that ask, *"What things do the laws of physics permit an infinitely advanced civilization to do, and what things do the laws forbid?"* (By an "infinitely advanced civilization," I mean one whose activities are limited only by the laws of physics, and not at all by ineptness, lack of know-how, or anything else.)

We physicists, I believe, have tended to avoid such questions because they are so close to science fiction. While many of us may enjoy reading science fiction or may even write some, we fear ridicule from our colleagues for working on research close to the science fiction fringe. We therefore have tended to focus on two other, less radical, types of questions: "What kinds of things *occur naturally* in the Universe?" (for example, do black holes occur naturally? and do wormholes occur naturally?). And "What kinds of things can we as humans, with our present or near-future technology, do?" (for example, can we produce new elements such as plutonium and use them to make atomic bombs? and can we produce high-temperature superconductors and use them to lower the power bills for levitated trains and Superconducting Supercollider magnets?).

By 1988 it seemed clear to me that we physicists had been much too conservative in our questions. Already, one *Sagan-type question* (as I shall call them) was beginning to bring a payoff. By asking, "Can an infinitely advanced civilization maintain wormholes for rapid interstellar travel?" Morris and I had identified exotic material as the key to wormhole maintenance, and we had triggered a somewhat fruitful effort to understand the circumstances under which the laws of physics do and do not permit exotic material to exist.

Suppose that our Universe was created (in the big bang) with no wormholes at all. Then eons later, when intelligent life has evolved and has produced a (hypothetical) infinitely advanced civilization, *can that infinitely advanced civilization construct wormholes for rapid interstellar travel?* Do the laws of physics permit wormholes to be constructed where previously there were none? Do the laws permit this type of change in the topology of our Universe's space?

These questions are the *second half* of Carl Sagan's interstellar transport problem. The *first half,* maintaining a wormhole once it has been constructed, Sagan solved with the help of exotic matter. The second half he finessed. In his novel, he describes the wormhole through which Eleanor Arroway traveled as now being maintained by exotic matter, but as having been created in the distant past by some infinitely advanced civilization, from which all records have been lost.

We physicists, of course, are not happy to relegate wormhole creation to prehistory. We want to know whether and how the Universe's topology can be changed *now,* within the confines of physical law.

We can imagine two strategies for constructing a wormhole where before there was none: a *quantum strategy,* and a *classical strategy.*

The quantum strategy relies on *gravitational vacuum fluctuations* (Box 12.4), that is, the gravitational analogue of the electromagnetic vacuum fluctuations discussed above: random, probabilistic fluctuations in the curvature of space caused by a tug-of-war in which adjacent regions of space are continually stealing energy from each other and then giving it back. Gravitational vacuum fluctuations are thought to be everywhere, but under ordinary circumstances they are so tiny that no experimenter has ever detected them.

Just as an electron's random degeneracy motions become more vigorous when one confines the electron to a smaller and smaller region (Chapter 4), so also gravitational vacuum fluctuations are more vigorous in small regions than in large, that is, for small wavelengths rather than for large. In 1955, John Wheeler, by combining the laws of quantum mechanics and the laws of general relativity in a tentative and crude way, deduced that in a region the size of the *Planck–Wheeler length,*[6] 1.62×10^{-33} centimeter or smaller, the vacuum fluctuations are so huge that space as we know it "boils" and becomes a froth of quantum foam—the same sort of quantum foam as makes up the core of a spacetime singularity (Chapter 13; Figure 14.3).

Quantum foam, therefore, is everywhere: inside black holes, in interstellar space, in the room where you sit, in your brain. But to see the quantum foam, one would have to zoom in with a (hypothetical) supermicroscope, looking at space and its contents on smaller and smaller scales. One would have to zoom in from the scale of you and me

---

6. The Planck–Wheeler length is the square root of the Planck–Wheeler area (which entered into the formula for the entropy of a black hole, Chapter 12); it is given by the formula $\sqrt{G\hbar/c^3}$, where $G = 6.670 \times 10^{-8}$ dyne-centimeter²/gram² is Newton's gravitation constant, $\hbar = 1.055 \times 10^{-27}$ erg-second is Planck's quantum mechanical constant, and $c = 2.998 \times 10^{10}$ centimeter/second is the speed of light.

(hundreds of centimeters) to the scale of an atom (10^{-8} centimeter), to the scale of an atomic nucleus (10^{-13} centimeter), and then on downward by *twenty* factors of 10 more, to 10^{-33} centimeter. At all the early, "large" scales, space would look completely smooth, with a very definite (but tiny) amount of curvature. As the microscopic zoom nears, then passes 10^{-32} centimeter, however, one would see space begin to writhe, ever so slightly at first, and then more and more strongly until, when a region just 10^{-33} centimeter in size fills the supermicroscope's entire eyepiece, space has become a froth of probabilistic quantum foam.

14.3   (Same as Figure 13.7.) Embedding diagrams illustrating quantum foam. The geometry and topology of space are not definite; instead, they are probabilistic. They might have, for example, a 0.1 percent probability for the form shown in (a), a 0.4 percent probability for (b), a 0.02 percent probability for (c), and so on.

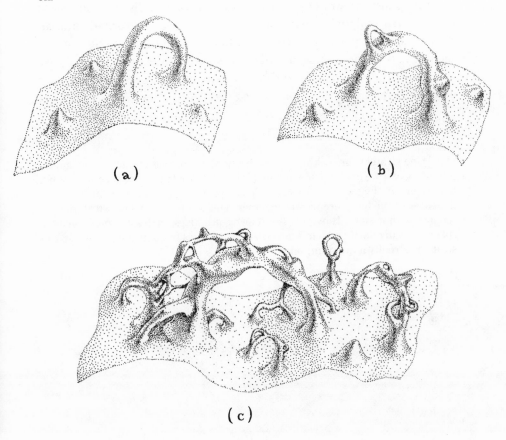

(a)

(b)

(c)

Since the quantum foam is everywhere, it is tempting to imagine an infinitely advanced civilization reaching down into the quantum foam, finding in it a wormhole (say, the "big" one in Figure 14.3b with its 0.4 percent probability), and trying to grab that wormhole and enlarge it to classical size. In 0.4 percent of such attempts, if the civilization were truly infinitely advanced, they might succeed. Or would they?

We do not yet understand the laws of quantum gravity well enough to know. One reason for our ignorance is that we do not understand the quantum foam itself very well. We aren't even 100 percent sure it exists. However, the challenge of this Sagan-type thought experiment—an advanced civilization pulling wormholes out of the quantum foam—might be of some conceptual help in the coming years, in efforts to firm up our understanding of quantum foam and quantum gravity.

So much for the *quantum strategy* of wormhole creation. What is the *classical strategy?*

In the classical strategy, our infinitely advanced civilization would try to warp and twist space on macroscopic scales (normal, human scales) so as to make a wormhole where previously none existed. It seems fairly obvious that, in order for such a strategy to succeed, *one must tear two holes in space and sew them together.* Figure 14.4 shows an example.

14.4   One strategy for making a wormhole. (a) A "sock" is created in the curvature of space. (b) Space outside the sock is gently folded in hyperspace. (c) A small hole is torn in the toe of the sock, a hole is torn in space just below the hole, and the edges of the holes are "sewn" together. This strategy looks classical (macroscopic) at first sight. However, the tearing produces, at least momentarily, a spacetime singularity which is governed by the laws of quantum gravity, so this strategy is really a quantum one.

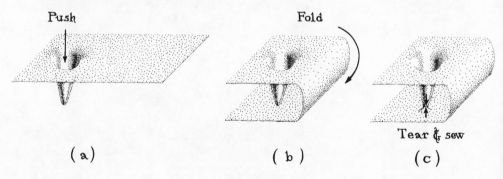

Push   Fold

Tear & sew

(a)   (b)   (c)

Now, any such tearing of space produces, momentarily, at the point of the tear, a singularity of spacetime, that is, a sharp boundary at which spacetime ends; and since singularities are governed by the laws of quantum gravity, such a strategy for making wormholes is actually quantum mechanical, not classical. We will not know whether it is permitted until we understand the laws of quantum gravity.

Is there no way out? Is there no way to make a wormhole without getting entangled with the ill-understood laws of quantum gravity— no *perfectly classical* way?

Somewhat surprisingly, there *is*—but only if one pays a severe price. In 1966, Robert Geroch (a student of Wheeler's at Princeton) used global methods to show that one *can* construct a wormhole by a smooth, singularity-free warping and twisting of spacetime, but one can do so only if, during the construction, time also becomes twisted up as seen in all reference frames.[7] More specifically, while the construction is going on, it must be possible to travel backward in time, as well as forward; the "machinery" that does the construction, whatever it might be, must function briefly as a time machine that carries things from late moments of the construction back to early moments (but not back to moments before the construction began).

The universal reaction to Geroch's theorem, in 1967, was *"Surely* the laws of physics forbid time machines, and thereby they will prevent a wormhole from ever being constructed classically, that is, without tearing holes in space."

In the decades since 1967, some things we thought were *sure* have been proved wrong. (For example, we would never have believed in 1967 that a black hole can evaporate.) This has taught us caution. As part of our caution, and triggered by Sagan-type questions, we began asking in the late 1980s, "Do the laws of physics *really* forbid time machines, and if so, *how?* How might the laws enforce such a prohibition?" To this question I shall return below.

L̲et us now pause and take stock. In 1993 our best understanding of wormholes is this:

If no wormholes were made in the big bang, then an infinitely advanced civilization might try to construct one by two methods, quantum (pulling it out of the quantum foam) or classical (twisting space-

---

7. I wish that I could draw a simple, clear picture to show how this smooth creation of a wormhole is accomplished; unfortunately, I cannot.

time without tearing it). We do *not* understand the laws of quantum gravity well enough to deduce, in 1993, whether the quantum construction of wormholes is possible. We *do* understand the laws of classical gravity (general relativity) well enough to know that the classical construction of wormholes is permitted only if the construction machinery, whatever it might be, twists time up so strongly, as seen in all reference frames, that it produces, at least briefly, a time machine.

We also know that, if an infinitely advanced civilization somehow acquires a wormhole, then the only way to hold the wormhole open (so it can be used for interstellar travel) is by threading it with exotic material. We know that vacuum fluctuations of the electromagnetic field are a promising form of exotic material: They can be exotic (have a negative average energy density as measured by a light beam) in curved spacetime under a wide variety of circumstances. However, we do *not* yet know whether they can be exotic inside a wormhole and thereby hold the wormhole open.

In the pages to come, I shall assume that an infinitely advanced civilization has somehow acquired a wormhole and is holding it open by means of some sort of exotic material; and I shall ask what other uses, besides interstellar travel, the civilization might find for its wormhole.

Time Machines

In December 1986, the fourteenth semi-annual Texas Symposium on Relativistic Astrophysics was held in Chicago, Illinois. These "Texas" symposia, patterned after the 1963 one in Dallas, Texas, where the mystery of quasars was first discussed (Chapters 7 and 9), had by now become a firmly established institution. I went to the symposium and lectured on dreams and plans for LIGO (Chapter 10). Mike Morris (my "wormhole" student) also went, to get his first full-blown exposure to the international community of relativity physicists and astrophysicists.

In the corridors between lectures, Morris became acquainted with Tom Roman, a young assistant professor from Central Connecticut State University who, several years earlier, had produced deep insights about exotic matter. Their conversation quickly turned to wormholes. "If a wormhole can really be held open, then it will permit one to travel over interstellar distances far faster than light," Roman noted.

"Doesn't this mean that one can also use a wormhole to travel backward in time?"

How stupid Mike and I felt! Of course; Roman was right. We, in fact, had learned about such time travel in our childhoods from a famous limerick:

> There once was a lady named Bright
> who traveled much faster than light.
> She departed one day in a relative way
> and came home the previous night.

With Roman's comment and the famous limerick to goad us, we easily figured out how to construct a time machine using two wormholes that move at high speeds relative to each other.[8] (I shall not describe that time machine here, because it is a bit complicated and there is a simpler, more easily described time machine to which I shall come shortly.)

I am a loner; I like to retreat to the mountains or an isolated seacoast, or even just into an attic, and think. New ideas come slowly and require large blocks of quiet, undisturbed time to gestate; and most worthwhile calculations require days or weeks of intense, steady concentration. A phone call at the wrong moment can knock my concentration off balance, setting me back by hours. So I hide from the world.

But hiding for too long is dangerous. I need, from time to time, the needle-pricking stimulus of conversations with people whose viewpoints and expertise are different from mine.

---

8. This time machine and others described later in this chapter are by no means the first time machine–type solutions to the Einstein field equation that people have found. In 1937, W. J. van Stockum in Edinburgh discovered a solution in which an infinitely long, rapidly spinning cylinder functions as a time machine. Physicists have long objected that nothing in the Universe can be infinitely long, and they have suspected (but nobody has proved) that, if the length of the cylinder were made finite, it would cease to be a time machine. In 1949, Kurt Gödel, at the Institute for Advanced Study in Princeton, New Jersey, found a solution to Einstein's equation that describes a whole universe which spins but does not expand or contract, and in which one can travel backward in time by simply going out to great distances from Earth and then returning. Physicists object, of course, that our real Universe does not at all resemble Gödel's solution: It is *not* spinning, at least not much, and it *is* expanding. In 1976 Frank Tipler used the Einstein field equation to prove that, in order to create a time machine in a finite-sized region of space, one must use exotic material as part of the machine. (Since any traversable wormhole must be threaded by exotic material, the wormhole-based time machines described in this chapter satisfy Tipler's requirement.)

In this chapter thus far I have described three examples. Without Carl Sagan's phone call and the challenge to make his novel scientifically correct, I would never have ventured into research on wormholes and time machines. Without Don Page's letter, Mike Morris and I would not have known that all wormholes, regardless of their shape, require exotic material to keep them open. And without Tom Roman's remark, Morris and I might have gone on blithely unaware that from wormholes an advanced civilization can easily make a time machine.

In the pages to come, I will describe other examples of the crucial role of needle-pricking interactions. However, not *all* ideas arise that way. Some arise from introspection. June 1987 was a case in point.

In early June 1987, emerging from several months of frenetic classroom teaching and interactions with my research group and the LIGO team, I retreated, exhausted, into isolation.

All spring long something had been gnawing at me, and I had been trying to ignore it, waiting for some days of quiet, to ponder. Those days, at last, had come. In isolation, I let the gnawing emerge from my subconscious and began to examine it: *"How does time decide how to hook itself up through a wormhole?"* That was the nub of the gnaw.

To make this question more concrete, I thought about an example: Suppose that I have a very short wormhole, one whose tunnel through hyperspace is only 30 centimeters long, and suppose that both mouths of the wormhole—two spheres, each 2 meters in diameter—are sitting in my Pasadena living room. And suppose that I climb through the wormhole, head first. From my viewpoint, I must emerge from the second mouth immediately after I enter the first, with no delay at all;

14.5 A picture of me crawling through a hypothetical, very short wormhole.

in fact, my head is coming out of the second mouth while my feet are still entering the first. Does this mean that my wife, Carolee, sitting there on the living room sofa, will also see my head emerging from the second mouth while my feet are still climbing into the first, as in Figure 14.5? If so, then time "hooks up *through* the wormhole" in the same manner as it hooks up *outside* the wormhole.

On the other hand, I asked myself, isn't it possible that, although the trip through the wormhole takes almost no time as seen by me, Carolee must wait an hour before she sees me emerge from the second mouth; and isn't it also possible that she sees me emerge an hour before I entered? If so, then time would be hooked up *through* the wormhole in a different manner than it hooks up *outside* the wormhole.

What could possibly make time behave so weirdly? I asked myself. On the other hand, why shouldn't it behave in this way? Only the laws of physics know the answer, I reasoned. Somehow, I ought to be able to deduce from the laws of physics just how time will behave.

As an aid to understanding how the laws of physics control time's hookup, I thought about a more complicated situation. Suppose that one mouth of the wormhole is at rest in my living room and the other is in interstellar space, traveling away from Earth at nearly the speed of light. And suppose that, despite this relative motion of its two mouths, the wormhole's length (the length of its tunnel through hyperspace) remains always fixed at 30 centimeters. (Figure 14.6 explains how it is

14.6 Explanation of how the mouths of a wormhole can move relative to each other as seen in the external Universe, while the length of the wormhole remains fixed. Each of the diagrams is an embedding diagram like that in Figure 14.1, seen in profile. The diagrams are a sequence of snapshots that depict motion of the Universe and the wormhole *relative to hyperspace*. (Recall, however, that hyperspace is just a useful figment of our imaginations; there is no way that we as humans can ever see or experience it in reality; see Figures 3.2 and 3.3.) Relative to hyperspace, the bottom part of our Universe is sliding rightward in the diagrams, while the wormhole and the top part of our Universe remain at rest. Correspondingly, as seen in our Universe, the mouths of the wormhole are moving relative to each other (they are getting farther apart), but as seen through the wormhole they are at rest with respect to each other; the wormhole's length does not change.

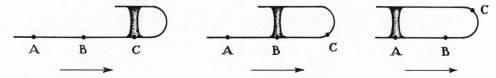

possible for the length of the wormhole to remain fixed while its mouths, as seen in the external Universe, move relative to each other.) Then, as seen in the external Universe, the two mouths are in different reference frames, frames that move at a high speed relative to each other; and *the mouths therefore must experience different flows of time.* On the other hand, as seen through the wormhole's interior, the mouths are at rest with respect to each other, so they share a common reference frame, which means that *the mouths must experience the same flow of time.* From the external viewpoint they experience different time flows, and from the internal viewpoint, the same time flow; how confusing!

Gradually, in my quiet isolation, the confusion subsided and all became clear. The laws of general relativity predict, unequivocally, the flow of time at the two mouths, and they predict, unequivocally, that the two time flows will be *the same* when compared through the wormhole, but will be *different* when compared outside the wormhole. Time, in this sense, hooks up to itself differently through the wormhole than through the external Universe, when the two mouths are moving relative to each other.

And this difference of hookup, I then realized, implies that *from a single wormhole, an infinitely advanced civilization can make a time machine.* There is no need for two wormholes. How? Easy, if you are infinitely advanced.

To explain how, I shall describe a thought experiment in which we humans are infinitely advanced beings. Carolee and I find a very short wormhole, and put one of its mouths in the living room of our home and the other in our family spacecraft, outside on the front lawn.

Now, as this thought experiment will show, the manner in which time is hooked up through any wormhole actually depends on the wormhole's past history. For simplicity, I shall assume that when Carolee and I first acquire the wormhole, it has the simplest possible hookup of time: the same hookup through the wormhole's interior as through the exterior Universe. In other words, if I climb through the wormhole, Carolee, I, and everyone on Earth will agree that I emerge from the mouth in the spacecraft at essentially the same moment as I entered the mouth in the living room.

Having checked that time is, indeed, hooked up through the wormhole in this way, Carolee and I then make a plan: I will stay at home in our living room with the one mouth, while Carolee in our spacecraft takes the other mouth on a very high speed trip out into the Universe

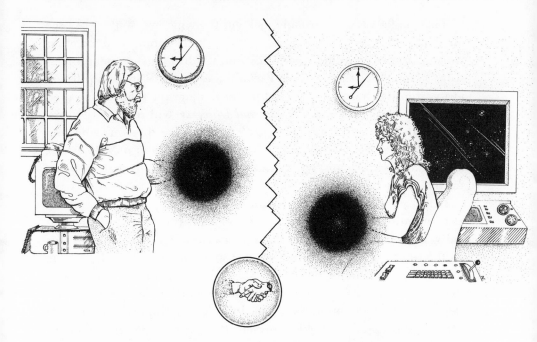

14.7 Carolee and I construct a time machine from a wormhole. *Left:* I stay at home in Pasadena with one mouth of the wormhole and hold hands with Carolee through the wormhole. *Right:* Carolee carries the other mouth on a high-speed trip through the Universe. *Inset:* Our hands inside the wormhole.

and back. Throughout the trip, we will hold hands through the wormhole; see Figure 14.7.

Carolee departs at 9:00 A.M. on 1 January 2000, as measured by herself, by me, and by everybody else on Earth. Carolee zooms away from Earth at nearly the speed of light for 6 hours as measured by her own time; then she reverses course and zooms back, arriving on the front lawn 12 hours after her departure as measured by her own time.[9] I hold hands with her and watch her through the wormhole throughout the trip, so obviously I agree, *while looking through the wormhole,* that she has returned after just 12 hours, at 9:00 P.M. on 1 January 2000. Looking through the wormhole at 9:00 P.M., I can see not only Carolee; I can also see, behind her, our front lawn and our house.

9. In reality, if Carolee were to accelerate up to the speed of light and then back down so quickly, the acceleration would be so great that it would kill her and mutilate her body. However, in the spirit of a physicist's thought experiment, I shall pretend that her body is made of such strong stuff that she can survive the acceleration comfortably.

Then, at 9:01 P.M., I turn and look out the window—and there I see an empty front lawn. The spaceship is not there; Carolee and the other wormhole mouth are not there. Instead, if I had a good enough telescope pointed out the window, I would see Carolee's spaceship flying away from Earth on its outbound journey, a journey that as measured on Earth, *looking through the external Universe*, will require 10 years. [This is the standard "twins paradox"; the high-speed "twin" who goes out and comes back (Carolee) measures a time lapse of only 12 hours, while the "twin" who stays behind on Earth (me) must wait 10 years for the trip to be completed.]

I then go about my daily routine of life. For day after day, month after month, year after year, I carry on with life, waiting—until finally, on 1 January 2010, Carolee returns from her journey and lands on the front lawn. I go out to meet her, and find, as expected, that she has aged just 12 hours, not 10 years. She is sitting there in the spaceship, her hand thrust into the wormhole mouth, holding hands with somebody. I stand behind her, look into the mouth, and see that the person whose hand she holds is myself, 10 years younger, sitting in our living room on 1 January 2000. The wormhole has become a time machine. If I now (on 1 January 2010) climb into the wormhole mouth in the spaceship, I will emerge through the other mouth in our living room on 1 January 2000, and there I will meet my younger self. Similarly, if my younger self climbs into the mouth in the living room on 1 January 2000, he will emerge from the mouth in the spaceship on 1 January 2010. Travel through the wormhole in one direction takes me backward 10 years in time; travel in the other direction takes me 10 years forward.

Neither I nor anyone else, however, can use the wormhole to travel back in time beyond 9:00 P.M., 1 January 2000. It is impossible to travel to a time earlier than when the wormhole first became a time machine.

The laws of general relativity are unequivocal. If wormholes can be held open by exotic material, then these are general relativity's predictions.

In summer 1987, a month or so after I arrived at these predictions, Richard Price telephoned Carolee. Richard—a close friend of mine and the man who sixteen years earlier had shown that a black hole radiates away all its "hair" (Chapter 7)—was worried about me. He had heard that I was working on the theory of time machines, and he feared I had gone a little crazy or senile or. . . Carolee tried to reassure him.

Richard's call shook me up a bit. Not because I doubted my own sanity; I had few doubts. However, if even my closest friends were worried, then (at least as a protection for Mike Morris and my other students, if not for myself) I would have to be careful about how we presented our research to the community of physicists and to the general public.

During the winter of 1987–88, as part of my caution, I decided to move slowly on publishing anything about time machines. Together with two students, Mike Morris and Ulvi Yurtsever, I focused on trying to understand everything I could about wormholes and time. Only after all issues were crystal clear did I want to publish.

Morris, Yurtsever, and I worked together by computer link and telephone, since I was hiding in isolation. Carolee had taken a two-year postdoctoral appointment in Madison, Wisconsin, and I had gone along as her "house husband" for the first seven months (January–July 1988). I had set up my computer and working tables in the attic of the house we rented in Madison; and I was spending most of my waking hours there in the attic, thinking, calculating, and writing—largely on other projects, but partly on wormholes and time.

For stimulus and to test my ideas against skilled "opponents," every few weeks I drove over to Milwaukee to talk with a superb group of relativity researchers led by John Friedman and Leonard Parker, and occasionally I drove down to Chicago to talk with another superb group led by Subrahmanyan Chandrasekhar, Robert Geroch, and Robert Wald.

On a March visit to Chicago, I got a jolt. I gave a seminar describing everything I understood about wormholes and time machines; and after the seminar, Geroch and Wald asked me (in effect), *"Won't a wormhole be automatically destroyed whenever an advanced civilization tries to convert it into a time machine?"*

Why? How? I wanted to know. They explained. Translated into the language of the Carolee-and-me story, their explanation was the following: Imagine that Carolee is zooming back to Earth with one wormhole mouth in her spacecraft and I am sitting at home on Earth with the other. When the spacecraft gets to within 10 light-years of Earth, it suddenly becomes possible for radiation (electromagnetic waves) to use the wormhole for time travel: Any random bit of radiation that leaves our home in Pasadena traveling at the speed of light toward the spacecraft can arrive at the spacecraft after 10 years' time (as seen on Earth), enter the wormhole mouth there, travel back in time by 10 years (as

seen on Earth), and emerge from the mouth on Earth at precisely the
same moment as it started its trip. The radiation piles right on top of its
previous self, not just in space but in spacetime, doubling its strength.
What's more, during the trip each quantum of radiation (each photon)
got boosted in energy due to the relative motion of the wormhole
mouths (a "Doppler-shift" boost).

After the radiation's next trip out to the spacecraft then back
through the wormhole, it again returns at the same time as it left and

14.8 (a) The Geroch–Wald suggestion for how a wormhole might get destroyed
when one tries to make it into a time machine. An intense beam of radiation
zooms between the two mouths and through the wormhole, piling up on and
reinforcing itself. The beam becomes infinitely energetic and destroys the wormhole. (b) What actually happens. The wormhole defocuses the beam, reducing
the amount of pileup. The beam remains weak; the wormhole is not destroyed.

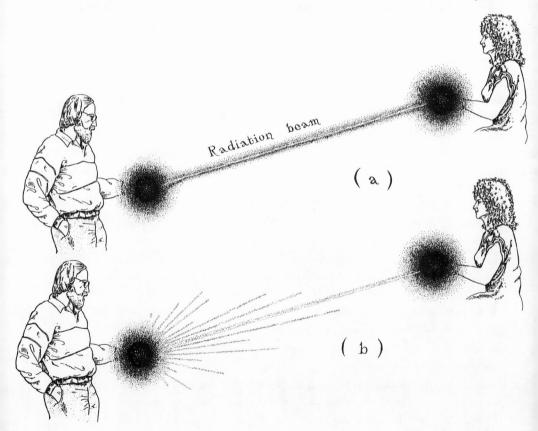

Radiation beam

(a)

(b)

again piles up on itself, again with a Doppler-boosted energy. Again and again this happens, making the beam of radiation infinitely strong (Figure 14.8a).

In this way, beginning with an arbitrarily tiny amount of radiation, a beam of infinite energy is created, coursing through space between the two wormhole mouths. As the beam passes through the wormhole, Geroch and Wald argued, it will produce infinite spacetime curvature and probably destroy the wormhole, thereby preventing the wormhole from becoming a time machine.

I drove away from Chicago and up Interstate 90 toward Madison in a daze. My mind was filled with geometric pictures of radiation beams shooting from one wormhole mouth to the other, as the mouths move toward each other. I was trying to compute, pictorially, just what *would* happen. I was trying to understand whether Geroch and Wald were right or wrong.

Gradually, as I neared the Wisconsin border, the pictures in my mind became clear. The wormhole would *not* be destroyed. Geroch and Wald had overlooked a crucial fact: Every time the beam of radiation passes through the wormhole, the wormhole *defocuses* it in the manner of Box 14.1 above. After the defocusing, the beam emerges from the mouth on Earth and spreads out over a wide swath of space, so that only a tiny fraction of it can get caught by the mouth on the spacecraft and transported through the wormhole back to Earth to "pile up" on itself (Figure 14.8b).

I could do the sum visually in my head, as I drove. By adding up all the radiation from all the trips through the wormhole (a tinier and tinier amount after each defocusing trip), I computed that the final beam would be weak; far too weak to destroy the wormhole.

My calculation turned out to be right; but, as I shall explain later, I should have been more cautious. This brush with wormhole destruction should have warned me that unexpected dangers await any maker of time machines.

W hen graduate students reach the final year of their research, they often give me great pleasure. They produce major insights on their own; they argue with me and win; they teach me unexpected things. Such was the case with Morris and Yurtsever as we gradually moved toward finalizing our manuscript for *Physical Review Letters*. Large portions of the manuscript's technical details and technical ideas were theirs.

As our work neared completion, I oscillated between worrying about tarnishing Morris's and Yurtsever's budding scientific reputations with a label of "crazy science fiction physicists" and waxing enthusiastic about the things we had learned and about our realization that Sagan-type questions can be powerful in physics research. At the last minute, as we finalized the paper, I suppressed my caution (which Morris and Yurtsever seemed not to share), and agreed with them to give our paper the title "Wormholes, Time Machines, and the Weak Energy Condition" ("weak energy condition" being the technical term associated with "exotic matter").

Despite the "time machines" in the title, our paper was accepted for publication without question. The two anonymous referees seemed to be sympathetic; I heaved a sigh of relief.

With the publication date nearing, caution took hold of me again; I asked the staff of the Caltech Public Relations Office to avoid and, indeed, try to suppress *any and all* publicity about our time machine research. A sensational splash in the press might brand our research as crazy in the eyes of many physicists, and I wanted our paper to be studied seriously by the physics community. The public relations staff acquiesced.

Our paper was published, and all went well. As I had hoped, the press missed it, but among physicists it generated interest and controversy. Letters trickled in, asking questions and challenging our claims; but we had done our homework. We had answers.

My friends' reactions were mixed. Richard Price continued to worry; he had decided I wasn't crazy or senile, but he feared I would sully my reputation. My Russian friend Igor Novikov, by contrast, was ecstatic. Telephoning from Santa Cruz, California, where he was visiting, Novikov said, "I'm so happy, Kip! You have broken the barrier. If *you* can publish research on time machines, then so can *I!*" And he proceeded to do so, forthwith.

The Matricide Paradox

Among the controversies stirred up by our paper, the most vigorous was over what I like to call the *matricide paradox*[10]: If I have a time

10. In most science fiction literature, the term "grandfather paradox" is used rather than "matricide paradox." Presumably, the chivalrous men who dominate the science fiction writing profession feel more comfortable pushing the murder back a generation and onto a male.

machine (wormhole-based or otherwise), I should be able to use it to go back in time and kill my mother before I was conceived, thereby preventing myself from being born and killing my mother.[11]

Central to the matricide paradox is the issue of *free will:* Do I, or do I not, as a human being, have the power to determine my own fate? Can I *really* kill my mother, after going backward in time, or (as in so many science fiction stories) will something inevitably stay my hand as I try to stab her in her sleep?

Now, even in a universe without time machines, free will is a terribly difficult thing for physicists to deal with. We usually try to avoid it. It just confuses issues that otherwise might be lucid. With time machines, all the more so. Accordingly, before publishing our paper (but after long discussions with our Milwaukee colleagues), Morris, Yurtsever, and I decided to avoid entirely the issue of free will. We insisted on not discussing at all, in print, human beings who go through a wormhole-based time machine. Instead, we dealt *only* with simple, inanimate time-traveling things, such as electromagnetic waves.

Before publishing, we thought a lot about waves that travel backward in time through a wormhole; we searched hard for unresolvable paradoxes in the waves' evolution. Ultimately (and with crucial proddings from John Friedman), we convinced ourselves that there probably will be *no unresolvable paradoxes,* and we conjectured so in our paper.[12] We even broadened our conjecture to suggest that there would never be unresolvable paradoxes for *any* inanimate object that passes through the wormhole. It was this conjecture that created the most controversy.

Of the letters we received, the most interesting was from Joe Polchinski, a professor of physics at the University of Texas in Austin. Polchinski wrote, "Dear Kip, . . . If I understand correctly, you are conjecturing that in your [wormhole-based time machine there will be no unresolvable paradoxes]. It seems to me that . . . this is not the case." He then posed an elegant and simple variant of the matricide paradox—a variant that is *not* entangled with free will and that we therefore felt competent to analyze:

11. I and my four siblings are very respectful and obedient toward our mother; see, for example, Footnote 2 in Chapter 7. Accordingly, I have sought and received permission from my mother to use this example.

12. Three years later, John Friedman and Mike Morris together managed to prove rigorously that, when waves travel backward in time through a wormhole, there indeed are no unresolvable paradoxes—provided the waves superimpose linearly on themselves in the manner of Box 10.3.

Take a wormhole that has been made into a time machine, and place its two mouths at rest near each other, out in interplanetary space (Figure 14.9). Then, if a billiard ball is launched toward the right mouth from an appropriate initial location and with an appropriate initial velocity, the ball will enter the right mouth, travel backward in time, and fly out of the left mouth before it entered the right (as seen by you and me outside the wormhole), and it will then hit its younger self, thereby preventing itself from ever entering the right mouth and hitting itself.

This situation, like the matricide paradox, entails going back in time and changing history. In the matricide paradox, I go back in time and, by killing my mother, prevent myself from being born. In Polchinski's paradox, the billiard ball goes back in time and, by hitting itself, prevents itself from ever going back in time.

Both situations are nonsensical. Just as the laws of physics must be logically consistent with each other, so also the evolution of the Universe, as governed by the laws of physics, must be fully consistent with itself—or at least it must be so when the Universe is behaving classically (non–quantum mechanically); the quantum mechanical realm is

14.9 Polchinski's billiard ball version of the matricide paradox. The wormhole is very short and has been made into a time machine, so that anything that enters the right mouth emerges, as measured on the outside, 30 minutes before it went in. The flow of time outside the mouth is denoted by the symbol t; the flow of time as experienced by the billiard ball itself is denoted by τ. The billiard ball is launched at t = 3:00 P.M. from the indicated location and with just the right velocity to enter the right mouth at t = 3:45. The ball emerges from the left mouth 30 minutes earlier, at t = 3:15, and then hits its younger self at t = 3:30 P.M., knocking itself off track so it cannot enter the right mouth and hit itself.

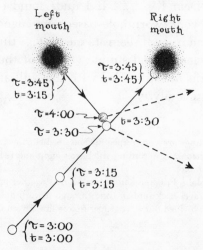

a little more subtle. Since both I and a billiard ball are highly classical objects (that is, we can exhibit quantum mechanical behavior only when one makes exceedingly accurate measurements on us; see Chapter 10), there is no way that either I or the billiard ball can go back in time and change our own histories.

So what happens to the billiard ball? To find out, Morris, Yurtsever, and I focused our attention on the ball's *initial conditions,* that is, its initial location and velocity. We asked ourselves, "For the same initial conditions as led to Polchinski's paradox, is there any *other* billiard ball trajectory that, unlike the one in Figure 14.9, is a *logically self-consistent* solution to the physical laws that govern classical billiard balls?" After much discussion, we agreed that the answer was probably "yes," but we were not absolutely sure—and there was no time for us to figure it out. Morris and Yurtsever had completed their Ph.D.s and were leaving Caltech to take up postdoctoral appointments in Milwaukee and Trieste.

Fortunately, Caltech continually draws great students. There were two new ones waiting in the wings: Fernando Echeverria and Gunnar Klinkhammer. Echeverria and Klinkhammer took Polchinski's paradox and ran with it: After some months of on-and-off mathematical struggle, they proved that there indeed *is* a fully self-consistent billiard ball trajectory that begins with Polchinski's initial data and satisfies all the laws of physics that govern classical billiard balls. In fact, there are *two* such trajectories. They are shown in Figure 14.10. I shall describe each of these trajectories in turn, from the viewpoint of the ball itself.

On trajectory (a) (left half of Figure 14.10), the ball, young, clean, and pristine, starts out at time t = 3:00 P.M., moving along precisely the same route as in Polchinski's paradox (Figure 14.9), a route taking it toward the wormhole's right mouth. A half hour later, at t = 3:30, the young, pristine ball gets hit on its *left, rear side,* by an older-looking, cracked ball (which will turn out to be its older self). The collision is gentle enough to deflect the young ball only slightly from its original course, but hard enough to crack it. The young ball, now cracked, continues onward along its slightly altered trajectory and enters the wormhole mouth at t = 3:45, travels backward in time by 30 minutes, and exits from the other mouth at t = 3:15. Because its trajectory has been altered slightly by comparison with Polchinski's paradoxical trajectory (Figure 14.9), the ball, now old and cracked, hits its younger self a gentle, glancing blow on the left, rear side at t =

3:30, instead of the vigorous, highly deflecting blow of Figure 14.9. The evolution thereby is made fully self-consistent.

Trajectory (b), the right half of Figure 14.10, is the same as (a), except that the geometry of the collision is slightly different, and correspondingly the trajectory between collisions is slightly different. In particular, the old, cracked ball emerges from the left mouth on a different trajectory than in (a), a trajectory that takes it in front of the young, pristine ball (instead of behind it), and produces a glancing blow on the young ball's *front, right side* (instead of left rear side).

Echeverria and Klinkhammer showed that both trajectories, (a) and (b), satisfy all the physical laws that govern classical billiard balls, so both are possible candidates to occur in the real Universe (*if* the real Universe can have wormhole-based time machines).

This is most disquieting. Such a situation can never occur in a universe without time machines. Without time machines, each set of initial conditions for a billiard ball gives rise to one and only one trajectory that satisfies all the classical laws of physics. There is a unique prediction for the ball's motion. The time machine has ruined this. There now are two, equally good predictions for the ball's motion.

14.10   The resolution of Polchinski's version of the matricide paradox (Figure 14.9): A billiard ball, starting out at 3:00 P.M. with the same initial conditions (same location and velocity) as in Polchinski's paradox, can move along either of the two trajectories shown here. Each of these trajectories is fully self-consistent and satisfies the classical laws of physics everywhere along the trajectory.

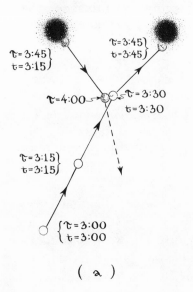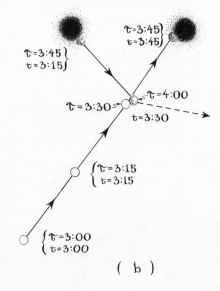

(a)   (b)

Actually, the situation is even worse than it looks at first sight: The time machine makes possible an *infinite number* of equally good predictions for the ball's motion, not just two. Box 14.2 shows a simple example.

Box 14.2

The Billiard Ball Crisis: An Infinity of Trajectories

One day, while sitting in San Francisco Airport waiting for a plane, it occurred to me that, if a billiard ball is fired between the two mouths of a wormhole-based time machine, there are two trajectories on which it can travel. On one (a), it hurtles between the mouths unscathed. On the other (b), as it is passing between the two mouths, it gets hit and knocked rightward, toward the right mouth; it then goes down the wormhole, emerges from the left mouth before it went down, hits itself, and flies away.

Some months later, Robert Forward [one of the pioneers of laser interferometer gravitational-wave detectors (Chapter 10) and also a science fiction writer] discovered a third trajectory that satisfies all the laws of physics, the trajectory (c) below: The collision, instead of occurring between the mouths, occurs before the ball reaches the mouths' vicinity. I then realized that the collision could be made to occur earlier and earlier, as in (d) and (e), if the ball travels through the wormhole several times between its two visits to the collision event. For example, in (e), the ball travels up route α, gets hit by its older self and knocked along β and into the right mouth; it then travels through the wormhole (and backward in

(continued next page)

(Box 14.2 continued)

time), emerging from the left mouth on γ, which takes it through the wormhole again (and still farther back in time), emerging along δ, which takes it through the wormhole yet again (and even farther back in time), emerging along ε, which takes it to the collision event, from which it is deflected down ζ.

Evidently, there are an infinite number of trajectories (each with a different number of wormhole traversals) that all satisfy the classical (non-quantum) laws of physics, and all begin with identically the same initial conditions (the same initial billiard ball location and velocity). One is left wondering whether physics has gone crazy, or whether, instead, the laws of physics can somehow tell us which trajectory the ball ought to take.

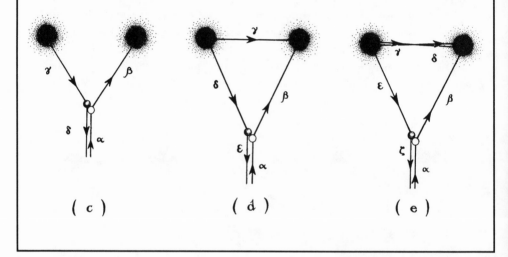

Do time machines make physics go crazy? Do they make it impossible to predict how things evolve? If not, then how do the laws of physics choose which trajectory, out of the infinite allowed set, a billiard ball will follow?

In search of an answer, Gunnar Klinkhammer and I in 1989 turned from the *classical* laws of physics to the *quantum* laws. Why the quantum laws? Because they are the Ultimate Rulers of our Universe.

For example, the laws of quantum gravity have ultimate control over gravitation and the structure of space and time. Einstein's classical, general relativistic laws of gravity are mere approximations to the quantum gravity laws—approximations with excellent accuracy when one is far from all singularities and looks at spacetime on scales far larger than 10^{-33} centimeter, but approximations nevertheless (Chapter 13).

Similarly, the classical laws of billiard ball physics, which my students and I had used in studying Polchinski's paradox, are mere approximations to the quantum mechanical laws. Since the classical laws seem to predict "nonsense" (an infinity of possible billiard ball trajectories), Klinkhammer and I turned to the quantum mechanical laws for deeper understanding.

The "rules of the game" are very different in quantum physics than in classical physics. When one provides the classical laws with initial conditions, they predict what will happen afterward (for example, what trajectory a ball will follow); and, if there are no time machines, their predictions are unique. The quantum laws, by contrast, predict only *probabilities* for what will happen, not certainties (for example, the probability that a ball will travel through this, that, or another region of space).

In light of these rules of the quantum mechanical game, the answer that Klinkhammer and I got from the quantum mechanical laws is not surprising. We learned that, if the ball starts out moving along Polchinski's paradoxical trajectory (Figures 14.9 and 14.10 at time $t =$ 3:00 P.M.), then there will be a certain quantum mechanical probability—say, 48 percent—for it subsequently to follow trajectory (a) in Figure 14.10, and a certain probability—say, also 48 percent—for trajectory (b), and a certain (far smaller) probability for each of the infinity of other classically allowed trajectories. In any one "experiment," the ball will follow just one of the trajectories that the classical laws allow; but if we perform a huge number of identical billiard ball experiments, in 48 percent of them the ball will follow trajectory (a), in 48 percent trajectory (b), and so forth.

This conclusion is somewhat satisfying. It suggests that the laws of physics might accommodate themselves to time machines fairly nicely. There are surprises, but there seem not to be any outrageous predictions, and there is no sign of any unresolvable paradox. Indeed, the *National Enquirer*, hearing of this, could easily display a banner headline: PHYSICISTS PROVE TIME MACHINES EXIST. (That kind of outrageous distortion, of course, has been my recurrent fear.)

In the autumn of 1988, three months after the publication of our paper "Wormholes, Time Machines, and the Weak Energy Condition,"

Keay Davidson, a reporter for the *San Francisco Examiner*, discovered it in *Physical Review Letters* and broke the story.

It could have been worse. At least the physics community had had three months of quiet in which to absorb our ideas without the blare of sensational headlines.

But the blare was unstoppable. PHYSICISTS INVENT TIME MACHINES, read a typical headline. *California* magazine, in an article on "The Man Who Invented Time Travel," even ran a photograph of me doing physics in the nude on Palomar Mountain. I was mortified—not by the photo, but by the totally outrageous claims that I had invented time machines and time travel. *If time machines are, in fact, allowed by the laws of physics (and, as will become clear at the end of the chapter, I doubt that they are), then they are probably much farther beyond the human race's present technological capabilities than space travel was beyond the capabilities of cavemen.*

After talking with two reporters, I abandoned all efforts to stem the tide and get the story told accurately, and went into hiding. My besieged administrative assistant, Pat Lyon, had to fend off the press with a firm "Professor Thorne believes it is too early in this research effort to communicate results to the general public. When he feels he has a better understanding of whether or not time machines are forbidden by the laws of physics, he will write an article for the public, explaining."

With this chapter of this book, I am making good on that promise.

Chronology Protection?

In February 1989, as the hoopla in the press was beginning to subside, and while Echeverria, Klinkhammer, and I were struggling with Polchinski's paradox, I flew to Bozeman, Montana, to give a lecture. There I ran into Bill Hiscock, a former student of Charles Misner's. As I have with so many colleagues, I pressed Hiscock for his views on wormholes and time machines. I was searching for cogent criticisms, new ideas, new viewpoints.

"Maybe you should study electromagnetic vacuum fluctuations," Hiscock told me. "Maybe they will destroy the wormhole when infinitely advanced beings try to turn it into a time machine." Hiscock had in mind the thought experiment in which my wife Carolee (assumed to be infinitely advanced) is flying back to Earth in the family spacecraft with one wormhole mouth, while I sit on Earth with the other mouth, and the wormhole is on the verge of becoming a time machine (Figures 14.7 and 14.8 above). Hiscock was speculating that electromagnetic vacuum fluctuations might circulate through the wormhole in the same manner as did bits of radiation in Figure 14.8; and, piling up on themselves, the fluctuations might become infinitely violent and destroy the wormhole.

I was skeptical. A year earlier, on my drive home from Chicago, I had convinced myself that bits of radiation, circulating through the wormhole, will *not* pile up on themselves, create an infinitely energetic beam, and destroy the wormhole. By defocusing the radiation, the wormhole saves itself. Surely, I thought, the wormhole will also defocus a circulating beam of electromagnetic vacuum fluctuations and thereby save itself.

On the other hand, I thought to myself, time machines are such a radical concept in physics that we must investigate anything which has any chance at all of destroying them. So, despite my skepticism, I set out with a postdoc in my group, Sung-Won Kim, to compute the behavior of circulating vacuum fluctuations.

Though we were helped greatly by mathematical tools and ideas that Hiscock and Deborah Konkowski had developed a few years earlier, Kim and I were hampered by our own ineptness. Neither of us was an expert on the laws that govern the circulating vacuum fluctuations: the laws of quantum fields in curved spacetime (Chapter 13). Finally, however, in February 1990, after a full year of false starts and mistakes, our calculations coalesced and gave an answer.

I was surprised and shocked. Despite the wormhole's attempt to defocus them, the vacuum fluctuations tended to refocus of their own accord (Figure 14.11). Defocused by the wormhole, they splayed out from the mouth on Earth as though they were going to miss the spacecraft; then of their own accord, as though being attracted by some mysterious force, they zeroed in on the wormhole mouth in Carolee's spacecraft. Returning to Earth through the wormhole, they then splayed out from the mouth on Earth again, and zeroed in once again on the mouth in the spacecraft. Over and over again they repeated this motion, building up an intense beam of fluctuational energy.

Will this beam of electromagnetic vacuum fluctuations be intense enough to destroy the wormhole? Kim and I asked ourselves. For eight months, February to September 1990, we struggled with this question. Finally, after several flip-flops, we concluded (incorrectly) "probably

not." Our reasoning seemed compelling to us and to the several colleagues we ran it past, so we laid it out in a manuscript and submitted it
to the *Physical Review*.

Our reasoning was this: Our calculations had shown that the circulating electromagnetic vacuum fluctuations are *infinitely intense
only for a vanishingly short period of time*. They rise to their peak at
precisely the instant when it is first possible to use the wormhole for
backward time travel (that is, at the moment when the wormhole first
becomes a time machine), and then they immediately start to die out;
see Figure 14.12.

Now, the (ill-understood) laws of quantum gravity seem to insist
that there is no such thing as a "vanishingly short period of time."
Rather, just as fluctuations of spacetime curvature make the concept of
length meaningless on scales smaller than the Planck–Wheeler length,
10^{-33} centimeter (Figure 14.3 and associated discussion), so also the
curvature fluctuations should make the concept of time meaningless on
scales smaller than 10^{-43} second (the "Planck–Wheeler time," which is
equal to the Planck–Wheeler length divided by the speed of light).
Time intervals shorter than this cannot exist, the laws of quantum

14.11 As Carolee and I try to convert a wormhole into a time machine by the
method of Figure 14.7, electromagnetic vacuum fluctuations zoom between the
two mouths and through the wormhole, piling up on themselves and creating a
beam of huge fluctuational energy.

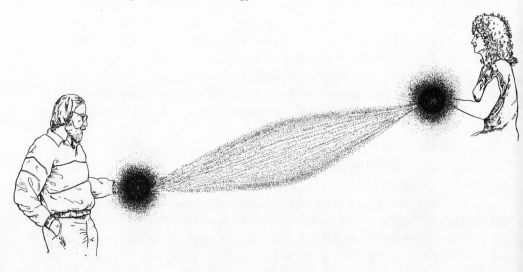

gravity seem to insist. The concepts of *before* and *after* and *evolution with time* make no sense during intervals so small.

Therefore, Kim and I reasoned, the circulating electromagnetic vacuum fluctuations must stop evolving with time, that is, must stop growing, 10^{-43} second before the wormhole becomes a time machine; the laws of quantum gravity must cut off the fluctuations' growth. And the quantum gravity laws will let the fluctuations continue their evolution again only 10^{-43} second after the time machine is born, which means after they have begun to die out. In between these times, there is no *time* and there is no evolution (Figure 14.12). The crucial issue, then, was *just how intense has the beam of circulating fluctuations become when quantum gravity cuts off their growth?* Our calculations were clear and unequivocal: The beam, when it stops growing, is far too weak to damage the wormhole, and therefore, in the words of our manuscript, it seemed likely that "vacuum fluctuations cannot prevent the formation of or existence of closed timelike curves." (As I mentioned earlier, *closed timelike curves* is physicists' jargon for "time ma-

14.12 Evolution of the intensity of the electromagnetic vacuum fluctuations that circulate through a wormhole just before and just after the wormhole becomes a time machine.

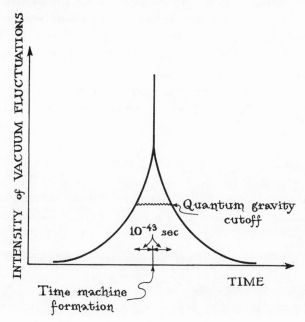

chines"; having been burned by the press, I had stopped using the phrase "time machines" in my papers; and the press, unfamiliar with physicists' jargon, was now unaware of the new time machine results I was publishing.)

In September 1990, when we submitted our manuscript to the *Physical Review*, Kim and I sent copies to a number of colleagues, including Stephen Hawking. Hawking read our manuscript with interest—and disagreed. Hawking had no quarrel with our calculation of the beam of circulating vacuum fluctuations (and, in fact, a similar calculation by Valery Frolov in Moscow had by then verified our results). Hawking's quarrel was with our analysis of quantum gravity's effects.

Hawking agreed that quantum gravity was likely to cut off the growth of the vacuum fluctuations 10^{-43} second before the time machine is created, that is, 10^{-43} second before they otherwise would become infinitely strong. "But 10^{-43} second as measured by whom? In whose reference frame?" he asked. Time is "relative," not absolute, Hawking reminded us; it depends on one's reference frame. Kim and I had assumed that the appropriate reference frame was that of somebody at rest in the wormhole throat. Hawking argued, instead (in effect), for a different choice of reference frame: that of the fluctuations themselves—or, stated more precisely, the reference frame of an observer who circulates, along with the fluctuations, from Earth to spacecraft and through the wormhole so rapidly that he sees the Earth–spacecraft distance contracted from 10 light-years (10^{19} centimeters) down to the Planck–Wheeler length (10^{-33} centimeter). The laws of quantum gravity can take over and stop the growth of the beam only 10^{-43} second before the wormhole becomes a time machine, *as seen by such a circulating observer,* Hawking conjectured.

Translating back to the viewpoint of an observer at rest in the wormhole (the observer that Kim and I had relied on), Hawking's conjecture meant that the quantum gravity cutoff occurs 10^{-95} second before the wormhole becomes a time machine, not 10^{-43} second—and by then, according to our calculations, the vacuum fluctuational beam is strong enough, but just barely so, that *it might indeed destroy the wormhole.*

Hawking's conjecture about the location of the quantum gravity cutoff was cogent. He might well be right, Kim and I concluded after much contemplation; and we managed to change our paper to say so before it got published.

The bottom line, however, was equivocal. Even if Hawking was right, it was far from clear whether the beam of vacuum fluctuations would destroy the wormhole or not—and to find out for certain would require understanding what quantum gravity does, when it takes hold in the 10^{-95}-second interval around the moment of time machine formation.

To put it succinctly, *the laws of quantum gravity are hiding from us the answer to whether wormholes can be converted successfully into time machines.* To learn the answer, we humans must first become experts on quantum gravity's laws.

Hawking has a firm opinion on time machines. He thinks that nature abhors them, and he has embodied that abhorence in a conjecture, the *chronology protection* conjecture, which says that *the laws of physics do not allow time machines.* (Hawking, in his characteristic off-the-wall humor, describes this as a conjecture that will "keep the world safe for historians.")

Hawking suspects that the growing beam of vacuum fluctuations is nature's way of enforcing chronology protection: *Whenever one tries to make a time machine, and no matter what kind of device one uses in one's attempt (a wormhole, a spinning cylinder,*[13] *a "cosmic string",*[14] *or whatever), just before one's device becomes a time machine, a beam of vacuum fluctuations will circulate through the device and destroy it.* Hawking seems ready to bet heavily on this outcome.

I am *not* willing to take the other side in such a bet. I *do* enjoy making bets with Hawking, but only bets that I have a reasonable chance of winning. My strong gut feeling is that I would lose this one. My own calculations with Kim, and unpublished calculations that Eanna Flanagan (a student of mine) has done more recently, suggest to me that Hawking is likely to be right. However, we cannot know for sure until physicists have fathomed in depth the laws of quantum gravity.

13. See Footnote 8 on page 499.
14. Recently Richard Gott of Princeton University has discovered that one can make a time machine by taking two infinitely long cosmic strings (hypothetical objects that might or might not exist in the real Universe) and moving them past each other at very high speed.

Epilogue

*an overview of Einstein's legacy,
past and future,
and an update on several central characters*

It now is nearly a full century since Einstein destroyed Newton's concept of space and time as absolute, and began laying the foundations for his own legacy. Over the intervening century, Einstein's legacy has grown to include, among many other things, a warpage of spacetime and a set of exotic objects made wholly and solely from that warpage: black holes, gravitational waves, singularities (clothed and naked), wormholes, and time machines.

At one epoch in history or another, physicists have regarded each of these objects as outrageous.

- We have met, in this book, Eddington's, Wheeler's, and even Einstein's vigorous skepticism about *black holes;* Eddington and Einstein died before they were firmly proved wrong, but Wheeler became a convert and black-hole advocate.
- During the 1940s and 1950s, a number of physicists, building on mistaken interpretations of the general relativistic mathematics they were studying, were highly skeptical of *gravitational waves* (ripples of curvature)—but that is a story for another book, and the skepticism long since has vanished.

- It was a horrendous shock to most physicists, and still is to many, to discover that singularities are an inevitable consequence of Einstein's general relativistic laws. Some physicists derive comfort from faith in Penrose's cosmic censorship conjecture (that all singularities are clothed; naked singularities are forbidden). But whether cosmic censorship is wrong or right, most physicists have accommodated to singularities and, like Wheeler, expect the ill-understood laws of quantum gravity to tame them—ruling and controlling them in just the same way as Newton's or Einstein's laws of gravity rule the planets and control their orbits around the Sun.

- Wormholes and time machines today are regarded as outrageous by most physicists, even though Einstein's general relativistic laws permit them to exist. Skeptical physicists can take comfort, however, in our newfound knowledge that the existence of wormholes and time machines is controlled not by Einstein's rather permissive laws, but rather by the more restrictive laws of quantum fields in curved spacetime, and quantum gravity. When we understand those laws better, *perhaps* they will teach us unequivocally that physical laws always protect the Universe against wormholes and time machines—or at least time machines. Perhaps.

What can we expect in the coming century, the second century of Einstein's legacy?

It seems likely that the revolution in our understanding of space, time, and objects built from spacetime warpage will be no less than in the first century. The seeds for revolution have been laid:

- Gravitational-wave detectors will soon bring us observational maps of black holes, and the symphonic sounds of black holes colliding—symphonies filled with rich, new information about how warped spacetime behaves when wildly vibrating. Supercomputer simulations will attempt to replicate the symphonies and tell us what they mean, and black holes thereby will become objects of detailed experimental scrutiny. What will that scrutiny teach us? There will be surprises.

- Ultimately, in the coming century, most likely sooner rather than later, some insightful physicist will discover and unveil the laws of quantum gravity, in all their intimate detail.

- With those quantum gravity laws in hand, we may figure out precisely how our Universe's spacetime came into being, how it emerged from the quantum foam and froth of the big bang singularity. We may learn for sure the meaning or the meaninglessness of the oft-asked question, "What preceded the big bang?" We may learn for sure whether quantum foam produces multiple universes with ease, and the full details of how spacetime gets destroyed in the singularity at the core of a black hole or in the big crunch, and how and whether and where spacetime gets re-created again. And we may learn whether the laws of quantum gravity permit or forbid time machines: Must time machines always self-destruct at the moment they are activated?

· The laws of quantum gravity are not the final set of physical laws along the route that has led from Newton to special relativity, to general relativity and quantum theory, and then to quantum gravity. The quantum gravity laws will still have to be married to (unified with) the laws that govern nature's other fundamental forces: the electromagnetic force, the weak force, and the strong force. We will probably learn the details of that unification in the coming century—and again, most likely sooner rather than later; and that unification may radically alter our view of the Universe. And what then? No human today can foresee beyond that point, I believe—and yet, that point may well come in my own lifetime, and in yours.

In Closing, November 1993

Albert Einstein spent most of his last twenty-five years in a fruitless quest to unify his general relativistic laws of physics with Maxwell's laws of electromagnetism; he did not know that the most important unification is with quantum mechanics. He died in Princeton, New Jersey, in 1955 at the age of seventy-six.

Subrahmanyan Chandrasekhar, now eighty-three years old, continues to plumb the secrets of Einstein's field equation, often in collaboration with much younger colleagues. In recent years he has taught us much about pulsations of stars and collisions of gravitational waves.

Fritz Zwicky became less a theorist and more an observational astronomer as he aged; and he continued to generate controversial, prescient ideas, though not on the topics of this book. He retired from his Caltech professorship in 1968 and moved to Switzerland, where he spent his final years promoting his own inside track to knowledge: the "morphological method." He died in 1974.

Lev Davidovich Landau recovered intellectually, but not emotionally, from his year in prison (1938–39) and then continued on as the dominant figure and most revered teacher among Soviet theoretical physicists. In 1962 he was critically injured in an automobile accident, which left him with brain damage that changed his personality and destroyed his ability to do physics. He died in 1968, but his closest friends said of him afterward, "For me, 'Dau died in 1962."

Yakov Borisovich Zel'dovich remained the world's most influential astrophysicist through the 1970s and into the 1980s. However, in 1978, in a tragic interpersonal explosion, he split off from most of his research group (the most powerful team of theoretical astrophysicists that the world has ever seen). He tried to rebuild with a fresh set of young colleagues, but was only partially successful, and then in the 1980s he became a guru for astrophysicists and cosmologists, worldwide. He died of a heart attack in Moscow in 1987, soon after Gorbachev's political changes made it possible for him to travel to America for the first time.

Igor Dmitrievich Novikov became the leader of the Zel'dovich/Novikov research group after the split with Zel'dovich. Through the 1980s he held the group together with the same kind of fire and stimulus as Zel'dovich had mustered in the old days. However, without Zel'dovich, his group was merely among the best in the world, and not far ahead of everyone else, as before. With the collapse of the Soviet Union in 1991, and following a heart operation that made him feel his finiteness, Novikov moved to the University of Copenhagen in Denmark, where he is now creating a new Theoretical Astrophysics Center.

Vitaly Lazarevich Ginzburg, at age seventy-seven, continues to do forefront research in several different branches of physics and astrophysics. During **Andrei Sakharov**'s exile to Gorky in 1980–86, Ginzburg, as Sakharov's official "boss" at the Lebedev Institute in Moscow, refused to fire him and acted as a sort of protector. Under Gorbachev's perestroika, Ginzburg and Sakharov were both elected members of the Chamber of People's Deputies of the U.S.S.R., where they pushed for reform. Sakharov died of a heart attack in 1989.

J. Robert Oppenheimer, though repudiated by the United States government in his 1954 security clearance hearings, became a hero to the majority of the physics community. He never returned to research, but he remained closely in touch with most all branches of physics, and served as a powerful foil off whom younger physicists could bounce their ideas, until his death from cancer in 1967.

John Wheeler, at age eighty-two, continues his quest to understand the marriage of quantum mechanics and general relativity—and continues to inspire younger generations with his lectures and writings, most notably his recent book *A Journey into Gravity and Spacetime* (Wheeler, 1990).

Roger Penrose, like Wheeler and many others, is obsessed with the marriage of general relativity and quantum mechanics and with the ill-understood laws of quantum gravity that should spring forth from that marriage. He has written about his unconventional ideas in a book for nonphysicists (*The Emperor's New Mind*, Penrose, 1989). Many physicists are skeptical of his views, but Penrose has been right so many times before . . .

Stephen Hawking also continues to be obsessed with the laws of quantum gravity, and most especially with the question of what those laws predict about the origin of the Universe. Like Penrose, he has written a book for nonphysicists, describing his ideas (*A Brief History of Time*, Hawking, 1988). His health holds strong, despite his ALS.

Acknowledgments

*my debts of gratitude
to friends and colleagues
who influenced this book*

Elaine Hawkes Watson, by her boundless curiosity about the Universe, inspired me to embark on this book. During my fifteen years of on-and-off writing, I received invaluable encouragement and support from several close friends and family: Linda Thorne, Kares Thorne, Bret Thorne, Alison Thorne, Estelle Gregory, Bonnie Schumaker, and most especially my wife, Carolee Winstein.

I am indebted to a number of my physicist, astrophysicist, and astronomer colleagues, who consented to be interviewed by me on tape about their recollections of the historical events and research efforts described in this book. Their names appear in the list of taped interviews at the beginning of the bibliography.

Four of my colleagues, Vladimir Braginsky, Stephen Hawking, Werner Israel, and Carl Sagan, were kind enough to read the entire manuscript and give me detailed critiques. Many others read individual chapters or several chapters and straightened me out on important historical and scientific details: Vladimir Belinsky, Roger Blandford, Carlton Caves, S. Chandrasekhar, Ronald Drever, Vitaly Ginzburg, Jesse Greenstein, Isaac Khalatnikov, Igor Novikov, Roger Penrose, Dennis Sciama, Robert Serber, Robert Spero, Alexi Starobinsky, Rochus Vogt, Robert Wald, John Wheeler, and Yakov Borisovich Zel'dovich. Without the advice of these colleagues, the book would be far less accurate than it is. However, one should not assume that my colleagues agree with me or approve of all my interpretations of our joint history. Inevitably there have been a few conflicts of viewpoint. In the text, for pedagogy's sake, I hew to my own viewpoint (often, but not always, significantly influenced by my colleagues' critiques). In the notes, for historical accuracy, I expose some of the conflicts.

Lynda Obst tore much of the first version of the book to shreds. I thank her. K. C. Cole tore the second version to shreds and then patiently gave me crucial advice on draft after draft, until the presentation was honed. To K. C. I am especially indebted. I also thank Debra Makay for meticulously cleaning up the final manuscript; she is even more of a perfectionist than I.

The book was significantly improved by critiques from several nonphysicist readers: Ludmila (Lily) Birladeanu, Doris Drücker, Linda Feferman, Rebecca Lewthwaite, Peter Lyman, Deanna Metzger, Phil Richman, Barrie Thorne, Alison Thorne, and Carolee Winstein. I thank them, and I thank Helen Knudsen for locating a number of references and facts—some unbelievably obscure.

I was fortunate to run across Matthew Zimet's delightful drawings in Heinz Pagel's book *The Cosmic Code*, and attract him to illustrate my book as well. His illustrations add so much.

Finally, I wish to thank the Commonwealth Fund Book Program and especially Alexander G. Bearn and Antonina W. Bouis—and also Ed Barber of W.W. Norton and Company—for their support, their patience, and their faith in me as a writer during the years that it took to bring this book to completion.

Characters

—————

*a list of characters
who appear significantly
at several different places in the book*

NOTE: The following descriptions are meant to serve solely as reminders of and cross-references to each person's various appearances in this book. These descriptions are *not* intended as biographical sketches. (Most of these people have made major contributions to science that are not relevant to this book and therefore are not listed here.) The principal criterion for inclusion in this section is *not* importance of contributions, but rather multiple appearances of the person at several different locations in the book.

Baade, Walter (1893–1960). German born, American optical astronomer; with Zwicky, developed the concept of a supernova and its connection to neutron stars (Chapter 5); identified the galaxies associated with cosmic radio sources (Chapter 9).

Bardeen, James Maxwell (b. 1939). American theoretical physicist; showed that many or most black holes in our Universe should be rapidly spinning and, with Petterson, predicted the influence of the holes' spins on surrounding accretion disks (Chapter 9); with Carter and Hawking, discovered the four laws of black-hole mechanics (the laws of evolution of black holes) (Chapter 12).

Bekenstein, Jacob (b. 1947). Israeli theoretical physicist; student of Wheeler's; with Hartle, showed that one cannot discern, by any external study of a black hole, what kinds of particles were among the material that formed it (Chapter 7); proposed that the surface area of a black hole is its entropy in disguise, and carried on a battle with Hawking over this idea, ultimately winning (Chapter 12).

Bohr, Niels Hendrik David (1885–1962). Danish theoretical physicist; Nobel laureate; one of the founders of quantum mechanics; mentor for many of the leading physicists of the middle twentieth century, including Lev Landau and John Wheeler; advised Chandrasekhar in his battle with Eddington (Chapter 4); tried to save Landau from prison (Chapter 5); with Wheeler developed the theory of nuclear fission (Chapter 6).

Braginsky, Vladimir Borisovich (b. 1931). Russian experimental physicist; discovered quantum mechanical limits on the precision of physical measurements, including those of gravitational-wave detectors (Chapter 10); inventor of the concept of "quantum nondemolition" devices, which circumvent those quantum limits (Chapter 10).

Carter, Brandon (b. 1942). Australian theoretical physicist; student of Dennis Sciama's in Cambridge, England; later moved to France; elucidated the properties of spinning black holes (Chapter 7); with others, proved that a black hole has no hair (Chapter 7); with Bardeen and Hawking discovered the four laws of black-hole mechanics (the laws of evolution of black holes) (Chapter 12).

Chandrasekhar, Subrahmanyan (b. 1910). Indian born, American astrophysicist; Nobel laureate; proved that there is a maximum mass for white-dwarf stars and fought a battle with Eddington over the correctness of his prediction (Chapter 4); developed much of the theory of how black holes respond to small perturbations (Chapter 7).

Eddington, Arthur Stanley (1882–1944). British astrophysicist; leading early exponent of Einstein's laws of general relativity (Chapter 3); vigorous opponent of the concept of a black hole and of Chandrasekhar's conclusion that white dwarfs have a maximum mass (Chapters 3 and 4).

Einstein, Albert (1879–1955). German born, Swiss/American theoretical physicist; Nobel laureate; formulated the laws of special relativity (Chapter 1) and general relativity (Chapter 2); showed that light is simultaneously a particle and a wave (Chapter 4); opposed the concept of a black hole (Chapter 3).

Geroch, Robert (b. 1942). American theoretical physicist; student of Wheeler's; with others, developed global methods for analyzing black holes (Chapter 13); showed that the topology of space can change (for example, when a wormhole forms) only if a time machine is produced in the process (Chapter 14); with Wald, gave the first argument suggesting that time machines might be destroyed whenever they try to form (Chapter 14).

Giacconi, Riccardo (b. 1931). Italian born, American experimental physicist and astrophysicist; led the team that discovered the first X-ray star, in 1962, using a detector flown on a rocket (Chapter 8); led the team that designed and built the Uhuru X-ray satellite, which produced the first strong X-ray evidence that Cygnus X-1 is a black hole (Chapter 8).

Ginzburg, Vitaly Lazarevich (b. 1916). Soviet theoretical physicist; invented the LiD fuel for the Soviet hydrogen bomb and then was separated from the bomb project (Chapter 6); with Landau, developed an explanation for the origin of superconductivity (Chapters 6 and 9); discovered the first evidence that a black hole has no hair (Chapter 7); developed the synchrotron radiation explanation for the origin of cosmic radio waves (Chapter 9).

Greenstein, Jesse L. (b. 1909). American optical astronomer; colleague of Zwicky's (Chapter 5); with Fred Whipple found it impossible to explain cosmic radio waves (Chapter 9); triggered the beginning of America's research effort in radio astronomy (Chapter 9); with Maarten Schmidt, discovered quasars (Chapter 9).

Hartle, James B. (b. 1939). Student of Wheeler's; with Bekenstein, showed that one cannot discern, by any external study of a black hole, what kinds of particles were among the material that formed it (Chapter 7); with Hawking, discovered the laws that govern the evolution of a black hole's horizon (Chapter 12); with Hawking, is developing insights into the laws of quantum gravity (Chapter 13).

Hawking, Stephen W. (b. 1942). British theoretical physicist; student of Sciama's; developed key parts of the proof that a black hole has no hair (Chapter 7); with Bardeen and Carter, discovered the four laws of black-hole mechanics (the laws of evolution of black holes) (Chapter 12); discovered that, if one ignores the laws of quantum mechanics, the surface areas of black holes can only increase, but quantum mechanics makes black holes evaporate and shrink (Chapter 12); showed that tiny black holes could have formed in the big bang and, with Page, placed observational limits on such primordial holes based on astronomers not seeing gamma rays produced by their evaporation (Chapter 12); developed global (topological) methods for analyzing black holes (Chapter 13); with Penrose, proved that the big bang contained a singularity (Chapter 13); formulated the chronology protection conjecture and argued that it is enforced by vacuum fluctuations destroying any time machine at the moment it is created (Chapter 14); made bets with Kip Thorne over whether Cygnus X-1 is a black hole (Chapter 8) and whether naked singularities can form in our Universe (Chapter 13).

Israel, Werner (b. 1931). South African born, Canadian theoretical physicist; proved that every nonspinning black hole must be spherical, and gave evidence that a black hole loses its "hair" by radiating it away (Chapter 7); discovered that the surface areas of black holes can only increase, but did not realize the significance of this conclusion (Chapter 12); with Poisson and Ori, showed that the tidal forces that surround a black hole's singularity become weaker as the hole ages (Chapter 13); developed insights into the early history of black-hole research (Chapter 3).

Kerr, Roy P. (b. 1934). New Zealander mathematician; discovered the solution to Einstein's field equation, which describes a spinning black hole: the "Kerr solution" (Chapter 7).

Landau, Lev Davidovich (1908–1968). Soviet theoretical physicist; Nobel laureate; transfused theoretical physics from Western Europe into the U.S.S.R. in the 1930s (Chapters 5 and 13); tried to explain stellar heat as produced by stellar material being captured onto a neutron core at the star's center, and thereby triggered Oppenheimer's research on neutron stars and black holes (Chapter 5); was imprisoned in Stalin's Great Terror and then released so he could develop the theory of superfluidity (Chapter 5); contributed to Soviet nuclear weapons research (Chapter 6).

Laplace, Pierre Simon (1749–1827). French natural philosopher; developed and popularized the concept of a dark star (black hole) as governed by Newton's laws of physics (Chapters 3 and 6).

Lorentz, Hendrik Antoon (1853–1928). Dutch theoretical physicist; Nobel laureate; developed key foundations for the laws of special relativity, the most important being the Lorentz–Fitzgerald contraction and time dilation (Chapter 1); friend and associate of Einstein when Einstein was developing his general relativistic laws of physics (Chapter 2).

Maxwell, James Clerk (1831–1879). British theoretical physicist; developed the laws of electricity and magnetism (Chapter 1).

Michell, John (1724–1793). British natural philosopher; developed and popularized the concept of a dark star (black hole) as governed by Newton's laws of physics (Chapters 3 and 6).

Michelson, Albert Abraham (1852–1931). German-born, American experimental physicist; Nobel laureate; invented the techniques of interferometry (Chapter 1); used those techniques to discover that the speed of light is independent of one's velocity through the Universe (Chapter 1).

Minkowski, Hermann (1864–1909). German theoretical physicist; teacher of Einstein (Chapter 1); discovered that space and time are unified into spacetime (Chapter 2).

Misner, Charles W. (b. 1932). American theoretical physicist; student of Wheeler's; developed an insightful embedding diagram description of how an imploding star produces a black hole (Chapter 6); created a research group that contributed significantly to the "golden age" of black-hole research (Chapter 7); discovered that electromagnetic and other waves propagating near a spinning black hole can extract rotational energy from the hole and use it to amplify themselves (Chapter 12); discovered the oscillatory, "mixmaster" oscillations of tidal gravity near singularities (Chapter 13).

Newton, Isaac (1642–1727). British natural philosopher; developed the foundations for the Newtonian laws of physics and for the concept of space and time as absolute (Chapter 1); developed the Newtonian laws of gravity (Chapter 2).

Novikov, Igor Dmitrievich (b. 1935). Soviet theoretical physicist and astrophysicist; student of Zel'dovich's; with Doroshkevich and Zel'dovich, developed some of the key initial evidence that a black hole has no hair (Chapter 7); with Zel'dovich, proposed the method for astronomical searches for black holes in our galaxy that seems to have finally succeeded (Chapter 8); with Thorne, developed the theory of the structures of accretion disks around black holes (Chapter 12); with Doroshkevich, predicted that the tidal forces inside a black hole must change as the hole ages (Chapter 13); carried out research on whether the laws of physics permit time machines (Chapter 14).

Oppenheimer, J. Robert (1904–1967). American theoretical physicist; transfused theoretical physics from Western Europe to the United States in the 1930s (Chapter 5); with Serber, disproved Landau's claim that stars might be kept hot by neutron cores, and with Volkoff, demonstrated that there is a maximum possible mass for neutron stars (Chapter 5); with Snyder, demonstrated, in a highly idealized model, that when massive stars die, they must implode to form black holes, and elucidated key features of the implosion (Chapter 6); led the American atomic bomb project, opposed the hydrogen bomb project early on and then endorsed it and lost his security clearance (Chapter 6); did battle with Wheeler over whether implosion produces black holes (Chapter 6).

Penrose, Roger (b. 1931). British mathematician and theoretical physicist; protégé of Sciama's; speculated that black holes lose their hair by radiating it away (Chapter 7); discovered that spinning black holes store huge amounts of energy in the swirl of space outside their horizons and that this energy can be extracted (Chapter 7); developed the concept of a black hole's apparent horizon (Chapters 12 and 13); discovered that the surface areas of black holes must increase, but did not realize the significance of that conclusion (Chapter 12); invented and developed global (topological) methods for analyzing black holes (Chapter 13); proved that black holes must have singularities in their cores and, with Hawking, proved that the big bang contained a singularity (Chapter 13); proposed the cosmic censorship conjecture, that the laws of physics prevent naked singularities from forming in our Universe (Chapter 13).

Press, William H. (b. 1948). American theoretical physicist and astrophysicist; student of Thorne's; with Teukolsky, proved that black holes are stable against small perturbations (Chapters 7 and 12); discovered that black holes can pulsate (Chapter 7); organized the funeral for the golden age of black-hole research (Chapter 7).

Price, Richard H. (b. 1943). American theoretical physicist and astrophysicist; student of Thorne's; gave the definitive proof that a black hole loses its hair by radiating the hair away and proved that anything which can be radiated will be radiated away completely (Chapter 7); saw evidence that black holes pulsate but did not recognize its significance (Chapter 7); with others developed the membrane paradigm for black holes (Chapter 11); worried about Thorne's sanity when Thorne initiated research on time machines (Chapter 14).

Rees, Martin (b. 1942). British astrophysicist; student of Sciama's; developed models that explain the observed features of binary systems in which a black hole accretes gas from a companion star (Chapter 8); proposed that the giant lobes of a radio galaxy are powered by beams of energy that travel from the galaxy's core to the lobes, and with Blandford developed detailed models for the beams (Chapter 9); with Blandford and others, developed models that explain how a supermassive black hole can energize radio galaxies, quasars, and active galactic nuclei (Chapter 9).

Sakharov, Andrei Dmitrievich (1921–1989). Soviet theoretical physicist; invented key ideas that underlie the Soviet hydrogen bomb (Chapter 6); close friend, associate, and competitor of Zel'dovich's (Chapters 6 and 7); later became the leading Soviet dissident and, after glasnost, Soviet saint.

Schwarzschild, Karl (1876–1916). German astrophysicist; discovered the Schwarzschild solution of the Einstein field equation, which describes the spacetime geometry of a nonspinning star that is either static or imploding, and also describes a nonspinning black hole (Chapter 3); discovered the solution of the Einstein equation for the interior of a constant-density star—a solution that Einstein used to argue that black holes cannot exist (Chapter 3).

Sciama, Dennis (b. 1926). British astrophysicist and mentor for British researchers on black holes (Chapters 7 and 13).

Teukolsky, Saul A. (b. 1947). South African born, American theoretical physicist; student of Thorne's; invented and developed the formalism by which perturbations of spinning black holes are analyzed and, with Press, used his formalism to show that black holes are stable against small perturbations (Chapters 7 and 12); with Shapiro, discovered evidence that the laws of physics might permit naked singularities to form in our Universe (Chapter 13).

Thorne, Kip S. (b. 1940). American theoretical physicist; student of Wheeler's; proposed the hoop conjecture which describes when black holes can form in an imploding star, and developed evidence for it (Chapter 7); made estimates of the gravitational waves from astrophysical sources and contributed to ideas and plans for the detection of those waves (Chapter 10); with others, developed the membrane paradigm for black holes (Chapter 11); developed ideas about the statistical origin of the entropy of a black hole (Chapter 12); probed the laws of physics by means of thought experiments about wormholes and time machines (Chapter 14).

Wald, Robert M. (b. 1947). American theoretical physicist; student of Wheeler's; contributed to the Teukolsky formalism for analyzing perturbations of black holes and its applications (Chapter 7); with others, developed an understanding of how electric fields behave outside a black hole—an understanding that underlies the membrane paradigm (Chapter 11); contributed to the theory of the evaporation of black holes and its implications for the origin of black-hole entropy (Chapter 12); with Geroch, gave the first argument suggesting that time machines might be destroyed whenever they try to form (Chapter 14).

Weber, Joseph (b. 1919). American experimental physicist; invented the world's first gravitational-wave detectors ("bar detectors") and co-invented interferometric detectors for gravitational waves (Chapter 10); universally regarded as the "father" of the field of gravitational-wave detection.

Wheeler, John Archibald (b. 1911). American theoretical physicist; mentor for American researchers on black holes and other aspects of general relativity (Chapters 7); with Harrison and Wakano, developed the equation of state for cold, dead matter and a complete catalog of cold, dead stars, thereby firming up evidence that when massive stars die they must form black holes (Chapter 5); with Niels Bohr, developed the theory of nuclear fission (Chapter 6); led a team that designed the first American hydrogen bombs (Chapter 6); argued in a battle with Oppenheimer that black holes cannot form, then retracted the argument and became the leading proponent of black holes (Chapter 6); coined the phrases "black hole" (Chapter 6) and "a black hole has no hair" (Chapter 7); argued that the "issue of the final state" of gravitationally imploding stars is a key to understanding the marriage between general relativity and quantum mechanics, and in this argument anticipated Hawking's discovery that black holes can evaporate (Chapters 6 and 13); developed foundations for the laws of quantum gravity and, most important, conceived and developed the concept of quantum foam, which we now suspect is the stuff of which singularities are made (Chapter 13); developed the concept of the Planck–Wheeler length and area (Chapters 12, 13, 14).

Zel'dovich, Yakov Borisovich (1914–1987). Soviet theoretical physicist and astrophysicist; mentor for Soviet astrophysicists (Chapter 7); developed the theory of nuclear chain reactions (Chapter 5); invented key ideas that underlie Soviet atomic and hydrogen bombs, and led a bomb design team (Chapter 6); with Doroshkevich and Novikov, developed early evidence that a black hole has no hair (Chapter 7); invented several methods for astronomical searches for black holes, one of which seems ultimately to have succeeded (Chapter 8); independently of Salpeter, proposed that supermassive black holes power quasars and radio galaxies (Chapter 9); conceived of the idea that the laws of quantum mechanics might cause spinning black holes to radiate and thereby lose their spin and, with Starobinsky, proved so, but then resisted Hawking's proof that even nonspinning holes can radiate and evaporate (Chapter 12).

Zwicky, Fritz (1898–1974). Swiss-born American theoretical physicist, astrophysicist, and optical astronomer; with Baade, identified supernovae as a class of astronomical objects and proposed that they are powered by energy released when a normal star becomes a neutron star (Chapter 5).

Chronology

*a chronology
of events, insights, and discoveries*

1687   Newton publishes his *Principia*, in which are formulated his concepts of absolute space and time, and his laws of motion and laws of gravity. [Ch. 1]

1783 & 1795   Michell and Laplace, using Newton's laws of motion, gravity, and light, formulate the concept of a Newtonian black hole. [Ch. 3]

1864   Maxwell formulates his unified laws of electromagnetism. [Ch. 1]

1887   Michelson and Morley show, experimentally, that the speed of light is independent of the velocity of the Earth through absolute space. [Ch. 1]

1905   Einstein shows that space and time are relative rather than absolute, and formulates the special relativistic laws of physics. [Ch. 1]

Einstein shows that electromagnetic waves behave under some circumstances like particles, thereby initiating the concept of wave/particle duality that underlies quantum mechanics. [Ch. 4]

1907   Einstein, taking his first steps toward general relativity, formulates the concept of a local inertial frame and the equivalence principle, and deduces the gravitational dilation of time. [Ch. 2]

1908   Hermann Minkowski unifies space and time into an absolute four-dimensional spacetime. [Ch. 2]

1912   Einstein realizes that spacetime is curved, and that tidal gravity is a manifestation of that curvature. [Ch. 2]

1915 Einstein and Hilbert independently formulate the Einstein field equation (which describes how mass curves spacetime), thereby completing the laws of general relativity. [Ch. 2]

1916 Karl Schwarzschild discovers the Schwarzschild solution of the Einstein field equation, which later will turn out to describe nonspinning, uncharged black holes. [Ch. 3]

Flamm discovers that, with an appropriate choice of topology, the Schwarzschild solution of the Einstein equation can describe a wormhole. [Ch. 14]

1916 & 1918 Reissner and Nordström discover their solution of the Einstein field equation, which later will describe nonspinning, charged black holes. [Ch. 7]

1926 Eddington poses the mystery of the white dwarfs and attacks the reality of black holes. [Ch. 4]

Schrödinger and Heisenberg, building on others' work, complete the formulation of the quantum mechanical laws of physics. [Ch. 4]

Fowler uses the quantum mechanical laws to show how electron degeneracy resolves the mystery of the white dwarfs. [Ch. 4]

1930 Chandrasekhar discovers that there is a maximum mass for white dwarfs. [Ch. 4]

1932 Chadwick discovers the neutron. [Ch. 5]

Jansky discovers cosmic radio waves. [Ch. 9]

1933 Landau creates his research group in the U.S.S.R. and begins to transfuse theoretical physics there from Western Europe. [Ch. 5, 13]

Baade and Zwicky identify supernovae, propose the concept of a neutron star, and suggest that supernovae are powered by the implosion of a stellar core to form a neutron star. [Ch. 5]

1935 Chandrasekhar makes more complete his demonstration of the maximum mass for white-dwarf stars, and Eddington attacks his work. [Ch. 4]

1935–1939 The Great Terror in the U.S.S.R. [Ch. 5, 6]

1937 Greenstein and Whipple demonstrate that Jansky's cosmic radio waves cannot be explained by then-known astrophysical processes. [Ch. 9]

Landau, in a desperate attempt to avoid prison and death, proposes that stars are kept hot by energy released when matter flows onto neutron cores at their centers. [Ch. 5]

1938 Landau is imprisoned in Moscow on charges of spying for Germany. [Ch. 5]

Oppenheimer and Serber disprove Landau's neutron core method for keeping stars hot; Oppenheimer and Volkoff show that there is a maximum mass for neutron stars. [Ch. 5]

Bethe and Critchfield show that the Sun and other stars are kept hot by burning nuclear fuel. [Ch. 5]

1939   Landau, near death, is released from prison. [Ch. 5]

Einstein argues that black holes cannot exist in the real Universe. [Ch. 4]

Oppenheimer and Snyder, in a highly idealized calculation, show that an imploding star forms a black hole, and (paradoxically) that the implosion appears to freeze at the horizon as seen from the outside but not as seen from the star's surface. [Ch. 6]

Reber discovers cosmic radio waves from distant galaxies, but does not know that is what he is seeing. [Ch. 9]

Bohr and Wheeler develop the theory of nuclear fission. [Ch. 6]

Khariton and Zel'dovich develop the theory of a chain reaction of nuclear fissions. [Ch. 6]

The German army invades Poland, setting off World War II.

1942   The U.S. launches a crash program to develop the atomic bomb, led by Oppenheimer. [Ch. 6]

1943   The U.S.S.R. launches a low-level effort to design nuclear reactors and atomic bombs, with Zel'dovich as a lead theorist. [Ch. 6]

1945   The U.S. drops atomic bombs on Hiroshima and Nagasaki. World War II ends. A low-level U.S. effort to develop the superbomb is begun. [Ch. 6]

The U.S.S.R. launches a crash program to develop the atomic bomb, with Zel'dovich as a lead theorist. [Ch. 6]

1946   Friedman and his team launch the first astronomical instrument above the Earth's atmosphere, on a captured German V-2 rocket. [Ch. 8]

Experimental physicists in England and Australia begin constructing radio telescopes and radio interferometers. [Ch. 9]

1948   Zel'dovich, Sakharov, Ginzburg, and others in the U.S.S.R. initiate design work for a superbomb (hydrogen bomb); Ginzburg invents the LiD fuel, Sakharov the layered-cake design. [Ch. 6]

1949   The U.S.S.R. explodes its first atomic bomb, setting off a debate in the U.S. about a crash program to develop the superbomb. The U.S.S.R. proceeds directly into a crash program for the superbomb, without debate. [Ch. 6]

1950   The U.S. launches a crash superbomb effort. [Ch. 6]

Kiepenheuer and Ginzburg realize that cosmic radio waves are produced by cosmic-ray electrons spiraling in interstellar magnetic fields. [Ch. 9]

Alexandrov and Pimenov initiate an ill-fated attempt to introduce topological tools into mathematical studies of curved spacetime. [Ch. 13]

1951   Teller and Ulam in the U.S. invent the idea for a "real" superbomb, one that can be arbitrarily powerful; Wheeler puts together a team to design a bomb based on the idea and simulate its explosion on computers. [Ch. 6]

Graham Smith provides Baade with a 1-arc-minute error box for the cosmic radio source Cyg A, and Baade discovers with an optical telescope that Cyg A is a distant galaxy—a "radio galaxy." [Ch. 9]

1952   The U.S. explodes its first superbomb device, one too massive to be delivered by an airplane or rocket, but using the Teller–Ulam invention and based on the Wheeler team's design work. [Ch. 6]

1953   Wheeler launches into research on general relativity. [Ch. 6]

Jennison and Das Gupta discover that the radio waves from galaxies are produced by two giant lobes on opposite sides of the galaxy. [Ch. 9]

Stalin dies. [Ch. 6]

The U.S.S.R. explodes its first hydrogen bomb, based on the Ginzburg and Sakharov ideas. It is claimed by U.S. scientists not to be a "real" superbomb because the design does not permit the bomb to be arbitrarily powerful. [Ch. 6]

1954   Sakharov and Zel'dovich invent the Teller–Ulam idea for a "real" super-bomb. [Ch. 6]

The U.S. explodes its first real superbomb, based on the Teller–Ulam/Sak-harov–Zel'dovich idea. [Ch. 6]

Teller testifies against Oppenheimer, and Oppenheimer's security clearance is revoked. [Ch. 6]

1955   The U.S.S.R. explodes its first real superbomb, based on the Teller–Ulam/ Sakharov–Zel'dovich idea. [Ch. 6]

Wheeler formulates the concept of gravitational vacuum fluctuations, identi-fies the Planck–Wheeler length as the scale on which they become huge, and suggests that on this scale the concept of spacetime gets replaced by quantum foam. [Ch. 12, 13, 14]

1957   Wheeler, Harrison, and Wakano formulate the concept of cold, dead matter and make a catalog of all possible cold, dead stars. Their catalog firms up the conclusion that massive stars must implode when they die. [Ch. 5]

Wheeler's group studies wormholes; Regge and Wheeler invent perturbation methods for analyzing small perturbations of wormholes; their formalism later will be used to study perturbations of black holes. [Ch. 7, 14]

Wheeler poses the issue of the final state of stellar implosion as a holy grail for research and, in a confrontation with Oppenheimer, opposes the idea that the final state will be hidden inside a black hole. [Ch. 6, 13]

1958   Finkelstein discovers a new reference frame for the Schwarzschild geometry, and it resolves the 1939 Oppenheimer–Snyder paradox of why an imploding star freezes at the critical circumference as seen from outside but implodes through the critical circumference as seen from inside. [Ch. 6]

1958–1960   Wheeler gradually embraces the concept of a black hole and becomes its leading proponent. [Ch. 6]

1959 Wheeler argues that spacetime singularities formed in the big crunch or inside a black hole are governed by the laws of quantum gravity, and may consist of quantum foam. [Ch. 13]

Burbidge shows that the giant lobes of radio galaxies contain magnetic and kinetic energy equivalent to that obtained by a perfect conversion of 10 million Suns into pure energy. [Ch. 9]

1960 Weber initiates construction of bar detectors for gravitational waves. [Ch. 10]

Kruskal shows that, if it is not threaded by any material, a spherical wormhole will pinch off so quickly that it cannot be traversed. [Ch. 14]

Graves and Brill discover that the Reissner–Nordström solution of Einstein's equation describes a spherical, electrically charged black hole and also a wormhole. [Ch. 7] Their work suggests (incorrectly) that it might be possible to travel from the interior of a black hole in our Universe through hyperspace and into some other universe. [Ch. 13]

1961 Khalatnikov and Lifshitz argue (incorrectly) that Einstein's field equation does not permit the existence of singularities with randomly deformed curvature, and therefore singularities cannot form inside real black holes or in the Universe's big crunch. [Ch. 13]

1961–1962 Zel'dovich begins research on astrophysics and general relativity, recruits Novikov, and begins to build his research team. [Ch. 6]

1962 Thorne begins research under Wheeler's guidance and initiates research that will lead to the hoop conjecture. [Ch. 7]

Giacconi and his team discover cosmic X-rays, using a Geiger counter flown above the Earth's atmosphere on an Aerobee rocket. [Ch. 8]

1963 Kerr discovers his solution of Einstein's field equation. [Ch. 7]

Schmidt, Greenstein, and Sandage discover quasars. [Ch. 9]

1964 The golden age of theoretical black-hole research begins. [Ch. 7]

Penrose introduces topology as a tool in relativity research, and uses it to prove that singularities must reside inside all black holes. [Ch. 13]

Ginzburg and then Doroshkevich, Novikov, and Zel'dovich discover the first evidence that a black hole has no "hair." [Ch. 7]

Colgate, May, and White in the U.S., and Podurets, Imshennik, and Nadezhin in the U.S.S.R., adapt bomb design computer codes to simulate realistic implosions of stellar cores; they confirm Zwicky's 1934 speculation that implosions with low mass will form a neutron star and trigger a supernova, and confirm the 1939 Oppenheimer–Snyder conclusion that implosions with larger mass will create a black hole. [Ch. 6]

Zel'dovich, Guseinov, and Salpeter make the first proposals for how to search for black holes in the real Universe. [Ch. 8]

Salpeter and Zel'dovich speculate (correctly) that supermassive black holes power quasars and radio galaxies. [Ch. 9]

Herbert Friedman and his team discover Cygnus X-1, using a Geiger counter flown on a rocket. [Ch. 8]

1965   Boyer and Lindquist, Carter, and Penrose discover that Kerr's solution of Einstein's field equation describes a spinning black hole. [Ch. 7]

1966   Zel'dovich and Novikov propose searching for black holes in binaries where one object emits X-rays and the other light; this method will succeed in the 1970s (probably). [Ch. 8]

Geroch shows that the topology of space can change (for example, a worm-hole can form) non–quantum mechanically only if a time machine is created in the process, at least momentarily. [Ch. 14]

1967   Wheeler coins the name *black hole*. [Ch. 7]

Israel proves rigorously the first piece of the black-hole, no-hair conjecture: A nonspinning black hole must be precisely spherical. [Ch. 7]

1968   Penrose argues that it is impossible to travel from the interior of a black hole in our Universe through hyperspace and into some other universe; others, in the 1970s, will confirm that his argument is correct. [Ch. 13]

Carter discovers the nature of the swirl of space around a spinning black hole and its influence on infalling particles. [Ch. 7]

Misner and independently Belinsky, Khalatnikov, and Lifshitz discover the oscillatory "mixmaster" singularity as a solution of Einstein's equation. [Ch. 13]

1969   Hawking and Penrose prove that our Universe must have had a singularity at the beginning of its big bang expansion. [Ch. 13]

Belinsky, Khalatnikov, and Lifshitz discover the oscillatory BKL singularity as a solution of Einstein's equation; they show that it has random deforma-tions of its spacetime curvature and argue that therefore it is the type of singularity that forms inside black holes and in the big crunch. [Ch. 13]

Penrose discovers that a spinning black hole stores enormous energy in the swirling motion of space around it, and that this rotational energy can be extracted. [Ch. 7]

Penrose proposes his cosmic censorship conjecture, that the laws of physics prevent naked singularities from forming. [Ch. 13]

Lynden-Bell proposes that gigantic black holes reside in the nuclei of galaxies and are surrounded by accretion disks. [Ch. 9]

Christodoulou notices a similarity between the evolution of a black hole when it slowly accretes matter and the laws of thermodynamics. [Ch. 12]

Weber announces tentative observational evidence for the existence of gravi-tational waves, triggering many other experimenters to start constructing bar detectors. By 1975 it will be clear he was not seeing waves. [Ch. 10]

Braginsky discovers evidence that there will be a quantum limit on the sensitivities of gravitational-wave detectors. [Ch. 10]

1970   Bardeen shows that the accretion of gas is likely to make typical black holes in our Universe spin very rapidly. [Ch. 9]

Price, building on work of Penrose, Novikov, and Chase, de la Cruz, and Israel, shows that black holes lose their hair by radiating it away, and he proves that anything which can be radiated will be radiated away completely. [Ch. 7]

Hawking formulates the concept of a black hole's absolute horizon and proves that the surface areas of absolute horizons always increase. [Ch. 12]

Giacconi's team constructs Uhuru, the first X-ray detector on a satellite; it is launched into orbit. [Ch. 8]

1971   Combined X-ray, radio-wave, and optical observations begin to bring strong evidence that Cygnus X-1 is a black hole orbiting a normal star. [Ch. 8]

Weiss at MIT and Forward at Hughes pioneer interferometric detectors for gravitational waves. [Ch. 10]

Rees proposes that a radio galaxy's giant lobes are powered by jets that shoot out of the galaxy's core. [Ch. 9]

Hanni and Ruffini formulate the concept of surface charge on a horizon, a foundation for the membrane paradigm. [Ch. 11]

Press discovers that black holes can pulsate. [Ch. 7]

Zel'dovich speculates that spinning black holes radiate, and Zel'dovich and Starobinsky use the laws of quantum fields in curved spacetime to justify Zel'dovich's speculation. [Ch. 12]

Hawking points out that tiny "primordial" black holes might have been created in the big bang. [Ch. 12]

1972   Carter, building on work by Hawking and Israel, proves the no-hair conjecture for spinning, uncharged black holes (except for some technical details filled in later by Robinson). He shows that such a black hole is always described by Kerr's solution of Einstein's equation. [Ch. 7]

Thorne proposes the hoop conjecture as a criterion for when black holes form. [Ch. 7]

Bekenstein conjectures that a black hole's surface area is its entropy in disguise, and conjectures that the hole's entropy is the logarithm of the number of ways the hole could have been made. Hawking argues vigorously against this conjecture. [Ch. 12]

Bardeen, Carter, and Hawking formulate the laws of evolution of black holes in a form that is identical to the laws of thermodynamics, but maintain that the horizon's surface area cannot be the hole's entropy in disguise. [Ch. 12]

Teukolsky develops perturbation methods to describe the pulsations of spinning black holes. [Ch. 7]

1973   Press and Teukolsky prove that the pulsations of a spinning black hole are stable; they do not grow by feeding off the hole's rotational energy. [Ch. 7]

1974   Hawking shows that *all* black holes, spinning or nonspinning, radiate precisely as though they had a temperature that is proportional to their surface gravity, and they thereby evaporate. He then recants his claim that the laws of black-hole mechanics are not the laws of thermodynamics in disguise and recants his critique of Bekenstein's conjecture that a hole's surface area is its entropy in disguise. [Ch. 12]

1974–1978   Blandford, Rees, and Lynden-Bell identify several methods by which supermassive black holes in the nuclei of galaxies and quasars can create jets. [Ch. 9]

1975   Bardeen and Petterson show that the swirl of space around a spinning black hole can act as a gyroscope to maintain the directions of jets. [Ch. 9]

Chandrasekhar embarks on a five-year quest to develop a complete mathematical description of perturbations of black holes. [Ch. 7]

Unruh and Davies infer that, as seen by accelerating observers just above a black hole's horizon, the hole is surrounded by a hot atmosphere of particles, whose gradual escape accounts for the hole's evaporation. [Ch. 12]

Page computes the spectrum of particles radiated by black holes. Hawking and Page, from observational data on cosmic gamma rays, infer that there can be no more than 300 tiny, primordial, evaporating black holes in each cubic light-year of space. [Ch. 12]

The golden age of theoretical black-hole research is declared finished by youthful researchers. [Ch. 7]

1977   Gibbons and Hawking verify Bekenstein's conjecture that a black hole's entropy is the logarithm of the number of ways it might have been made. [Ch. 12]

Radio astronomers use interferometers to discover the jets that feed power from a galaxy's central black-hole engine to its giant radio-emitting lobes. [Ch. 9]

Blandford and Znajek show that magnetic fields, threading the horizon of a spinning black hole, can extract the hole's spin energy, and that the extracted energy can power quasars and radio galaxies. [Ch. 9]

Znajek and Damour formulate the membrane description of a black-hole horizon. [Ch. 11]

Braginsky and colleagues, and Caves, Thorne, and colleagues, devise quantum nondemolition sensors for circumventing the quantum limit on bar detectors of gravitational waves. [Ch. 10]

1978   Giacconi's group completes construction of the first high-resolution X-ray telescope, called "Einstein," and it is launched into orbit. [Ch. 8]

1979   Townes and others discover evidence for a 3-million-solar-mass black hole at the center of our galaxy. [Ch. 9]

Drever initiates an interferometric gravitational-wave detection project at Caltech. [Ch. 10]

1982   Bunting and Mazur prove the no-hair conjecture for spinning, electrically charged black holes. [Ch. 7]

1983–1988   Phinney and others develop comprehensive black-hole-based models to explain the full details of quasars and radio galaxies. [Ch. 9]

1984   The National Science Foundation forges a shotgun marriage between the Caltech and MIT gravitational-wave detection efforts, giving rise to the LIGO Project. [Ch. 10]

Redmount (building on earlier work by Eardley) shows that radiation falling into an empty, spherical wormhole gets accelerated to high energy and greatly speeds up the wormhole's pinch-off. [Ch. 14]

1985–1993   Thorne, Morris, Yurtsever, Friedman, Novikov, and others probe the laws of physics by asking whether they permit traversable wormholes and time machines. [Ch. 14]

1987   Vogt becomes director of the LIGO Project, and it then begins to move forward vigorously. [Ch. 10]

1990   Kim and Thorne show that, whenever one tries to create a time machine, by any method whatsoever, an intense beam of vacuum fluctuations circulates through the machine at the moment it is first created. [Ch. 14]

1991   Hawking proposes the chronology protection conjecture (that the laws of physics forbid time machines) and argues that it will be enforced by the circulating beam of vacuum fluctuations destroying any time machine at its moment of formation. [Ch. 14]

Israel, Poisson, and Ori, building on work by Doroshkevich and Novikov, show that the singularity inside a black hole ages; Ori shows that when the hole is old and quiescent, infalling objects do not get strongly deformed by the singularity's tidal gravity until the moment they hit its quantum gravity core. [Ch. 13]

Shapiro and Teukolsky discover evidence, in supercomputer simulations, that the cosmic censorship conjecture might be wrong: Naked singularities might be able to form when highly nonspherical stars implode. [Ch. 13]

1993   Hulse and Taylor are awarded the Nobel Prize for demonstrating, by measurements of a binary pulsar, that gravitational waves exist. [Ch. 10]

Glossary

absolute. Independent of one's reference frame; the same as measured in each and every reference frame.

absolute horizon. The surface of a black hole. See *horizon*.

absolute space. Newton's conception of the three-dimensional space in which we live as having a notion of absolute rest, and as having the property that the lengths of objects are independent of the motion of the reference frame in which they are measured.

absolute time. Newton's conception of time as being universal, with a unique, universally agreed upon notion of simultaneity of events and a unique, universally agreed upon time interval between any two events.

accelerated observer. An observer who does not fall freely.

accretion disk. A disk of gas that surrounds a black hole or neutron star. Friction in the disk makes the gas gradually spiral inward and *accrete* onto the hole or star.

adiabatic index. Same as *resistance to compression*.

aether. The hypothetical medium which (according to nineteenth-century thinking) oscillates when electromagnetic waves go by, and by its oscillations, makes the waves possible. The aether was believed to be at rest in absolute space.

angular momentum. A measure of the amount of rotation that a body has. In this book the word *spin* is often used in place of "angular momentum."

antimatter. A form of material that is "anathema" to ordinary matter. To each type of particle of ordinary matter (for example, an electron or proton or neutron) there corresponds an almost identical antiparticle of antimatter (the positron or antiproton or antineutron). When a particle of matter meets its corresponding antiparticle of antimatter, they annihilate each other.

apparent horizon. The outermost location around a black hole, where photons, trying to escape, get pulled inward by gravity. This is the same as the *(absolute) horizon* only when the hole is in a quiescent, unchanging state.

astronomer. A scientist who specializes in observing cosmic objects using telescopes.

astrophysicist. A physicist (usually a theoretical physicist) who specializes in using the laws of physics to try to understand how cosmic objects behave.

astrophysics. The branch of physics that deals with cosmic objects and the laws of physics that govern them.

atom. The basic building block of matter. Each atom consists of a nucleus with positive electric charge and a surrounding cloud of electrons with negative charge. Electric forces bind the electron cloud to the nucleus.

atomic bomb. A bomb whose explosive energy comes from a chain reaction of fissions of uranium-235 or plutonium-239 nuclei.

band. A range of frequencies.

bandwidth. The range of frequencies over which an instrument can detect a wave.

bar detector. A gravitational-wave detector in which the waves squeeze and stretch a large metal bar, and a sensor monitors the bar's vibrations.

beam splitter. A device used to split a light beam into two parts going in different directions, and to combine two light beams that come from different directions.

big bang. The explosion in which the Universe began.

big crunch. The final stage of recollapse of the Universe (assuming the Universe does ultimately recollapse; we don't know whether it will or not).

binary system. Two objects in orbit around each other; the objects may be stars or black holes or a star and a black hole.

BKL singularity. A singularity near which tidal gravity oscillates chaotically both in time and space. This is the type of singularity that probably forms at the centers of black holes and in the big crunch of our Universe.

black hole. An object (created by the implosion of a star) down which things can fall but out of which nothing can ever escape.

black-hole binary. A binary system made of two black holes.

Blandford–Znajek process. The extraction of rotational energy from a spinning black hole by magnetic fields that thread through the hole.

boosted atomic bomb. An atomic bomb whose explosive power is increased by one or more layers of fusion fuel.

chain reaction. A sequence of fissions of atomic nuclei in which neutrons from one fission trigger additional fissions, and neutrons from those trigger still more fissions, and so on.

Chandrasekhar limit. The maximum mass that a white-dwarf star can have.

chronology protection conjecture. Hawking's conjecture that the laws of physics do not allow time machines.

classical. Subject to the laws of physics that govern macroscopic objects; non–quantum mechanical.

cold, dead matter. Cold matter in which all nuclear reactions have gone to completion, expelling from the matter all the nuclear energy that can possibly be removed.

collapsed star. The name used for a black hole in the West in the 1960s.

conservation law. Any law of physics that says some specific quantity can never change. Examples are conservation of mass and energy (taken together as a single entity via Einstein's $E = Mc^2$), conservation of total electric charge, and conservation of angular momentum (total amount of spin).

corpuscle. The name used for a particle of light in the seventeenth and eighteenth centuries.

cosmic censorship conjecture. The conjecture that the laws of physics prevent naked singularities from forming when an object implodes.

cosmic ray. A particle of matter or antimatter that bombards the Earth from space. Some cosmic rays are produced by the Sun, but most are created in distant regions of our Milky Way galaxy, perhaps in hot clouds of gas that are ejected into interstellar space by supernovae.

cosmic string. A hypothetical one-dimensional, string-like object that is made from a warpage of space. The string has no ends (either it is closed on itself like a rubber band or it extends on and on forever), and its space warpage causes any circle around it to have a circumference divided by diameter slightly less than π.

critical circumference. The circumference of the horizon of a black hole; the circumference inside which an object must shrink in order for it to form a black hole around itself. The value of the critical circumference is 18.5 kilometers times the mass of the hole or object in units of the mass of the Sun.

curvature of space or spacetime. The property of space or spacetime that makes it violate Euclid's or Minkowski's notions of geometry; that is, the property that enables straight lines that are initially parallel to cross.

Cyg A. Cygnus A; a radio galaxy that looks like (but is not) two colliding galaxies. The first radio galaxy to be firmly identified.

Cyg X-1. Cygnus X-1; a massive object in our galaxy that is probably a black hole. Hot gas falling toward the object emits X-rays observed on Earth.

dark star. A phrase used in the late eighteenth and early nineteenth centuries to describe what we now call a black hole.

degeneracy pressure. Pressure inside high-density matter, produced by erratic, high-speed, wave/particle-duality-induced motions of electrons or neutrons. This type of pressure remains strong when matter is cooled to absolute zero temperature.

deuterium nuclei, or deuterons. Atomic nuclei made from a single proton and a single neutron held together by the nuclear force. Also called "heavy hydrogen" because atoms of deuterium have almost the same chemical properties as hydrogen.

differential equation. An equation that combines in a single formula various functions and their rates of change; that is, the functions and their "derivatives." By "solve a differential equation" is meant "compute the functions themselves from the differential equation."

Doppler shift. The shift of a wave to a higher frequency (shorter wavelength, higher energy) when its source is moving toward a receiver, and to a lower frequency (longer wavelength, lower energy) when the source is moving away from the receiver.

electric charge. The property of a particle or matter by which it produces and feels electric forces.

electric field. The force field around an electric charge, which pulls and pushes on other electric charges.

electric field lines. Lines that point in the direction of the force that an electric field exerts on charged particles. Electric analogue of magnetic field lines.

electromagnetic waves. Waves of electric and magnetic forces. These include, depending on the wavelength, radio waves, microwaves, infrared radiation, light, ultraviolet radiation, X-rays, and gamma rays.

electron. A fundamental particle of matter, with negative electric charge, which populates the outer regions of atoms.

electron degeneracy. The behavior of electrons at high densities, in which they move erratically with high speeds as a result of quantum mechanical wave/particle duality.

elementary particle. A subatomic particle of matter or antimatter. Among the elementary particles are electrons, protons, neutrons, positrons, antiprotons, and antineutrons.

embedding diagram. A diagram in which one visualizes the curvature of a two-dimensional surface by embedding it in a flat, three-dimensional space.

entropy. A measure of the amount of randomness in large collections of atoms, molecules, and other particles; equal to the logarithm of the number of ways that the particles could be distributed without changing their macroscopic appearance.

equation of state. The manner in which the pressure of matter (or matter's resistance to compression) depends on its density.

equivalence principle. See *principle of equivalence.*

error box. The region of the sky in which observations suggest that a specific star or other object is located. It is called an error box because the larger are the uncertainties (errors) of the observations, the larger will be this region.

escape velocity. The speed with which an object must be launched from the surface of a gravitating body in order for it to escape the body's gravitational pull.

event. A point in spacetime; that is, a location in space at a specific moment of time. Alternatively, something that happens at a point in spacetime, for example, the explosion of a firecracker.

exotic material. Material that has a *negative* average energy density, as measured by someone moving through it at nearly the speed of light.

field. Something that is distributed continuously and smoothly in space. Examples are the electric field, the magnetic field, the curvature of spacetime, and a gravitational wave.

fission, nuclear. The breakup of a large atomic nucleus to form several smaller ones. The fission of uranium or plutonium nuclei is the source of the energy that drives the explosion of an atomic bomb, and fission is the energy source in nuclear reactors.

freely falling object. An object on which no forces act except gravity.

free particle. A particle on which no forces act; that is, a particle that moves solely under the influence of its own inertia. When gravity is present: A particle on which no forces act *except* gravity.

frequency. The rate at which a wave oscillates; that is, its number of cycles of oscillation per second.

frozen star. The name used for a black hole in the U.S.S.R. during the 1960s.

function. A mathematical expression that tells how one quantity, for example, the circumference of a black hole's horizon, depends on some other quantity, for example, the black hole's mass; in this example, the function is $C = 4\pi GM/c^2$, where C is the circumference, M is the mass, G is Newton's gravitation constant, and c is the speed of light.

fusion, nuclear. The merger of two small atomic nuclei to form a larger one. The Sun is kept hot and hydrogen bombs are driven by the fusion of hydrogen, deuterium, and tritium nuclei to form helium nuclei.

galaxy. A collection of between 1 billion and 1 trillion stars that all orbit around a common center. Galaxies are typically about 100,000 light-years in diameter.

gamma rays. Electromagnetic waves with extremely short wavelengths; see Figure P.2 on page 25.

Geiger counter. A simple instrument for detecting X-rays; also called a "proportional counter."

general relativity. Einstein's laws of physics in which gravity is described by a curvature of spacetime.

geodesic. A straight line in a curved space or curved spacetime. On the Earth's surface the geodesics are the great circles.

gigantic black hole. A black hole that weighs as much as a million Suns, or more. Such holes are thought to inhabit the cores of galaxies and quasars.

global methods. Mathematical techniques, based on a combination of topology and geometry, for analyzing the structure of spacetime.

gravitational cutoff. Oppenheimer's phrase for the formation of a black hole around an imploding star.

gravitational lens. The role of a gravitating body, such as a black hole or a galaxy, to focus light from a distant source by deflecting the light rays; see *light deflection*.

gravitational redshift of light. The lengthening of the wavelength of light (the reddening of its color) as it propagates upward through a gravitational field.

gravitational time dilation. The slowing of the flow of time near a gravitating body.

gravitational wave. A ripple of spacetime curvature that travels with the speed of light.

graviton. The particle which, according to wave/particle duality, is associated with gravitational waves.

gyroscope. A rapidly spinning object which holds its spin axis steadily fixed for a very long time.

"hair." Any property that a black hole can radiate away and thus cannot hold on to; for example, a magnetic field or a mountain on its horizon.

hoop conjecture. The conjecture that a black hole forms when and only when a body gets compressed so small that a hoop with the critical circumference can be placed around it and twisted in all directions.

horizon. The surface of a black hole; the point of no return, out of which nothing can emerge. Also called the *absolute horizon* to distinguish it from the *apparent horizon*.

hydrogen bomb. A bomb whose explosive energy comes from the fusion of hydrogen, deuterium, and tritium nuclei to form helium nuclei. See also *superbomb*.

hyperspace. A fictitious flat space in which one imagines pieces of our Universe's curved space as embedded.

implosion. The high-speed shrinkage of a star produced by the pull of its own gravity.

inertia. A body's resistance to being accelerated by forces that act on it.

inertial reference frame. A reference frame that does not rotate and on which no external forces push or pull. The motion of such a reference frame is driven solely by its own inertia. See also *local inertial reference frame*.

infrared radiation. Electromagnetic waves with wavelength a little longer than light; see Figure P.2 on page 25.

interference. The manner in which two waves, superimposing on each other and adding linearly, reinforce each other when their crests match with crests and troughs with troughs (constructive interference), and cancel each other when crests match up with troughs (destructive interference).

interferometer. A device based on the interference of waves. See *radio interferometer* and *interferometric detector*.

interferometric detector. A detector of gravitational waves in which the waves' tidal forces wiggle masses that hang from wires, and the interference of laser beams is used to monitor the masses' motions. Also called *interferometer*.

interferometry. The process of interfering two or more waves with each other.

intergalactic space. The space between the galaxies.

interstellar space. The space between the stars of our Milky Way galaxy.

inverse square law of gravity. Newton's law of gravity, which says that between every pair of objects in the Universe there acts a gravitational force that pulls the objects toward each other, and the force is proportional to the product of the objects' masses and inversely proportional to the square of the distance between them.

ion. An atom that has lost some of its orbital electrons and therefore has a net positive charge.

ionized gas. Gas in which a large fraction of the atoms have lost orbital electrons.

jet. A beam of gas that carries power from the central engine of a radio galaxy or quasar to a distant, radio-emitting lobe.

laws of physics. Fundamental principles from which one can deduce, by logical and mathematical calculations, how our Universe behaves.

length contraction. The contraction of an object's length as a result of its motion past the person who measures the length. The contraction occurs only along the direction of motion.

light. The type of electromagnetic waves that can be seen by the human eye; see Figure P.2 on page 25.

light deflection. The deflection of the direction of propagation of light and other electromagnetic waves, as they pass near the Sun or any other gravitating body. This deflection is produced by the curvature of spacetime around the body.

LIGO. The Laser Interferometer Gravitational-Wave Observatory.

linear. The property of combining together by simple addition.

lobe. A huge radio-emitting cloud of gas outside a galaxy or quasar.

local inertial reference frame. A reference frame on which no forces except gravity act, that falls freely in response to gravity's pull, and that is small enough for tidal gravitational accelerations to be negligible inside it.

magnetic field. The field that produces magnetic forces.

magnetic field lines. Lines that point along the direction of a magnetic field (that is, along the direction that a compass needle would point if it were placed in the magnetic field). These field lines can be made evident around a bar magnet by placing a sheet of paper above the magnet and scattering bits of iron on the paper.

mass. A measure of the amount of matter in an object. (The object's inertia is proportional to its mass, and Einstein showed that mass is actually a very compact form of energy.) The word "mass" is also used to mean "an object made of mass," in contexts where the inertia of the object is important.

Maxwell's laws of electromagnetism. The set of laws of physics by which James Clerk Maxwell unified all electromagnetic phenomena. From these laws one can predict, by mathematical calculations, the behaviors of electricity, magnetism, and electromagnetic waves.

metaprinciple. A principle that all physical laws should obey. The principle of relativity is an example of a metaprinciple.

microsecond. One-millionth of a second.

microwaves. Electromagnetic radiation with wavelength a little shorter than radio waves; see Figure P.2 on page 25.

Milky Way. The galaxy in which we live.

mixmaster singularity. A singularity near which tidal gravity oscillates chaotically with time, but does not necessarily vary in space. See also *BKL singularity*.

molecule. An entity made of several atoms that share their electron clouds with each other. Water is a molecule made in this way from two hydrogen atoms and one oxygen.

mouth. An entrance to a wormhole. There is a mouth at each of the two ends of the wormhole.

naked singularity. A singularity that is not inside a black hole (not surrounded by a black-hole horizon), and that therefore can be seen and studied by someone outside it. See *cosmic censorship conjecture*.

National Science Foundation (NSF). The agency of the United States government charged with the support of basic scientific research.

natural philosopher. A phrase widely used in the seventeenth, eighteenth, and nineteenth centuries to describe what we now call a scientist.

nebula. A cloud of brightly shining gas in interstellar space. Before the 1930s, galaxies were generally mistaken for nebulas.

neutrino. A very light particle that resembles the photon, except that it interacts hardly at all with matter. Neutrinos produced in the Sun's center, for example, fly out through the Sun's surrounding matter without being absorbed or scattered hardly at all.

neutron. A subatomic particle. Neutrons and protons, held together by the nuclear force, make up the nuclei of atoms.

neutron core. Oppenheimer's name for a neutron star. Also a neutron star at the center of a normal star.

neutron star. A star, about as massive as the Sun but only 50 to 1000 kilometers in circumference, and made from neutrons packed tightly together by the force of gravity.

new quantum mechanics. The final version of the laws of quantum mechanics, formulated in 1926.

Newtonian laws of physics. The laws of physics, built on Newton's conception of space and time as absolute, which were the centerpiece of nineteenth-century thinking about the Universe.

Newton's law of gravity. See *inverse square law of gravity*.

no-hair conjecture. The conjecture in the 1960s and 1970s (which was proved to be true in the 1970s and 1980s) that all the properties of a black hole are determined uniquely by its mass, electric charge, and spin.

nonlinear. The property of combining together in a more complicated way than simple addition.

nova. A brilliant outburst of light from an old star, now known to be caused by a nuclear explosion in the star's outer layers.

nuclear burning. Nuclear fusion reactions that keep stars hot and power hydrogen bombs.

nuclear force. Also called the "strong interaction." The force between protons and protons, protons and neutrons, and neutrons and neutrons, which holds atomic nuclei together. When the particles are somewhat far from each other, the nuclear force is attractive; when they are closer it becomes repulsive. The nuclear force is responsible for much of the pressure near the center of a neutron star.

nuclear reaction. The merging of several atomic nuclei to form a larger one (fusion), or the breakup of a larger one to form several smaller ones (fission).

nuclear reactor. A device in which a chain reaction of nuclear fissions is used to generate energy, produce plutonium, and in some cases produce electricity.

nucleon. Neutron or proton.

nucleus, atomic. The dense core of an atom. Atomic nuclei have positive electric charge, are made of neutrons and protons, and are held together by the nuclear force.

observer. A (usually hypothetical) person or being who makes a measurement.

old quantum mechanics. The early version of the laws of quantum mechanics, developed in the first two decades of the twentieth century.

optical astronomer. An astronomer who observes the Universe using visible light (light that can be seen by the human eye).

orbital period. The time it takes for one object, in orbit around another, to encircle its companion once.

paradigm. A set of tools that a community of scientists uses in its research on a given topic, and in communicating the results of its research to others.

particle. A tiny object; one of the building blocks of matter (such as an electron, proton, photon, or graviton).

perihelion. The location, on a planet's orbit around the Sun, at which it is closest to the Sun.

perihelion shift of Mercury. The tiny failure of Mercury's elliptical orbit to close on itself, which results in its perihelion shifting in position each time Mercury passes through the perihelion.

perturbation. A small distortion (from its normal shape) of an object or of the spacetime curvature around an object.

perturbation methods. Methods of analyzing, mathematically, the behaviors of small perturbations of an object, for example, a black hole.

photon. A particle of light or of any other type of electromagnetic radiation (radio, microwave, infrared, ultraviolet, X-ray, gamma ray); the particle which, according to wave/particle duality, is associated with electromagnetic waves.

piezoelectric crystal. A crystal that produces a voltage when squeezed or stretched.

Planck's constant. A fundamental constant, denoted $\hbar$, that enters into the laws of quantum mechanics; the ratio of the energy of a photon to its angular frequency (that is, to 2π times its frequency); 1.055×10^{-27} erg-second.

Planck–Wheeler length, area, and time. Quantities associated with the laws of quantum gravity. The Planck–Wheeler length, $\sqrt{G\hbar/c^3} = 1.62 \times 10^{-33}$ centimeter, is the length scale below which space as we know it ceases to exist and becomes quantum foam. The Planck–Wheeler time ($1/c$ times the Planck–Wheeler length or about 10^{-43} second) is the shortest time interval that can exist; if two events are separated by less than this, one cannot say which comes before and which after. The Planck–Wheeler area (the square of the Planck–Wheeler length, that is, 2.61×10^{-66} square centimeter) plays a key role in black-hole entropy. In the above formulas, $G = 6.670 \times 10^{-8}$ dyne-centimeter^2/gram^2 is Newton's gravitation constant, $\hbar = 1.055 \times 10^{-27}$ erg-second is Planck's quantum mechanical constant, and $c = 2.998 \times 10^{10}$ centimeter/second is the speed of light.

plasma. Hot, ionized, electrically conducting gas.

plutonium-239. A specific type of plutonium atomic nucleus which contains 239 protons and neutrons (94 protons and 145 neutrons).

polarization. The property that electromagnetic and gravitational waves have of consisting of two components, one that oscillates in one direction or set of directions, and the other in a different direction or set of directions. The two components are called the waves' two polarizations.

polarized body. A body with negative electric charge concentrated in one region and positive charge concentrated in another region.

polarized light; polarized gravitational waves. Light or gravitational waves in which one of the two polarizations is completely absent (vanishes).

postdoc. Postdoctoral fellow; a person who has recently received the Ph.D. degree and is continuing his or her training in how to do research, usually under the guidance of a more senior researcher.

pressure. The amount of outward force that matter produces when it is squeezed.

Price's theorem. The theorem that all properties of a black hole that can be converted into radiation will be converted into radiation and will be radiated away completely, thereby making the hole "hairless."

primordial black hole. A black hole typically far less massive than the Sun that was created in the big bang.

principle of absoluteness of the speed of light. Einstein's principle that the speed of light is a universal constant, the same in all directions and the same in every inertial reference frame, independent of the frame's motion.

principle of equivalence. The principle that in a local inertial reference frame in the presence of gravity, all the laws of physics should take the same form as they do in an inertial reference frame in the absence of gravity.

principle of relativity. Einstein's principle that the laws of physics should not be able to distinguish one inertial reference frame from another; that is, that they should take on the same form in every inertial reference frame. When gravity is present: this same principle, but with local inertial reference frames playing the role of the inertial reference frames.

pulsar. A magnetized, spinning neutron star that emits a beam of radiation (radio waves and sometimes also light and X-rays). As the star spins, its beam sweeps around like the beam of a turning spotlight; each time the beam sweeps past Earth, astronomers receive a pulse of radiation.

pulsation. The vibration or oscillation of an object, for example, a black hole or a star or a bell.

quantum field. A field that is governed by the laws of quantum mechanics. All fields, when measured with sufficient accuracy, turn out to be quantum fields; but when measured with modest accuracy, they may behave classically (that is, they do not exhibit wave/particle duality or vacuum fluctuations).

quantum fields in curved spacetime, the laws of. A partial marriage of general relativity (curved spacetime) with the laws of quantum fields, in which gravitational waves and nongravitational fields are regarded as quantum mechanical, while the curved spacetime in which they reside is regarded as classical.

quantum foam. A probabilistic foamlike structure of space that probably makes up the cores of singularities, and that probably occurs in ordinary space on scales of the Planck–Wheeler length and less.

quantum gravity. The laws of physics that are obtained by merging ("marrying") general relativity with quantum mechanics.

quantum mechanics. The laws of physics that govern the realm of the small (atoms, molecules, electron, protons), and that also underlie the realm of the large, but rarely show themselves there. Among the phenomena that quantum mechanics predicts are the *uncertainty principle, wave/particle duality*, and *vacuum fluctuations.*

quantum nondemolition. A method of measurement that circumvents the standard quantum limit.

quantum theory. The same as *quantum mechanics.*

quasar. A compact, highly luminous object in the distant Universe, believed to be powered by a gigantic black hole.

radiation. Any form of high-speed particles or waves.

radio astronomer. An astronomer who studies the Universe using radio waves.

radio galaxy. A galaxy that emits strong radio waves.

radio interferometer. A device consisting of several radio telescopes linked together, which simulates a single much larger radio telescope.

radio source. Any astronomical object that emits radio waves.

radio telescope. A telescope that observes the Universe using radio waves.

radio waves. Electromagnetic waves of very low frequency, used by humans to transmit radio signals and used by astronomers to study distant astronomical objects; see Figure P.2 on page 25.

redshift. A shifting of electromagnetic waves to longer wavelengths, that is, a "reddening" of the waves.

reference frame. A (possibly imaginary) laboratory for making physical measurements, which moves through the Universe in some particular manner.

relative. Dependent on one's reference frame; different, as measured in one frame which moves through the Universe in one manner, than as measured in another frame which moves in another manner.

resistance to compression, or simply **resistance.** Also called *adiabatic index.* The percentage by which the pressure inside matter increases when the density is increased by 1 percent.

rigor; rigorous. A high degree of precision, exactness, and reliability (a term applied to mathematical calculations and arguments).

rotational energy. The energy associated with the spin of a black hole or a star or some other object.

Schwarzschild geometry. The geometry of spacetime around and inside a spherical, nonspinning hole.

Schwarzschild singularity. The phrase used between 1916 and about 1958 to describe what we now call a black hole.

Sco X-1. Scorpius X-1, the brightest X-ray star in the sky.

second law of thermodynamics. The law that entropy can never decrease and almost always increases.

sensitivity. The weakest signal that can be measured by some device. Alternatively, the ability of a device to measure signals.

sensor. A device for monitoring the vibrations of a bar or motions of a mass.

shocked gas. Gas that has been heated and compressed in a shock front.

shock front. A place, in flowing gas, where the density and temperature of the gas suddenly jump upward by a large amount.

simultaneity breakdown. The fact that events which are simultaneous as measured in one reference frame are not simultaneous as measured in another frame that moves relative to the first.

singularity. A region of spacetime where spacetime curvature becomes so strong that the general relativistic laws break down and the laws of quantum gravity take over. If one tries to describe a singularity using general relativity alone, one finds (incorrectly) that tidal gravity and spacetime curvature are infinitely strong there. Quantum gravity probably replaces these infinities by quantum foam.

Sirius B. The white-dwarf star that orbits around the star Sirius.

spacetime. The four-dimensional "fabric" that results when space and time are unified.

spacetime curvature. The property of spacetime that causes freely falling particles that are initially moving along parallel world lines to subsequently move together or apart. Spacetime curvature and *tidal gravity* are different names for the same thing.

spacetime diagram. A diagram with time plotted upward and space plotted horizontally.

special relativity. Einstein's laws of physics in the absence of gravity.

spectral lines. Sharp features in the spectrum of the light emitted by some source. These features are due to strong emission at specific wavelengths, emission produced by specific atoms or molecules.

spectograph. A sophisticated version of a prism, for separating the various colors (wavelengths) of light and thereby measuring the light's spectrum.

spectrum. The range of wavelengths or frequencies over which electromagnetic waves can exist, running from extremely low-frequency radio waves up through light to extremely high-frequency gamma rays; see Figure P.2 in the prologue. Also, a picture of the distribution of light as a function of frequency (or wavelength), obtained by sending the light through a prism.

spin. Rotation. See *angular momentum.*

stability. The issue of whether an object is unstable or not. See also *unstable.*

standard quantum limit. A limit, due to the uncertainty principle, on how accurately certain quantities can be measured using standard methods. This limit can be circumvented using quantum nondemolition methods.

stroboscopic measurement. A specific kind of quantum nondemolition measurement in which one makes a sequence of very quick measurements of a vibrating bar, each measurement separated by one vibration period.

structure of a star. The details of how a star's pressure, density, temperature, and gravity change as one goes inward from its surface to its center.

superbomb. A hydrogen bomb that uses a principle by which one can produce an arbitrarily large explosion.

superconductor. A material that conducts electricity perfectly, without any resistance.

supermassive star. A hypothetical star that weighs as much as or more than 10,000 Suns.

supernova. A gigantic explosion of a dying star. The explosion of the star's outer layers is powered by energy that is released when the star's inner core implodes to form a neutron star.

surface gravity. Roughly speaking, the strength of the gravitational pull felt by an observer at rest just above a black hole's horizon. (More precisely: that gravitational pull multiplied by the amount of gravitational time dilation at the observer's location.)

synchrotron radiation. Electromagnetic waves emitted by high-speed electrons that are spiraling around and around magnetic field lines.

thermal pressure. Pressure created by the heat-induced, random motions of atoms, molecules, electrons, and/or other particles.

thermodynamics. The set of physical laws that govern the random, statistical behavior of large numbers of atoms and molecules, including their heat.

thermonuclear reactions. Heat-induced nuclear reactions.

tidal gravity. Gravitational accelerations that squeeze objects along some directions and stretch them along others. Tidal gravity produced by the Moon and Sun is responsible for the tides on the Earth's oceans.

time dilation. A slowing of the flow of time.

time machine. A device for traveling backward in time. In physicists' jargon, a "closed timelike curve."

topology. The branch of mathematics that deals with the qualitative ways that objects are connected to each other or to themselves. For example, topology distinguishes a sphere (which has no hole) from a doughnut (which has one).

tritium. Atomic nuclei made of one proton and two neutrons bound together by the nuclear force.

ultraviolet radiation. Electromagnetic radiation with a wavelength a little shorter than light; see Figure P.2 on page 25.

uncertainty principle. A quantum mechanical law which states that, if one measures the position of an object or the strength of a field with high precision, one's measurement must necessarily perturb the object's velocity or the field's rate of change by an unpredictable amount.

universe. A region of space that is disconnected from all other regions of space, much as an island is disconnected from all other pieces of land.

Universe. Our universe.

unstable. The property of an object that if one perturbs it slightly, the perturbation will grow large, thereby changing the object greatly and perhaps even destroying it. Also called, in more complete terminology, "unstable against small perturbations."

uranium-235. A specific type of uranium nucleus which contains 235 protons and neutrons (92 protons and 143 neutrons).

vacuum. A region of spacetime from which have been removed all the particles and fields and energy that one can remove; the only things left are the irremovable vacuum fluctuations.

vacuum fluctuations. Random, unpredictable, irremovable oscillations of a field (for example, an electromagnetic or gravitational field), which are caused by a tug-of-war in which small regions of space momentarily steal energy from adjacent regions and then give it back. See also *vacuum* and *virtual particles*.

virtual particles. Particles that are created in pairs using energy borrowed from a nearby region of space. The laws of quantum mechanics require that the energy be given back quickly, so the virtual particles annihilate quickly and cannot be captured. Virtual particles are the particle aspect of vacuum fluctuations, as seen by freely falling observers. Virtual photons and virtual gravitons are the particle aspects of electromagnetic vacuum fluctuations and gravitational vacuum fluctuations, respectively. See also *wave/particle duality*.

warpage of spacetime. Same as *curvature of spacetime*.

wave. An oscillation in some field (for example, the electromagnetic field or spacetime curvature) that propagates through spacetime.

waveform. A curve showing the details of the oscillations of a wave.

wavelength. The distance between the crests of a wave.

wave/particle duality. The fact that all waves sometimes behave like particles, and all particles sometimes behave like waves.

white-dwarf star. A star with roughly the circumference of the Earth but the mass of the Sun, which has exhausted all its nuclear fuel and is gradually cooling off. It supports itself against the squeeze of its own gravity by means of electron degeneracy pressure.

world line. The path of an object through spacetime or through a spacetime diagram.

wormhole. A "handle" in the topology of space, connecting two widely separated locations in our Universe.

X-rays. Electromagnetic waves with wavelength between that of ultraviolet radiation and gamma rays; see Figure P.2 on page 25.

Notes

what makes me confident
of what I say?

SOURCES AND ABBREVIATIONS

Sources cited in these notes are listed in the bibliography.
Abbreviations used in these notes are:

ECP-1—The Collected Papers of Albert Einstein, Volume 1, cited in bibliography as ECP-1.

ECP-2—The Collected Papers of Albert Einstein, Volume 2, cited in bibliography as ECP-2.

INT—Interviews by the author, listed at beginning of bibliography.

MTW—Misner, Thorne, and Wheeler (1973).

PROLOGUE

Page

23 [Of all the conceptions . . . finding them.] This paragraph is adapted from Thorne
(1974).

26 [From the orbital period . . . ("10 solar masses").] Newton's formula is $M_h = C_o{}^3/(2\pi GP_o{}^2)$, where M_h is the mass of the hole (or any other gravitating body),
C_o and P_o are the circumference and period of any circular orbit around the hole,

π is 3.14159..., and G is Newton's gravitation constant, 1.327×10^{11} kilometers^3 per second^2 per solar mass. See note to page 61, below. Inserting into this formula the starship's orbital period $P_0 = 5$ minutes 46 seconds, and its orbital circumference $C_0 = 10^6$ kilometers, one obtains a mass $M_h = 10$ solar masses. (One solar mass is 1.989×10^{30} kilograms.)

28 [As for size, ... Sun's mass.] The formula for the horizon circumference is $C_h = 4\pi G M_h/c^2 = 18.5$ kilometers $\times (M_h/M_\odot)$, where M_h is the hole's mass, G is Newton's gravitation constant (see above), $c = 2.998 \times 10^5$ kilometers per second is the speed of light, and $M_\odot = 1.989 \times 10^{30}$ kilograms is the mass of the Sun. See, e.g., Chapters 31 and 32 of MTW.

35 [In honor of those tides, ... *tidal force.*] The tidal force, expressed as a relative acceleration between your head and feet (or between any other two objects), is $\triangle a = 16\pi^3 G(M_h/C^3)L$, where G is Newton's gravitation constant (see above), M_h is the black-hole mass, C is the circumference at which you are located, and L is the distance between your head and feet. Note that 1 Earth gravity is 9.81 meters per second^2. See, e.g., page 29 of MTW.

37 [General relativity predicts, ... actually decreases.] The above formula (note to page 35) gives for the tidal force $\triangle a \propto M_h/C^3$. When the circumference is nearly that of the horizon, $C \propto M_h$ (note to page 35), so $\triangle a \propto 1/M_h^2$.

37 [The entire trip of 30,100 light-years ... only 11 years.] Starship time T_{ship}, Earth time T_E, and distance D traveled are related by $T_E = (2c/G)\sinh(gT_{\text{ship}}/2c)$ and $D = (2c^2/g)[\cosh(gT_{\text{ship}}/2c) - 1]$, where g is the ship's acceleration ("one Earth gravity," 9.81 meters per second^2), c is the speed of light, and cosh and sinh are the hyperbolic cosine and hyperbolic sine functions. See, e.g., Chapter 6 of MTW. For trips that last much more than one year, these formulas become, approximately, $T_E = D/c$ and $T_{\text{ship}} = (2c/g)\ln(gD/c^2)$, where ln is the natural logarithm.

39 [To remain in a circular orbit, ... hurled you inward.] For a mathematical analysis of circular (and other) orbits around a nonspinning black hole, see, e.g., Chapter 25 of MTW, and especially Box 25.6.

40 [Your calculations show ... 1.0001 horizon circumferences.] The acceleration force you will feel, hovering at a circumference C above a black hole of mass M_h and horizon circumference C_h, is $a = 4\pi^2 G(M_h/C^2) \times (1/\sqrt{1 - C/C_h})$, where G is Newton's gravitation constant. If you are very close to the horizon, then $C \simeq C_h \propto M_h$, which implies $a \propto 1/M_h$.

40 [Using the usual 1-g acceleration ... crew in the starship.] See the second note for page 37 above.

43 [The spot is small ... seen from Earth.] When one hovers at a circumference C slightly above a horizon with circumference C_h, one sees all the light from the external Universe concentrated in a bright disk with angular diameter $\alpha \simeq 3\sqrt{3}\sqrt{1 - C_h/C}$ radians $\simeq 300\sqrt{1 - C_h/C}$ degrees. See, e.g., Box 25.7 of MTW.

44 [Equally peculiar, the colors ... 5×10^{-7} meter light.] When one hovers at a circumference C slightly above a horizon with circumference C_h, one sees the wavelengths λ of all light from the external Universe gravitationally blue-shifted (the inverse of the gravitational redshift) by $\lambda_{\text{received}}/\lambda_{\text{emitted}} = 1/\sqrt{1 - C/C_h}$. See, e.g., page 657 of MTW.

49 [Inserting these numbers . . . coalesce seven days from now.] When two black holes, each with mass M_h, orbit each other with separation D, they have an orbital period $2\pi\sqrt{D^3/2GM_h}$, and their gravitational-wave recoil forces them to spiral together and coalesce after a time $(5/512) \times (c^5/G^3)(D^4/M_h^3)$. G is Newton's gravitation constant and c is the speed of light; see above. See, e.g., Equation (36.17b) of MTW.

53 [The ring has a circumference of 5 million kilometers, . . . curvature of spacetime.] A person on the girder-work ring at a distance L from its central layer feels an acceleration $a = (32\pi^3 GM_h/C^3)L$ toward the central layer, caused half by the rotating ring's centrifugal force and half by the hole's tidal force. G is Newton's gravitation constant, M_h is the hole's mass, and C is the circumference of the ring's central layer. For comparison, 1 Earth gravity of acceleration is 9.81 meters per second². See the note for page 37 above.

55 [The laws of quantum gravity . . . usable for time travel.] 10^{-33} centimeter = $\sqrt{G\hbar/c^3}$ is the "Planck-Wheeler length," with G = Newton's gravitation constant, c = the speed of light, and $\hbar$ = Planck's constant (1.055×10^{-34} kilogram-meter² per second). See page 494 of Chapter 14.

57–58 [Another is the fact that, . . . flying colors.] See, e.g., Will (1986).

CHAPTER 1

59 General comment about Chapter 1: Most of this chapter's material about Einstein's life comes from the standard biographies of him: Pais (1982), Hoffman (1972), Clark (1971), Einstein (1949), and Frank (1947). For most of the historical perspective and quotations in Chapter 1, which I have gleaned from these standard biographies, I do not give individual citations below. Much new historical material is becoming available with the gradual publication of Einstein's collected papers, ECP-1, ECP-2, and Einstein and Marić (1992). I do cite, below, material from these sources.

59–60 [Professor Wilhelm Ostwald . . . Hermann Einstein.] Document 99 in ECP-1.

60 ["Unthinking respect . . . enemy of truth,"] Document 115 of ECP-1, as translated on page xix of Renn and Schulmann (1992).

61 Footnote 1: The following example illustrates what is meant by "mathematically manipulating" the laws of physics.

Early in the seventeenth century, Johannes Kepler deduced, from Tycho Brahe's observations of the planets, that the cube of the circumference C of a planet's orbit divided by the square of its orbital period P, i.e., C^3/P^2, was the same for all the planets then known: Mercury, Venus, Earth, Mars, Jupiter, Saturn. A half century later, Isaac Newton explained Kepler's discovery by a mathematical manipulation of the Newtonian laws of motion and gravity (the laws listed on page 61 of the text):

1. From the following diagram and a fair amount of sweat, one deduces that, as a planet encircles the Sun, the planet's velocity changes at a rate given by the formula, (rate of change of velocity) $= 2\pi C/P^2$, where $\pi = 3.14159\ldots$. This rate of change of velocity is sometimes called the *centrifugal acceleration* that the orbiting planet experiences.

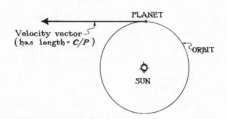

2. Newton's second law of motion tells us that this rate of change of velocity (centrifugal acceleration) must be equal to the gravitational force, F_{grav}, exerted by the Sun on the planet, divided by the planet's mass, M_{planet}; in other words, $2\pi C/P^2 = F_{grav}/M_{planet}$.

3. Newton's gravitational law tells us that the gravitational force F_{grav} is proportional to the Sun's mass M_{Sun} times the planet's mass M_{planet} divided by the square of the planet's orbital circumference. Stated as an equality rather than a proportionality, $F_{grav} = 4\pi^2 G M_{Sun} M_{planet}/C^2$. Here G is Newton's constant of gravitation, equal to 6.670×10^{-20} kilometer^3 per second^2 per kilogram, or equivalently 1.327×10^{11} kilometers^3 per second^2 per solar mass.

4. By inserting this expression for the gravitational force F_{grav} into Newton's second law of motion (Step 2 above), we obtain $2\pi C/P^2 = 4\pi^2 G M_{Sun}/C^2$. By then multiplying both sides of this equation by $C^2/2\pi$, we obtain $C^3/P^2 = 2\pi G M_{Sun}$.

Thus, Newton's laws of motion and gravity explain—in fact they enforce—the relationship discovered by Kepler: C^3/P^2 is the same for all planets; it depends only on Newton's gravitation constant and the Sun's mass.

As an illustration of the power of the laws of physics, the above manipulations not only explain Kepler's discovery, they also offer us a method to weigh the Sun. By dividing the final equation in Step 4 by $2\pi G$, we obtain an equation for the Sun's mass, $M_{Sun} = C^3/(2\pi G P^2)$. By inserting into this equation the circumference C and period P of any planet's orbit as measured by astronomers and the value of Newton's gravitation constant G as measured in Earth-bound laboratories by physicists, we infer that the mass of the Sun is 1.989×10^{30} kilograms.

62 ["Weber lectured . . . his every class."] Document 39 in ECP-1; Document 2 in Einstein and Marić (1992).

63 [And since the aether . . . at rest in absolute space,] In this chapter, I ignore the speculations by some physicists in the late nineteenth century that in the vicinity of the Earth the aether might be dragged along by the motion of the Earth through absolute space. There in fact was strong experimental evidence against such dragging: If, near the Earth's surface, the aether was at rest with respect to the Earth, then there should be no aberration of starlight; but aberration due to the Earth's motion around the Sun was a well-established fact. For a brief discussion of the history of ideas about the aether, see Chapter 6 of Pais (1982); for more detailed discussions, see references cited therein.

64 [Albert Michelson . . . had invented.] The technology of Michelson's time was not capable of comparing *one-way* light speeds in various directions with sufficient accuracy (1 part in 10^4) to test the Newtonian prediction. However, there was a

similar prediction of a difference in *round-trip* light speeds (about 5 parts in 10^9 difference between a round-trip parallel to the Earth's motion through the aether and one perpendicular). Michelson's new technique was ideally suited to measuring such round-trip differences; they were what Michelson searched for and could not find.

65 [By contrast, Heinrich Weber . . . mislead young minds.] I do not know *for certain* that Weber was confident of this, or that he in particular took the attitude that it would be inappropriate to mention the Michelson–Morley experiment in his lectures. This passage is speculation based on the absence of any sign that Weber discussed the experiment, or the issues raised by the experiment, in his lectures; see the detailed notes on his lectures taken by Einstein (Document 37 in ECP-1) and the brief description (page 62 of ECP-1) of the only other existing set of notes from Weber's lectures.

65 [By comparing it with other experiments,] The other experiments were those, such as measurements of the aberration of starlight, which implied that the aether is not dragged along by the Earth; see note to page 63, above.

65 [A tiny (five parts in a billion) . . . Michelson–Morley experiment.] Recall (note to page 64) that Michelson was actually measuring round-trip light speeds and looking for variations with direction of about five parts in a billion.

66 [If one expressed . . . (see Figure 1.1c).] This discussion of the "no ends on magnetic field lines" law, and the more detailed discussion in Figure 1.1, is my own translation, into modern pictorial language, of one aspect of the Maxwell's equations issue with which Lorentz, Larmor, and Poincaré struggled. For a more precise discussion of this issue and their struggle, see pages 123–130 of Pais (1982).

66 [If the Fitzgerald contraction . . . "dilates" time.] To make the laws beautiful required not only the contraction of moving objects and the dilation of their time, it also required pretending that the concept of simultaneity is relative, i.e., that simultaneity depends on one's state of motion; and Lorentz, Larmor, and Poincaré paid considerable attention to this as well as to length contraction and time dilation. However, for pedagogical simplicity, I ignore this in the text and take up the issue of simultaneity somewhat later in Chapter 1.

68 ["I am more and more convinced . . . not correct."] Document 52 in ECP-1; Document 8 in Einstein and Marić (1992).

68 [Over the next six years, . . . dilation of time.] Here I am speculating. It is not really known to what extent Einstein's mind focused on these issues during 1899–1905. As Pais (1982, Section 6b) makes clear, during these six years Einstein was unaware of the Lorentz–Poincaré–Larmor deduction of length contraction and time dilation from Maxwell's laws. Stated more technically, he *was* aware of Lorentz's derivation of the Lorentz transformation up to first order in velocity (including simultaneity breakdown), but not to second order where length contraction and time dilation occur. On the other hand, he presumably was aware of the Fitzgerald–Lorentz inference of length contraction from the Michelson–Morley experiment; and we do know that in his 1905 paper on special relativity he gives his own derivation of the full Lorentz transformation, accurate to all orders, and of length contraction, time dilation, and simultaneity breakdown.

69 [To the saucy . . . Mileva Marić,] For a description of Marić's personality based largely on the love letters between her and Einstein, see Renn and Schulmann (1992); for the love letters see ECP-1 or Einstein and Marić (1992).

69 ["I'm absolutely convinced . . . bad recommendation."] Document 94 of
   ECP-1; Document 95 of Einstein and Marić (1992).

69 ["I could have found . . . thick hide."] Document 100 of ECP-1.

69 ["This Miss Marić . . . dislike her."] Document 138 of ECP-1.

69 ["That lady seems . . . wicked people!"] Document 125 of ECP-1.

69 ["I am beside myself . . . former teachers."] Document 104 of ECP-1.

70 [an illegitimate child . . . staid Switzerland;] ECP-1; Renn and Schulmann
   (1992); Einstein and Marić (1992).

70 [Most of these he spent studying and thinking] I am speculating, based on
   various biographies of Einstein, that he spent most of his free hours in this
   way.

70 ["He was sitting in his study . . . went on working."] Seelig (1956), as quoted
   by Clark (1971).

70–71 [Sometimes it helped . . . "I could not have found . . . whole of Europe."] But
   see the discussion, on page xxvi of Renn and Schulmann (1992), of the contributions that Besso made to Einstein's work.

77 [This proof is essentially . . . devised by Einstein in 1905.] Section 2 of
   Document 23 of ECP-2.

78 [Indeed, a wide variety . . . in just this way.] See, e.g., the appendix in Will
   (1986).

79 [*Having deduced that space . . . to his principle of relativity:*] As Pais (1982,
   Section 6b.6) makes clear, Henri Poincaré formulated a primitive version of
   the principle of relativity (calling it the "relativity principle") one year before
   Einstein, but was unaware of its power.

83 [Einstein's article . . . was published.] Document 23 of ECP-2.

CHAPTER 2

87 General comments about Chapter 2: Most of this chapter's material about
   Einstein's life comes from the standard biographies of him: Pais (1982),
   Hoffman (1972), Clark (1971), Einstein (1949), and Frank (1947). For most of
   the historical perspective and quotations in Chapter 2, which I have gleaned
   from these standard biographies, I do not give individual citations below.
   Much new historical material will become available in the next few years,
   with the gradual publication of Einstein's collected papers: the volumes that
   follow the already published ECP-1 and ECP-2.

   The intellectual route that Einstein followed to get from special relativity
   to general relativity was basically that described in this chapter. However, of
   necessity I have simplified his route substantially; and for clarity, I have
   described the route in modern language rather than in the language that
   Einstein used. For a careful historical reconstruction of Einstein's intellectual
   route, see Pais (1982).

87 [The views of space and time . . . independent reality.] Hermann Minkowski's
   address was delivered at the 80th Assembly of German Natural Scientists and
   Physicians, at Cologne, 21 September 1908. An English translation has been
   published in Lorentz, Einstein, Minkowski, and Weyl (1923).

94 [The other, a peculiarity in the Moon's . . . misinterpretation of the astronomers' measurements.] The Moon *appeared* to be speeding up ever so slightly
   in its motion around the Earth, an effect that Newton's gravitational law

could not explain. In 1920 G. I. Taylor and H. Jeffries realized that, in fact, the Moon was *not* speeding up. Rather, the Earth's spin was slowing down due to the gravitational pull of the Moon on high-tide water in the Earth's oceans. By comparing the Moon's steady motion to the Earth's slowing spin, astronomers had incorrectly inferred a lunar speedup. See Smart (1953).

96 [review article . . . *Radioaktivität und Elektronik*] An English translation of Einstein's beautiful review article is published as Document 47 of ECP-2.

100 [Einstein discovered gravitational time dilation . . . presented in Box 2.4,] Einstein's argument as presented in Box 2.4 was originally published in Einstein (1911).

100 [When starting to write his 1907 review article, . . . *Radioaktivität und Elektronik*] Document 47 of ECP-2.

103 [Einstein's life as a professor . . . he was brilliant.] See Frank (1947), pages 89–91.

117 [These conclusions . . . on 25 November.] Einstein (1915).

118–119 Box 2.6: Remark for readers who are familiar with the mathematical formulation of general relativity: The description of the Einstein field equation given in this box corresponds to the mathematical relation $R_{tt} = 4\pi G(T_{tt} + T_{xx} + T_{yy} + T_{zz})$, where R_{tt} is the time–time component of the Ricci curvature tensor, G is Newton's gravitation constant, T_{tt} is the density of mass expressed in energy units (see Box 5.2), and $T_{xx} + T_{yy} + T_{zz}$ is the sum of the principal pressures along three orthogonal directions. See page 406 of MTW. This "time–time" component of the Einstein field equation, when imposed in *all* reference frames, guarantees that the other nine components of the field equation are satisfied.

119 [As I browse . . . (a browsing which, . . . into English!)] Einstein's personal papers and the rights to some of his published papers were tied up in a legal battle for several decades. The Russian edition of his collected works was produced and published at a time when the Soviet Union did not adhere to the International Copyright Convention. The far more complete English edition is now being published, very gradually; the first two volumes are ECP-1 and ECP-2.

CHAPTER 3

121 ["The essential result of this investigation . . . reality."] Einstein (1939).

122 [In 1783 John Michell . . . should look like.] Michell (1784). For discussions of this work see Gibbons (1979), Schaffer (1979), Israel (1987), and Eisenstaedt (1991).

123 [Thirteen years later, . . . subsequent editions of his book.] Laplace (1796, 1799). For discussions of Laplace's publications on dark stars, see Israel (1987) and Eisenstaedt (1991). Eisenstaedt discusses the attempts and failure to verify, observationally, Michell's prediction that light emitted by massive stars is affected by their gravitational pull, and the contribution that this failure might have had to Laplace's deletion of dark stars from the third edition of his book.

124 [Schwarzschild mailed to Einstein . . . curvature *inside* the star.] Schwarzschild (1916a,b).

131 [Jim Brault . . . Einstein's prediction.] Brault (1962). For a detailed discussion of tests of Einstein's general relativistic laws of gravity see Will (1986).

131–132 [However, few were . . . highly compact stars.] For a detailed discussion of the early history of people's reaction to the Schwarzschild geometry and research on it, see Eisenstaedt (1982). A broader-brushed history that covers the period from 1916 to 1974 will be found in Israel (1987).

135 [In 1939, Einstein published . . . cannot exist.] Einstein (1939).

136 [As backing for . . . Einstein believed.] Schwarzschild (1916b).

138 ["I am sure . . . to that faith."] Israel (1990).

139 ["There is a curious . . . dream of."] Israel (1990).

CHAPTER 4

140 General comment about Chapter 4: The historical aspects of this chapter are based largely on (i) personal conversations with S. Chandrasekhar over the past twenty-five years, (ii) a taped interview with him (INT-Chandrasekhar), (iii) a book about Eddington by him (Chandrasekhar, 1983a), and (iv) a beautiful biography of him (Wali, 1991). I do not cite specific sources for specific items, except in special cases. Chandrasekhar's scientific publications on white dwarfs are collected together in Chandrasekhar (1989).

141–142 [Especially interesting was . . . *Royal Astronomical Society*.] Fowler (1926).

142 [Fowler's article pointed . . . Arthur S. Eddington,] Eddington (1926).

143 Footnote 2: For a detailed discussion of the difficulties Adams faced and the errors that he made in his measurements, see Greenstein, Oke, and Shipman (1985). This reference also gives information about observational studies of Sirius B up to 1985.

150 [Chandrasekhar worked out . . . in pressure.] Here I have taken literary license in two ways. First, Fowler (1926) had already computed the resistance to compression, so Chandrasekhar was merely checking Fowler's calculation. Second, this is not the route by which Chandrasekhar carried out his computation (INT-Chandrasekhar), though it is mathematically equivalent to the true route. This route is the one that is easiest for me to explain; the true route entailed computing the electrons' pressure as an integral over their momentum space.

152 [Finally, a full year . . . published.] Chandrasekhar (1931).

152 Footnote 4: Stoner (1930). This contribution by Stoner is briefly mentioned by Chandrasekhar (1931). For a discussion of the work of Stoner and related work by Wilhelm Anderson, see Israel (1987).

153 [In late 1934 . . . in Estonia.] Anderson (1929), Stoner (1930).

154 Figure 4.3: The masses and circumferences of white dwarfs as shown in this figure, and Chandrasekhar's results for the interior structures of white-dwarf stars, were later published in Chandrasekhar (1935).

160 ["The star has to go on radiating . . . in this absurd way!"] Eddington (1935a). For further details of Eddington's specious arguments see Eddington (1935b).

161–162 [To Leon Rosenfeld . . . "If Eddington is right, . . . Eddington's statements"]
Wali (1991).

162 [in Paris in 1939, . . . "Out there we don't believe in Eddington."] Wali
(1991).

162 [If nature provided no law . . . white-dwarf grave.] I was told this in an
authoritative fashion by an eminent Caltech astronomy professor when I
was an undergraduate in 1958–62. It is my strong personal impression from
that era that most astronomers were taking this view and had done so since
the early 1940s, but I cannot be sure.

163 ["I felt that . . . into something else."] Quoted by Wali (1991).

163 [To Eddington, the treatment . . . astronomical establishment.] This interpretation of Eddington's behavior was suggested to me by Werner Israel, in
a critique of an early version of this chapter; I believe it accords well with
the historical record.

Chapter 5

164 General comment about Chapter 5: The historical aspects of this chapter
are based in large part (i) on my interviews with participants in the events
described, or with their scientist colleagues and friends (INT-Baym, INT-
Braginsky, INT-Eggen, INT-Fowler, INT-Ginzburg, INT-Greenstein,
INT-Harrison, INT-Khalatnikov, INT-Lifshitz, INT-Sandage, INT-Serber,
INT-Volkoff, INT-Wheeler), and (ii) on my reading of the scientific papers
the participants wrote. For general background on the history of physics in
the 1920s and 1930s, I have relied somewhat on Kevles (1971), and for
background on the history of Soviet physics, on Medvedev (1978). Useful
information and background about Landau came from Livanova (1980)
and Gamow (1970), about Oppenheimer from Rabi et al. (1969) and Smith
and Weiner (1980), and about the development of Wheeler's ideas from his
research notebooks, Wheeler (1988). In some places I have relied on other
sources cited below.

164 ["By the time I knew Fritz . . . wrong,"] INT-Fowler.

164 [Jesse Greenstein . . . "a self-proclaimed genius . . . other people."] INT-
Greenstein, and Greenstein (1982).

165 [He even went on the air . . . popularize his neutron stars.] Zwicky
(1935).

166 ["Zwicky called Baade . . . same room," recalls Jesse Greenstein.] INT-
Greenstein.

168 [(Today we know, . . . factor of 10,] Baade (1952).

168 [By combining Baade's knowledge . . . larger factor, 10 million,] These are
Baade and Zwicky's numbers, as they appear in the abstract of a talk that is
replicated in Figure 5.2 (Baade and Zwicky, 1934a), except for the "10,000
and perhaps 10 million," which come from their more detailed paper on
the issue (Baade and Zwicky, 1934b). Their error resulted from assuming
that when the supernova is brightest, the circumference of its hot, radiating
gas is in the range of 1 to 100 solar circumferences. In fact, the circumference is far larger than this, and when one traces through their argument,
this results in far less ultraviolet light and X-rays.

171 [The neutron arrived . . . it seemed to Zwicky.] In this section and throughout Chapter 5, I attribute to Zwicky the concept of a neutron star and its
consequences for supernovae and cosmic rays, although the publication of
the ideas was joint with Baade. Giving Zwicky the credit for the ideas (and
Baade the credit for the key understanding of the observational data) is an
informed speculation based on my discussions with their scientist colleagues: INT-Eggen, INT-Fowler, INT-Greenstein, INT-Sandage.

174 Figure 5.2: Baade and Zwicky (1934a). For some justification of the numbers in the abstract see the more detailed presentation in Baade and Zwicky
(1934b).

178 [Landau's publication . . . cry for help:] This interpretation of Landau's
publication was explained to me by his closest, lifelong friend, Evgeny
Michailovich Lifshitz (INT-Lifshitz).

180 [A fellow postdoctoral . . . "I vividly remember . . . of the paper."] Quoted in
Livanova (1980).

180–181 ["All the nice girls . . . are left,"] Quoted in Livanova (1980).

181 [As George Gamow, . . . "Russian science . . . capitalistic countries."]
Gamow (1970).

181 [In 1936 Stalin, . . . were destroyed.] The statistics on imprisonments and
deaths under Stalin are somewhat uncertain. Medvedev (1978) gives what
are perhaps the most reliable numbers available in the 1970s. However, in
the late 1980s glasnost made possible the public dissemination of information that drove the numbers upward. The numbers I quote are an overall
assessment made by Russian friends of mine who have studied the issue in
some depth in the light of the glasnost revelations.

182 [Arthur Eddington . . . *nuclear fusion*;] Chapter 11 of Eddington (1926) and
references therein.

184 [Landau had actually . . . fail in atomic nuclei.] Landau (1932).

184 [In late 1937, Landau wrote a manuscript] Landau's manuscript was published in Landau (1938). Unbeknownst to Landau, his close friend George
Gamow had already published the same idea (Gamow, 1937). Gamow had
escaped from the U.S.S.R. in 1933, shortly after Stalin's iron curtain descended (see Gamow, 1970), but before escaping he had learned Landau's
original pre-neutron idea of keeping a star hot by a dense central core. After
the neutron was discovered, it was natural that Gamow and Landau (now
out of contact with each other) would independently reinterpret Landau's
1931 core as a neutron core.

185–186 [Landau sent Bohr . . . "The new idea of L. Landau is excellent and very
promising."] Landau's closest personal friend, Evgeny Michailovich Lifshitz, called my attention to this correspondence in 1982 (INT-Lifshitz) and
explained to me the history behind it, as recounted here. After Lifshitz's
death, the full correspondence—including that between Kapitsa and Molotov, Kapitsa and Stalin, and Kapitsa and Beria, which ultimately produced
Landau's release from prison—was published in Khalatnikov (1988). The
excerpts quoted here are my own translation from the Russian.

186 [—though in Landau's case, . . . KGB files:] Gorelik (1991).

186–187 [Landau was lucky . . . (Superfluidity had been discovered . . . power of
Soviet science.)] See note to pages 185–186.

188 ["Well, Robert, . . . damned word."] Quoted in Royal (1969).

188–189 ["Oppie . . . twenty-five dollars a month."] Serber (1969).

191 [(As of the early 1990s, . . . Landau's mechanism.)] These giant stars are thought to be created in binary star systems when one star implodes to become a neutron star, and then, much later, spirals into the core of its companion star and takes up residence there. These peculiar beasts have come to be called "Thorne–Żytkow objects" because Anna Żytkow and I were the first to compute their structures in detail. See Thorne and Żytkow (1977); also Cannon et al. (1992).

191 [they submitted their critique . . . "An estimate of Landau . . . of the Sun."] Oppenheimer and Serber (1938).

192 [In the 1990s, . . . 3 solar masses,] Shapiro and Teukolsky (1983), Hartle and Sabbadini (1977).

193–196 Box 5.4: In this box, most of my description of the sequence of steps by which the research was done is informed speculation, based on an interview with Volkoff (INT-Volkoff), the Tolman archives (Tolman, 1948), and the participants' publications (Oppenheimer and Volkoff, 1939; Tolman, 1939).

195 [On 19 October, . . . more formulas.] The correspondence between Tolman and Oppenheimer is archived in Tolman (1948).

195–196 ["I remember being . . . my calculations."] INT-Volkoff.

196 [There must still be . . . several solar masses.] This conclusion was published in Oppenheimer and Volkoff (1939). Tolman's analytic analyses, on which Oppenheimer and Volkoff relied for their estimates of the effect of nuclear forces, were published in Tolman (1939).

197 [In March 1956, Wheeler . . . and Oppenheimer and Volkoff.] Volume 4, pages 33–40 of Wheeler (1988).

199 [Wheeler was superbly prepared . . . hydrogen bomb] For details of Wheeler's background and earlier work see Wheeler (1979) and Thorne and Zurek (1986).

200–202 Box 5.5: This equation of state (the fruit of the work of Harrison and Wheeler) was published in Harrison, Wakano, and Wheeler (1958), and in greater detail in Harrison, Thorne, Wakano, and Wheeler (1965). The more recent, solid curve at and above nuclear densities (10^{14} grams per cubic centimeter) is an approximation to various modern equations of state as reviewed by Shapiro and Teukolsky (1983).

203 Figure 5.5: From Harrison, Wakano, and Wheeler (1958) and Harrison, Thorne, Wakano, and Wheeler (1965). The solid neutron-star curve is an approximation to various modern computations as reviewed by Shapiro and Teukolsky (1983).

206 [Thus, his article with Volkoff . . . "On Massive Neutron Cores."] Oppenheimer and Volkoff (1939).

207 [His best effort . . . "On the Theory and Observation of Highly Collapsed Stars."] Zwicky (1939).

208 [Isidore I. Rabi, . . . "[I]t seems to me . . . had already gone."] Rabi et al. (1969).

CHAPTER 6

209 General comment about Chapter 6: The historical aspects of this chapter are based in large part on the following: (i) my interviews with participants

in the events described, or with their scientist colleagues and friends (INT-Braginsky, INT-Finkelstein, INT-Fowler, INT-Ginzburg, INT-Harrison, INT-Lifshitz, INT-Misner, INT-Serber, INT-Wheeler, INT-Zel'dovich), (ii) my own participation in a small portion of the history, (iii) my reading of the scientific papers the participants wrote, (iv) the descriptions of the American nuclear weapons projects in Bethe (1982), Rhodes (1986), Teller (1955), and York (1976), (v) the descriptions of the Soviet nuclear weapons projects and other events in the U.S.S.R. in Golovin (1973), Medvedev (1978), Ritus (1990), Romanov (1990), and Sakharov (1990), and (vi) John Wheeler's research notebooks (Wheeler, 1988).

209–211 [It was Tuesday, 10 June 1958 . . . "It is very difficult to believe 'gravitational cutoff' is a satisfactory answer,"] A written version of Wheeler's lecture and the interchange of comments between Wheeler and Oppenheimer are published in Solvay (1958).

210 ["there seems no escape . . . [below about 2 Suns]"] This quote is paraphrased from Harrison, Wakano, and Wheeler (1958), with minor changes of detail to fit the phraseology and conventions of this book.

212 ["Hartland pooh-poohed . . . liberal politics."] INT-Serber.

212 ["Oppie was extremely cultured; . . . most independent."] INT-Fowler.

212 ["Hartland had more talent . . . rest of us did."] INT-Serber.

212 [Before embarking . . . quick survey of the problem.] Here I am speculating; I do not know for sure that he carried out such a quick survey, but based on my understanding of Oppenheimer and the contents of the paper he wrote when the research was finished (Oppenheimer and Snyder, 1939), I strongly suspect that he did.

216–217 [By scrutinizing those formulas, . . . looks on the star's surface,] Oppenheimer and Snyder published the results of their research in Oppenheimer and Snyder (1939).

219 [At Caltech, for example, . . . was very convinced.] INT-Fowler.

219 [There Lev Landau, . . . human mind to comprehend.] INT-Lifshitz.

220 ["In personality they . . . I chose Breit."] Wheeler (1979). This reference is an autobiographical account of Wheeler's research in nuclear physics.

220, 222 [Wheeler and Bohr at Princeton . . . The Bohr–Wheeler article . . . *Physical Review*] Bohr and Wheeler (1939), Wheeler (1979). Bohr and Wheeler did not name plutonium-239 by name in their paper, but Louis A. Turner inferred directly from their Figure 4 that it was an ideal nucleus for sustaining chain reactions, and proposed in a famous classified memorandum that it be used as the fuel for the atomic bomb (Wheeler, 1985).

223 [Zel'dovich and a close friend, . . . for all the world to see.] INT-Zel'dovich, Zel'dovich and Khariton (1939).

223 [Wheeler was the lead scientist . . . Nagasaki bomb.] For some details of Wheeler's key role, see pages 2–5 of Klauder (1972).

223 ["If atomic bombs . . . at Los Alamos and Hiroshima."] From a speech by Oppenheimer at Los Alamos, New Mexico, on 16 October 1945; see page 172 of Goodchild (1980).

223 ["In some sort of crude sense . . . cannot lose."] Page 174 of Goodchild (1980).

223–224 ["As I look back . . . August 6, 1943."] Wheeler (1979).

225 [While this massive effort . . . over to the American design.] These details were revealed by Khariton in a lecture in Moscow, which was reported in the *New York Times* of Thursday, 14 January 1993, page A5.

225 [accumulation of waste . . . square miles of countryside.] Medvedev (1979).

226–227 ["We base our recommendations . . . genocide."] Report of 30 October 1949 from the General Advisory Committee to the U.S. Atomic Energy Commission. Reproduced in the appendix of York (1976).

227 ["Nine out of ten . . . stroke of genius."] Bethe (1982).

227 [As Wheeler recalls, "We did an immense amount . . . get things out."] INT-Wheeler.

228 [Wheeler recalls, "While I was starting . . . on the project."] INT-Wheeler.

229 ["The program we had in 1949 . . . once you had it."] USAEC (1984), p. 251.

229 ["I'm told . . . thermonuclear devices"] INT-Wheeler.

229 [In spring 1948, fifteen months before] There seems to be some confusion over the date on which the Soviet H-bomb design work was initiated. Sakharov (1990) dates it as spring 1948, but Ginzburg (1990) dates it as 1947.

229 [In June 1948, a second superbomb team] This is the date given by Sakharov (1990); Ginzburg (1990) places the date in 1947.

229 Footnote 3: Sakharov's speculation is outlined in Sakharov (1990). Zel'dovich's assertion was made verbally to close Russian friends, who transmitted it to me.

230 ["Our job is to lick Zel'dovich's anus."] Quoted to me by Vitaly Ginzburg, who was present. Sakharov was also present; in the English version of his memoirs (Sakharov, 1990), the quotation is expressed as "Our job is to kiss Zel'dovich's ass." For some of my own views on the complex relationship between Zel'dovich and Sakharov, see Thorne (1991).

230 ["that bitch, Zel'dovich."] This quote by Landau has been passed on to me independently by several Soviet theoretical physicists.

230 [Sakharov proposed . . . lithium deuteride (LiD).] Romanov (1990).

231 [it was 800 times more powerful . . . Hiroshima.] The numbers I cite for the energy release in various bomb explosions are taken from York (1976).

231 ["I am under the influence . . . his humanity."] Sakharov (1990).

232 [In March 1954, Sakharov . . . Teller–Ulam idea,] Romanov (1990), Sakharov (1990). Romanov, in an article in honor of Sakharov, attributes the discovery jointly to Sakharov and Zel'dovich. Sakharov says that "[s]everal of us in the theoretical departments came up with [this idea] at about the same time," and he then leaves the impression that he himself deserves the greater share of the credit but says that "Zel'dovich, Yuri Trutnev, and others undoubtedly made significant contributions."

235 ["In a great number of cases . . . not to grant clearance."] USAEC (1954).

235 [Teller had "had the courage . . . deserved consideration,"] J. A. Wheeler, telephone conversation with K. S. Thorne, July 1991.

235 [Andrei Sakharov, . . . came to agree.] Sakharov (1990).

239 [At Livermore . . . produced a black hole.] The motivation for this research,

a quest to understand supernovae and their roles as sources of cosmic rays, is described in Colgate and Johnson (1960). Colgate and White (1963, 1966) carried out the small-mass, supernova-forming simulations, using Newton's description of gravity rather than Einstein's. May and White (1965, 1966) did the large-mass, black-hole–forming simulations, using Einstein's general relativistic description of gravity.

240–241 [To puzzle out the details . . . nearly identical to the Americans'.] Imshennik and Nadezhin (1964), Podurets (1964).

244 ["You cannot appreciate . . . true simultaneously,"] INT-Lifshitz.

244 [Then one day in 1958, . . . David Finkelstein,] Finkelstein (1958).

244 [Footnote 5: See, e.g., the discussions in Box 31.1 and Chapter 31 of MTW.

245 [Finkelstein discovered, quite by chance . . . and stellar implosion.] For Finkelstein's description of how the discovery was made, see Finkelstein (1993).

246 [an article in *Scientific American.*] Thorne (1967).

254 [In 1964 and 1965 . . . stellar implosion.] Harrison, Thorne, Wakano, and Wheeler (1965).

256 [He tried it out at a conference . . . "By reason . . . increase its gravitational attraction."] Wheeler (1968).

CHAPTER 7

258 General comment about Chapter 7: The historical aspects of this chapter are based on (i) my own personal experience as a participant, (ii) my interviews with other participants (INT-Carter, INT-Chandrasekhar, INT-Detweiler, INT-Eardley, INT-Ellis, INT-Geroch, INT-Ginzburg, INT-Hartle, INT-Ipser, INT-Israel, INT-Misner, INT-Novikov, INT-Penrose, INT-Press, INT-Price, INT-Rees, INT-Sciama, INT-Smarr, INT-Teukolsky, INT-Wald, INT-Wheeler, INT-Zel'dovich), and (iii) my reading of the scientific papers the participants wrote.

262 ["There have been few occasions . . . consummated."] Wheeler (1964b).

266 [*hoop conjecture:*] I first published the concept of the hoop conjecture in a Festschrift volume in honor of Wheeler (Thorne, 1972), and in Box 32.3 of MTW.

268 [the idea that the implosion of a star . . . run backward.] This idea was called by Novikov and Zel'dovich the *semiclosed universe.* They ultimately published separate papers describing it: Zel'dovich (1962) and Novikov (1963).

269 ["Maybe you *now* don't . . . but you *will* want to."] INT-Novikov.

269 ["Yakov Boris'ch would . . . next day."] INT-Novikov.

274 [To test this speculation . . . no magnetic field whatsoever.] The key ideas and initial calculations of this research were published in Ginzburg (1964); more complete mathematical details were worked out by Ginzburg and a young colleague, Leonid Moiseevich Ozernoy (Ginzburg and Ozernoy, 1964).

275 [Doroshkevich, Novikov, and Zel'dovich quickly . . . no protrusion] They published their analysis and conclusions in Doroshkevich, Zel'dovich, and Novikov (1965). (The order of the authors is alphabetic in the Russian language.)

278 [In London, Novikov presented . . . anything like it.] Readers can see the flavor of Novikov's lecture in the influential review articles that he and Zel'dovich wrote shortly before the conference: Zel'dovich and Novikov (1964, 1965).

278 [The written version . . . in Russian.] Doroshkevich, Zel'dovich, and Novikov (1965); see note to page 275.

279 [The first was Werner Israel, . . . will become clear below] Israel's analysis was published in Israel (1967).

281 [He got there third, after Novikov and after Israel,] Novikov (1969), de la Cruz, Chase, and Israel (1970), Price (1972).

283–284 [(The mechanism, . . . Ted Chase.)] de la Cruz, Chase, and Israel (1970).

284 [The field now threads the horizon, . . . leaving the hole unmagnetized] For a more detailed and complete discussion of the interaction of magnetic fields with a black hole, see Figures 10, 11, and 36 of Thorne, Price, and Macdonald (1986).

285 [The lion's share . . . Mazur.] For a review and references, see Section 6.7 of Carter (1979); the subsequent, final stage was published in Mazur (1982) and Bunting (1983).

288 [John Graves and Dieter Brill, . . . charged black hole.] Graves and Brill (1960) and references therein.

289 [Roy Kerr had . . . outside a spinning star.] Kerr (1963).

290 [Within a year Carter . . . Richard Lindquist,] Carter (1966), Boyer and Lindquist (1967).

290 [Carter and others . . . possibly exist.] Carter (1979) and earlier references therein.

290 [Carter, by plumbing that mathematics, . . . should be.] Carter (1968).

293 [Werner Israel showed . . . always fail.] Israel (1986).

294 [In 1969, Roger Penrose . . . marvelous discovery.] Penrose (1969).

295 [Ted Newman . . . Robert Torrence.] Newman et al. (1965).

295 [In autumn 1971, Bill Press, . . . black hole itself.] Press (1971).

297 [The winner was Saul Teukolsky,] Teukolsky (1972).

298 [Teukolsky recalls vividly . . . "Sometimes when you play with mathematics, . . . terms together."] INT-Teukolsky.

298 [Teukolsky himself, . . . its pulsations are stable.] Press and Teukolsky (1973).

299 [*The Mathematical Theory of Black Holes*] Chandrasekhar (1983b).

Chapter 8

300 [General comment about Chapter 8: The historical aspects of this chapter are based on (i) my own personal experience as a participant, (ii) my interviews with other participants (INT-Giacconi, INT-Novikov, INT-Rees, INT-Van Allen, INT-Zel'dovich), (iii) my reading of the scientific

papers the participants wrote, and (iv) the following published accounts of
the history: Friedman (1972), Giacconi and Gursky (1974), Hirsh (1979),
and Uhuru (1981).

301 ["Such an object . . . another star"] Wheeler (1964a).

301 [If you are Zel'dovich . . . stellar implosion.] Twenty-two years later, in
1986, Zel'dovich expressed to me regret that he had not been more open-
minded about the issue of what goes on inside black holes; INT-Zel'dovich.

306 [Together, Guseinov and Zel'dovich . . . the catalogs.] Zel'dovich and
Guseinov (1965).

307 [By searching through . . . eight black-hole candidates.] Trimble and
Thorne (1969).

307 [Fortunately, his brainstorming . . . New York.] Salpeter (1964), Zel'dovich
(1964).

308 [Zel'dovich and Novikov together . . . infalling gas idea] Novikov and
Zel'dovich (1966).

309 ["the rocket returned . . . on impact."] Friedman (1972).

311 [they announced their discovery: . . . *had predicted.*] Giacconi, Gursky,
Paolini, and Rossi (1962).

318 [(suggested in 1972 by Rashid Sunyaev, . . . Zel'dovich's team)] Sunyaev
(1972).

Chapter 9

322 General comment about Chapter 9: The historical aspects of this chapter
are based on (i) my own personal experience as a peripheral participant
from 1962 onward, (ii) my interviews with several participants (INT-Ginzburg, INT-Greenstein, INT-Rees, INT-Zel'dovich), (iii) my reading of the
scientific papers the participants wrote, and (iv) the following published
and unpublished accounts of the history: Hey (1973), Greenstein (1982),
Kellermann and Sheets (1983), Struve and Zebergs (1962), and Sullivan
(1982, 1984).

323 [Cosmic radio waves . . . 1932 by Karl Jansky,] Jansky (1932).

323–324 [The two exceptions . . . Jansky was seeing.] Whipple and Greenstein
(1937).

324 ["I never met . . . not one astronomer,"] INT-Greenstein.

324 [So uninterested . . . call number W9GFZ.] For Reber's own historical
description of his work, see Reber (1958).

327 [In 1940, having made . . . paper for publication.] Reber (1940).

327 [Greenstein describes Reber as "the ideal American inventor . . . a million
dollars."] INT-Greenstein.

327 ["The University didn't want . . . independent cuss,"] INT-Greenstein.

330 [The first crucial benchmark, . . . radio sources must lie.] Bolton, Stanley,
and Slee (1949).

331 [When Baade developed . . . two galaxies colliding with each other] Baade
and Minkowski (1954).

333 [R. C. Jennison and M. K. Das Gupta . . . opposite sides of the "colliding
galaxies."] Jennison and Das Gupta (1953).

334 [Greenstein organized . . . 5 and 6 January 1954.] The proceedings of this
conference are published in Washington (1954).

335 [The mental block . . . Maarten Schmidt,] Schmidt (1963).

336 [Greenstein turned, . . . 37 percent of the speed of light.] Greenstein (1963).

337 [Harlan Smith . . . as short as a month.] Smith (1965).

339 [Building on seminal ideas . . . fill interstellar space] Alfvén and Herlofson (1950), Kiepenheuer (1950), Ginzburg (1951). For a discussion of the history of this work see Ginzburg (1984).

339 [Geoffrey Burbidge . . . 100 percent efficiency.] Burbidge (1959).

341 [To foster dialogue . . . Dallas, Texas.] The proceedings of this conference are published in Robinson, Schild, and Shucking (1965).

342 [So, as Kerr got up to speak, . . . picked up pace.] This description is from my own vivid memory of the conference.

343 [In 1971, this suggested . . . that powers quasars.] Rees (1971).

343 [Malcolm Longair, . . . electromagnetic waves.] Longair, Ryle, and Scheuer (1973).

346 [The idea that gigantic black holes . . . Edwin Salpeter and Yakov Borisovich Zel'dovich] Salpeter (1964), Zel'dovich (1964).

346 [A more complete . . . by Donald Lynden-Bell,] Lynden-Bell (1969).

346 [How can a black hole . . . answer in 1975:] Bardeen and Petterson (1975).

347–348 [How strong will the swirl of space be . . . nearly its maximum possible rate] Bardeen (1970).

348 [First, Blandford and Rees realized,] Blandford and Rees (1974).

348 [Second, . . . Lynden-Bell pointed out,] Lynden-Bell (1978).

348 [Third, Blandford realized,] Blandford (1976).

350 [The fourth method . . .*Blandford–Znajek process.*] Blandford and Znajek (1977).

351 [If quasars and radio galaxies are powered by the same kind of black-hole engine,] For more detailed discussions of the present state of our understanding of quasars, radio galaxies, jets, and the roles of black holes and their accretion disks as the central engines that power them, see, e.g., Begelman, Blandford, and Rees (1984) and Blandford (1987).

354 [The evidence for such a hole . . . far from firm.] See, e.g., Phinney (1989).

Chapter 10

357 General comment about Chapter 10: The historical aspects of this chapter are based on (i) my own personal experience as a participant, (ii) my interviews with several participants (INT-Braginsky, INT-Drever, INT-Forward, INT-Grishchuk, INT-Weber, INT-Weiss), and (iii) my reading of scientific papers the participants wrote. For more technical overviews of gravitational radiation and efforts to detect it, see, e.g., Blair (1991) and Thorne (1987).

366 [While Weber was publishing his concept,] Weber (1953).

366–367 [Through late 1957, . . . broadside to the incoming waves] The fruits of Weber's work were published in Weber (1960, 1961).

367 [His sole guide . . . near the critical circumference.] Letter from Weber to me, dated 1 October 1992; Weber did not publish this argument at the time. Weber's colleague Freeman Dyson was the first to show that nature is likely to produce gravitational-wave bursts near the frequencies Weber had chosen (Dyson, 1963).

369 [However, in the early 1970s, . . . a reality.] Weber's announcement of observational evidence for gravitational waves was made in Weber (1969). The ensuing experimental activity and controversy over whether waves had really been detected are documented, e.g., in deSabbata and Weber (1977) and papers cited therein. For a sociological study of the controversy see Collins (1975, 1981).

369 [two-month summer school] The lectures presented at the summer school, including Weber's, were published in DeWitt and DeWitt (1964).

372 [During our 1969 meeting, . . . ultimate limitation.] This initial version of Braginsky's warning was published in Braginsky (1967).

372 [However, in 1976, . . . *uncertainty principle.*] The clarified warnings were published in Braginsky (1977) and Giffard (1976), and the uncertainty principle origin of the limit was explained in Thorne, Drever, Caves, Zimmermann, and Sandberg (1978).

375 [Roughly 10^{-21} was the answer,] See, e.g., the quasi-transcript of a 1978 conference discussion in Epstein and Clark (1979).

375 [We both found the answer . . . different routes.] Braginsky, Vorontsov, and Khalili (1978); Thorne, Drever, Caves, Zimmermann, and Sandberg (1978).

378 [In principle it would be possible to widen the bars' bandwidths] Michelson and Taber (1984).

383 [Interferometers for gravitational-wave detection . . . as did Robert Forward and colleagues] Gertsenshtein and Pustovoit (1962), Weber (1964), Weiss (1972), Moss, Miller, and Forward (1971).

383 [and Drever had added . . . to their design.] See, e.g., Drever (1991) and references therein.

387 [he redirected most of his own team's efforts . . . and modest funds.] See Braginsky and Khalili (1992).

391 [A key to success in our endeavor . . . or *LIGO.*] For an overview of the plans for LIGO see Abramovici et al. (1992).

CHAPTER 11

397 General comment about Chapter 11: The (rather minor) historical aspects of this chapter are based on (i) my own personal experience as a participant, (ii) my interviews with two other participants (INT-Damour, INT-Wald), (iii) my reading of scientific papers the participants wrote, and (iv) my experience as a student in a course on paradigms and scientific revolutions taught by Thomas Kuhn at Princeton University in 1965.

401 [*The Structure of Scientific Revolutions*] Kuhn (1962).

403 [This freedom carries power.] Richard Feynman, one of the greatest physicists of our century, described beautifully the power of having several different paradigms at one's fingertips in his lovely little book *The Character of Physical Law* (Feynman, 1965). Note, however, that he never uses the word "paradigm," and I suspect that he never read Thomas Kuhn's writings. Kuhn described how people like Feynman operate; Feynman just operated that way.

403 [That is why physicists . . . supplement to it.] The flat spacetime paradigm was devised more or less independently by a number of different people; it is known, technically, as a "field theory in flat spacetime formulation of general relativity." For an overview of its history and concepts, see the following passages in MTW: Sections 7.1 and 18.1; Boxes 7.1, 17.2, and 18.1; Exercise 7.3. For an

elegant generalization of it, which elucidates its relationship to the curved space-time paradigm, see Grishchuk, Petrov, and Popova (1984).

406 [In 1971 Hanni and Ruffini, . . . Jeff Cohen] Cohen and Wald (1971), Hanni and Ruffini (1973).

407 [Five years later Roger Blandford . . . power jets] Blandford and Znajek (1977).

409 [During 1977 and 1978, Znajek and . . . pictorial interpretation:] Znajek (1978), Damour (1978).

409 [*Black Holes: The Membrane Paradigm.*] Thorne, Price, and Macdonald (1986). See also Price and Thorne (1988).

CHAPTER 12

412 General comment about Chapter 12: The historical aspects of this chapter are based on (i) my own personal experience as a participant, (ii) my interviews with other participants (INT-DeWitt, INT-Eardley, INT-Hartle, INT-Hawking, INT-Israel, INT-Penrose, INT-Unruh, INT-Wald, INT-Wheeler, INT-Zel'-dovich), (iii) my reading of scientific papers the participants wrote, and (iv) the following published accounts of the history: Bekenstein (1980), Hawking (1988), Israel (1987).

412 [The Idea hit . . . so quickly.] This and the subsequent description of how Hawking arrived at the idea come from INT-Hawking and Hawking (1988). Hawking published the details and consequences of his idea, as sketched in the first section of this chapter, "Black Holes Grow," in Hawking (1971b, 1972, 1973).

414 [following Roger Penrose's lead,] Penrose (1965).

414 Box 12.1: Hawking (1972, 1973).

417 [Stephen Hawking was not the first . . . Werner Israel] INT-Israel, INT-Penrose, INT-Hawking.

417 [Penrose's 1964 discovery . . . singularity at its center.] Penrose (1965).

418 Box 12.2: Hawking (1972, 1973).

419 [Hawking and James Hartle . . . gravity of other bodies.] Hawking and Hartle (1972).

422 [Demetrios Christodoulou . . . equations of thermodynamics.] Christodoulou (1970).

425 [Jacob Bekenstein was not persuaded.] Bekenstein describes this and the controversy with Hawking that followed in Bekenstein (1980). Bekenstein published his black-hole entropy conjecture and his arguments for it in Bekenstein (1972, 1973).

426 [Les Houches summer school,] The proceedings of the 1972 summer school were published in DeWitt and DeWitt (1973).

427 [Bardeen, Carter, and Hawking . . . *laws of black-hole mechanics*] Bardeen, Carter, and Hawking (1973).

428 [Zel'dovich had brought me to Moscow . . .] Charles Misner and John Wheeler accompanied me on my June 1971 visit to Moscow, but they were not with me at Zel'dovich's apartment during the discussion described in the following paragraphs.

429 [Zel'dovich, his eyes dancing, . . .] I have reconstructed the following conversation from memory, and have translated it into less technical language than we actually used.

433 [Zel'dovich, however, did not forget; . . . his paper was published] Zel'-
dovich (1971).

434–435 [Starobinsky described Zel'dovich's conjecture . . . does, indeed, radiate.]
Zel'dovich and Starobinsky (1971).

435 [Then came a bombshell.] Hawking describes, in Hawking (1988), how he
arrived at his "bombshell" discovery that all black holes radiate. He published the discovery and its implications in Hawking (1974, 1975, 1976).

437 [This and the demand for a perfect mesh, . . . almost completely.] See, e.g.,
Wald (1977).

437 Footnote 11: Wald (1977).

439 [Perhaps the simplest . . . particles rather than waves:] Hawking (1988).

442 [Gradually . . . new understanding embodied in Figure 12.3.] Chapter 8 of
Thorne, Price, and Macdonald (1986), and references therein.

444 Box 12.5: Davies (1975), Unruh (1976), Unruh and Wald (1982, 1984).

446 [a highly abstract proof . . . in 1977.] Gibbons and Hawking (1977).

446 [The total lifetime, . . . Don Page] Page (1976).

447 [Detailed calculations by Hawking, . . . to produce tiny holes.] E.g., Hawking (1971a); Novikov, Polnarev, Starobinsky, and Zel'dovich (1979).

447 [The absence of excess gamma rays . . . soft equation of state.] Page and
Hawking (1975); Novikov, Polnarev, Starobinsky, and Zel'dovich (1979).

CHAPTER 13

449 General comment about Chapter 13: The historical aspects of this chapter
are based on (i) my own personal experience (though as an observer rather
than a participant), (ii) my interviews with participants (INT-Belinsky,
INT-DeWitt, INT-Geroch, INT-Khalatnikov, INT-Lifshitz, INT-MacCal-
lum, INT-Misner, INT-Penrose, INT-Sciama, INT-Wheeler), and (iii) my
reading of scientific papers the participants wrote.

449 [John Archibald Wheeler taught . . . outside the horizon.] Harrison,
Wakano, and Wheeler (1958); Wheeler (1960).

450 [Wheeler retained his conviction . . . pursuing.] Wheeler (1964a,b); Harrison, Thorne, Wakano, and Wheeler (1965).

450 [J. Robert Oppenheimer and Hartland Snyder,] Oppenheimer and Snyder
(1939).

450 [Perhaps Oppenheimer's unwillingness to speculate,] See the last several
pages of Chapter 5.

451 [The singularity predicted by the Oppenheimer–Snyder calculations] The
singularity as described here is that in the vacuum outside the imploding
star, and since the vacuum region is described by the Schwarzschild solution of Einstein's equations, this singularity is often referred to as the
singularity of the Schwarzschild geometry. It is analyzed quantitatively, e.g.,
in Chapter 32 of MTW.

451 Figure 13.1: Ibid.

453 [One group, . . . general relativity fails] Wheeler (1960, 1964a,b); Harrison,
Thorne, Wakano, and Wheeler (1965).

453 [A second group, . . . Khalatnikov and Evgeny Michailovich Lifshitz . . .

could not be trusted.] This viewpoint and the calculations that led Khalatnikov and Lifshitz to it were published in Lifshitz and Khalatnikov (1960, 1963) and in Landau and Lifshitz (1962).

454 [Khalatnikov and Lifshitz ... *small perturbations.*] Ibid.

456 [*The Classical Theory of Fields.*] Landau and Lifshitz (1962).

457 Figure 13.4: It was obvious in the early 1960s to students in Wheeler's group, where the Graves–Brill (1960) research had been done, that there must exist a solution to Einstein's equations of the sort depicted here. However, I gather from a discussion with Penrose that researchers in most other groups did not become aware of it until the late 1960s. It was difficult to construct such solutions explicitly, and we in Wheeler's group did not try, and did not publish anything on the issue. The first publication of the idea and the first attempt at an explicit solution, so far as I know, were by Novikov (1966).

458 [Hans Reissner and Gunnar Nordström ... Dieter Brill and John Graves,] Graves and Brill (1960) and references therein.

459 [Roger Penrose grew up in a British ...] This biographical discussion of Penrose comes largely from INT-Penrose and INT-Sciama.

460 [The seduction began in 1952,] Ibid.

462 [One day in late autumn of 1964, ...] INT-Penrose; Penrose (1989).

462 ["My conversation with Robinson ... crossing the street."] Penrose (1989).

462 [a short article for ... *Physical Review Letters,*] Penrose (1965).

465 [*global methods.*] The global methods were codified in a classic book by Hawking and Ellis (1973).

465 [Hawking and Penrose in 1970 proved ... big crunch.] Hawking and Penrose (1970).

466 [Lifshitz, though Jewish, ... 1976.] From my private discussions with Lifshitz in the 1970s.

466 [Khalatnikov had two strikes against him; ... come to London.] Letter from Khalatnikov to me, 18 June 1990.

466 [As he spoke in the packed London lecture hall, ... Penrose, they asserted, was probably wrong.] From my own memory of the meeting and its aftermath.

468 ["Please, ... submit it to *Physical Review Letters,*"] Khalatnikov and Lifshitz (1970). See also Belinsky, Khalatnikov, and Lifshitz (1970, 1982).

469 [I carried the manuscript ... published.] Ibid.

470 [Lev Davidovich Landau ... great physics discoveries.] INT-Lifshitz, Livanova (1980).

471 [Curiously, topological techniques ... Pimenov.] I learned this from Penrose.

471 [In 1950–59, Aleksandrov ... that cannot.] Aleksandrov (1955, 1959).

471 [picked up and pushed further by Pimenov,] Pimenov (1968).

473 [(due to Khalatnikov and Lifshitz) ... "unstable against small perturbations."] Lifshitz and Khalatnikov (1960, 1963).

473 [The Reissner–Nordström ... large universe] e.g., Novikov (1966).

473–474 [it is unstable ... many different physicists.] In technical language, it is the *inner Cauchy horizon* of the Reissner–Nordström solution that is unstable.

The conjecture is in Penrose (1968); the proofs are in Chandrasekhar and Hartle (1982) and earlier references cited therein.

474 [Belinsky, Khalatnikov, and Lifshitz . . . (This is the kind . . . holes.)] Belinsky, Khalatnikov, and Lifshitz (1970, 1982).

474 [Charles Misner . . . *mixmaster oscillation*] Misner (1969).

476 [Just when does quantum gravity take over, . . . or less.] This was first deduced by Wheeler (1960), building on his own earlier ideas of vacuum fluctuations of the geometry of spacetime (Wheeler, 1955, 1957).

476 Footnote 2: The *Planck–Wheeler time* was introduced and its physical significance deduced by Wheeler (1955, 1957).

476–477 [Quantum gravity then radically changes . . . random, probabilistic froth,] This was first suggested by Wheeler (1960), and has been made more quantitative since via what is now called the "Wheeler–DeWitt equation." See, e.g., the discussion in Hawking (1987).

477 [John Wheeler, . . . *quantum foam.*] Wheeler (1957, 1960).

479 [Clear answers . . . DeWitt.] See, e.g., Hawking (1987, 1988).

479 [The tidal forces . . . and gradually disappear.] Doroshkevich and Novikov (1978) showed that the singularity ages; Poisson and Israel (1990) and Ori (1991) deduced the details of the aging in idealized models; and Ori (1992) has tentatively shown that these models are good guides to the behavior of singularities in real black holes.

481 [Some implosions, . . . might actually create naked singularities.] For details of these simulations see Shapiro and Teukolsky (1991).

482 [Just four months . . . tiny naked singularity.] Hawking's evidence was published in Hawking (1992a).

CHAPTER 14

483 General comment about Chapter 14: The historical aspects of this chapter are based almost entirely on my own experiences as a participant.

485–486 [Wormholes are not mere figments . . . in 1916,] Ludwig Flamm (1916) discovered that, with an appropriate choice of topology, the Schwarzschild (1916a) solution of Einstein's equation describes an empty, spherical wormhole.

487 Figure 14.2: Kruskal (1960).

490 [We wrote slowly . . . *American Journal of Physics,*] Morris and Thorne (1988).

490 [(the topic of the Hawking and Ellis book)] Hawking and Ellis (1973).

491 [*vacuum fluctuations near a hole's horizon are exotic:*] Hawking inferred this only very indirectly and somewhat tentatively from his discovery of black-hole evaporation. It was firmly demonstrated to be so six years later, by Candelas (1980).

492 [The answer has not come easily, . . . thereby makes them exotic.] See Wald and Yurtsever (1991) and other references cited therein.

494 [In 1955, John Wheeler, . . . quantum foam—] Wheeler (1955, 1957, 1960).

497 [In 1966, Robert Geroch . . . to travel backward in time,] Geroch (1967). Friedman, Papastamatiou, Parker, and Zhang (1988) have given an explicit example of wormhole creation of the sort envisioned by Geroch's theorem.

499 Footnote 8: van Stockum (1937), Gödel (1949), Tipler (1976).

508 [Our paper was published,] Morris, Thorne, and Yurtsever (1988).

509 [we conjectured so in our paper.] Morris, Thorne, and Yurtsever (1988).

509 Footnote 12: Friedman and Morris (1991).

511 [Echeverria and Klinkhammer... *two* such trajectories.] Echeverria, Klinkhammer, and Thorne (1991).

513–514 Box 14.2: Echeverria, Klinkhammer, and Thorne (1991).

513 [Robert Forward ... discovered a third trajectory] Forward (1992).

515 [but there seem not to be ... unresolvable paradox.] For a careful and fairly thorough technical discussion of the issue of paradoxes when one has a wormhole-based time machine, see Friedman et al. (1990).

516 [*California* magazine, ... on Palomar Mountain.] Hall (1989).

517 [Though we were helped ... Konkowski] Hiscock and Konkowski (1982).

520 [a similar calculation by Valery Frolov ... our results] Frolov (1991).

521 [we managed to change ... got published] Kim and Thorne (1991).

521 [*the chronology protection conjecture,*] Hawking (1992b).

521 Footnote 14: Gott (1991).

521 [I am *not* willing to take ... the laws of quantum gravity.] For a somewhat technical description of my 1993 reasons for skepticism about time machines, and a detailed overview of research on time machines up to spring 1993, see Thorne (1993).

Bibliography

TAPED INTERVIEWS

Baym, Gordon. 5 September 1985, Champaign/Urbana, Illinois.

Belinsky, Vladimir. 27 March 1986, Moscow, U.S.S.R.

Braginsky, Vladimir Borisovich. 20 December 1982, Moscow, U.S.S.R.; 27 March 1986, Moscow, U.S.S.R.

Carter, Brandon. 6 July 1983, Padova, Italy.

Chandrasekhar, Subrahmanyan. 3 April 1982, Chicago, Illinois.

Damour, Thibault. 26 July 1986, Cargese, Corsica.

Detweiler, Steven. December 1980, Baltimore, Maryland.

DeWitt, Bryce. December 1980, Baltimore, Maryland.

Drever, Ronald W. P. 21 June 1982, Les Houches, France.

Eardley, Doug M. December 1980, Baltimore, Maryland.

Eggen, Olin. 13 September 1985, Pasadena, California.

Ellis, George. December 1980, Baltimore, Maryland.

Finkelstein, David. 8 July 1983, Padova, Italy.

Forward, Robert. 31 August 1982, Oxnard, California.

Fowler, William A. 6 August 1985, Pasadena, California.

Geroch, Robert. 2 April 1982, Chicago, Illinois.

Giacconi, Riccardo. 29 April 1983, Greenbelt, Maryland.

Ginzburg, Vitaly Lazarevich. December 1982, Moscow, U.S.S.R.; 3 February 1989, Pasadena, California.

Greenstein, Jesse L. 9 August 1985, Pasadena, California.

Grishchuk, Leonid P. 26 March 1986, Moscow, U.S.S.R.

Harrison, B. Kent. 5 September 1985, Provo, Utah.

Hartle, James B. December 1980, Baltimore, Maryland; 2 April 1982, Chicago, Illinois.

Hawking, Stephen W. July 1980, Cambridge, England (not taped).

Ipser, James R. December 1980, Baltimore, Maryland.

Israel, Werner. June 1982, Les Houches, France.

Khalatnikov, Isaac Markovich. 27 March 1986, Moscow, U.S.S.R.

Lifshitz, Evgeny Michailovich. December 1982, Moscow, U.S.S.R.

MacCallum, Malcolm. 30 August 1982, Santa Barbara, California.

Misner, Charles W. 10 May 1981, Pasadena, California.

Novikov, Igor Dmitrievich. December 1982, Moscow, U.S.S.R.; 28 March 1986, Moscow, U.S.S.R.

Penrose, Roger. 7 July 1983, Padova, Italy.

Press, William H. December 1980, Baltimore, Maryland.

Price, Richard. December 1980, Baltimore, Maryland.

Rees, Martin. December 1980, Baltimore, Maryland.

Sandage, Allan. 13 September 1985, Baltimore, Maryland.

Sciama, Dennis. 8 July 1983, Padova, Italy.

Serber, Robert. 5 August 1985, New York City.

Smarr, Larry. December 1980, Baltimore, Maryland.

Teukolsky, Saul A. 27 January 1985, Ithaca, New York.

Unruh, William. December 1980, Baltimore, Maryland.

Van Allen, James. 29 April 1973, Greenbelt, Maryland.

Volkoff, George. 11 September 1985, Vancouver, British Columbia.

Wald, Robert M. December 1980, Baltimore, Maryland; 2 April 1982, Chicago, Illinois.

Weber, Joseph. 20 July 1982, College Park, Maryland.

Weiss, Rainer. 7 July 1983, Padova, Italy.

Wheeler, John. December 1980, Baltimore, Maryland.

Zel'dovich, Yakov Borisovich. 17 December 1982, Moscow, U.S.S.R.; 22 and 27 March 1986, Moscow, U.S.S.R.

References

Abramovici, A., Althouse, W. E., Drever, R. W. P., Gürsel, Y., Kawamura, S., Raab, F. J., Shoemaker, D., Sievers, L., Spero, R. E., Thorne, K. S., Vogt, R. E., Weiss, R., Whitcomb, S. E., and Zucker, M. E. (1992). "LIGO: The Laser Interferometer Gravitational-Wave Observatory," *Science*, **256**, 325–333.

Aleksandrov, A. D. (1955). "The Space-Time of the Theory of Relativity," *Helvetica Physica Acta, Supplement*, **4**, 4.

Aleksandrov, A. D. (1959). "The Philosophical Implication and Significance of the Theory of Relativity," *Voprosy Filosofii*, No. 1, 67.

Alfvén, H., and Herlofson, N. (1950). "Cosmic Radiation and Radio Stars," *Physical Review*, **78**, 738.

Anderson, W. (1929). "Über die Grenzdichte der Materie und der Energie," *Zeitschrift für Physik*, **56**, 851.

Baade, W. (1952). "Report of the Commission on Extragalactic Nebulae," *Transactions of the International Astronomical Union*, **8**, 397.

Baade, W., and Minkowski, R. (1954). "Identification of the Radio Sources in Cassiopeia, Cygnus A, and Puppis," *Astrophysical Journal,* **119,** 206.

Baade, W., and Zwicky, F. (1934a). "Supernovae and Cosmic Rays," *Physical Review,* **45,** 138.

Baade, W., and Zwicky, F. (1934b). "On Super-Novae," *Proceedings of the National Academy of Sciences,* **20,** 254.

Bardeen, J. M. (1970). "Kerr Metric Black Holes," *Nature,* **226,** 64.

Bardeen, J. M., Carter, B., and Hawking, S. W. (1973). "The Four Laws of Black Hole Mechanics," *Communications in Mathematical Physics,* **31,** 161.

Bardeen, J. M., and Petterson, J. A. (1975). "The Lense–Thirring Effect and Accretion Disks around Kerr Black Holes," *Astrophysical Journal (Letters),* **195,** L65.

Begelman, M. C., Blandford, R. D., and Rees, M. J. (1984). "Theory of Extragalactic Radio Sources," *Reviews of Modern Physics,* **56,** 255.

Bekenstein, J. D. (1972). "Black Holes and the Second Law," *Lettere al Nuovo Cimento,* **4,** 737.

Bekenstein, J. D. (1973). "Black Holes and Entropy," *Physical Review D,* **7,** 2333.

Bekenstein, J. D. (1980). "Black Hole Thermodynamics," *Physics Today,* January 24.

Belinsky, V. A., Khalatnikov, I. M., and Lifshitz, E. M. (1970). "Oscillatory Approach to a Singular Point in the Relativistic Cosmology," *Advances in Physics,* **19,** 525.

Belinsky, V. A., Khalatnikov, I. M., and Lifshitz, E. M. (1982). "Solution of the Einstein Equations with a Time Singularity," *Advances in Physics,* **31,** 639.

Bethe, H. A. (1982). "Comments on the History of the H-Bomb," *Los Alamos Science,* Fall 1982, 43.

Bethe, H. A. (1990). "Sakharov's H-Bomb," *Bulletin of the Atomic Scientists,* October 1990. Reprinted in Drell and Kapitsa (1991), p. 149.

Blair, D., ed. (1991). *The Detection of Gravitational Waves* (Cambridge University Press, Cambridge, England).

Blandford, R. D. (1976). "Accretion Disc Electrodynamics—A Model for Double Radio Sources," *Monthly Notices of the Royal Astronomical Society,* **176,** 465.

Blandford, R. D. (1987). "Astrophysical Black Holes," in *300 Years of Gravitation,* edited by S. W. Hawking and W. Israel (Cambridge University Press, Cambridge, England), p. 277.

Blandford, R. D., and Rees, M. (1974). "A Twin-Exhaust Model for Double Radio Sources," *Monthly Notices of the Royal Astronomical Society,* **169,** 395.

Blandford, R. D., and Znajek, R. L. (1977). "Electromagnetic Extraction of Energy from Kerr Black Holes," *Monthly Notices of the Royal Astronomical Society,* **179,** 433.

Bohr, N., and Wheeler, J. A. (1939). "The Mechanism of Nuclear Fission," *Physical Review,* **56,** 426.

Bolton, J. G., Stanley, G. J., and Slee, O. B. (1949). "Positions of Three Discrete Sources of Galactic Radio-Frequency Radiation," *Nature,* **164,** 101.

Boyer, R. H., and Lindquist, R. W. (1967). "Maximal Analytic Extension of the Kerr Metric," *Journal of Mathematical Physics,* **8,** 265.

Braginsky, V. B. (1967). "Classical and Quantum Restrictions on the Detection of Weak Disturbances of a Macroscopic Oscillator," *Zhurnal Eksperimentalnoi i Teoreticheskoi Fiziki,* **53,** 1434. English translation in *Soviet Physics—JETP,* **26,** 831 (1968).

Braginsky, V. B. (1977). "The Detection of Gravitational Waves and Quantum Nondisturbtive Measurements," in *Topics in Theoretical and Experimental Gravi-*

tation Physics, edited by V. de Sabbata and J. Weber (Plenum, London), p. 105.

Braginsky, V. B., and Khalili, F. Ya. (1992). *Quantum Measurements* (Cambridge University Press, Cambridge, England).

Braginsky, V. B., Vorontsov, Yu. I., and Khalili, F. Ya. (1978). "Optimal Quantum Measurements in Detectors of Gravitational Radiation," *Pis'ma v Redaktsiyu Zhurnal Eksperimentalnoi i Teoreticheskoi Fiziki*, **27**, 296. English translation in *JETP Letters*, **27**, 276 (1978).

Braginsky, V. B., Vorontsov, Yu. I., and Thorne, K. S. (1980). "Quantum Nondemolition Measurements," *Science*, **209**, 547.

Brault, J. W. (1962). "The Gravitational Redshift in the Solar Spectrum," unpublished doctoral dissertation, Princeton University; available from University Microfilms, Ann Arbor, Michigan.

Brown, A. C., ed. (1978). *DROPSHOT: The American Plan for World War III against Russia in 1957* (Dial Press/James Wade, New York).

Bunting, G. (1983). "Proof of the Uniqueness Conjecture for Black Holes," unpublished Ph.D. dissertation, Department of Mathematics, University of New England, Armidale, N.S.W. Australia.

Burbidge, G. R. (1959). "The Theoretical Explanation of Radio Emission," in *Paris Symposium on Radio Astronomy*, edited by R. N. Bracewell (Stanford University Press, Stanford, California).

Candelas, P. (1980). "Vacuum Polarization in Schwarzschild Spacetime," *Physical Review D*, **21**, 2185.

Cannon, R. C., Eggleton, P. P., Żytkow, A. N., and Podsiadlowski, P. (1992). "The Structure and Evolution of Thorne–Żytkow Objects," *Astrophysical Journal*, **386**, 206–214.

Carter, B. (1966). "Complete Analytic Extension of the Symmetry Axis of Kerr's Solution of Einstein's Equations," *Physical Review*, **141**, 1242.

Carter, B. (1968). "Global Structure of the Kerr Family of Gravitational Fields," *Physical Review*, **174**, 1559.

Carter, B. (1979). "The General Theory of the Mechanical, Electromagnetic and Thermodynamic Properties of Black Holes," in *General Relativity: An Einstein Centenary Survey*, edited by S. W. Hawking and W. Israel (Cambridge University Press, Cambridge, England), p. 294.

Caves, C. M., Thorne, K. S., Drever, R. W. P., Sandberg, V. D., and Zimmermann, M. (1980). "On the Measurement of a Weak Classical Force Coupled to a Quantum-Mechanical Oscillator. I. Issues of Principle," *Reviews of Modern Physics*, **52**, 341.

Chandrasekhar, S. (1931). "The Maximum Mass of Ideal White Dwarfs," *Astrophysical Journal*, **74**, 81.

Chandrasekhar, S. (1935). "The Highly Collapsed Configurations of a Stellar Mass (Second Paper)," *Monthly Notices of the Royal Astronomical Society*, **95**, 207.

Chandrasekhar, S. (1983a). *Eddington: The Most Distinguished Astrophysicist of His Time* (Cambridge University Press, Cambridge, England).

Chandrasekhar, S. (1983b). *The Mathematical Theory of Black Holes* (Oxford University Press, New York).

Chandrasekhar, S. (1989). *Selected Papers of S. Chandrasekhar*. Volume I: *Stellar Structure and Stellar Atmospheres* (University of Chicago Press, Chicago).

Chandrasekhar, S., and Hartle, J. M. (1982). "On Crossing the Cauchy Horizon of a

Reissner-Nordström Black Hole," *Proceedings of the Royal Society of London,* **A384,** 301.

Christodoulou, D. (1970). "Reversible and Irreversible Transformations in Black-Hole Physics," *Physical Review Letters,* **25,** 1596.

Clark, R. W. (1971). *Einstein: The Life and Times* (World Publishing Co., New York).

Cohen, J. M., and Wald, R. M. (1971). "Point Charge in the Vicinity of a Schwarzschild Black Hole," *Journal of Mathematical Physics,* **12,** 1845.

Colgate, S. A., and Johnson, M. H. (1960). "Hydrodynamic Origin of Cosmic Rays," *Physical Review Letters,* **5,** 235.

Colgate, S. A., and White, R. H. (1963). "Dynamics of a Supernova Explosion," *Bulletin of the American Physical Society,* **8,** 306.

Colgate, S. A., and White, R. H. (1966). "The Hydrodynamic Behavior of Supernova Explosions," *Astrophysical Journal,* **143,** 626.

Collins, H. M. (1975). "The Seven Sexes: A Study in the Sociology of a Phenomenon, or the Replication of Experiments in Physics," *Sociology,* **9,** 205.

Collins, H. M. (1981). "Son of Seven Sexes: The Social Destruction of a Physical Phenomenon," *Social Studies of Science* (SAGE, London and Beverly Hills), **11,** 33.

Damour, T. (1978). "Black-Hole Eddy Currents," *Physical Review D,* **18,** 3598.

Davies, P. C. W. (1975). "Scalar Particle Production in Schwarzschild and Rindler Metrics," *Journal of Physics A,* **8,** 609.

de la Cruz, V., Chase, J. E., and Israel, W. (1970). "Gravitational Collapse with Asymmetries," *Physical Review Letters,* **24,** 423.

de Sabbata, V., and Weber, J., eds. (1977). *Topics in Theoretical and Experimental Gravitation Physics* (Plenum, New York).

DeWitt, C., and DeWitt, B. S., eds. (1964). *Relativity, Groups, and Topology* (Gordon and Breach, New York).

DeWitt, C., and DeWitt, B. S., eds. (1973). *Black Holes* (Gordon and Breach, New York).

Doroshkevich, A. D., and Novikov, I. D. (1978). "Space-Time and Physical Fields in Black Holes," *Zhurnal Eksperimentalnoi i Teoreticheskii Fiziki,* **74,** 3. English translation in *Soviet Physics—JETP,* **47,** 1 (1978).

Doroshkevich, A. D., Zel'dovich, Ya. B., and Novikov, I. D. (1965). "Gravitational Collapse of Nonsymmetric and Rotating Masses," *Zhurnal Eksperimentalnoi i Teoreticheskii Fiziki,* **49,** 170. English translation in *Soviet Physics—JETP,* **22,** 122 (1966).

Drell, S., and Kapitsa, S., eds. (1991). *Sakharov Remembered: A Tribute by Friends and Colleagues* (American Institute of Physics, New York).

Drever, R. W. P. (1991). "Fabry–Perot Cavity Gravity-Wave Detectors," in *The Detection of Gravitational Waves,* edited by D. Blair (Cambridge University Press, Cambridge, England), p. 306.

Dyson, F. J. (1963). "Gravitational Machines," in *The Search for Extraterrestrial Life,* edited by A. G. W. Cameron (W. A. Benjamin, New York), p. 115.

Echeverria, F., Klinkhammer, G., and Thorne, K. S. (1991). "Billiard Balls in Wormhole Spacetimes with Closed Timelike Curves. I. Classical Theory," *Physical Review D,* **44,** 1077.

ECP-1: Einstein, A. (1987). *The Collected Papers of Albert Einstein.* Volume 1: *The Early Years, 1879–1902,* edited by John Stachel (Princeton University Press,

Princeton, New Jersey). English translation by Anna Beck in a companion volume of the same title.

ECP-2: Einstein, A. (1989). *The Collected Papers of Albert Einstein.* Volume 2: *The Swiss Years: Writings, 1900–1909,* edited by John Stachel (Princeton University Press, Princeton, New Jersey). English translation by Anna Beck in a companion volume of the same title.

Eddington, A. S. (1926). *The Internal Constitution of the Stars* (Cambridge University Press, Cambridge, England).

Eddington, A. S. (1935a). "Relativistic Degeneracy," *Observatory,* **58,** 37.

Eddington, A. S. (1935b). "On Relativistic Degeneracy," *Monthly Notices of the Royal Astronomical Society,* **95,** 194.

Einstein, A. (1911). "On the Influence of Gravity on the Propagation of Light," *Annalen der Physik,* **35,** 898.

Einstein, A. (1915). "The Field Equations for Gravitation," *Sitzungsberichte der Deutschen Akademie der Wissenschaften zu Berlin, Klasse fur Mathematik, Physik, und Technik,* **1915,** 844.

Einstein, A. (1939). "On a Stationary System with Spherical Symmetry Consisting of Many Gravitating Masses," *Annals of Mathematics,* **40,** 922.

Einstein, A. (1949). "Autobiographical Notes," in *Albert Einstein: Philosopher-Scientist,* edited by Paul A. Schilpp (Library of Living Philosophers, Evanston, Illinois).

Einstein, A., and Marić, M. (1992). *Albert Einstein/Mileva Marić: The Love Letters,* edited by Jürgen Renn and Robert Schulman (Princeton University Press, Princeton, New Jersey).

Eisenstaedt, J. (1982). "Histoire et Singularités de la Solution de Schwarzschild," *Archive for History of Exact Sciences,* **27,** 157.

Eisenstaedt, J. (1991). "De l'Influence de la Gravitation sur la Propagation de la Lumière en Théorie Newtonienne. L'Archéologie des Trous Noirs," *Archive for History of Exact Sciences,* **42,** 315.

Epstein, R., and Clark, J. P. A. (1979). "Discussion Session II: Sources of Gravitational Radiation," in *Sources of Gravitational Radiation,* edited by L. Smarr (Cambridge University Press, Cambridge, England), p. 477.

Feynman, R. P. (1965). *The Character of Physical Law* (British Broadcasting Corporation, London; paperback edition: MIT Press, Cambridge, Massachusetts).

Finkelstein, D. (1958). "Past–Future Asymmetry of the Gravitational Field of a Point Particle," *Physical Review,* **110,** 965.

Finkelstein, D. (1993). "Misner, Kinks, and Black Holes," in *Directions in General Relativity.* Volume 1: *Papers in Honor of Charles Misner,* edited by B. L. Hu, M. P. Ryan Jr., and C. V. Vishveshwara (Cambridge University Press, Cambridge, England), p. 99.

Flamm, L. (1916). "Beitrage zur Einsteinschen Gravitationstheorie," *Physik Zeitschrift,* **17,** 448.

Forward, R. L. (1992). *Timemaster.* (Tor Books, New York).

Fowler, R. H. (1926). "On Dense Matter," *Monthly Notices of the Royal Astronomical Society,* **87,** 114.

Frank, P. (1947). *Einstein: His Life and Times* (Alfred A. Knopf, New York).

Friedman, H. (1972). "Rocket Astronomy," *Annals of the New York Academy of Sciences,* **198,** 267.

Friedman, J., and Morris, M. S. (1991). "The Cauchy Problem for the Scalar Wave

Equation Is Well Defined on a Class of Spacetimes with Closed Timelike Curves," *Physical Review Letters,* **66,** 401.

Friedman, J., Morris, M. S., Novikov, I. D., Echeverria, F., Klinkhammer, G., Thorne, K. S., and Yurtsever, U. (1990). "Cauchy Problem in Spacetimes with Closed Timelike Curves," *Physical Review D,* **42,** 1915.

Friedman, J., Papastamatiou, N., Parker, L., and Zhang, H. (1988). "Non-orientable Foam and an Effective Planck Mass for Point-like Fermions," *Nuclear Physics,* **B309,** 533; appendix.

Frolov, V. P. (1991). "Vacuum Polarization in a Locally Static Multiply Connected Spacetime and a Time-Machine Problem," *Physical Review D,* **43,** 3878.

Gamow, G. (1937). *Structure of Atomic Nuclei and Nuclear Transformations* (Clarendon Press, Oxford, England), pp. 234–238.

Gamow, G. (1970). *My World Line* (Viking Press, New York).

Geroch, R. P. (1967). "Topology in General Relativity," *Journal of Mathematical Physics,* **8,** 782.

Gertsenshtein, M. E., and Pustovoit, V. I. (1962). "On the Detection of Low-Frequency Gravitational Waves," *Zhurnal Eksperimentalnoi i Teoreticheskoi Fiziki,* **43,** 605. English translation in *Soviet Physics—JETP,* **16,** 433 (1963).

Giacconi, R., and Gursky, H., eds. (1974). *X-Ray Astronomy* (Reidel, Dordrecht, Holland).

Giacconi, R., Gursky, H., Paolini, F. R., and Rossi, B. B. (1962). "Evidence for X-Rays from Sources Outside the Solar System," *Physical Review Letters,* **9,** 439.

Gibbons, G. (1979). "The Man Who Invented Black Holes," *New Scientist,* **28,** 1101 (29 June).

Gibbons, G. W., and Hawking, S. W. (1977). "Action Integrals and Partition Functions in Quantum Gravity," *Physical Review D,* **15,** 2752.

Giffard, R. (1976). "Ultimate Sensitivity Limit of a Resonant Gravitational Wave Antenna Using a Linear Motion Detector," *Physical Review D,* **14,** 2478.

Ginzburg, V. L. (1951). "Cosmic Rays as the Source of Galactic Radio Waves," *Doklady Akademii Nauk SSSR,* **76,** 377.

Ginzburg, V. L. (1964). "The Magnetic Fields of Collapsing Masses and the Nature of Superstars," *Doklady Akademii Nauk SSSR,* **156,** 43. English translation in *Soviet Physics—Doklady,* **9,** 329 (1964).

Ginzburg, V. L. (1984). "Some Remarks on the History of the Development of Radio Astronomy," in *The Early Years of Radio Astronomy,* edited by W. J. Sullivan (Cambridge University Press, Cambridge, England).

Ginzburg, V. L. (1990). Private communication to K. S. Thorne.

Ginzburg, V. L., and Ozernoy, L. M. (1964). "On Gravitational Collapse of Magnetic Stars," *Zhurnal Eksperimentalnoi i Teoreticheskoi Fiziki,* **47,** 1030. English translation in *Soviet Physics—JETP,* **20,** 689 (1965).

Gleick, J. (1987). *Chaos: Making a New Science* (Viking/Penguin, New York).

Gödel, K. (1949). "An Example of a New Type of Cosmological Solution of Einstein's Field Equations of Gravitation," *Reviews of Modern Physics,* **21,** 447.

Golovin, I. N. (1973). *I. V. Kurchatov* (Atomizdat, Moscow), 2nd edition. An English translation of the earlier and less complete first edition was published as *Academician Igor Kurchatov* (Mir Publishers, Moscow, 1969; also, Selbstverlag Press, Bloomington, Indiana, 1968).

Goodchild, P. (1980). *J Robert Oppenheimer, Shatterer of Worlds* (British Broadcasting Company, London).

Gorelik, G. E. (1991). " 'My Anti-Soviet Activities . . .' One Year in the Life of L. D. Landau," *Priroda*, November issue, p. 93; in Russian.

Gott, J. R. (1991). "Closed Timelike Curves Produced by Pairs of Moving Cosmic Strings: Exact Solutions," *Physical Review Letters*, **66**, 1126.

Graves, J. C., and Brill, D. R. (1960). "Oscillitory Character of the Reissner–Nordström Metric for an Ideal Charged Wormhole," *Physical Review*, **120**, 1507.

Greenstein, J. L. (1963). "Red-shift of the Unusual Radio Source: 3C48," *Nature*, **197**, 1041.

Greenstein, J. L. (1982). Oral history interview by Rachel Prud'homme, February and March 1982, Archives, California Institute of Technology.

Greenstein, J. L., Oke, J. B., and Shipman, H. (1985). "On the Redshift of Sirius B," *Quarterly Journal of the Royal Astronomical Society*, **26**, 279.

Grishchuk, L. P., Petrov, A. N., and Popova, A. D. (1984). "Exact Theory of the Einstein Gravitational Field in an Arbitrary Background Space-Time," *Communications in Mathematical Physics*, **94**, 379.

Hall, S. S. (1989). "The Man Who Invented Time Travel: The Astounding World of Kip Thorne," *California*, October, p. 68.

Hanni, R. S., and Ruffini, R. (1973). "Lines of Force of a Point Charge Near a Schwarzschild Black Hole," *Physical Review D*, **8**, 3259.

Harrison, B. K., Thorne, K. S., Wakano, M., and Wheeler, J. A. (1965). *Gravitation Theory and Gravitational Collapse* (University of Chicago Press, Chicago).

Harrison, B. K., Wakano, M., and Wheeler, J. A. (1958). "Matter–Energy at High Density: End Point of Thermonuclear Evolution," in *La Structure et l'Evolution de l'Univers*, Onzième Conseil de Physique Solvay (Stoops, Brussels), p. 124.

Hartle, J. B., and Sabbadini, A. G. (1977). "The Equation of State and Bounds on the Mass of Nonrotating Neutron Stars," *Astrophysical Journal*, **213**, 831.

Hawking, S. W. (1971a). "Gravitationally Collapsed Objects of Very Low Mass," *Monthly Notices of the Royal Astronomical Society*, **152**, 75.

Hawking, S. W. (1971b). "Gravitational Radiation from Colliding Black Holes," *Physical Review Letters*, **26**, 1344.

Hawking, S. W. (1972). "Black Holes in General Relativity," *Communications in Mathematical Physics*, **25**, 152.

Hawking, S. W. (1973). "The Event Horizon," in *Black Holes*, edited by C. DeWitt and B. S. DeWitt (Gordon and Breach, New York), p. 1.

Hawking, S. W. (1974). "Black Hole Explosions?" *Nature*, **248**, 30.

Hawking, S. W. (1975). "Particle Creation by Black Holes," *Communications in Mathematical Physics*, **43**, 199.

Hawking, S. W. (1976). "Black Holes and Thermodynamics," *Physical Review D*, **13**, 191.

Hawking, S. W. (1987). "Quantum Cosmology," in *300 Years of Gravitation*, edited by S. W. Hawking and W. Israel (Cambridge University Press, Cambridge, England), p. 631.

Hawking, S. W. (1988). *A Brief History of Time* (Bantam Books, Toronto, New York).

Hawking, S. W. (1992a). "The Chronology Protection Conjecture," *Physical Review D*, **46**, 603.

Hawking, S. W. (1992b). "Evaporation of Two-Dimensional Black Holes," *Physical Review Letters*, **69**, 406.

Hawking, S. W., and Ellis, G. F. R. (1973). *The Large Scale Structure of Space-Time* (Cambridge University Press, Cambridge, England).

Hawking, S. W., and Hartle, J. B. (1972). "Energy and Angular Momentum Flow into a Black Hole," *Communications in Mathematical Physics*, **27**, 283.

Hawking, S. W., and Penrose, R. (1970). "The Singularities of Gravitational Collapse and Cosmology," *Proceedings of the Royal Society of London*, **A314**, 529.

Hey, J. S. (1973). *The Evolution of Radio Astronomy* (Neale Watson Academic Publications, Inc., New York).

Hirsh, R. F. (1979). "Science, Technology, and Public Policy: The Case of X-Ray Astronomy, 1959 to 1972," unpublished Ph.D. dissertation, University of Wisconsin–Madison; available from University Microfilms, Ann Arbor, Michigan.

Hiscock, W. A., and Konkowski, D. A. (1982). "Quantum Vacuum Energy in Taub–NUT (Newman–Unti–Tamburino)-Type Cosmologies," *Physical Review D*, **6**, 1225.

Hoffman, B. (1972). In collaboration with H. Dukas, *Albert Einstein: Creator and Rebel* (Viking, New York).

Imshennik, V. S., and Nadezhin, D. K. (1964). "Gas Dynamical Model of a Type II Supernova Outburst," *Astronomicheskii Zhurnal*, **41**, 829. English translation in *Soviet Astronomy—AJ*, **8**, 664 (1965).

Israel, W. (1967). "Event Horizons in Static Vacuum Spacetimes," *Physical Review*, **164**, 1776.

Israel, W. (1986). "Third Law of Black Hole Dynamics—A Formulation and Proof," *Physical Review Letters*, **57**, 397.

Israel, W. (1987). "Dark Stars: The Evolution of an Idea," in *300 Years of Gravitation*, edited by S. W. Hawking and W. Israel (Cambridge University Press, Cambridge, England), p. 199.

Israel, W. (1990). Letter to K. S. Thorne, dated 28 May 1990, commenting on the semifinal draft of this book.

Jansky, K. (1932). "Directional Studies of Atmospherics at High Frequencies," *Proceedings of the Institute of Radio Engineers*, **20**, 1920.

Jennison, R. C., and Das Gupta, M. K. (1953). "Fine Structure of the Extra-terrestrial Radio Source Cygnus 1," *Nature*, **172**, 996.

Kellermann, K., and Sheets, B. (1983). *Serendipitous Discoveries in Radio Astronomy* (National Radio Astronomy Observatory, Green Bank, West Virginia).

Kerr, R. P. (1963). "Gravitational Field of a Spinning Mass as an Example of Algebraically Special Metrics," *Physical Review Letters*, **11**, 237.

Kevles, D. J. (1971). *The Physicists* (Random House, New York).

Khalatnikov, I. M., ed. (1988). *Vospominaniya o L. D. Landau* (Nauka, Moscow). English translation: *Landau, the Physicist and the Man: Recollections of L. D. Landau* (Pergamon Press, Oxford, England, 1989).

Khalatnikov, I. M., and Lifshitz, E. M. (1970). "The General Cosmological Solution of the Gravitational Equations with a Singularity in Time," *Physical Review Letters*, **24**, 76.

Kiepenheuer, K. O. (1950). "Cosmic Rays as the Source of General Galactic Radio Emission," *Physical Review*, **79**, 738.

Kim, S.-W., and Thorne, K. S. (1991). "Do Vacuum Fluctuations Prevent the Creation of Closed Timelike Curves?" *Physical Review D*, **43**, 3939.

Klauder, J. R., ed. (1972). *Magic without Magic: John Archibald Wheeler* (W. H. Freeman, San Francisco).

Kruskal, M. D. (1960). "Minimal Extension of the Schwarzschild Metric," *Physical Review*, **119**, 1743.

Kuhn, T. (1962). *The Structure of Scientific Revolutions* (University of Chicago Press, Chicago).

Landau, L. D. (1932). "On the Theory of Stars," *Physikalische Zeitschrift Sowjetunion,* **1,** 285.

Landau, L. D. (1938). "Origin of Stellar Energy," *Nature,* **141,** 333.

Landau, L. D., and Lifshitz, E. M. (1962). *Teoriya Polya* (Gosudarstvennoye Izdatel'stvo Fiziko-Matematicheskoi Literaturi, Moscow), Section 108. English translation: *The Classical Theory of Fields* (Pergamon Press, Oxford, England, 1962), Section 110.

Laplace, P. S. (1796). *Exposition du Système du Monde.* Volume II: *Des Mouvements Réels des Corps Célestes* (Paris). Published in English as *The System of the World* (W. Flint, London, 1809).

Laplace, P. S. (1799). "Proof of the Theorem, that the Attractive Force of a Heavenly Body Could Be So Large, that Light Could Not Flow Out of It," *Allgemeine Geographische Ephemeriden,* verfasset von Einer Gesellschaft Gelehrten. 8vo Weimer, IV, Bd I St. English translation in Appendix A of Hawking and Ellis (1973).

Lifshitz, E. M., and Khalatnikov, I. M. (1960). "On the Singularities of Cosmological Solutions of the Gravitational Equations. I." *Zhurnal Eksperimentalnoi i Teoreticheskoi Fiziki,* **39,** 149. English translation in *Soviet Physics—JETP,* **12,** 108 and 558 (1961).

Lifshitz, E. M., and Khalatnikov, I. M. (1963). "Investigations in Relativistic Cosmology," *Advances in Physics,* **12,** 185.

Livanova, A. (1980). *Landau: A Great Physicist and Teacher* (Pergamon Press, Oxford, England).

Longair, M. S., Ryle, M., and Scheuer, P. A. G. (1973). "Models of Extended Radio Sources," *Monthly Notices of the Royal Astronomical Society,* **164,** 243.

Lorentz, H. A., Einstein, A., Minkowski, H., and Weyl, H. (1923). *The Principle of Relativity: A Collection of Original Memoirs on the Special and General Theory of Relativity* (Dover, New York).

Lynden-Bell, D. (1969). "Galactic Nuclei as Collapsed Old Quasars," *Nature,* **223,** 690.

Lynden-Bell, D. (1978). "Gravity Power," *Physica Scripta,* **17,** 185.

Mazur, P. (1982). "Proof of Uniqueness of the Kerr–Newman Black Hole Solution," *Journal of Physics A,* **15,** 3173.

May, M. M., and White, R. H. (1965). "Hydrodynamical Calculation of General Relativistic Collapse," *Bulletin of the American Physical Society,* **10,** 15.

May, M. M., and White, R. H. (1966). "Hydrodynamic Calculations of General Relativistic Collapse," *Physical Review,* **141,** 1232.

Medvedev, Z. A. (1978). *Soviet Science* (W. W. Norton, New York).

Medvedev, Z. A. (1979). *Nuclear Disaster in the Urals* (W. W. Norton, New York).

Michell, J. (1784). "On the Means of Discovering the Distance, Magnitude, Etc., of the Fixed Stars, in Consequence of the Diminution of Their Light, in Case Such a Diminution Should Be Found to Take Place in Any of Them, and Such Other Data Should Be Procured from Observations, as Would Be Further Necessary for That Purpose," in *Philosophical Transactions of the Royal Society of London,* **74,** 35; presented to the Royal Society on 27 November 1783.

Michelson, P. F., and Taber, R. C. (1984). "Can a Resonant-Mass Gravitational-Wave Detector Have Wideband Sensitivity?" *Physical Review D,* **29,** 2149.

Misner, C. W. (1969). "Mixmaster Universe," *Physical Review Letters*, **22**, 1071.

Misner, C. W., Thorne, K. S., and Wheeler, J. A. (1973). *Gravitation* (W. H. Freeman, San Francisco).

Mitton, S., and Ryle, M. (1969). "High Resolution Observations of Cygnus A at 2.7 GHz and 5 GHz," *Monthly Notices of the Royal Astronomical Society*, **146**, 221.

Morris, M. S., and Thorne, K. S. (1988). "Wormholes in Spacetime and Their Use for Interstellar Travel: A Tool for Teaching General Relativity," *American Journal of Physics*, **56**, 395.

Morris, M. S., Thorne, K. S., and Yurtsever, U. (1988). "Wormholes, Time Machines, and the Weak Energy Condition," *Physical Review Letters*, **61**, 1446.

Moss, G. E., Miller, L. R., and Forward, R. L. (1971). "Photon Noise Limited Laser Transducer for Gravitational Antenna," *Applied Optics*, **10**, 2495.

MTW: Misner, Thorne, and Wheeler (1973).

Newman, E. T., Couch, E., Chinnapared, K., Exton, A., Prakash, A., and Torrence, R. (1965). "Metric of a Rotating, Charged Mass," *Journal of Mathematical Physics*, **6**, 918.

Novikov, I. D. (1963). "The Evolution of the Semi-Closed World," *Astronomicheskii Zhurnal*, **40**, 772. English translation in *Soviet Astronomy—AJ*, **7**, 587 (1964).

Novikov, I. D. (1966). "Change of Relativistic Collapse into Anticollapse and Kinematics of a Charged Sphere," *Pis'ma v Redaktsiyu Zhurnal Eksperimentalnoi i Teoreticheskoi Fiziki*, **3**, 223. English translation in *JETP Letters*, **3**, 142 (1966).

Novikov, I. D. (1969). "Metric Perturbations When Crossing the Schwarzschild Sphere," *Zhurnal Eksperimentalnoi i Teoreticheskoi Fiziki*, **57**, 949. English translation in *Soviet Physics—JETP*, **30**, 518 (1970).

Novikov, I. D., Polnarev, A. G., Starobinsky, A. A., and Zel'dovich, Ya. B. (1979). "Primordial Black Holes," *Astronomy and Astrophysics*, **80**, 104.

Novikov, I. D., and Zel'dovich, Ya. B. (1966). "Physics of Relativistic Collapse," *Supplemento al Nuovo Cimento*, **4**, 810; Addendum 2.

Oppenheimer, J. R., and Serber, R. (1938). "On the Stability of Stellar Neutron Cores," *Physical Review*, **54**, 608.

Oppenheimer, J. R., and Snyder, H. (1939). "On Continued Gravitational Contraction," *Physical Review*, **56**, 455.

Oppenheimer, J. R., and Volkoff, G. (1939). "On Massive Neutron Cores," *Physical Review*, **54**, 540.

Ori, A. (1991). "The Inner Structure of a Charged Black Hole: An Exact Mass Inflation Solution," *Physical Review Letters*, **67**, 789.

Ori, A. (1992). "Structure of the Singularity Inside a Realistic Rotating Black Hole," *Physical Review Letters*, **68**, 2117.

Page, D. N. (1976). "Particle Emission Rates from a Black Hole," *Physical Review D*, **13**, 198, and **14**, 3260.

Page, D. N., and Hawking, S. W. (1975). "Gamma Rays from Primordial Black Holes," *Astrophysical Journal*, **206**, 1.

Pagels, H. (1982). *The Cosmic Code* (Simon and Schuster, New York).

Pais, A. (1982). *"Subtle Is the Lord . . ." The Science and the Life of Albert Einstein* (Oxford University Press, Oxford, England).

Penrose, R. (1965). "Gravitational Collapse and Spacetime Singularities," *Physical Review Letters*, **14**, 57.

Penrose, R. (1968). "The Structure of Spacetime," in *Battelle Rencontres: 1967 Lec-*

tures in Mathematics and Physics, edited by C. M. DeWitt and J. A. Wheeler (Benjamin, New York), p. 565.

Penrose, R. (1969). "Gravitational Collapse: The Role of General Relativity," *Rivista Nuovo Cimento*, **1**, 252.

Penrose, R. (1989). *The Emperor's New Mind* (Oxford University Press, New York), pp. 419–421.

Phinney, E. S. (1989). "Manifestations of a Massive Black Hole in the Galactic Center," in *The Center of the Galaxy: Proceedings of IAU Symposium 136*, edited by M. Morris (Reidel, Dordrecht, Holland), p. 543.

Pimenov, R. I. (1968). *Prostranstva Kinimaticheskovo Tipa [Seminars in Mathematics]*, Vol. 6 (V. A. Steklov Mathematical Institute, Leningrad). English translation: *Kinematic Spaces* (Consultants Bureau, New York, 1970).

Podurets, M. A. (1964). "The Collapse of a Star with Back Pressure Taken Into Account," *Doklady Akademi Nauk*, **154**, 300. English translation in *Soviet Physics—Doklady*, **9**, 1 (1964).

Poisson, E., and Israel, W. (1990). "Internal Structure of Black Holes," *Physical Review D*, **41**, 1796.

Press, W. H. (1971). "Long Wave Trains of Gravitational Waves from a Vibrating Black Hole," *Astrophysical Journal Letters*, **170**, 105.

Press, W. H., and Teukolsky, S. A. (1973). "Perturbations of a Rotating Black Hole. II. Dynamical Stability of the Kerr Metric," *Astrophysical Journal*, **185**, 649.

Price, R. H. (1972). "Nonspherical Perturbations of Relativistic Gravitational Collapse," *Physical Review D*, **5**, 2419 and 2439.

Price, R. H., and Thorne, K. S. (1988). "The Membrane Paradigm for Black Holes," *Scientific American*, **258** (No. 4), 69.

Rabi, I. I., Serber, R., Weisskopf, V. F., Pais, A., and Seaborg, G. T. (1969). *Oppenheimer* (Scribners, New York).

Reber, G. (1940). "Cosmic Static," *Astrophysical Journal*, **91**, 621.

Reber, G. (1944). "Cosmic Static," *Astrophysical Journal*, **100**, 279.

Reber, G. (1958). "Early Radio Astronomy at Wheaton, Illinois," *Proceedings of the Institute of Radio Engineers*, **46**, 15.

Rees, M. (1971). "New Interpretation of Extragalactic Radio Sources," *Nature*, **229**, 312 and 510.

Renn, J., and Schulman, R. (1992). Introduction to *Albert Einstein/Mileva Marić: The Love Letters*, edited by Jürgen Renn and Robert Schulman (Princeton University Press, Princeton, New Jersey).

Rhodes, R. (1986). *The Making of the Atomic Bomb* (Simon and Schuster, New York).

Ritus V. I. (1990). "If Not I, Then Who?" *Priroda*, August issue. English translation in Drell and Kapitsa, eds. (1991).

Robinson, I., Schild, A., and Schucking, E. L., eds. (1965). *Quasi-Stellar Sources and Gravitational Collapse* (University of Chicago Press, Chicago).

Romanov, Yu. A. (1990). "The Father of the Soviet Hydrogen Bomb," *Priroda*, August issue. English translation in Drell and Kapitsa, eds. (1991).

Royal, D. (1969). *The Story of J. Robert Oppenheimer* (St. Martin's Press, New York).

Sagan, C. (1985). *Contact* (Simon and Schuster, New York).

Sakharov, A. (1990). *Memoirs* (Alfred A. Knopf, New York).

Salpeter, E. E. (1964). "Accretion of Interstellar Matter by Massive Objects," *Astrophysical Journal*, **140**, 796.

Schaffer, S. (1979). "John Michell and Black Holes," *Journal for the History of Astronomy*, **10**, 42.

Schmidt, M. (1963). "3C273: A Star-like Object with Large Red-shift," *Nature*, **197**, 1040.

Schwarzschild, K. (1916a). "Uber das Gravitationsfeld eines Massenpunktes nach der Einsteinschen Theorie," *Sitzungsberichte der Deutschen Akademie der Wissenschaften zu Berlin, Klasse fur Mathematik, Physik, und Technik*, **1916**, 189.

Schwarzschild, K. (1916b). "Uber das Gravitationsfeld einer Kugel aus inkompressibler Flussigkeit nach der Einsteinschen Theorie," *Sitzungsberichte der Deutschen Akademie der Wissenschaften zu Berlin, Klasse fur Mathematik, Physik, und Technik*, **1916**, 424.

Seelig, C. (1956). *Albert Einstein: A Documentary Biography* (Staples Press, London), p. 104.

Serber, R. (1969). "The Early Years," in Rabi et al. (1969); also published in *Physics Today*, October 1967, p. 35.

Shapiro, S. L., and Teukolsky, S. A. (1983). *Black Holes, White Dwarfs, and Neutron Stars* (Wiley, New York).

Shapiro, S. L., and Teukolsky, S. A. (1991). "Formation of Naked Singularities—The Violation of Cosmic Censorship," *Physical Review Letters*, **66**, 994.

Smart, W. M. (1953). *Celestial Mechanics* (Longmans, Green and Co., London), Section 19.03.

Smith, A. K., and Weiner, C. (1980). *Robert Oppenheimer: Letters and Recollections* (Harvard University Press, Cambridge, Massachusetts).

Smith, H. J. (1965). "Light Variations of 3C273," in *Quasi-Stellar Sources and Gravitational Collapse*, edited by I. Robinson, A. Schild, and E. L. Schucking (University of Chicago Press, Chicago), p. 221.

Solvay (1958). Onzième Conseil de Physique Solvay, *La Structure et l'Evolution de l'Univers* (Editions R. Stoops, Brussels).

Stoner, E. C. (1930). "The Equilibrium of Dense Stars," *Philosophical Magazine*, **9**, 944.

Struve, O., and Zebergs, V. (1962). *Astronomy of the 20th Century* (Macmillan, New York).

Sullivan, W. J., ed. (1982). *Classics in Radio Astronomy* (Reidel, Dordrecht, Holland).

Sullivan, W. J., ed. (1984). *The Early Years of Radio Astronomy* (Cambridge University Press, Cambridge, England).

Sunyaev, R. A. (1972). "Variability of X Rays from Black Holes with Accretion Disks," *Astronomicheskii Zhurnal*, **49**, 1153. English translation in *Soviet Astronomy—AJ*, **16**, 941 (1973).

Taylor, E. F., and Wheeler, J. A. (1992). *Spacetime Physics: Introduction to Special Relativity* (W. H. Freeman, San Francisco).

Teller, E. (1955). "The Work of Many People," *Science*, **121**, 268.

Teukolsky, S. A. (1972). "Rotating Black Holes: Separable Wave Equations for Gravitational and Electromagnetic Perturbations," *Physical Review Letters*, **29**, 1115.

Thorne, K. S. (1967). "Gravitational Collapse," *Scientific American*, **217** (No. 5), 96.

Thorne, K. S. (1972). "Nonspherical Gravitational Collapse—A Short Review," in *Magic without Magic: John Archibald Wheeler*, edited by J. R. Klauder (W. H. Freeman, San Francisco), p. 231.

Thorne, K. S. (1974). "The Search for Black Holes," *Scientific American*, **231** (No. 6), 32.

Thorne, K. S. (1987). "Gravitational Radiation," in *300 Years of Gravitation*, edited by S. W. Hawking and W. Israel (Cambridge University Press, Cambridge, England), p. 330.

Thorne, K. S. (1991). "An American's Glimpses of Sakharov," *Priroda*, May issue; in Russian. English translation in Drell and Kapitsa, eds. (1991), p. 74.

Thorne, K. S. (1993). "Closed Timelike Curves," in *General Relativity and Gravitation 1992*, edited by R. J. Gleiser, C. N. Kozameh, and D. M. Moreschi (Institute of Physics Publishing, Bristol, England), p. 295.

Thorne, K. S., Drever, R. W. P., Caves, C. M., Zimmermann, M., and Sandberg, V. D. (1978). "Quantum Nondemolition Measurements of Harmonic Oscillators," *Physical Review Letters*, **40**, 667.

Thorne, K. S., Price, R. H., and Macdonald, D. A., eds. (1986). *Black Holes: The Membrane Paradigm* (Yale University Press, New Haven, Connecticut).

Thorne, K. S., and Zurek, W. (1986). "John Archibald Wheeler: A Few Highlights of His Contributions to Physics," *Foundations of Physics*, **16**, 79.

Thorne, K. S., and Żytkow, A. N. (1977). "Stars with Degenerate Neutron Cores. I. Structure of Equilibrium Models," *Astrophysical Journal*, **212**, 832.

Tipler, F. J. (1976). "Causality Violation in Asymptotically Flat Space-Times," *Physical Review Letters*, **37**, 879.

Tolman, R. C. (1939). "Static Solutions of Einstein's Field Equations for Spheres of Fluid," *Physical Review*, **55**, 364.

Tolman, R. C. (1948). The Richard Chace Tolman Papers, archived in the California Institute of Technology Archives.

Trimble, V. L., and Thorne, K. S. (1969). "Spectroscopic Binaries and Collapsed Stars," *Astrophysical Journal*, **56**, 1013.

Uhuru (1981). "Proceedings of the Uhuru Memorial Symposium: The Past, Present, and Future of X-Ray Astronomy," *Journal of the Washington Academy of Sciences*, **71** (No. 1).

Unruh, W. G. (1976). "Notes on Black-Hole Evaporation," *Physical Review D*, **14**, 870.

Unruh, W. G., and Wald, R. M. (1982). "Acceleration Radiation and the Generalized Second Law of Thermodynamics," *Physical Review D*, **25**, 942.

Unruh, W. G., and Wald, R. M. (1984). "What Happens When an Accelerating Observer Detects a Rindler Particle," *Physical Review D*, **29**, 1047.

USAEC [United States Atomic Energy Commission] (1954). *In the Matter of J. Robert Oppenheimer; Transcript of Hearing before Personnel Security Board, Washington, D.C., April 12, 1954, through May 6, 1954* (U.S. Government Printing Office, Washington, D.C.).

van Stockum, W. J. (1937). "The Gravitational Field of a Distribution of Particles Rotating about an Axis of Symmetry," *Proceedings of the Royal Society of Edinburgh*, **57**, 135.

Wald, R. M. (1977). "The Back Reaction Effect in Particle Creation in Curved Space-time," *Communications in Mathematical Physics*, **54**, 1.

Wald, R. M., and Yurtsever, U. (1991). "General Proof of the Averaged Null Energy Condition for a Massless Scalar Field in Two-Dimensional Curved Space-time," *Physical Review D*, **44**, 403.

Wali, K. C. (1991). *Chandra: A Biography of S. Chandrasekhar* (University of Chicago Press, Chicago).

Washington (1954). "Washington Conference on Radio Astronomy—1954," *Journal of Geophysical Research*, **59**, 1–204.

Weber, J. (1953). "Amplification of Microwave Radiation by Substances Not in Thermal Equilibrium," *Transactions of the IEEE, PG Electron Devices—3*, 1 (June).

Weber, J. (1960). "Detection and Generation of Gravitational Waves," *Physical Review*, **117,** 306.

Weber, J. (1961). *General Relativity and Gravitational Waves* (Wiley–Interscience, New York).

Weber, J. (1964). Unpublished research notebooks; also documented in Robert Forward's unpublished Personal Journal No. C1338, page 66, 13 September 1964.

Weber, J. (1969). "Evidence for Discovery of Gravitational Radiation," *Physical Review Letters*, **22,** 1320.

Weiss, R. (1972). "Electromagnetically Coupled Broadband Gravitational Antenna," *Quarterly Progress Report of the Research Laboratory of Electronics, M.I.T.,* **105,** 54.

Wheeler, J. A. (1955). "Geons," *Physical Review*, **97,** 511. Reprinted in Wheeler (1962), p. 131.

Wheeler, J. A. (1957). "On the Nature of Quantum Geometrodynamics," *Annals of Physics*, **2,** 604.

Wheeler, J. A. (1960). "Neutrinos, Gravitation and Geometry," in *Proceedings of the International School of Physics, "Enrico Fermi," Course XI* (Zanichelli, Bologna). Reprinted in Wheeler (1962), p. 1.

Wheeler, J. A. (1962). *Geometrodynamics* (Academic Press, New York).

Wheeler, J. A. (1964a). "The Superdense Star and the Critical Nucleon Number," in *Gravitation and Relativity,* edited by H. Y. Chiu and W. F. Hoffman (Benjamin, New York), p. 10.

Wheeler, J. A. (1964b). "Geometrodynamics and the Issue of the Final State," in *Relativity, Groups, and Topology,* edited by C. DeWitt and B. S. DeWitt (Gordon and Breach, New York), p. 315.

Wheeler, J. A. (1968). "Our Universe: The Known and the Unknown," *American Scientist,* **56,** 1.

Wheeler, J. A. (1979). "Some Men and Moments in the History of Nuclear Physics: The Interplay of Colleagues and Motivations," in *Nuclear Physics in Retrospect,* edited by Roger H. Stuewer (University of Minnesota, Minneapolis).

Wheeler, J. A. (1985). Letter to K. S. Thorne dated 3 December.

Wheeler, J. A. (1988). Notebooks in which Wheeler recorded his research work and ideas as they developed; now archived at the American Philosophical Society Library, Philadelphia, Pennsylvania.

Wheeler, J. A. (1990). *A Journey into Gravity and Spacetime* (Scientific American Library, New York).

Whipple, F. L., and Greenstein, J. L. (1937). "On the Origin of Interstellar Radio Disturbances," *Proceedings of the National Academy of Sciences,* **23,** 177.

White, T. H. (1939). *The Once and Future King* (Collins, London), Chapter 13 of Part I, "The Sword in the Stone."

Will, C. M. (1986). *Was Einstein Right?* (Basic Books, New York).

York, H. (1976). *The Advisors: Oppenheimer, Teller and the Superbomb* (W. H. Freeman, San Francisco).

Zel'dovich, Ya. B. (1962). "Semi-closed Worlds in the General Theory of Relativity," *Zhurnal Eksperimentalnoi i Teoreticheskoi Fiziki,* **43,** 1037. English translation in *Soviet Physics—JETP,* **16,** 732 (1963).

Zel'dovich, Ya. B. (1964). "The Fate of a Star and the Evolution of Gravitational Energy upon Accretion," *Doklady Akademii Nauk,* **155,** 67. English translation in *Soviet Physics—Doklady,* **9,** 195 (1964).

Zel'dovich, Ya. B. (1971). "The Generation of Waves by a Rotating Body," *Pis'ma v Redaktsiyu Zhurnal Eksperimentalnoi i Teoreticheskoi Fiziki,* **14,** 270. English translation in *JETP Letters,* **14,** 180 (1971).

Zel'dovich, Ya. B. (1985). *Collected Works: Particles, Nuclei, and the Universe* (Nauka, Moscow); in Russian. English translation: *Selected Works of Yakov Borisovich Zel'dovich.* Volume II: *Particles, Nuclei, and the Universe* (Princeton University Press, Princeton, 1993).

Zel'dovich, Ya. B., and Guseinov, O. Kh. (1965). "Collapsed Stars in Binaries," *Astrophysical Journal,* **144,** 840.

Zel'dovich, Ya. B., and Khariton, Yu. B. (1939). "On the Issue of a Chain Reaction Based on an Isotope of Uranium," *Zhurnal Eksperimentalnoi i Teoreticheskoi Fiziki,* **9,** 1425; see also the follow-up papers by the same authors in the same journal, **10,** 29 (1940) and **10,** 477 (1940). Reprinted as the first three papers in Volume II of Zel'dovich's collected works, Zel'dovich (1985).

Zel'dovich, Ya. B., and Novikov, I. D. (1964). "Relativistic Astrophysics, Part I," *Uspekhi Fizicheskikh Nauk,* **84,** 877. English translation in *Soviet Physics—Uspekhi,* **7,** 763 (1965).

Zel'dovich, Ya. B., and Novikov, I. D. (1965). "Relativistic Astrophysics, Part II," *Uspekhi Fizicheskikh Nauk,* **86,** 447. English translation in *Soviet Physics—Uspekhi,* **8,** 522 (1966).

Zel'dovich, Ya. B., and Starobinsky, A. A. (1971). "Particle Production and Vacuum Polarization in an Anisotropic Gravitational Field," *Zhurnal Eksperimentalnoi i Teoreticheskoi Fiziki,* **61,** 2161. English translation in *Soviet Physics—JETP,* **34,** 1159 (1972).

Znajek, R. (1978). "The Electric and Magnetic Conductivity of a Kerr Hole," *Monthly Notices of the Royal Astronomical Society,* **185,** 833.

Zwicky, F. (1935). "Stellar Guests," *Scientific Monthly,* **40,** 461.

Zwicky, F. (1939). "On the Theory and Observation of Highly Collapsed Stars," *Physical Review,* **55,** 726.

Subject Index

People Index

COVERAGE AND ABBREVIATIONS

This index covers the Prologue, Chapters, Epilogue, and Notes.
Additional information about people will be found in the Characters section (pages 531–36) and the Bibliography section (pages 585–600).

Letters appended to page numbers have the following meanings:
    b—box
    f—figure or photograph
    n—footnote